SYMBOL	MEANING	
$\mathscr{P}(M)$	Power set of the set M	53
$\cup$	Set union	56
$\cap$	Set intersection	56
$\setminus$	Relative complement	56
$\# M$	The number of elements, or the cardinal number, of the set M	62
$\dot\cup$	Disjoint union	64
Δ	Symmetric difference	67
C_r^n	Number of combinations of n things taken r at a time	69
$(M_i)_{i \in I}$	Indexed family of sets	70
$\mathscr{E}$	Set of sets	71
$\bigcup_{M \in \mathscr{E}} M$	Set consisting of all elements that belong to at least one of the sets M in $\mathscr{E}$	71
$\bigcup_{i \in I} M_i$	Union of the family of sets M_i, i running through the index I	72
$\bigcap_{M \in \mathscr{E}} M$	Set consisting of all elements that belong to every one of the sets M in $\mathscr{E}$	74
$\bigcap_{i \in I} M_i$	Intersection of the family of sets M_i, i running through the index I	75
xRy	Assertion that the object x is related by R to the object y	91
$\mathrm{dom}(R)$	Domain of R	92
$\mathrm{codom}(R)$	Codomain of R	92
R^{-1}	Converse, or inverse, of the relation R	92
Δ_X	Diagonal relation on the set X	93
$(X, \leq)$	Partially ordered set or poset	97
$\mathrm{LUB}(M)$	Least upper bound of the set M	99
$\mathrm{GLB}(M)$	Greatest lower bound of the set M	99
$\sup(M)$	Supremum of the set M	99
$\inf(M)$	Infimum of the set M	99
$[x]_E$	E-equivalence class determined by x	107
X/E	X modulo E	109
$f:X \to Y$	f is a mapping from X to Y	118
$X \xrightarrow{f} Y$	f is a mapping from X to Y	118
$x \mapsto y$	x is mapped or transformed into y	119
χ_M	Characteristic function of M	128
$g \circ f$	The composite of the functions g and f	130
$R \circ S$	The composite of the relations R and S	139
$*:S \times S \to S$	A binary operation or binary composition on S	143
$x * y$	Infix notation for $*((x, y))$	143
$(S, *)$	An algebraic system with a single binary operation $*$	144
$\mathscr{B}(X)$	Set of all bijections $f:X \to X$	145
$x * y * z$	Iterated composition of x, y, and z in an algebraic system $(S, *)$	146

AFTER CALCULUS: ALGEBRA

AFTER CALCULUS: ALGEBRA

DAVID J. FOULIS
UNIVERSITY OF MASSACHUSETTS

MUSTAFA A. MUNEM
MACOMB COLLEGE

DELLEN PUBLISHING COMPANY
SAN FRANCISCO, CALIFORNIA

COLLIER MACMILLAN PUBLISHERS
LONDON

DIVISIONS OF
MACMILLAN, INC.

On the cover: "Hot Plate" by Sam Tchakalian, 1986; oil on canvas, 42 × 60 inches. Working spontaneously and with much physical involvement, Sam Tchakalian's "attitude" rather than formalized structural elements becomes the content of his work. His work may be seen at the San Francisco Museum of Modern Art; the Oakland Museum, Oakland, California; the Albright Knox Art Gallery, Buffalo, New York; and the University Art Museum of the University of California, Berkeley. Tchakalian is represented by Modernism in San Francisco, California.

Printed in the United States of America

Permissions: Dellen Publishing Company
400 Pacific Avenue
San Francisco, California 94133

Orders: Dellen Publishing Company
c/o Macmillan Publishing Company
Front and Brown Streets
Riverside, New Jersey 08075

Collier Macmillan Canada, Inc.

LIBRARY OF CONGRESS CATALOGING-IN-PUBLICATION DATA

Foulis, David J., 1930–
After calculus—algebra.

Bibliography
Includes index.
1. Algebra. I. Munem, Mustafa A. II. Title.
QA154.2.F67 1988 512 87-33044
ISBN 0-02-384790-5

Printing: 1 2 3 4 5 6 7 8 9 *Year:* 8 9 0 1 2

ISBN 0-02-384790-5

CONTENTS

3

GROUPS, RINGS, FIELDS, AND VECTOR SPACES 178

4

NUMBER SYSTEMS 252

*5

SPECIAL TOPICS 353

PREFACE FOR THE INSTRUCTOR

This textbook is written for students who have completed at least one semester of calculus and who intend to take more advanced courses in mathematics. It is designed to provide a gentle transition from "cookbook" mathematics to the more rigorous and sophisticated material in post-calculus courses. Our primary concern is to acquaint the student with the language and general methods of contemporary mathematics; however, since this cannot be done in a vacuum, we have chosen to present this material in a predominantly algebraic context.

This is not a monograph on the foundations of mathematics, nor is it a cultural survey for liberal arts students, nor is it a vehicle for mathematical proselytization. We assume that our readers are people who already have a strong attraction to mathematics (e.g., mathematics majors) or who are going into a field (e.g., engineering, physics, or economics) in which more advanced mathematical techniques are used. We believe that these people—who are already highly motivated to study mathematics—are best served by a straightforward exposition without frills. Therefore, although we do include a few historical vignettes and annotated bibliographies for suggested further reading, we have indulged in a minimum of motivational rhetoric.

There are no prerequisites for an intelligent reading of this textbook other than competence in precalculus mathematics. However, some of the examples are based on material usually covered in the first semester of calculus, and some of the (optional) problems involve a bit of elementary matrix algebra.

This book includes well over 1200 problems, ranging from simple drill-type exercises to problems that probe for deep understanding and insight. The odd-numbered problems at the end of each section are usually quite straightforward and should present little difficulty for the student who has carefully read the section. Numerous examples, given in an example–

solution format, show the student how these problems are to be handled; answers or hints for selected odd-numbered problems can be found at the end of the book. Some of the even-numbered problems, particularly those toward the ends of the problem sets, are considerably more challenging. In general, *even-numbered problems should not be assigned without careful prior consideration* since the more difficult of these problems can be frustrating for all but the best students.

We have written this textbook using the so-called spiral approach. Thus, at the outset, we assume that the reader is familiar with some elementary facts about counting, real numbers, real-valued functions of real variables, and so forth, and we make free use of these ideas in our examples and problems. Later, we return to these concepts and show how they can be handled more carefully. For instance, we assume in Section 5 of Chapter 1 that the reader knows what is meant by the number of elements in a (finite) set, and we make free use of this idea. Later, in Sections 6 and 7 of Chapter 4, we state the formal definition of a cardinal number and give a careful mathematical development of the properties of cardinal numbers.

In those cases for which there is no universal agreement regarding notation or terminology, we have made our choice (when possible) on the basis of what seems to be most common in the standard algebraic literature. For example, since the majority of algebraists and computer scientists use *injective*, *surjective*, and *bijective* in preference to *one-to-one*, *onto*, and *one-to-one correspondence*, we emphasize the former terminology rather than the latter. In those instances for which there is no clear basis for a decision regarding notation or terminology, we have made an arbitrary choice, but we caution our readers about the alternatives used by other authors.

In our experience, even students with otherwise strong backgrounds in mathematics have difficulty with mathematical induction. For this reason, we have included a *review of mathematical induction* in an appendix. Some instructors may prefer to organize their first lesson around this appendix; others may suggest that students refer to the appendix for help with problems that require mathematical induction.

Other special features of this book are as follows:

1. Propositional calculus and predicate calculus are presented informally in the first two sections of Chapter 1. It is emphasized that this machinery will be used only when necessary to help clarify portions of mathematical arguments.
2. In the third section of Chapter 1, we discuss the notion of a mathematical proof based on axioms, definitions, and rules of procedure. This section also includes an indication of why the implication connective $\Rightarrow$ works so well in mathematical arguments in spite of some of its (perhaps) unusual properties.

3. Prior to a brief study of the algebra of sets, we suggest that a mathematical set is a concept, never a physical entity. This helps to clarify the idea of a singleton set and to dispel skepticism about the "existence" of the empty set.
4. In Chapter 2, relations are defined as sets of ordered pairs, and functions are introduced as special kinds of relations. Here we discuss alternative ways of thinking about a function, for example, as a rule of correspondence, as a mapping, or as a relation between dependent and independent variables.
5. The entire third section of Chapter 2 is devoted to the important topic of partitions and equivalence relations.
6. Binary operations are introduced in the sixth section of Chapter 2 in preparation for the study of groups, rings, fields, and vector spaces in Chapter 3.
7. To lay the groundwork for our study of the field $\mathbb{R}$ of real numbers, the notion of an ordered field is first introduced in Chapter 3.
8. In Chapter 4, we present a top-down development of the standard number systems; that is, we begin by postulating the existence of a complete ordered field, and we extract from this field the systems of natural numbers, integers, and rational numbers. We feel that this approach provides an excellent preparation for those students who go on to study the constructive development of $\mathbb{R}$.
9. The field $\mathbb{C}$ of complex numbers is constructed in the usual way from the field of real numbers in the fifth section of Chapter 3, and a brief indication of number systems "beyond $\mathbb{C}$" is given in the final (optional) section of this chapter.

Several sections in this textbook, including all of the sections in Chapter 5, are marked with asterisks to indicate that they are *optional* in the sense that they can be omitted without compromising the reader's understanding of the remaining sections. Material in these sections can be covered if time permits, or it can be assigned as extra credit work. Because many students will use this book for reference in connection with more advanced mathematics courses, some of the optional material is included in the interest of completeness.

Most of the nonoptional material in this textbook can be covered in a one-term course on the *fundamental concepts of mathematics.* If time constraints make it necessary, the first chapter can be treated with dispatch, emphasizing only the peculiarities of the implication connective, the use of quantifiers, and the elements of set algebra. On the other hand, the entire textbook, including most of the optional material, can be used as the basis of a two-term undergraduate course in *abstract algebra.* In the second term of such a course, the instructor may wish to supplement the material in this book with selected topics from the standard literature, such as principal ideal domains, fields of quotients, Sylow theorems, solvable groups, or field extensions.

It is a pleasure to acknowledge the valuable advice of the following reviewers of our manuscript:

David Barnette, University of California, Davis
Sherralyn Craven, Central Missouri State University
David Ellenbogen, University of Vermont
Michael Evans, North Carolina State University
Robert Gamble, Winthrop College
Juan Gatica, University of Iowa
Robert Hunter, Pennsylvania State University
Jack Johnson, Brigham Young University, Hawaii Campus
Daniel Kocan, Potsdam College of the State University of New York
Joseph H. Oppenheim, San Francisco State University
Robert Piziak, Baylor University
Daniel Sweet, University of Maryland
Howard Taylor, West Georgia College

Special thanks are also given to Al Bednarek of the University of Florida for his many helpful suggestions, to Gerald L. Alexanderson of Santa Clara University for providing the annotated bibliographies, and to William C. Schulz of Northern Arizona University for supplying the historical notes. Finally, we would like to express our gratitude to Hyla Gold Foulis and Steve Fasbinder for assisting with the proofreading of the pages.

David J. Foulis
Mustafa A. Munem

INTRODUCTION

From ancient times until nearly the end of the nineteenth century, mathematics was generally viewed as the study of number (algebra) and form (geometry). This view was reinforced in the seventeenth century by the great synthesis of algebra and geometry (analytic geometry) created by the French mathematicians René Descartes and Pierre de Fermat. By the early part of the twentieth century, however, it had become clear that this perception of mathematics was overly restrictive, that the study of algebraic and geometric structure is only part of the mathematical enterprise, and that the true concern of mathematics is the study of abstract structure in general. It is because sets are natural carriers of this structure that set theory has played such a central role in twentieth century mathematics.

The unification effected by the modern structural view of mathematics has been even more profound than the synthesis wrought by Descartes and Fermat; and, as a consequence, all branches of mathematics now share a common core of ideas and methods. The purpose of this textbook is to acquaint you with these fundamental mathematical concepts and techniques. In your study of this material, you may find the following suggestions helpful:

1. Your previous mathematical experience may have been with "cookbook" courses requiring only that you memorize formulas and procedures and apply them to solving problems, usually by simply "plugging in" suitable values for variables. If so, be forewarned that the material in this texbook is a departure from this approach. You will find that this material has some of the flavor of a foreign language, with its own alphabet, vocabulary, orthography, grammar, and idiomatic expressions. Indeed, you will be learning the language used by mathematicians in communicating their ideas and discoveries.
2. Try to understand every word in every sentence in every paragraph in every section of this textbook. In mathematics, new ideas almost always depend strongly on previously introduced concepts, and a lack of understanding of earlier material often interferes with comprehension of

later ideas. (The sections marked with asterisks can be omitted without loss of continuity.) Since you probably will not grasp everything the first time around, we recommend a spiral approach, in which you repeatedly reconsider earlier ideas after acquiring a better notion of where they lead.

3. In reading this textbook (or mathematics in general) it is *imperative* that you have a pencil and paper at hand. You must *write out* the definitions, formulas, theorems, and key ideas that you encounter as you go along. Some students like to use a highlighting pen to color this material in the text, but this is not nearly as effective as the act of transcribing them to a sheet of paper. *Doing the writing* focuses your attention. Considerable effort is required to learn mathematics, and it is impossible to do so sitting in an easy chair, reading your textbook as if it were a novel.
4. Draw pictures and diagrams when possible to illustrate the concepts under consideration.
5. The problem sets at the ends of the sections are integral parts of this textbook. Although some of the even-numbered problems, particularly those toward the ends of the problem sets, are quite challenging, most of the odd-numbered problems are straightforward, and working them will enhance your grasp of the material. An odd-numbered problem with which you cannot cope may indicate a lack of understanding of the concepts involved. Answers and hints for most of the odd-numbered problems are provided at the end of the book.
6. In working the problems (and in writing mathematics in general) cultivate the habit of writing complete sentences. An isolated expression such as $x^2 + 2x + 1$ means very little. What about it? Do you want to find the values of x for which the expression is zero? If so, say so. Are you writing this as an example of a polynomial with integer coefficients? If so, say so.
7. Every mathematical concept has to be understood on two levels: the *formal* level and the *intuitive* level. To understand the concept on the formal level, you need to know its mathematical definition. Although it might be possible to get through a beginning calculus course without knowing the definition of, say, the Riemann integral, it is inconceivable to do more advanced mathematics without paying careful attention to definitions. Some people prefer to memorize definitions outright; others would rather look them up as they need them in working problems; this is a matter of individual preference. An intuitive grasp of a mathematical concept may be more difficult to come by; it generally develops only after you deal with the concept over a period of time. To build your intuition, you might begin by collecting pertinent examples of the concept. Perhaps, as you go along, you will want to keep a notebook of interesting and suggestive examples. After you have a suitable stock

of examples at hand, you should begin to use the concept, and this is where the problems are of importance. Finally, you should think about the concept, mull it over, see how it relates to other ideas, and see if it helps to clarify notions that might otherwise be hazy or imprecise.

8. From time to time you will encounter an idea that you find exciting. When you do, why not decide to become a "local expert" on this idea and its ramifications? Go to the library and read everything you can find that pertains to the idea. Start a notebook for your own use, devoted to this and related ideas. Collect pertinent examples, exercises, definitions, theorems, diagrams, remarks, historical lore—anything you can find that is germane to the idea. Explain the idea to your friends and tell them why you find it so fascinating. If you persist in this, you will come to the realization that mathematics is remarkably holistic and that, in learning all you can about one apparently isolated concept, you will automatically learn considerably more mathematics than you might expect.

This textbook is organized into five chapters, and each chapter is divided into numbered sections. Problem Set 3.2, for instance, refers to the problem set at the end of Section 2 in Chapter 3. Major definitions, theorems, and examples are numbered consecutively through each section; for instance, Theorem 2.13 refers to the thirteenth numbered item (i.e., definition, theorem, or example) in the second section of a chapter. If no chapter is mentioned, it is understood that the item appears in the current chapter; otherwise, the chapter is always indicated.

The most important definitions and theorems are in boxes, and all major theorems have names (as well as numbers) for ease of reference. Important words, phrases, and remarks are *italicized*, and terms are written in **boldface** type when they are defined for the first time. Separate, but related, parts of definitions or theorems are labeled with small Roman numerals (i), (ii), (iii), and so on; separate, but related, parts of problems are labeled with small letters (a), (b), (c), and so on. The symbol □ is used to signal the end of a proof or solution.

AFTER CALCULUS: ALGEBRA

1

INTRODUCTION TO LOGIC AND SETS

In 1666, the German mathematician and philosopher G. W. Leibniz (one of the codiscoverers of the infinitesimal calculus) conceived of a "calculus of reasoning" that could be used to make complicated logical deductions merely by mechanically manipulating suitable symbols. The formulas of such a **symbolic logic** would be handled in much the same way as the formulas of ordinary algebra; and logical errors, which would arise only through mistakes in calculation, could quickly be detected and corrected. Reasoning would be reduced to computation!

Between 1666 and the 1690s, Leibniz himself made an effort to lay the foundations for symbolic logic, but it remained for a nineteenth-century British mathematician, George Boole, to take the first really decisive steps in this direction. In the hands of Boole and his successors, symbolic logic developed into a powerful tool that not only helped to clarify mathematical thought but also had a number of unforeseen applications. For instance, in the late 1930s, the American electrical engineer Claude Shannon showed that Boolean algebra, a branch of symbolic logic, could be used to analyze and design electrical switching circuits.

In relation to mathematics, symbolic logic plays a dual role: First, it functions as a precise *language* that is tailor-made for clarifying complex logical deductions; second, in connection with a discipline called *metamathematics*, it is a tool for studying the potentialities and limitations of mathematics itself. However, symbolic logic is not a miraculous machine for grinding out mathematical theorems; and the would-be provers of such theorems must be prepared for some hard work no matter how much symbolic logic they happen to know.

Although it is possible to approach symbolic logic in a formal axiomatic way, we propose to present it informally as a language with its own alphabet, grammar, and idioms. You should expect to learn this material in much the same way that you would learn any foreign language; however, you will find that it is a language capable of exquisite precision.

Although most of our mathematical deliberations will be conducted in ordinary English, we intend to use symbolic logic when necessary to clear up arguments that might otherwise cause difficulty.

1 ELEMENTS OF PROPOSITIONAL CALCULUS

The Latin word *calculus* refers to a small stone or pebble. In ancient times, pebbles were used as counters to aid in doing arithmetic, and so the word calculus has come to be used in mathematics to refer to a systematic method of calculating or reasoning. Thus, *differential calculus* is the study of calculating with differentials or derivatives, whereas *integral calculus* is the study of calculating with integrals. One of the important branches of symbolic logic is **propositional calculus**, a systematic method of reasoning or "computing" with statements or propositions. We shall launch our study of propositional calculus with the following basic definition.

1.1 DEFINITION

PROPOSITION By a **proposition** we mean a complete declarative sentence that is either true or false. If a proposition is true, we say that its **truth value** is 1, and if it is false, we say that its truth value is 0.

Some people use the letters T (true) and F (false) to denote the truth values of propositions. However, since most computer languages denote truth values by 1 and 0, we prefer to use 1 and 0 rather than T and F.

Our definition of a proposition is not intended as a formal mathematical definition but merely as an informal working definition that will serve our present purposes. We often use the word **statement** as a synonym for a proposition. Note that there is no condition on the language in which a proposition is expressed; it may be, for instance, English, German, Esperanto, or mathematical symbolism. Thus, *Two is smaller than three*, *Zwei ist kleiner als Drei*, and $2 < 3$ are propositions. Also, in testing to see whether an expression is a proposition, you are not required to determine its truth value. Thus, the statement *The trillionth digit after the decimal point in the decimal expansion of pi is 7* is a proposition even though no one may know whether it is true or false.

1.2 Example Which of the following are propositions?

(a) Hoping to evaluate the integral
(b) $2 + 2 = 4$
(c) Is pi equal to 22/7?
(d) Please go to the blackboard and solve the equation $x^2 + 1 = 0$

SOLUTION Only (b) is a proposition; (a) is not a complete sentence; (c) is a question, not a declarative sentence; and (d) is a request, not a statement of fact. □

In ordinary algebra, we use symbols such as $x, y, z, \ldots$, called *number variables* to denote arbitrary numbers. Likewise, in propositional calculus, we use symbols such as $P, Q, R, \ldots$, called **propositional variables** to stand for arbitrary propositions. In ordinary algebra, we have operations such as addition, subtraction, multiplication, division, negation, and reciprocation, which can be used with number variables to form new number expressions $x + y$, $x - y$, $x \cdot y$, $x \div y$, $-x$, and x^{-1}. *Propositional connectives*, which we now define, play a similar role in propositional calculus.

1.3 DEFINITION

BINARY PROPOSITIONAL CONNECTIVES A **binary propositional connective** is an operation that combines two propositions to yield a new proposition whose truth value depends only on the truth values of the two original propositions.

The English word *and* provides a good example of a binary propositional connective. Indeed, if P, Q are propositions, then

P *and* Q

is a new proposition declaring that both P and Q are true. Its truth value is 1 (true) if and only if both P and Q have truth value 1; otherwise, its truth value is 0 (false).

More generally, if the symbol c represents some binary propositional connective, we write

$P\ c\ Q$

for the new proposition obtained by combining the propositions P and Q with the use of the connective c. It is important to understand that the truth value of the new proposition $P\ c\ Q$ does not depend on the *content* or *meaning* of the original propositions P and Q, but *only on their truth values.* For instance, to determine the truth value of

P *and* Q

you do not need to know what either P or Q happens to say; you need only to know their truth values. Propositions built up by combining propositional variables with the aid of propositional connectives are called **compound** propositions and statements.

1.4 DEFINITION

TRUTH COMBINATION Let $P, Q, R, \ldots$ be propositional variables. Any definite assignment of truth values to each of these propositional variables is called a **truth combination** for $P, Q, R, \ldots$.

For instance, one possible truth combination for the triple P, Q, R of propositional variables would be obtained by assigning, say, the truth value 1 to P, the truth value 0 to Q, and the truth value 1 to R. This particular truth combination is displayed in the following table:

P	Q	R
1	0	1

How many different truth combinations can be formed for n different propositional variables

$$P_1, P_2, P_3, \ldots, P_n?$$

Any one of the two possible truth values (0 or 1) can be assigned to P_1; then, any one of two possible truth values (0 or 1) can be assigned to P_2, for a total of $2 \times 2 = 4$ different ways of assigning truth combinations to P_1 and P_2. Continuing in this way, we double the number of truth combinations for each additional propositional variable, and it follows that there are

$$\overbrace{2 \times 2 \times \cdots \times 2}^{n \text{ factors}} = 2^n$$

different ways of assigning truth combinations to

$$P_1, P_2, P_3, \ldots, P_n.$$

Thus:

> *For n different propositional variables, there are exactly 2^n different truth combinations.*

1.5 Example List in tabular form all of the $2^3 = 8$ possible truth combinations for the three propositional variables P, Q, and R.

SOLUTION

P	Q	R
1	1	1
0	1	1
1	0	1
0	0	1
1	1	0
0	1	0
1	0	0
0	0	0

Notice that we have obtained all possible truth combinations by alternating 1's and 0's in a systematic way: singly in the first column, in pairs in the second column, and in sets of four in the third column. □

Now, let's use a table to show exactly how the truth value of the compound proposition

P and Q

depends on the truth values of the two propositional variables P, Q. There are $2^2 = 4$ different truth combinations for these two propositional variables; and, for each such truth combination, the compound proposition

P and Q

will have the corresponding truth value, shown in the following table:

P	Q	P *and* Q
1	1	1
0	1	0
1	0	0
0	0	0

As this table shows, the compound proposition

P and Q

is true only if both P and Q are true; otherwise, it is false.

More generally, we can make a **truth table** for any binary propositional connective c by showing the truth values of $P \, c \, Q$ that correspond to the four possible truth combinations for the two propositional variables P and Q:

P	Q	$P \, c \, Q$
1	1	v_1
0	1	v_2
1	0	v_3
0	0	v_4

In this table the entries v_1, v_2, v_3, and v_4 represent the truth values of the compound proposition $P \, c \, Q$ corresponding to the four possible truth combinations for P and Q. Thus, v_1 is the truth value of $P \, c \, Q$ when P and Q are both true, v_2 is the truth value of $P \, c \, Q$ when P is false and Q is true, and so forth. Notice that a *propositional connective is completely determined by its truth table.* Since each of the entries v_1, v_2, v_3, and v_4 has two possible values (1 or 0), it follows that there are only 16 possible ways to assign these values (see Problem 43). Therefore, *there are only 16 possible binary propositional connectives.*

We are going to focus our attention on four of the sixteen possible binary propositional connectives. Special names and symbols are used for these four connectives.

1.6 DEFINITION

$\wedge$, $\vee$, $\Rightarrow$, AND $\Leftrightarrow$ The binary propositional connectives $\wedge$, $\vee$, $\Rightarrow$, and $\Leftrightarrow$, called **conjunction**, **disjunction**, **implication**, and **equivalence**, respectively, are defined by the following truth tables:

P	Q	$P \wedge Q$
1	1	1
0	1	0
1	0	0
0	0	0

P	Q	$P \vee Q$
1	1	1
0	1	1
1	0	1
0	0	0

P	Q	$P \Rightarrow Q$
1	1	1
0	1	1
1	0	0
0	0	1

P	Q	$P \Leftrightarrow Q$
1	1	1
0	1	0
1	0	0
0	0	1

Notice that the conjunction connective $\wedge$ is just alternative notation for the connective *and*. These four connectives are usually read (in English) as follows:

$P \wedge Q$ is read "*P and Q*,"

$P \vee Q$ is read "*P or Q*,"

$P \Rightarrow Q$ is read "*P implies Q*,"

and

$P \Leftrightarrow Q$ is read "*P is equivalent to Q*."

Sometimes,

$P \Rightarrow Q$ is read "*if P then Q*,"

and sometimes,

$P \Leftrightarrow Q$ is read "*P if and only if Q*."

Although the connective $\wedge$ is used in propositional calculus in exactly the same way the conjunction *and* is used in English, the connective $\vee$ is not always used in the same way as the English word *or*. Notice that, in the statement

John will marry Marsha or Dolores,

the word *or* is used in the **exclusive** sense of *one or the other but not both.* On the other hand, in the statement

Most families in our town have a cat or a dog,

the word *or* is used in the **inclusive** sense of *at least one and possibly both.*

This ambiguity in the meaning of *or* normally does not cause any difficulty because we can tell the intended meaning from the context; however, to avoid any possible misunderstandings in carefully written documents, *and/or* is often used for *or* in the inclusive sense. As you can see from the table for $\vee$ in Definition 1.6:

> *The connective* $\vee$ *is used in mathematics as a symbol for* or *in the inclusive sense of* and/or.

Thus, in mathematical writing, to say that $P \vee Q$ is true means that *at least one and possibly both* of the propositions P, Q are true.

The implication connective $\Rightarrow$ is usually a source of difficulty for beginning students of propositional calculus. As the truth table for $\Rightarrow$ in Definition 1.6 shows, *whenever the proposition P is false, the proposition* $P \Rightarrow Q$ is true! Thus, for instance, the compound proposition

$1 = 2 \Rightarrow$ *all pigs have wings*

is true because $1 = 2$ is false. This situation understandably upsets many people; they feel that the propositional connective $\Rightarrow$ does not correspond to their intuitive idea of logical implication. Nevertheless, it has been found that the connective $\Rightarrow$, as defined by the truth table in Definition 1.6, is precisely the version of logical implication appropriate for the construction of mathematical proofs. After studying Section 1.3, you will understand why this is so.

If you look at the truth table for the equivalence connective $\Leftrightarrow$ in Definition 1.6, you will see that the compound proposition $P \Leftrightarrow Q$ is true precisely when the two propositions P, Q have the same truth value—both true or both false. For instance, the compound proposition

$1 = 2 \Leftrightarrow$ *all pigs have wings*

is true because both of the propositions from which it is formed have the same truth value (namely, 0).

In ordinary conversation, we often have occasion to deny that something is true. The same thing is accomplished in propositional calculus by the use of the *negation connective.*

1.7 DEFINITION

> **NEGATION CONNECTIVE** If P is a propositional variable, we define the expression $\sim P$ to be a new proposition that is true if P is false and false if P is true.

The expression $\sim P$ is usually read (in English) as *not* P and the symbol $\sim$ is called the **denial connective** or the **negation connective**. Note that $\sim$ is a **unary** connective rather than a binary connective; that is, it "acts on" a *single* propositional variable rather than on a pair of propositional variables. The truth table for $\sim$ is simple enough:

P	$\sim P$
1	0
0	1

By a *formula* of propositional calculus we mean, roughly, any expression built up in a meaningful way with the use of a finite number of propositional variables and the five connectives $\wedge$, $\vee$, $\sim$, $\Rightarrow$, and $\Leftrightarrow$. More precisely:

1.8 DEFINITION

FORMULA Propositional variables are defined to be **formulas**. If $\mathscr{F}$ and $\mathscr{G}$ are formulas, then the expressions

$$\mathscr{F} \wedge \mathscr{G},\ \mathscr{F} \vee \mathscr{G},\ \sim\mathscr{F},\ \mathscr{F} \Rightarrow \mathscr{G},\quad \text{and} \quad \mathscr{F} \Leftrightarrow \mathscr{G}$$

are **formulas**. Only those expressions that can be obtained in these ways are **formulas**.

In building up formulas by the use of this definition, it may be necessary to enclose previously constructed formulas in parentheses. Formulas are also called **compound** propositions or statements.

1.9 Example Show that the expression $(P \vee Q) \Leftrightarrow (Q \Rightarrow (\sim R))$ is a formula.

SOLUTION The propositional variables P and Q are formulas; hence, $P \vee Q$ is a formula. The propositional variable R is a formula; so $\sim R$ is a formula. Because Q and $\sim R$ are formulas, it follows that $Q \Rightarrow (\sim R)$ is a formula. Finally, because $P \vee Q$ and $Q \Rightarrow (\sim R)$ are formulas, $(P \vee Q) \Leftrightarrow (Q \Rightarrow (\sim R))$ is a formula. Notice that we enclosed the previously constructed formulas $\sim R$, $P \vee Q$, and $Q \Rightarrow (\sim R)$ in parentheses before forming the final compound proposition. □

Evidently:

Every formula of propositional calculus is a proposition whose truth value depends in a definite way on the truth values of the propositional variables from which it is formed.

The dependence of the truth value of a formula on its propositional variables can be shown by means of a truth table.

1.10 Example Construct the truth table for the formula

$$(P \vee Q) \Leftrightarrow (Q \Rightarrow (\sim R)).$$

SOLUTION Because there are three propositional variables, P, Q, and R, involved in this formula, we need a truth table showing all of the $2^3 = 8$ truth combinations for these variables:

P	Q	R	$(P \vee Q) \Leftrightarrow (Q \Rightarrow (\sim R))$
1	1	1	1 1 1 **0** 1 0 0
0	1	1	0 1 1 **0** 1 0 0
1	0	1	1 1 0 **1** 0 1 0
0	0	1	0 0 0 **0** 0 1 0
1	1	0	1 1 1 **1** 1 1 1
0	1	0	0 1 1 **1** 1 1 1
1	0	0	1 1 0 **1** 0 1 1
0	0	0	0 0 0 **0** 0 1 1

In forming this truth table, we work from the "inside out." For instance, in the first row, we begin with $P \vee Q$, noting that when P and Q both have truth value 1, $P \vee Q$ has truth value 1. Thus, in the first row, we place a 1 under the symbol $\vee$ in the middle of $P \vee Q$. Likewise, in the first row, R has truth value 1, so $\sim R$ has truth value 0; and since Q has truth value 1, the implication $Q \Rightarrow (\sim R)$ has truth value 0, which we indicate by placing a 0 under the symbol $\Rightarrow$ in the middle of $Q \Rightarrow (\sim R)$. Finally, since the propositions $P \vee Q$ and $Q \Rightarrow \sim R$ on both sides of the equivalence $\Leftrightarrow$ have opposite truth values (0 and 1), we place a 0 under the symbol $\Leftrightarrow$ in the first row. The 0 under the symbol $\Leftrightarrow$ denotes the truth value (false) of the compound statement $(P \vee Q) \Leftrightarrow (Q \Rightarrow (\sim R))$ for the case in which P, Q, and R are all true. Continuing in this way, we work out the rest of the truth table as shown. □

1.11 DEFINITION

TAUTOLOGY Let $\mathcal{T}$ be a formula built up from the propositional variables $P, Q, \ldots, R$, and suppose that $\mathcal{T}$ is always true for every possible truth combination for these variables. Then we call $\mathcal{T}$ a **tautology**.

Tautologies are the theorems of propositional calculus. It is an easy matter to decide whether a formula of propositional calculus is a tautology: Just construct its truth table and look to see if an unbroken column of 1's appears beneath the formula.

1.12 Example Show that $P \vee (\sim P)$ is a tautology.

SOLUTION The truth table

P	$P \vee (\sim P)$
1	1 **1** 0
0	0 **1** 1

shows that $P \vee (\sim P)$ is a tautology. □

The formula $P \vee (\sim P)$, which is perhaps the simplest of all tautologies, is called the **law of the excluded middle** because it states P *or not* P, that is, P is either true or false—there is no "middle" possibility. The formula

$$((\sim P) \vee Q) \Leftrightarrow (P \Rightarrow Q)$$

is a somewhat less trivial example of a tautology.

1.13 Example Show that $((\sim P) \vee Q) \Leftrightarrow (P \Rightarrow Q)$ is a tautology.

SOLUTION

P	Q	$((\sim P) \vee Q) \Leftrightarrow (P \Rightarrow Q)$
1	1	0 1 1 **1** 1 1 1
0	1	1 1 1 **1** 0 1 1
1	0	0 0 0 **1** 1 0 0
0	0	1 1 0 **1** 0 1 0

□

Here is a list of some of the more useful tautologies:

1. $P \Leftrightarrow \sim(\sim P)$ — Double denial
2. $(P \wedge Q) \Leftrightarrow (Q \wedge P)$ — Commutative law for conjunction
3. $(P \vee Q) \Leftrightarrow (Q \vee P)$ — Commutative law for disjunction
4. $(P \wedge Q) \wedge R \Leftrightarrow P \wedge (Q \wedge R)$ — Associative law for conjunction
5. $(P \vee Q) \vee R \Leftrightarrow P \vee (Q \vee R)$ — Associative law for disjunction
6. $P \vee (\sim P)$ — Law of the excluded middle
7. $\sim(P \wedge (\sim P))$ — Law of noncontradiction
8. $P \wedge (Q \vee R) \Leftrightarrow (P \wedge Q) \vee (P \wedge R)$ — Distributive law for $\wedge$ over $\vee$
9. $P \vee (Q \wedge R) \Leftrightarrow (P \vee Q) \wedge (P \vee R)$ — Distributive law for $\vee$ over $\wedge$
10. $\sim(P \wedge Q) \Leftrightarrow (\sim P) \vee (\sim Q)$ — De Morgan's law for $\wedge$
11. $\sim(P \vee Q) \Leftrightarrow (\sim P) \wedge (\sim Q)$ — De Morgan's law for $\vee$
12. $(P \Rightarrow Q) \Leftrightarrow ((\sim Q) \Rightarrow (\sim P))$ — Law of contraposition
13. $P \Leftrightarrow P$ — Reflexivity of equivalence
14. $(P \Leftrightarrow Q) \Leftrightarrow (Q \Leftrightarrow P)$ — Symmetry of equivalence
15. $(P \Leftrightarrow Q) \wedge (Q \Leftrightarrow R) \Rightarrow (P \Leftrightarrow R)$ — Transitivity of equivalence
16. $(P \Rightarrow Q) \wedge (Q \Rightarrow R) \Rightarrow (P \Rightarrow R)$ — Transitivity of implication
17. $((\sim P) \Rightarrow (Q \wedge (\sim Q))) \Rightarrow P$ — Reductio ad absurdum
18. $(P \wedge Q) \Rightarrow P$ — Simplification
19. $P \Rightarrow (Q \Rightarrow (P \wedge Q))$ — Conjunction
20. $P \Rightarrow (P \vee Q)$ — Disjunction

21. $(P \Rightarrow Q) \Leftrightarrow ((\sim P) \vee Q)$
22. $(P \Leftrightarrow Q) \Leftrightarrow ((P \Rightarrow Q) \wedge (Q \Rightarrow P))$
23. $Q \Rightarrow (P \Rightarrow Q)$
24. $(\sim P) \Rightarrow (P \Rightarrow Q)$
25. $(P \Rightarrow (Q \Rightarrow R)) \Leftrightarrow ((P \wedge Q) \Rightarrow R)$
26. $((P \Rightarrow R) \wedge (Q \Rightarrow R)) \Leftrightarrow ((P \vee Q) \Rightarrow R)$
27. $((P \wedge Q) \Rightarrow R) \Leftrightarrow ((P \wedge (\sim R)) \Rightarrow (\sim Q))$
28. $(P \Leftrightarrow Q) \Leftrightarrow ((\sim P) \Leftrightarrow (\sim Q))$
29. $(P \wedge (Q \Rightarrow R)) \Rightarrow ((P \wedge Q) \Rightarrow (P \wedge R))$
30. $(P \Rightarrow Q) \vee (Q \Rightarrow P)$

If $\mathscr{T}$ is a tautology, then $\sim \mathscr{T}$ is a formula whose truth value is always 0 (false), no matter what truth combination is assigned to its propositional variables. Such a formula is called an **antitautology**. The simplest example of an antitautology is the formula

$$P \wedge (\sim P)$$

which says, in words, that P is both true and at the same time false—an obvious impossibility. A compound statement of the form

$$P \wedge (\sim P)$$

is called a **contradiction**. Note that tautology 17 above says, in words, that *if the denial of P leads to a contradiction, then P must be true.* This is an important tautology, for it supplies the basis for the indirect proofs to be discussed in Section 1.3.

In the following definition, we set forth some terminology often used in connection with propositions of the form $P \Rightarrow Q$.

1.14 DEFINITION

HYPOTHESIS, CONCLUSION, CONVERSE, CONTRAPOSITIVE In an implication of the form

$$P \Rightarrow Q,$$

the proposition P is called the **hypothesis** or the **antecedent**, and the proposition Q is called the **conclusion** or the **consequent**. By interchanging the hypothesis and conclusion, we obtain a new implication

$$Q \Rightarrow P$$

called the **converse** of the original implication. By negating both the hypothesis and the conclusion in this converse implication, we obtain yet another implication

$$(\sim Q) \Rightarrow (\sim P)$$

called the **contrapositive** of the original implication.

According to the truth table

P	Q	$P \Rightarrow Q$
1	1	1
0	1	1
1	0	0
0	0	1

we have the following:

If the hypothesis of an implication is false, or if its conclusion is true, then the implication is true. Indeed, an implication is false only in the case in which its hypothesis is true and its conclusion is false.

As a consequence of tautology 12 in the list above:

An implication is always logically equivalent to its own contrapositive implication.

For instance, if P is the proposition *today is Thanksgiving* and Q is the proposition *tomorrow is Friday*, then the implication

$$P \Rightarrow Q$$

is certainly true. Indeed, if today is Thanksgiving, then tomorrow is Friday. The contrapositive implication

$$(\sim Q) \Rightarrow (\sim P)$$

is also true, because, if tomorrow is not Friday, then today cannot possibly be Thanksgiving. However, the converse of $P \Rightarrow Q$, the implication

$$Q \Rightarrow P,$$

claims that, if tomorrow is Friday, then today is Thanksgiving! Because not every Thursday is Thanksgiving (a pity, isn't it?), the converse implication $Q \Rightarrow P$ is false.

From the fact that an implication $P \Rightarrow Q$ is true, we can draw no conclusion about the truth or falsity of its converse implication $Q \Rightarrow P$.

When a mathematician proves a theorem having the form $P \Rightarrow Q$, the question of whether the converse $Q \Rightarrow P$ is also true naturally presents itself. If the converse implication also proves to be true, then tautology 22 can be used to conclude that $P \Leftrightarrow Q$; that is, P and Q are logically equivalent (have the same truth value).

Advances in our mathematical knowledge often begin with educated guesses or conjectures concerning the truth value of various mathematical propositions. Suppose that a mathematician is somehow led to conjecture that a certain implication $P \Rightarrow Q$ is true. Such a conjecture could be defeated by finding a particular example for which the hypothesis P is true and the conclusion Q is false. An example used to show that a guess or conjecture is false is called a **counterexample**.

1.15 Example Let x, y, and z represent arbitrary real numbers. Show by means of a counterexample that the implication

$$xy = xz \Rightarrow y = z$$

is false.

SOLUTION At first glance, this implication looks plausible enough: It seems that the conclusion $y = z$ follows from the hypothesis $xy = xz$ if we just "cancel the x on both sides." However, consider the case in which $x = 0$, $y = 1$, and $z = 2$. In this case, the hypothesis $xy = xz$ is true (it just says that $0 = 0$), whereas the conclusion $y = z$ is false (it says that $1 = 2$). □

In working with mathematical statements, it is often necessary to find a formula that is equivalent to the denial of a given formula, but in which no denial symbol appears in front of a compound proposition. For instance, by De Morgan's law for $\wedge$ (tautology 10), the denial of the formula $P \wedge Q$ is equivalent to the formula $(\sim P) \vee (\sim Q)$. Since every formula is built up from our five basic connectives $\wedge$, $\vee$, $\Rightarrow$, $\Leftrightarrow$, and $\sim$, it suffices to know how to deny each of the formulas $P \wedge Q$, $P \vee Q$, $P \Rightarrow Q$, $P \Leftrightarrow Q$, and $\sim P$. The following tautologies show how this is done:

(i) $\sim(P \wedge Q) \Leftrightarrow (\sim P) \vee (\sim Q)$
(ii) $\sim(P \vee Q) \Leftrightarrow (\sim P) \wedge (\sim Q)$
(iii) $\sim(P \Rightarrow Q) \Leftrightarrow P \wedge (\sim Q)$
(iv) $\sim(P \Leftrightarrow Q) \Leftrightarrow (P \wedge (\sim Q)) \vee (Q \wedge (\sim P))$
(v) $\sim(\sim P) \Leftrightarrow P$

Tautologies (i) and (ii) are De Morgan's laws (tautologies 10 and 11), and (v) is the double denial tautology (tautology 1). We leave it to you to verify (iii) and (iv) by means of truth tables (Problem 36).

1.16 Example Write out, in English, the denial of the following statement:

If Alfie is taking chemistry, then he is smart.

SOLUTION Let P be the statement *Alfie is taking chemistry*, and let Q be the statement *Alfie is smart*. The statement to be denied has the form $P \Rightarrow Q$; hence, by (iii) above, its denial is equivalent to $P \wedge (\sim Q)$, that is:

Alfie is taking chemistry and he is not smart.

Note that in translating from English to the symbolism of propositional calculus and vice versa, we have freely interchanged the pronoun *he* and the noun *Alfie* to which it refers. □

You will find that the following "dictionary" is handy for translating certain English expressions into the formulas of propositional calculus.

English expression	Translation
1. P or Q or both	$P \vee Q$
2. P and Q	$P \wedge Q$
3. P, but Q	$P \wedge Q$
4. P, however Q	$P \wedge Q$
5. P is not true.	$\sim P$
6. P implies Q.	$P \Rightarrow Q$
7. If P, then Q	$P \Rightarrow Q$
8. P is a sufficient condition for Q.	$P \Rightarrow Q$
9. Q if P	$P \Rightarrow Q$
10. P is a necessary condition for Q.	$Q \Rightarrow P$
11. Q only if P	$Q \Rightarrow P$
12. P is implied by Q.	$Q \Rightarrow P$
13. P if and only if Q	$P \Leftrightarrow Q$
14. P is a necessary and sufficient condition for Q.	$P \Leftrightarrow Q$
15. P is equivalent to Q.	$P \Leftrightarrow Q$
16. Either P or Q, but not both	$\sim(P \Leftrightarrow Q)$
17. P, but not Q	$P \wedge (\sim Q)$
18. P unless Q	$(\sim Q) \Rightarrow P$

PROBLEM SET 1.1

1. Which of the following expressions are propositions? [Note: Our definition of *proposition* is not sufficiently precise to allow for clear-cut answers in every case, so you might want to label some cases as indeterminate.]
   (a) If only I could make an A in this course.
   (b) Statement (b) in the present list is false.
   (c) Somewhere in the decimal expansion of pi the sequence of digits 123456789 occurs.
   (d) If p is a prime number, then $2^p - 1$ is also a prime number.
   (e) $57x^2 + \sin x - x^{-1}$.

(f) The present king of France is bald.
(g) How many solutions does a quadratic equation have?
(h) $57x^2 + \sin x - x^{-1} = 243$.
(i) If pigs have wings, then some winged animals are good to eat.
(j) Every boojum is a snark.
(k) It is raining.
(l) If a function has a derivative, then it is continuous.
(m) All squares are rectangles.
(n) All rectangles are squares.
(o) Let $x < 2$.

2. List in tabular form all of the $2^4 = 16$ possible truth combinations for the four propositional variables P, Q, R, and S.

3. Suppose P, Q, and R are propositions with $P \Leftrightarrow Q$ and $Q \Leftrightarrow R$.
(a) Explain in words why it follows that $P \Leftrightarrow R$.
(b) Which tautology in the list on pages 10–11 confirms the conclusion in Part (a).

4. Let P be the proposition $4 > 3$, and let Q be the proposition $4 > 2$. Find the truth value of each of the following:
(a) $P \vee Q$ (b) $P \wedge Q$ (c) $P \Rightarrow Q$ (d) $P \Leftrightarrow Q$ (e) $P \vee (\sim Q)$

5. Suppose that two baseball teams A and B are scheduled to play each other with the understanding that extra innings will be used, if necessary, to break a tie. Let P be the proposition *team A will win the game*, and let Q be the proposition *team B will win the game*. Find the truth value of each of the following:
(a) $P \vee Q$ (b) $P \wedge Q$ (c) $P \Leftrightarrow Q$
(d) $(\sim P) \Rightarrow Q$ (e) $P \Rightarrow (\sim Q)$ (f) $(\sim P) \Leftrightarrow (\sim Q)$

6. Professor Grumbles promises Alfie that, if he makes a C on the final exam in calculus, then he will pass the course. Let P be the statement *Alfie passes the course*, and let Q be the statement *Alfie makes a C on the final exam in calculus.*
(a) Express Professor Grumbles' promise in symbolic form as an implication.
(b) For each of the possible truth combinations for P and Q, determine whether or not Professor Grumbles' promise is kept.

7. Let P be the proposition *interest rates are high*, and let Q be the proposition *the inflation rate is low.* Interpret the word *low* to mean *not high.* Write each of the following propositions in symbolic form:
(a) Interest rates are high, and the inflation rate is high.
(b) Interest rates are low, and the inflation rate is low.
(c) If the inflation rate is low, then interest rates are low.
(d) If the inflation rate is high, then interest rates are high.
(e) Neither the inflation rate nor the interest rate is high.

8. If P and Q are propositions, let $P \dashv Q$ denote the proposition asserting that statement Q can be deduced from statement P. Is $\dashv$ a propositional connective? Why or why not?

9. Using De Morgan's laws, write out (in English) the denial of each of the following statements:
(a) This equation has a solution, or I am stupid.
(b) Alfie is studying hard and doing well in his classes.

10. Write out (in English) the denial of the statement *if it is not raining, then it is snowing.*

11. Let P, Q, and R denote the statements *it is snowing, I am cold*, and *I am going home*, respectively. Write out in English the propositions represented symbolically by the following:
(a) $\sim Q$ (b) $Q \Rightarrow (P \wedge (\sim R))$ (c) $P \wedge (\sim R)$
(d) $Q \Leftrightarrow R$ (e) $R \Rightarrow Q$ (f) $\sim(Q \Leftrightarrow R)$
(g) $(P \vee Q) \Rightarrow R$ (h) $\sim(R \Rightarrow Q)$

12. Complete the tautology $(P \Rightarrow Q) \Leftrightarrow ?$ by replacing the question mark with a formula involving only P, Q, and the connectives $\sim$ and $\wedge$.

13. Continuing with the notation of Problem 11, translate each of the following statements into the symbolism of the propositional calculus:
(a) If it is snowing, then I am going home.
(b) If it is snowing and I am cold, then I am going home.
(c) I am not going home unless I am cold.
(d) A necessary condition that I go home is that I am cold.
(e) If I am going home, then it is snowing or I am cold.
(f) It is either snowing or I am cold, but not both.

14. Complete the tautology $(P \Leftrightarrow Q) \Leftrightarrow ?$ by replacing the question mark with a formula involving only P, Q, and the connectives $\sim$ and $\wedge$.

15. Complete the following truth tables:

P	Q	R	$P \vee R$	$Q \vee R$	$P \wedge R$	$Q \wedge R$	$P \wedge (Q \wedge R)$	$P \wedge (Q \vee R)$	$P \vee (Q \wedge R)$
1	1	1							
0	1	1							
1	0	1							
0	0	1							
1	1	0							
0	1	0							
1	0	0							
0	0	0							

16. Show that the eight truth combinations for the three propositional variables P, Q, and R correspond in a natural way to the representations of the numbers 0, 1, 2, 3, 4, 5, 6, and 7 in the binary (base 2) system.

17. Complete the truth table:

P	Q	R	$(P \wedge ((\sim Q) \vee R)) \vee Q$
1	1	1	
0	1	1	
1	0	1	
0	0	1	
1	1	0	
0	1	0	
1	0	0	
0	0	0	

18. Construct a truth table for the formula $(P \Rightarrow Q) \Rightarrow (Q \Rightarrow P)$.

19. Check tautology 1 (double denial) in the list on pages 10–11 by a truth table.

20. Professor Grumbles says, "I can't afford not to do without a new car."
(a) Can he afford the new car?
(b) Justify your reasoning in Part (a) by forming a suitable tautology.

21. Check the commutative and associative laws (tautologies 2, 3, 4, and 5) by truth tables.

22. Check each of the following tautologies by truth tables:
(a) The law of noncontradiction (tautology 7).
(b) The reductio ad absurdum tautology (tautology 17).

23. Check the distributive laws (tautologies 8 and 9) by truth tables.

24. Check tautologies 13 and 14 by truth tables.

25. Check De Morgan's laws (tautologies 10 and 11) by truth tables.

26. Check tautologies 15 and 16 by truth tables.

27. Check the law of contraposition (tautology 12) by a truth table.

28. Check tautologies 23 and 24 by truth tables, and give verbal explanations of their meanings.

29. Check tautologies 21 and 22 by truth tables.

30. Check tautologies 25 through 30 by truth tables.

31. Check the simplification, conjunction, and disjunction tautologies (tautologies 18, 19, and 20) by truth tables.

32. Make up a tautology that is not in the list on pages 10–11 and check it by a truth table.

33. The tautology $(P \wedge (P \Rightarrow Q)) \Rightarrow Q$ is called *modus ponens.* Check this tautology by a truth table.

34. (a) Without using a truth table, explain why you would expect the formula $((\sim P) \wedge (P \vee Q)) \Rightarrow Q$ to be a tautology.
(b) Check that this formula is a tautology by using a truth table.

35. Write the contrapositive of each implication:
(a) If n is an integer, then $2n + 1$ is an odd integer.
(b) You can take Math 212 only if you passed Math 211.
(c) If the series $\sum_{k=1}^{\infty} a_k$ converges, then $\lim_{n \to \infty} a_n = 0$.
(d) If a citizen is of age 18 or over, then he or she can vote.

36. Check tautologies (iii) and (iv) on page 13 by truth tables.

37. Show by means of a counterexample that the following statement is false: If p is a prime number, then p is odd.

38. Show by means of a counterexample that the following statement is false: If a series converges, then it converges absolutely.

39. True or false: If $x^2 > 9$, then $x > 3$?

40. Give truth tables for each of the 16 possible binary connectives, calling them $c_1, c_2, c_3, \ldots, c_{16}$. For each of these binary connectives c_j, $j = 1, 2, 3, \ldots, 16$, find a formula $\mathscr{F}_j$ involving only the propositional variables P and Q such that $P\ c_j\ Q$ is equivalent to $\mathscr{F}_j$.

41. Let $P_1, P_2, P_3, \ldots, P_n$ represent n different propositional variables. Using mathematical induction (see the Appendix on Mathematical Induction), prove that there are exactly 2^n different truth combinations for these n propositional variables.

42. Explain why any formula $\mathscr{F}$ of the propositional calculus is equivalent to a formula $\mathscr{G}$ of the propositional calculus in which $\mathscr{G}$ involves the same propositional variables as $\mathscr{F}$ but only the propositional connectives $\sim$ and $\vee$. [Hint: Note that $P \wedge Q$ is equivalent to $\sim((\sim P) \vee (\sim Q))$, $P \Rightarrow Q$ is equivalent to $(\sim P) \vee Q$, and so on.]

43. Using mathematical induction (see the Appendix on Mathematical Induction), give a careful proof showing that there are only 16 different possible binary connectives.

44. Let $P_1, P_2, P_3, \ldots, P_n$ represent n different propositional variables. Show that a truth table exhibiting all of the 2^n possible truth combinations for these variables can be constructed as follows: Head n columns with the symbols $P_1, P_2, P_3, \ldots,$ and P_n. Alternate 1's and 0's in the first column (under P_1) until 2^n rows are used. Alternate 1's and 0's in pairs in the second column (under P_2) until 2^n rows are used. Continue in this way, alternating 1's and 0's in the jth column in sets of 2^{j-1} for $j = 1, 2, 3, \ldots, n$.

2 QUANTIFIERS

Phrases of the form *for every x* and *there exists an x such that* are called **quantifiers**. The expression

for every x

is the **universal quantifier**, and the expression

there exists an x such that

is the **existential quantifier**. Although these quantifiers are indispensable in advanced mathematics, they are not often used in elementary algebra and geometry. Why? One reason is that in elementary textbooks the universal quantifier is often suppressed in the interest of brevity. For instance, if you read

$$(x + 1)^2 = x^2 + 2x + 1$$

in an elementary algebra book, you are supposed to understand that this equation holds for every value of x. Another reason is that theorems asserting the existence of something (*existence theorems*) are usually too difficult to be included in elementary mathematics courses. For instance, in elementary algebra, you learned how to handle square roots of positive numbers, but there probably was little, if any, discussion of whether these square roots actually exist.

In more advanced mathematics courses, the suppression of universal quantifiers can lead to hopeless confusion. Also, some of the more important theorems of advanced mathematics are existence theorems. Thus, to progress from elementary to more advanced mathematics, it is necessary to learn to understand and deal with quantifiers.

In any mathematical discussion, we usually have in mind some particular "universe" or **domain of discourse** that consists of those things whose properties and relationships are under consideration. For instance, in elementary algebra our domain of discourse is the set of all real numbers; in elementary geometry it is the set of all points in the plane.

In what follows, U represents some particular domain of discourse. Assuming that there is at least one object in U, we introduce symbols

$$x, y, z, \ldots$$

to stand for arbitrary objects in U, and we call these symbols **object variables**. For instance, in elementary algebra, U is the set of all real numbers, and the object variables $x, y, z, \ldots$ are just the usual numerical variables or "unknowns."

2.1 DEFINITION

PREDICATE OR PROPOSITIONAL FUNCTION By a **predicate** or **propositional function**, we mean a complete declarative sentence $P(x)$ that makes a statement about the object variable x. We call x the **argument** of $P(x)$. Thus, if x is assigned a particular value in U, then $P(x)$ becomes a proposition with a definite truth value.

2.2 Example Give several examples of propositional functions in elementary algebra.

SOLUTION We take the domain of discourse U to be the set of all real numbers. Here are some examples of propositional functions:

(a) Let $P(x)$ be the statement $x > 0$.
(b) Let $Q(x)$ be the statement $x^2 - 5x + 6 = 0$.
(c) Let $W(x)$ be the statement x *is a whole number.*

In this example, for instance, $P(3)$ is true, $P(-2)$ is false, $Q(-1)$ is false, $Q(2)$ is true, $W(7)$ is true, and $W(\pi)$ is false. □

We are now ready to introduce some useful symbolism for universal and existential quantifiers.

2.3 DEFINITION

UNIVERSAL QUANTIFIER Let $P(x)$ be a propositional function whose argument x ranges over the domain of discourse U. Then the expression

$$(\forall x)(P(x))$$

stands for the proposition asserting that, no matter what value (in U) is assigned to x, the resulting proposition $P(x)$ is true.

The proposition $(\forall x)(P(x))$ is read in English as

for all x, $P(x)$

or

for every x, $P(x)$,

and the symbol $\forall x$ denotes the universal quantifier. We note that if y is an object in U, then

$$(\forall x)(P(x)) \Rightarrow P(y).$$

2.4 Example If U is the set of all real numbers, determine the truth value of each of the following quantified propositions:

(a) $(\forall x)((x + 1)^2 = x^2 + 2x + 1)$
(b) $(\forall x)(x > 2)$

SOLUTION Proposition (a) is true because, indeed, $(x + 1)^2 = x^2 + 2x + 1$ is a true statement, no matter what real number is put in place of x. However, (b) is false because not every real number is greater than 2; for instance, if the value 1 is assigned to x, the resulting statement $1 > 2$ is false. □

2.5 DEFINITION

EXISTENTIAL QUANTIFIER Let $P(x)$ be a propositional function whose argument x ranges over the domain of discourse U. Then the expression

$$(\exists x)(P(x))$$

will stand for the proposition asserting that there exists at least one value of x (in U) for which the resulting proposition $P(x)$ is true.

The proposition $(\exists x)(P(x))$ is read in English as

there exists an x such that P(x)

or

for some x, P(x),

and the symbol $\exists x$ denotes the existential quantifier. We note that if y is an object in U, then

$$P(y) \Rightarrow (\exists x)(P(x)).$$

2.6 Example If U is the set of all real numbers, determine the truth value of each of the following quantified propositions:

(a) $(\exists x)(x^2 - 5x + 6 = 0)$
(b) $(\exists x)(x^2 + 2x + 6 = 0)$

SOLUTION Proposition (a) is true because there does exist at least one real number x (for instance, $x = 2$ or $x = 3$) such that $x^2 - 5x + 6 = 0$. However, (b) is false because the discriminant of the quadratic equation $x^2 + 2x + 6 = 0$ is negative; so this equation has no real solution. □

If $P(x)$ is a propositional function, then $(\forall x)(P(x))$ and $(\exists x)(P(x))$ are actually propositions, *not propositional functions.* Thus, even though the expressions $(\forall x)(P(x))$ and $(\exists x)(P(x))$ contain the object variable x, their truth value does not depend in any way on this variable, but only on the propositional function $P(x)$ and the domain of discourse U. For instance, if U is the set of all real numbers, then

$$(\exists x)(x > 2)$$

merely says that there is a real number greater than 2; hence, it is a true proposition having to do only with real numbers and the relation of being greater than. It really has nothing to do with x at all! For this reason, we say that the symbol x that occurs in the quantified expressions $(\forall x)(P(x))$ and $(\exists x)(P(x))$ is a **dummy variable** or an **apparent variable** because it only *appears* (to the uninformed) that these propositions depend on x in some way. Logicians describe this by saying that the variable x occurs as a **free variable** in the propositional function $P(x)$, so we can freely assign it any value in U; but it is a **bound variable** in the quantified expressions $(\forall x)(P(x))$ and $(\exists x)(P(x))$ and, as such, is no longer available for the assignment of values.

For ordinary numerical-valued functions $f(x)$, a similar situation occurs in the integral calculus. Indeed, we can assign x any value in the domain of such a function and the result is a real number, for instance $f(-7)$ or $f(1.3)$ or $f(\pi)$. However, the definite integral

$$\int_a^b f(x)\,dx,$$

which appears (again to the uninformed) to depend in some way on x, is actually a number depending only on the function being integrated and the interval $[a, b]$ over which the integration takes place. In the expression for the definite integral, x is a dummy variable which cannot be assigned particular numerical values.

The fact that x is a dummy variable in the definite integral can be expressed by the equation

$$\int_a^b f(x)\,dx = \int_a^b f(y)\,dy.$$

Likewise, the fact that x is a dummy variable in the quantified expressions $(\forall x)(P(x))$ and $(\exists x)(P(x))$ can be expressed by the rules

$$(\forall x)(P(x)) \Leftrightarrow (\forall y)(P(y))$$

and

$$(\exists x)(P(x)) \Leftrightarrow (\exists y)(P(y)).$$

If $P(x)$ is a propositional function for the domain of discourse U, then the proposition

$$(\forall x)(P(x))$$

asserts that $P(x)$ is true no matter what value (in U) is assigned to x; hence, its denial

$$\sim(\forall x)(P(x))$$

is equivalent to the statement that there exists at least one value of x (in U) for which $P(x)$ is false. Therefore, we have the important rule

$$\sim(\forall x)(P(x)) \Leftrightarrow (\exists x)(\sim P(x)).$$

Similarly, the proposition

$$(\exists x)(P(x))$$

asserts that $P(x)$ is true for at least one value of x (in U); hence, its denial

$$\sim(\exists x)(P(x))$$

is equivalent to the statement that $P(x)$ is false for all values of x (in U). Therefore, we also have the rule

$$\sim(\exists x)(P(x)) \Leftrightarrow (\forall x)(\sim P(x)).$$

These two rules for forming the denial of a quantified statement can be summarized as follows:

The denial symbol ~ can be "pushed through" a quantifier at the expense of changing the quantifier from universal to existential and vice versa.

A person untrained in logic who is asked to deny the proposition *all men are mortal* might respond with *all men are not mortal.* If interpreted literally, the last statement asserts that *all men are immortal*, which is not the denial of the original proposition.

2.7 Example Write the denial of the proposition *all men are mortal.*

SOLUTION Let U, the domain of discourse, be the set of all men, and let $M(x)$ be the propositional function asserting that x is mortal. Thus, the proposition *all men are mortal* is expressed by the quantified expression

$$(\forall x)(M(x)),$$

and its denial

$$\sim(\forall x)(M(x))$$

is equivalent to the proposition

$$(\exists x)(\sim M(x)),$$

which translates into the statement *there exists at least one man who is immortal.* □

The formal study of propositional functions and quantifiers is called **predicate calculus** or **functional calculus**. Predicate (or functional) calculus has roughly the same relationship to propositional calculus as the theory of numerical functions has to ordinary algebra. In multivariate calculus, we consider functions $f(x, y)$, $g(x, y, z)$, and so on, which have more than one independent variable. Likewise, in predicate calculus, we have to deal with propositional functions $P(x, y)$, $Q(x, y, z)$, and so on, which involve more than one object variable. Here are some examples of such propositional functions of more than one variable:

Domain of discourse	Number of variables	Propositional function
1. U = all living humans	2	x is older than y
2. U = all real numbers	2	$x^2 + y^2 = 1$
3. U = all real numbers	2	$x^2 + y^2 \leq 1$
4. U = all integers	2	x is an exact divisor of y
5. U = all real numbers	3	$x - \sin y > z$
6. U = all points in the plane	3	x is on the line segment between y and z
7. U = all real numbers	3	$(0 < \lvert x - 3 \rvert < \varepsilon) \Rightarrow (\lvert x^2 - 9 \rvert < \delta)$
8. U = all real numbers	n	$x_1 > x_2 > x_3 > \cdots > x_n$

If a quantifier is applied to a propositional function of several variables, the result is a propositional function of one fewer variable.

2.8 Example Let the domain of discourse U consist of all real numbers, and consider the propositional function $x < y$ of the two variables x and y. Quantify this expression:

(a) Universally with respect to y
(b) Existentially with respect to x

Interpret the results.

SOLUTION **(a)** If we quantify $x < y$ universally with respect to y, we obtain the expression

$$(\forall y)(x < y),$$

which is a propositional function of the single variable x (since y has become an apparent variable). The propositional function $(\forall y)(x < y)$ states that x is less than every number in U.

(b) If we quantify $x < y$ existentially with respect to x, we obtain the expression

$$(\exists x)(x < y),$$

which is a propositional function of the single variable y (since x has become an apparent variable). The propositional function $(\exists x)(x < y)$ states that there exists at least one number that is less than y. □

After a propositional function of several variables has been quantified with respect to one of these variables, the resulting expression can then be quantified with respect to any one of the remaining free variables.

2.9 Example Consider the propositional functions $(\forall y)(x < y)$ and $(\exists x)(x < y)$ obtained in Example 2.8.

(a) Quantify $(\forall y)(x < y)$ existentially with respect to x.
(b) Quantify $(\exists x)(x < y)$ universally with respect to y.

Interpret the results.

SOLUTION **(a)** Quantifying $(\forall y)(x < y)$ existentially with respect to x, we obtain

$$(\exists x)(\forall y)(x < y),$$

which states that there exists a real number x that is less than every real number. Thus, $(\exists x)(\forall y)(x < y)$ is a false statement.

(b) Quantifying $(\exists x)(x < y)$ universally with respect to y, we obtain

$$(\forall y)(\exists x)(x < y),$$

which states that for every real number y there is at least one smaller real number x. Thus, $(\forall y)(\exists x)(x < y)$ is a true statement. □

2.10 Example Rewrite the denial of the proposition

$$(\exists x)(\forall y)(x < y)$$

in such a way that no denial connective appears explicitly, and interpret the result.

SOLUTION Starting with

$$\sim(\exists x)(\forall y)(x < y)$$

we "push the denial connective $\sim$ through the quantifiers," changing existential to universal and universal to existential, to obtain the equivalent proposition

$$(\forall x)(\exists y)(\sim(x < y)).$$

The statement $\sim(x < y)$ says that the real number x is not less than the real number y; hence, it is equivalent to the statement that $x \geq y$. Therefore, the denial of the proposition

$$(\exists x)(\forall y)(x < y)$$

is equivalent to the proposition

$$(\forall x)(\exists y)(x \geq y).$$

In words: To deny that there is a real number x that is smaller than every real number y is to assert that, for every real number x, there is a real number y that is less than or equal to x. □

Example 2.9 shows that you must pay attention to the *order* in which quantifiers are written! Although the propositions

$$(\exists x)(\forall y)(x < y)$$

and

$$(\forall y)(\exists x)(x < y)$$

differ only in the order in which the quantifiers appear, they are *not equivalent*.

More generally, suppose that $P(x, y)$ is a propositional function of two variables, and consider the two propositions $(\exists x)(\forall y)(P(x, y))$ and $(\forall y)(\exists x)(P(x, y))$, which differ only in the order in which the quantifiers are written. On the one hand:

> The proposition $(\exists x)(\forall y)(P(x, y))$ states that there exists a fixed x in U such that $P(x, y)$ is true for every choice of y in U.

On the other hand:

> The proposition $(\forall y)(\exists x)(P(x, y))$ states that for every choice of y in U, there exists some x (possibly depending on the choice of y), such that $P(x, y)$ is true.

Consequently, we have the rule

$$(\exists x)(\forall y)(P(x, y)) \Rightarrow (\forall y)(\exists x)(P(x, y))$$

but, in general, the converse implication is false.

It turns out that:

Two adjacent quantifiers of the same type (universal or existential) can always be transposed.

Thus, we have the rules

$$(\forall x)(\forall y)(P(x, y)) \Leftrightarrow (\forall y)(\forall x)(P(x, y))$$

and

$$(\exists x)(\exists y)(P(x, y)) \Leftrightarrow (\exists y)(\exists x)(P(x, y))$$

For instance, if U consists of the real numbers, the proposition

$$(\forall x)(\forall y)((x + y)^2 = x^2 + 2xy + y^2)$$

asserts that the equation

$$(x + y)^2 = x^2 + 2xy + y^2$$

is true for all choices of the real numbers x and y, and this is exactly the same as the assertion made by the proposition

$$(\forall y)(\forall x)((x + y)^2 = x^2 + 2xy + y^2).$$

Similarly, the proposition

$$(\exists x)(\exists y)(x < y)$$

asserts the existence of a pair of real numbers, one of which is less than the other, and this is exactly the same as the assertion made by the proposition

$$(\exists y)(\exists x)(x < y).$$

It often happens that some, but not all, of the objects in the domain of discourse U have a particularly interesting or desirable property. In this case, it may be convenient to reserve certain symbols to represent objects in U that have this property. For instance, in studying differential and integral calculus, our domain of discourse U is likely to be the set of all real numbers. Here the property of being a *positive* number is often of significance, and the Greek letters ε (epsilon) and δ (delta) are often understood to represent only positive numbers. If this agreement is made, then the symbol $\forall\varepsilon$ would correspond to the phrase *for every positive* ε, and the symbol $\exists\delta$ would correspond to the phrase *there exists a positive* δ.

2.11 Example Suppose that $f(x)$ is a real-valued function of the real variable x. Write out, in the symbolism of predicate calculus, the definition of

$$\lim_{x \to a} f(x) = L.$$

SOLUTION Let us use the convention that ε and δ stand for positive real numbers. By definition, the statement

$$\lim_{x \to a} f(x) = L$$

means that, for every positive number ε, there exists a positive number δ such that, if $0 < |x - a| < \delta$, then $|f(x) - L| < \varepsilon$. Thus, by definition,

$$\lim_{x \to a} f(x) = L \Leftrightarrow (\forall \varepsilon)(\exists \delta)((0 < |x - a| < \delta) \Rightarrow (|f(x) - L| < \varepsilon)) \qquad \square$$

Predicate calculus is an important branch of symbolic logic and can be developed with great rigor and precision. Here we have given only a brief and informal introduction to this subject. Every serious student of mathematics should, at some time, work through the formal theory of the predicate calculus as it is found in any standard textbook of symbolic logic. For future reference, we present the following list of some of the basic theorems of the predicate calculus. In this list, x and y denote object variables ranging over a fixed universe of discourse U; $P(x)$, $Q(x)$, $P(x, y)$, and $Q(x, y)$ denote propositional functions; and R denotes a proposition or a propositional function that does not contain x as a free variable.

1. $(\forall x)(P(x)) \Rightarrow P(y)$
2. $P(y) \Rightarrow (\exists x)(P(x))$
3. $(\forall x)(P(x)) \Rightarrow (\exists x)(P(x))$
4. $(\forall x)(P(x)) \Leftrightarrow (\forall y)(P(y))$
5. $(\exists x)(P(x)) \Leftrightarrow (\exists y)(P(y))$
6. $\sim(\forall x)(P(x)) \Leftrightarrow (\exists x)(\sim P(x))$
7. $\sim(\exists x)(P(x)) \Leftrightarrow (\forall x)(\sim P(x))$
8. $(\exists x)(\forall y)(P(x, y)) \Rightarrow (\forall y)(\exists x)(P(x, y))$
9. $(\forall x)(\forall y)(P(x, y)) \Leftrightarrow (\forall y)(\forall x)(P(x, y))$
10. $(\exists x)(\exists y)(P(x, y)) \Leftrightarrow (\exists y)(\exists x)(P(x, y))$
11. $(\forall x)(P(x) \wedge Q(x)) \Leftrightarrow [(\forall x)(P(x)) \wedge (\forall x)(Q(x))]$
12. $(\exists x)(P(x) \vee Q(x)) \Leftrightarrow [(\exists x)(P(x)) \vee (\exists x)(Q(x))]$
13. $[(\forall x)(P(x)) \vee (\forall x)(Q(x))] \Rightarrow (\forall x)(P(x) \vee Q(x))$
14. $(\exists x)(P(x) \wedge Q(x)) \Rightarrow [(\exists x)(P(x)) \wedge (\exists x)(Q(x))]$
15. $(\forall x)(R \wedge Q(x)) \Leftrightarrow R \wedge (\forall x)(Q(x))$
16. $(\forall x)(R \vee Q(x)) \Leftrightarrow R \vee (\forall x)(Q(x))$
17. $(\exists x)(R \wedge Q(x)) \Leftrightarrow R \wedge (\exists x)(Q(x))$
18. $(\exists x)(R \vee Q(x)) \Leftrightarrow R \vee (\exists x)(Q(x))$

We have stated these theorems without proof, but we do ask you to translate their statements into English and convince yourself of their reasonableness.

2.12 Example **(a)** Give a convincing argument for theorem number 14 in the list above.
(b) Show by means of a counterexample that the converse of the implication in theorem number 14 is not a theorem of the predicate calculus.

SOLUTION **(a)** The hypothesis $(\exists x)(P(x) \wedge Q(x))$ of theorem number 14 says that there is at least one object x in U such that both $P(x)$ and $Q(x)$ are true. If this is so, then there is an object in U, namely this same x, such that $P(x)$ is true; that is, $(\exists x)(P(x))$. Likewise, $(\exists x)(Q(x))$ follows from the hypotheses; hence, $(\exists x)(P(x)) \wedge (\exists x)(Q(x))$ is a consequence of the hypothesis.
(b) The converse of the implication in theorem number 14 would be the statement

$$[(\exists x)(P(x)) \wedge (\exists x)(Q(x))] \Rightarrow (\exists x)(P(x) \wedge Q(x)). \quad (*)$$

Let U, the domain of discourse, consist of all positive integers. Let $P(x)$ be the statement x *is even*, and let $Q(x)$ be the statement x *is odd*. Since there exists an even positive integer, for instance 2, the statement $(\exists x)(P(x))$ is true. Since there exists an odd positive integer, for instance 1, the statement $(\exists x)(Q(x))$ is also true. Thus, the hypothesis $(\exists x)(P(x)) \wedge (\exists x)(Q(x))$ of $(*)$ is true. However, the conclusion $(\exists x)(P(x) \wedge Q(x))$ of $(*)$ is false because it says that there exists a positive integer x that is both even and odd. □

The propositional function

$$x = y$$

of the two variables x and y has a special role to play in the predicate calculus. This propositional function, called **equality** or **logical identity**, asserts that x and y are *identical*. In symbolic logic, the idea of logical identity is understood as follows:

> *Two objects a and b in the universe of discourse U are said to be **equal**, and we write*
>
> $$a = b,$$
>
> *provided that, for every propositional function $P(x)$,*
>
> $$P(a) \Leftrightarrow P(b).$$

In other words, *two objects are equal if and only if anything that can be said about the one object can be said about the other and vice versa.* If $\sim(a = b)$, then we say that a and b are **different** or **distinct** objects in U, and we write

$$a \neq b.$$

In elementary geometry, two line segments are sometimes said to be "equal" if they have the same length; however, according to the discussion above, this is a misuse of the idea of equality unless the line segments happen to be identical. Even in more advanced mathematics, an occasional abuse of the notion of equality is sometimes tolerated in the interest of avoiding more awkward notation. There is nothing wrong with an occasional abuse of notation, *provided that those who indulge in it know exactly what they are doing and can, upon demand, reformulate their statements with precision!*

The notion of logical identity can be used to help formalize the important idea of *unique existence.* If $P(x)$ is a propositional function, then the expression

$$(\exists! x)(P(x))$$

is understood to be the proposition asserting that there exists *one and only one* object x in U for which $P(x)$ is true. This symbolism is read in English as

there exists a unique x *such that* $P(x)$.

The following definition formalizes this idea.

2.13 DEFINITION

UNIQUE EXISTENCE The proposition $(\exists! x)(P(x))$ is true if and only if the following two conditions hold:

(i) $(\exists x)(P(x))$
(ii) $(\forall x)(\forall y)(P(x) \wedge P(y) \Rightarrow x = y)$

According to this definition, if you want to show that

$$(\exists! x)(P(x))$$

is true, you must do two things: First, you must show that there exists an x in U such that $P(x)$ is true. Second, you must show that there cannot be two different things x and y in U such that both $P(x)$ and $P(y)$ are true.

2.14 Example Let U be the set of all real numbers. Show that there exists a unique positive real number x such that $x^2 = 4$.

SOLUTION First, because $2^2 = 4$, there exists an x in U such that x is positive and $x^2 = 4$. Second, suppose that x and y are positive numbers such that $x^2 = 4$ and $y^2 = 4$. Then

$$x^2 - y^2 = 0$$

and therefore

$$(x - y)(x + y) = 0.$$

Since x and y are positive, $x + y$ is also positive; hence, $x + y \neq 0$, and it follows that

$$x - y = 0.$$

Therefore, $x = y$. □

PROBLEM SET 1.2

In Problems 1–4, let the domain of discourse U consist of all real numbers. Determine the truth value of each proposition.

1. $P(x)$ is the statement $x < 5$.
(a) $P(-1)$ (b) $P(6)$ (c) $P(5)$

2. $Q(x)$ is the statement $x^2 - x - 6 = 0$.
(a) $Q(-3)$ (b) $Q(3)$ (c) $Q(2)$

3. $R(x)$ is the statement x *is a nonnegative integer.*
(a) $R(2)$ (b) $R(0)$ (c) $R(\pi)$

4. $S(x)$ is the statement $|2x + 7| \geq 11$.
(a) $S(-9)$ (b) $S(3)$ (c) $S(0)$

In Problems 5 and 6, let the domain of discourse U be the set of all real numbers. Determine the truth value of each quantified proposition.

5. (a) $(\forall x)((x + 2)^2 = x^2 + 4x + 4)$ (b) $(\forall y)(|7 + 2y| \geq 7)$
(c) $(\exists x)(2x^2 - x - 1 = 0)$ (d) $(\exists y)(2y^2 + 3y + 2 = 0)$

6. (a) $(\forall \theta)(\cos^2 \theta + \sin^2 \theta = 1)$ (b) $(\forall z)(5 \leq 4z + 1 \leq 17)$
(c) $(\exists w)(18w^2 + 61w - 7 = 0)$ (d) $(\exists t)(\cos t \geq 1 + \sin t)$

In Problems 7–16, (a) translate each English statement into a symbolic proposition with quantifiers by using the indicated domain of discourse U and introducing suitable propositional functions: T(x) for x is a teacher, *S(x) for x* is a sadist, *and so on, and (b) write out, in English, the denial of each statement.*

7. All teachers are sadists. U = all people.

8. No teachers are sadists. U = all people.

9. All teachers are sadists. U = all teachers.

10. Some teachers are sadists and some are not. U = all teachers.

11. Not all lawyers are judges. U = all people.

12. Some actors are egoists. U = all people.

13. Not all lawyers are judges. U = all lawyers.

14. No pig has wings. U = all animals.

15. There is a prime number that is exactly divisible by 3. U = all positive integers.

16. All prime numbers are greater than 1. U = all positive integers.

In Problems 17–20, let the domain of discourse U be the real numbers. Quantify each expression: (a) with $(\forall y)$, *(b) with* $(\exists x)$, *(c) with* $(\exists x)(\forall y)$, *and (d) with* $(\forall y)(\exists x)$. *In each case, interpret the result.*

17. $x > y$

18. $x^2 + y^2 \leq 1$

19. $x + y = 0$

20. $x^2 + y^2 = 4$

21. Let the domain of discourse U consist of all integers, $0, \pm 1, \pm 2, \pm 3, \ldots$, and define the propositional function $D(x, y)$ of two variables by $D(x, y) \Leftrightarrow (\exists z)(zx = y)$. Read $D(x, y)$ as x *divides* y. Find the truth value of each of the following:
(a) $D(3, -12)$ (b) $D(3, 0)$ (c) $D(0, 3)$
(d) $D(0, 0)$ (e) $(\exists x)(D(0, x))$ (f) $(\forall x)(D(x, x))$
(g) $(\exists x)(\forall y)(D(x, y))$ (h) $(\forall y)(\exists x)(D(x, y))$ (i) $(\exists y)(\forall x)(D(x, y))$
(j) $(\forall x)(\exists y)(D(x, y))$ (k) $(\forall x)(\forall y)(D(x, y))$ (l) $(\exists x)(\exists y)(D(x, y))$

22. Let the domain of discourse U consist of all possible *events* in space-time, and let $C(x, y)$ be interpreted to mean that the event x *causes* the event y. Discuss the meanings of the following assertions:
(a) $(\exists x)(\forall y)(C(x, y))$ (b) $(\forall x)(\forall y)(C(x, y))$ (c) $(\forall y)(\exists x)(C(x, y))$
(d) $(\exists x)(\exists y)(C(x, y))$ (e) $(\exists x)(C(x, x))$
If we interpret (a)–(e) as philosophical doctrines, which doctrine is strongest in the sense that it implies all the others? Which doctrines would a classical physicist probably subscribe to? Which doctrines would you describe as being mystical?

In Problems 23 and 24, let the domain of discourse U be the set of all real numbers, and let n denote an integer. Write the denial of each proposition in such a way that no denial connective appears explicitly and interpret the result.

23. (a) $(\forall x)(\exists n)(n \leq x < n + 1)$ (b) $(\exists x)(\forall n)(x < n + 1 \Rightarrow x < n)$

24. (a) $(\forall x)(\exists n)(0 \leq n \wedge x < n/10)$ (b) $(\forall x)(\forall n)(x \leq n \vee n < x)$

25. Simplify each expression by writing an equivalent expression that does not involve the denial connective:
(a) $\sim(\exists x)(\sim P(x))$ (b) $\sim(\forall x)(\sim P(x))$

26. Explain why it is possible to develop predicate calculus using only the existential quantifier, that is, without using the universal quantifier at all. [Hint: See Problem 25.]

27. Let the universe of discourse U be all real numbers, and use the convention that ε and δ stand for positive real numbers. Suppose that $f(x)$ is a real-valued function of the real variable x.

(a) If a is a real number, write out, in the symbolism of predicate calculus, the definition of the statement *$f(x)$ is continuous at $x = a$.*

(b) Write out, in the symbolism of predicate calculus, but without using the denial connective, a statement equivalent to the condition that *$f(x)$ is not continuous at $x = a$.*

28. (a) Continuing with the notation of Problem 27, write out, in the symbolism of predicate calculus, the condition for $f(x)$ to be *continuous*, that is, continuous at every real number.

(b) By definition, $f(x)$ is said to be *uniformly continuous* if and only if

$$(\forall\varepsilon)(\exists\delta)(\forall x)(\forall y)(|x - y| < \delta \Rightarrow |f(x) - f(y)| < \varepsilon).$$

Show that, if $f(x)$ is uniformly continuous, then it is continuous.

29. Let the universe of discourse U be all real numbers, and let $f(x)$ be a real-valued function of the real variable x. Write out, in the symbolism of predicate calculus, the condition that $f(x)$ is a *strictly increasing function of x.*

30. By means of a counterexample, defeat the following conjecture: A continuous function is uniformly continuous. [See Problem 28.]

In Problems 31—37, we refer to the theorems of propositional calculus numbered 1–18 in the list on page 28.

31. Let U be the set of all people. Replace the propositional functions $P(x)$ and $Q(x)$ in Theorems 1–18 by particular propositional functions such as *x is a politician*, *x is honest*, and so on, and translate the resulting statements into English.

32. Some of the Theorems 1–18 are implications rather than equivalences. For those that are implications, show by means of counterexamples that the converse implications are not theorems. [This has already been done in Part (b) of Example 2.12 for Theorem 14.]

33. Give a convincing argument for Theorem 11.

34. Give a convincing argument for Theorem 12.

35. Give a convincing argument for Theorem 13.

36. Using Theorems 6 and 18, give an argument to show that, if R denotes a proposition or a propositional function that does not contain x as a free variable, then

$$(\exists x)(Q(x) \Rightarrow R) \Leftrightarrow ((\forall x)(Q(x)) \Rightarrow R)$$

is a theorem of predicate calculus. [Hint: Use the fact that $P \Rightarrow Q$ is equivalent to $(\sim P) \vee Q$.]

37. Give convincing arguments for Theorems 15, 16, 17, and 18.

38. Give an argument to show that, if R denotes a proposition or a propositional function that does not contain x as a free variable, then

$$(\forall x)(Q(x) \Rightarrow R) \Leftrightarrow ((\exists x)(Q(x)) \Rightarrow R)$$

is a theorem of predicate calculus. [Hint: See Problem 36.]

39. Give several examples of instances in which the notion of *equal* is not used in strict accord with the notion of *logical identity*.

40. The *principle of substitution of equals for equals* is stated in 3.10 on page 38.
(a) Explain why this principle is reasonable in view of the idea that *equal* means *logical identity*.
(b) Does this principle work for the other uses of the word *equal* that you cite in Problem 39?

41. One of the important principles governing the use of the notion of equality is that *things equal to the same thing are equal to each other*.
(a) Express this principle in symbolic form with the aid of quantifiers.
(b) Explain why this principle is reasonable in view of the idea that *equal* means *logical identity*.

42. Suppose that an experimental physicist adopts the following operational definition of *equal*: Two physical objects are equal if and only if every measurement of the one object yields the same result (within the limits of accuracy of the measuring instruments) as the corresponding measurement of the other object. Explain why, in this case, the principle that *things equal to the same thing are equal to each other* may fail.

43. Show that there exists a unique real number x such that $x^3 = 8$.

44. Show that there exists a unique even positive integer x such that x is a prime number.

45. Show that the proposition $(\forall x)(P(x)) \Rightarrow (\exists y)(P(y) \wedge x \neq y)$ is equivalent to the denial of the proposition $(\exists! x)(P(x))$.

46. A mathematician, arriving in a certain city to attend a convention, is amused by a sign stating that *this is America's most unique city*. Why does this amuse her?

In Problems 47–49, suppose that the universe of discourse U consists only of the numbers 1 *and* 2.

47. (a) Explain why the proposition $(\forall x)(P(x))$ is equivalent to $P(1) \wedge P(2)$.
(b) Explain why the proposition $(\exists x)(P(x))$ is equivalent to $P(1) \vee P(2)$.

48. Reasoning as in Problem 47, translate the proposition $(\exists x)(\forall y)(P(x, y))$ into a formula of propositional calculus.

49. Using the results of Problem 47, show that Theorems 6 and 7 in the list on page 28 are consequences of De Morgan's laws.

50. Show that the following is a tautology:

$$((P \wedge Q) \Rightarrow R) \Leftrightarrow ((P \Rightarrow R) \vee (Q \Rightarrow R))$$

Let U, our domain of discourse, consist of all straight lines in Euclidean three-dimensional space. In the tautology above, let P be the statement *x and y lie in the same plane*, let Q be the statement *x and y do not meet*, and let R be the statement *x and y are parallel*. Evidently, the left side of the tautology is true, since two straight lines that lie in the same plane and do not meet are indeed parallel. The statement $P \Rightarrow R$ says that lines that lie in the same plane are parallel, and this is false. The statement $Q \Rightarrow R$ says that lines that do not meet are parallel, and this also is false. Therefore, the right side of the tautology is false. What is the trouble here?

3
PROOFS AND RULES OF PROCEDURE

In its most general sense, the word *proof* means *an argument to convince a person (or persons) that some proposition is true.* Even an argument such as "you'd better believe it or I'll punch you in the nose" is a proof in this most general sense. (This barbaric argument is called the *argumentum baculinum.*) Obviously, we need a more restricted notion of proof that is better suited to the requirements of mathematicians.

It is not easy to give a definition of *mathematical proof* that will be acceptable to all mathematicians—in fact, it is probably impossible. Currently, for instance, there are even some doubts about the validity of certain proofs that involve the extensive use of high-speed computers. Fortunately, for our purposes in this textbook, we can settle for an informal working definition of a mathematical proof.

We begin by introducing some **rules of procedure**, or **rules of inference**, which state conditions under which we agree to say that a proposition is **justified**. The word *justified* should be interpreted as meaning something like "true in context." The most important rule of procedure, called **modus ponens** (Latin for *method of asserting*) is as follows:

3.1 RULE OF PROCEDURE

MODUS PONENS If P is a justified proposition and if $P \Rightarrow Q$ is a justified proposition, then we can infer that Q is a justified proposition.

The rationale for modus ponens is simple enough. If you examine the truth table

P	Q	$P \Rightarrow Q$
1	1	1
0	1	1
1	0	0
0	0	1

for $P \Rightarrow Q$, you see that it is true only for the first, second, and fourth truth combinations. Therefore, if $P \Rightarrow Q$ is true, one of these three truth combinations must hold. But, if P is true, only the first truth combination can hold; hence, Q must be true.

3.2 Example Consider the following argument: *Edna has good sense. If Edna has good sense, then she will not interrupt Professor Twit's lecture. Therefore, Edna will not interrupt the lecture.* Why is this a valid argument?

SOLUTION Let P denote the proposition

Edna has good sense,

and let Q denote the proposition

she will not interrupt Professor Twit's lecture.

We are given both P and $P \Rightarrow Q$; hence, by modus ponens, we can infer Q. □

3.3 DEFINITION

MATHEMATICAL PROOF

(i) A **mathematical proof** is a finite sequence of justified propositions. Each such justified proposition is called a *step* of the proof.
(ii) A **formal mathematical proof** is a mathematical proof in which each step is accompanied by the rule or rules of procedure that justify it.
(iii) By a **mathematical proof of a proposition** P, we mean a mathematical proof whose last step is the proposition P.

A proposition P that can be shown to have a mathematical proof is called a **theorem**. To **prove** a theorem means to exhibit its proof. An **abbreviation** or **outline** of a mathematical proof is an indication of some of its more important or less obvious steps. To prove a theorem **informally** means to exhibit an abbreviation or outline of its proof.

Working mathematicians usually give informal proofs of their theorems, replacing omitted portions of their arguments with words such as

it can easily be seen that . . .

or

it is clear that . . . .

Also, they often leave out justifications for steps when they feel that these are "obvious." If the person or persons to whom the proof is being shown do not accept or understand the argument, the mathematician is obliged to fill in the gaps, that is, restore some of the omitted steps and justifications. The process of filling in the gaps usually continues until a consensus is reached that the theorem can be proved formally.

A **lemma** is a theorem that we state and prove because we intend to use it to help us in the proof of a subsequent (and, perhaps, more important) theorem. Thus, a lemma is a helper-theorem. A **corollary** is a theorem whose proof follows easily if we make use of a previously proved theorem. That previously proved theorems can be used in the proofs of theorems is guaranteed by the following rule of procedure.

3.4 RULE OF PROCEDURE

PREVIOUSLY PROVED THEOREMS A theorem that has already been proved may be inserted in the proof of a subsequent theorem as a justified step.

Because mathematical theorems often have the form $P \Rightarrow Q$, the following two rules of procedure are especially useful.

3.5 RULE OF PROCEDURE

ASSERTION OF HYPOTHESIS In the proof of a theorem of the form $P \Rightarrow Q$, the hypothesis P may be written down as a justified step.

The rationale for 3.5 is easy to supply. Recall that an implication $P \Rightarrow Q$ is automatically true if its hypothesis P is false. Therefore, if we want to prove that $P \Rightarrow Q$ is true, we need consider only the case in which P is true.

3.6 RULE OF PROCEDURE

JUSTIFICATION OF CONCLUSION If Q is a justified proposition, then we can infer that $P \Rightarrow Q$ is a justified proposition.

Again, the rationale for 3.6 is simple. Just recall that an implication $P \Rightarrow Q$ is automatically true if its conclusion Q is true.

As we mentioned in Section 1, students of mathematics are often troubled by some of the apparently peculiar features of the implication connective $\Rightarrow$ as defined by its truth table. Note, however, that 3.1, 3.5, and 3.6 are the only rules of procedure pertaining to the use of this connective; and, as we have seen, they can be substantiated on the basis of this truth table. Thus, although the connective $\Rightarrow$ may not conform in every respect to our usual idea of implication, it works perfectly for the creation of mathematical proofs.

We now give several more rules of procedure and introduce a few more concepts relating to the idea of a mathematical proof.

3.7 RULE OF PROCEDURE

REPLACEMENT INSTANCES OF TAUTOLOGIES If P is a propositional variable occurring in a tautology $\mathscr{T}$, and if $\mathscr{R}$ is the proposition that results when P is replaced in its every occurrence in $\mathscr{T}$ by a particular proposition, then the proposition $\mathscr{R}$ is justified.

In 3.7, the proposition $\mathscr{R}$ is called a **replacement instance** of the tautology $\mathscr{T}$.

3.8 Example Justify the proposition $(1 = 0) \vee (1 \neq 0)$.

SOLUTION In the tautology $P \vee (\sim P)$, replace P by the proposition $1 = 0$. □

3.9 RULE OF PROCEDURE

IDENTITY If a is any object in the domain of discourse U, then $a = a$ is a justified proposition.

3.10 RULE OF PROCEDURE

SUBSTITUTION OF EQUALS Let P be a justified proposition about a certain object a in the domain of discourse U. Suppose that $a = b$ is also a justified proposition. Then, the proposition Q that results when b is substituted for a in any or all of its occurrences in P is also a justified proposition.

3.11 Example If a and b are objects in the domain of discourse U, prove that $a = b \Rightarrow b = a$.

SOLUTION The theorem to be proved has the form $P \Rightarrow Q$, where P is the proposition $a = b$ and Q is the proposition $b = a$. By 3.5 (assertion of hypotheses), we can assume that the hypothesis $a = b$ is justified. By 3.9, $a = a$ is also justified. By 3.10, we can substitute b for a in any or all of its occurrences in the proposition $a = a$. We choose to substitute b for a only on the left of the equal sign, and thus we obtain $b = a$ as a justified statement. By 3.6 (justification of conclusion), we can infer that

$$a = b \Rightarrow b = a$$

is justified, and the theorem is proved. □

At this point, you may be wondering, "Why all the fuss? It's obvious that $a = b$ implies $b = a$, so why do we bother to prove it?" The answer is that we are testing our theorem-proving tools. If these tools are incapable of proving simple facts, they can hardly be trusted for more complicated tasks. In testing a new jigsaw, you would begin by making rather simple cuts, not by doing elaborate scrollwork.

Among the more important theorem-proving tools are *axioms*, or *postulates*, and *definitions*. In mathematics, the word **axiom** is usually taken to mean a proposition whose truth we *assume* for the purpose of studying its consequences. The theorems that follow from the assumption of a set of axioms form what is called a **mathematical theory**. There is nothing sacred about axioms; it is not even necessary to believe in them. They are posited only as starting points for the development of theories. The only question is whether a theory is of interest to the mathematical community.

A **definition** in mathematics is usually a notational device in which a certain meaningful cluster of symbols is, for simplicity, represented by a simpler cluster of symbols or by a single symbol. For instance, in calculus, we define

$$e = \lim_{n \to \infty} \left(1 + \frac{1}{n}\right)^n.$$

Here the complicated expression on the right, presumably containing previously defined terms, is set equal to e by definition. Superficially, definitions are nothing but devices of convenience: It is much easier to write e than to write the more complicated expression for which it stands. In fact, however, definitions play a decisive role in the creation of a mathematical theory because they focus our attention on those particular clusters of symbols that the creator of the theory regards as being particularly significant. In other words, definitions establish the fundamental *concepts* with which the theory deals.

3.12 RULE OF PROCEDURE

AXIOMS AND DEFINITIONS Any previously introduced axiom or definition may be considered to be a justified proposition.

Occasionally, the word **postulate** is used in mathematics as a synonym for the word *axiom*. (This does not coincide with older usage in which a technical distinction was made between axioms and postulates, as, for example, in Euclid's geometry.) It should also be mentioned that the individual propositions that together make up a definition are often spoken of as the *axioms* or the *postulates* for the concept being defined. (For instance, one speaks of the *postulates for a group* or the *axioms for a topological space*.)

In mathematics, one of the most useful (and also one of the most controversial) rules of procedure is the rule permitting the formation of **indirect proofs**, or **proofs by contradiction**.

3.13 RULE OF PROCEDURE

INDIRECT PROOF, OR PROOF BY CONTRADICTION For a proposition P, if the assumption that the denial $\sim P$ is justified leads to a contradiction, then we can infer that P is justified.

Tautology 17 on page 10,

$$((\sim P) \Rightarrow (Q \wedge (\sim Q))) \Rightarrow P,$$

provides the rationale for proofs by contradiction using 3.13.

3.14 Example Prove that there is no smallest positive real number.

SOLUTION We make a proof by contradiction. Let P denote the proposition

there is no smallest positive real number.

Then $\sim P$ is the proposition

there is a smallest positive real number.

Assume $\sim P$; that is, assume there is a smallest positive real number. Call this number s. Let $t = s/2$, noting that $t < s$ and that t is positive. But, because s is the smallest positive real number, $t \not< s$. Thus, $\sim P$ leads to the contradiction $(t < s) \wedge (t \not< s)$; hence, P is true by rule of procedure 3.13. □

In practice, the proof by contradiction given in Example 3.14 might be abbreviated as follows:

Theorem *There is no smallest positive real number.*

PROOF Suppose there were such a number. Call it s. Then $s/2 < s$, contradicting the supposition that s is the smallest positive real number. □

The following two rules of procedure pertain to the use of existential and universal quantifiers.

3.15 RULE OF PROCEDURE

EXISTENTIAL QUANTIFICATION Let $P(x)$ be a propositional function. If b is an object in the domain of discourse U, and if $P(b)$ is justified, then we may infer that $(\exists x)(P(x))$ is justified. Conversely, if $(\exists x)(P(x))$ is justified, then we may infer that, for some suitable choice of an object c in U, $P(c)$ is justified.

3.16 RULE OF PROCEDURE

UNIVERSAL QUANTIFICATION Let $P(x)$ be a propositional function. If $P(x)$ is justified, where x represents an arbitrary object in the domain of discourse U, then we may infer that $(\forall x)(P(x))$ is justified. Conversely, if $(\forall x)(P(x))$ is justified and if b is an object in U, we may infer that $P(b)$ is justified.

Although we have given no formal definition of the word *justified*, it should be plain that the rules of procedure are actually agreements about the ways in which we intend to use this word. In this sense, the rules of procedure establish an operational definition of the word *justified*. We make no claim for the completeness of the rules of procedure given above; however, they are sufficiently comprehensive to provide a reasonably secure foundation for most of the proofs in this textbook.

The ideas discussed above have given rise to a general notion of a **formal mathematical system**. Such a system begins with a list of *primitive*, or *undefined, symbols*. Ordinarily, some attempt is made to indicate just what these symbols might stand for, but this indication is not to be regarded as part of the formal system itself. Next comes a list of *axioms*, or *postulates*, that govern the undefined symbols. These axioms are out-and-out assumptions about relationships among the undefined symbols, and as such, they constitute a contextual or operational definition of these symbols. *Rules of procedure* are then set forth, which determine the specific conditions under which propositions, expressed in terms of the primitive symbols, are considered justified. These rules of procedure enable the person who is constructing the formal theory to prove the *theorems* that form the body of the formal mathematical system. From time to time during the construction of the formal system, it may be convenient to introduce new symbols by *definitions* that fix their meanings in terms of the primitive symbols and previously defined symbols.

The concept of a formal mathematical system has so fascinated logicians and mathematicians that they have created a new discipline called

metamathematics, the study of formal mathematical systems *in general.* One of the most intriguing metamathematical results is a theorem, published by Kurt Gödel in 1931, which states that within a formal mathematical system sufficiently rich to enable the arithmetic of whole numbers, there are propositions that are *undecidable* in the sense that, by working within the system, *they can neither be proved nor disproved.* This astonishing theorem put an end to nearly a century of efforts to create a formal axiomatic basis for all of mathematics. Although Gödel's proof is highly technical, a very readable popular account of this and related matters can be found in Douglas R. Hofstadter, *Gödel, Escher, Bach: An Eternal Golden Braid* (New York: Basic Books), 1979.

We close this section with three remarks:

First: There is nothing inviolable about the rules of procedure given above. Although these rules, or close counterparts, are used (implicitly or explicitly) by most contemporary mathematicians, there are certain people (such as the so-called intuitionists) who find that they cannot subscribe to these particular rules and who do indeed replace them by substantially different rules.

Second: In practice, mathematicians rarely work explicitly within the confines of a formal mathematical system. Most mathematical proofs are given somewhat informally, and the use of rules of procedure is often implicit or tacit. However, when necessary, most practicing mathematicians can reformulate their work in the context of some suitable formal mathematical system.

Third: There is nothing magic about a mathematical theory: No more can be squeezed out of it than is implicit in its axioms and rules of procedure. In a sense, a mathematical theorem is nothing but a (perhaps) inobvious tautology. A person who does not understand and appreciate this fact is subject to being imposed on by charlatans who claim to be able to prove outlandish things "mathematically."

PROBLEM SET 1.3

1. By using the truth table for $P \Rightarrow Q$, explain why, if Q is a justified proposition and if $P \Rightarrow Q$ is a justified proposition, then (in general) we cannot infer that P is a justified proposition.
2. Give a specific example showing that $P \Rightarrow Q$ may be true and Q may be true, but P may be false.
3. If $P \Rightarrow Q$ is justified and $\sim Q$ is justified, explain how $\sim P$ can be justified. [Hint: Use the law of contraposition (tautology number 12 on page 10) and modus ponens.]

4. If $P \vee Q$ is justified and $\sim P$ is justified, explain how Q can be justified. [Hint: Show that $(P \vee Q) \Rightarrow ((\sim P) \Rightarrow Q)$ is a tautology, and then use modus ponens.]

In Problems 5–10, indicate which arguments are valid, which are invalid, and say why.

5. If Joe embezzled the union funds, then he is guilty of a felony. Joe did embezzle the union funds. Therefore, he is guilty of a felony.

6. If Maria can solve the problem, then she has promise as a mathematician. Maria cannot solve the problem. Therefore, she has no promise as a mathematician.

7. If Gwen is a brain surgeon, then she is highly trained. Gwen is highly trained. Therefore, she is a brain surgeon.

8. Carlos is either a genius or he is crazy. Carlos is not crazy. Therefore, he is a genius.

9. If Rodney cheated on the exam, then he passed his calculus class. Rodney passed his calculus class. Therefore, he cheated on the exam.

10. If $f(x)$ is a differentiable function, then it is continuous. The function $f(x)$ is not differentiable. Therefore, it is not continuous.

11. Professor Grumbles wants to prove that if a series $\sum_{j=1}^{\infty} a_j$ converges, then $\lim_{n\to\infty} a_n = 0$. He begins his proof by saying, "Suppose that the series $\sum_{j=1}^{\infty} a_j$ converges." Is this step justified? Why or why not?

12. Professor Twit wants to prove that if a series $\sum_{j=1}^{\infty} a_j$ converges, then $\lim_{n\to\infty} a_n = 0$. He begins his proof by saying, "Suppose that $\lim_{n\to\infty} a_n = 0$." Is this step justified? Why or why not?

13. Professor Keen wants to prove that, if a series $\sum_{j=1}^{\infty} a_j$ converges, then $\lim_{n\to\infty} a_n = 0$. After an argument in which each step is justified, she is able to justify the statement $\lim_{n\to\infty} a_n = 0$. Has she proved the theorem? Why or why not?

14. Archibald knows that if a series $\sum_{j=1}^{\infty} a_j$ converges, then $\lim_{n\to\infty} a_n = 0$. On an exam, he is asked whether the harmonic series $\sum_{j=1}^{\infty}(1/j)$ converges. He argues that, since $\lim_{n\to\infty}(1/n) = 0$, the harmonic series must converge. Criticize Archibald's reasoning.

In Problems 15–18, justify each proposition by showing that it is a replacement instance of a tautology.

15. $\cos 0 = 1 \Leftrightarrow \sim(\cos 0 \neq 1)$

16. Boris is either a spy, or he is not a spy.

17. $((\cos 0 = 1) \wedge ((\cos 0 = 1) \Rightarrow \sin 0 = 0)) \Rightarrow \sin 0 = 0$

18. $(((\sqrt{4} = 2) \vee (\sqrt{4} \neq 2)) \Rightarrow \sqrt{1} = 2) \Rightarrow (\sqrt{1} = 2)$

19. Let x and y denote real numbers. Show that the proposition

$$(x = 1) \wedge ((x = 1) \Rightarrow (y = 2)) \wedge (y \neq 2)$$

is equivalent to a contradiction.

20. Give a rationale for the Rule of Procedure 3.7.

21. If P and Q are propositions, and if $P \wedge Q$ is a justified step in a formal mathematical proof, explain how P can be obtained as a justified step in this proof. [Hint: Use tautology 18 in the list on page 10.]

22. Give a rationale for the Rule of Procedure 3.10.

23. If a, b, and c are objects in the domain of discourse U, prove that

$$((a = b) \wedge (b = c)) \Rightarrow a = c.$$

24. Although the Rule of Procedure 3.10, Substitution of Equals, seems reasonable and works in connection with the usual propositions encountered in mathematics, its unrestricted use for propositions of ordinary discourse can lead to difficulties. For instance, suppose that the proposition

Superman is so called because of his superhuman abilities

is justified. Suppose, also, that the proposition

Superman = Clark Kent

is justified. Using rule 3.10, we infer that

Clark Kent is so called because of his superhuman abilities,

which is absurd. Can you resolve this paradox? Why do such difficulties rarely arise in mathematical proofs?

25. In ordinary conversation, the word *axiom* means a self-evident or universally recognized truth. Contrast this with the *mathematical* use of the word.

26. (a) Explain the sense in which definitions in a dictionary are essentially *circular.*
(b) Explain how circular definitions are avoided in a formal mathematical theory.

27. Prove by contradiction (that is, make an indirect proof) that there is no largest positive integer.

28. Let the domain of discourse U consist of the integers.
(a) Prove that, if n is odd, then n^2 is odd. [Hint: If n is odd, it can be written in the form $n = 2k + 1$ for some integer k.]
(b) Using the result in Part (a), make an indirect proof to show that if n^2 is even, then n is even. [Hint: Use the fact that *n is even if and only if it is not odd.*]

29. Professor Keen wants to prove a theorem of the form $P \Rightarrow Q$. She elects to use an indirect proof. As her first step, she says, "Assume that P is true, but that Q is false." Explain.

30. Let the domain of discourse U consist of the real numbers. Take it as a previously proved theorem that

$$(0 \le x \le y) \Rightarrow x^2 \le y^2.$$

Using this fact, make an indirect proof showing that

$$(0 < x < y) \Rightarrow (\sqrt{x} < \sqrt{y}).$$

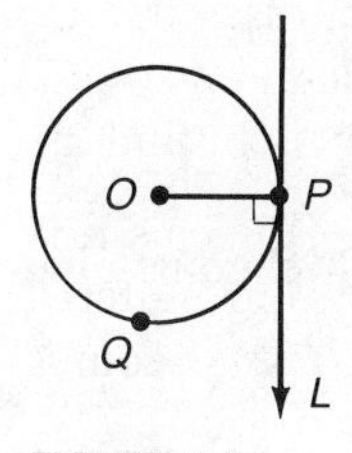

FIGURE 1-1

31. Prove by contradiction that if a straight line L intersects a circle with center O at a point P, and if L is perpendicular to the radius $\overline{OP}$ (Figure 1-1), then L does not intersect the circle at another point $Q \neq P$.

32. If P and Q are propositions, denote by the symbolism $P \dashv Q$ the statement that P implies Q in the usual intuitive sense, namely, that Q can be inferred or deduced from P.

(a) Does it appear to you that

$$(P \dashv Q) \dashv (P \Rightarrow Q)$$

should be a theorem? Why or why not?

(b) Does it appear to you that

$$(P \Rightarrow Q) \dashv (P \dashv Q)$$

should be a theorem? Why or why not?

33. Find some examples of indirect proofs in your calculus textbook.

34. A person claims to have a mathematical proof that the world is flat. What are some of the pertinent questions that you might ask in connection with the alleged proof?

4 SETS

The idea of a *set* permeates all of modern mathematics. Between the years 1873 and 1895, the basic elements of set theory were created by the German mathematician Georg Cantor (1845–1918), who became aware of the need for such a theory during his research on trigonometric (Fourier) series. Although Cantor's revolutionary ideas were initially scorned by other mathematicians, set theory soon came to be recognized as a legitimate branch of mathematics. Then, in 1902, the English philosopher Bertrand Russell (1872–1970) discovered that the methods used by Cantor and his followers in reasoning about sets lead to a contradiction—the *Russell paradox*. Using a so-called theory of types, Russell and Alfred North

Whitehead (1861–1947) attempted to recast set theory to avoid this paradox in their monumental *Principia Mathematica.* Later, other mathematicians and philosophers formulated several competing brands of set theory, each designed to avoid the paradoxes inherent in Cantor's original work. In this textbook, we ignore these difficulties and treat set theory intuitively and more or less informally. After you have learned the basics of this "naive" set theory, you will be better prepared to study some brand of formal axiomatic set theory, if you are so inclined.

Mathematicians use the word **set** to mean much the same thing that it means in ordinary English—a collection of things. The things comprising a set are understood to be objects from a fixed domain of discourse U. These objects may be physical things (trees, books, molecules, and so on), or they may be nonphysical entities (numbers, geometric points, continuous functions, and so on). In one of the influential early books on set theory, the German mathematician Felix Hausdorff wrote, "eine Menge ist eine Vielheit, als Einheit gedacht," which translates roughly as, "a set is a multiplicity thought of as a unity." The significant words here are *thought of*. Thus, in mathematics, a set should be regarded as a *concept* or an *idea in our minds* and not as a physical collection of objects in space and time. For instance, we may have a bag of marbles, and we may speak of the set of marbles in this bag, but this set is an idea in our minds. It can be spoken about but not exhibited, and it must not be confused with the bag of marbles itself.

Just as we use symbols (numerals) to denote numbers, we can use symbols to denote sets. And just as we use symbols in algebra to denote arbitrary numbers, we shall use symbols to denote arbitrary sets. In this book, we usually use capital letters—such as M or N—to represent arbitrary sets. If M is a set and b is an object in our domain of discourse U, then the symbolism

$$b \in M$$

stands for the proposition stating that the object b **belongs to**, or **is a member of**, the set M. If $b \in M$, we also say that b is an **element** of the set M. The symbol $\in$ is called the **set membership symbol**. Similarly, the symbolism

$$b \notin M$$

is understood to stand for the proposition stating that the object b does not belong to the set M. Therefore, by definition,

$$b \notin M \Leftrightarrow \sim(b \in M).$$

In mathematics, it is always understood that *a set is completely determined by its members.* Thus, a club or syndicate is not an example of a set in the mathematical sense. There is more to a club or syndicate than the mere totality of its members thought of as a unity; after all, even though these members are banded together for some purpose, it may not be possible to determine this purpose from the membership list alone. The requirement that a set is completely determined by its members is made precise by the following axiom.

4.1 AXIOM

AXIOM OF EXTENT Let M and N be sets. Then

$$M = N \Leftrightarrow (\forall x)(x \in M \Leftrightarrow x \in N).$$

In words, the axiom of extent says that *two sets are equal if and only if they have the same members.*

Because a set is completely determined by its members or elements, it can be specified simply by listing these elements. There is special notation for this. We write

$$M = \{a, b, c, \ldots, k\}$$

to indicate that *M is the set whose elements are $a, b, c, \ldots, k$, and no others.* When a set is specified in this way, we say that it is described **explicitly**.

4.2 Example Rewrite the set E of all even integers between 1 and 9 in explicit form.

SOLUTION $E = \{2, 4, 6, 8\}$. □

If M is a set and b is an object in the domain of discourse U, then b either belongs to M or it does not. If b does belong to M, then (as a consequence of the axiom of extent) the number of ways in which it qualifies for membership in M is of no significance. *We can ask only whether or not a given object belongs to a set; it makes no sense to ask how many "times" it belongs to the set.* For instance, if A is the set consisting of all people who hold a doctoral degree in astrophysics or who have orbited the earth, then Sally Ride is a member of A. But we do not say that she belongs to A twice, even though she qualifies for membership in A in two different ways.

4.3 Example Let $B = \{1, 3, 5, 7\}$ and let $C = \{2, 3, 7, 8\}$. Give an explicit description of the set D consisting of all numbers that belong to B or to C (or to both).

SOLUTION $D = \{1, 2, 3, 5, 7, 8\}$. □

Note that, in spite of the fact that 3 and 7 qualify in two different ways for membership in D, *we do not write* $D = \{1, 2, 3, 3, 5, 7, 7, 8\}$.

Another consequence of the axiom of extent is that *the elements of a set M should not be regarded as belonging to M in any particular order.* For instance,

$$\{1, 2, 3, 4, 5, 6\} = \{3, 6, 1, 5, 4, 2\},$$

because the set on the left and the set on the right have exactly the same members. It is an accident of our system of notation that when we write the elements of a set explicitly, we must put them down in some particular order; however, this order is not to be regarded as part of the structure of the set.

A set that has exactly one member is called a **singleton set**. If b is an object in the domain of discourse U, then the set $\{b\}$, whose only member is b, is called **singleton** b. *You must be careful to distinguish between the object b and the singleton set* $\{b\}$. The set $\{b\}$ is an idea or concept that certainly concerns the object b, but it is not the *same* as the object b. For instance, Detroit is a city, but {Detroit} is a set and not a city.

Because a set is a concept, there is no reason why we cannot form the concept of a set with no members whatsoever. By the axiom of extent, there can be only one such set. We call it the **null set**, or the **empty set**, and we symbolize it by $\varnothing$. Although it is not usually done, we could write the description of the null set in explicit form as

$$\varnothing = \{\ \}.$$

Do not make the mistake of supposing that the null set is nothing—after all, it is a set.

4.4 Example Is it true that $\{\varnothing\} = \varnothing$?

SOLUTION No! The set $\{\varnothing\}$ is not the null set because it does have a member: $\varnothing \in \{\varnothing\}$. In fact, $\{\varnothing\}$ is the singleton set whose only member is the null set. □

Until now, we have only one way to specify sets—namely, by means of explicit description. Of course, explicit description is restricted to sets that have a finite number of elements, and even then it is restricted in practice to small sets. But there is a second way to specify sets that overcomes this limitation. If $P(x)$ is a propositional function, and M is a set, then by the symbolism

$$\{x \in M \mid P(x)\}$$

we mean

the set of all objects x in M such that P(x) is true.

Thus, if

$$N = \{x \in M \mid P(x)\},$$

then, for all x in U,

$$x \in N \Leftrightarrow (x \in M) \wedge P(x).$$

The following basic axiom of set theory provides for the existence of such a set N.

4.5 AXIOM **AXIOM OF SPECIFICATION** If M is a set and if $P(x)$ is a propositional function, then there exists a set N such that

$$(\forall x)(x \in N \Leftrightarrow (x \in M) \wedge P(x)).$$

Notice that the axiom of extent guarantees the uniqueness of the set N in Axiom 4.5. The symbolism $\{x \in M \mid P(x)\}$ is sometimes called **set builder notation** for this unique set N.

4.6 Example If M is the set of all politicians, what are the members of the set $H = \{x \in M \mid x \text{ is honest}\}$?

SOLUTION H is the set of all honest politicians. (Some cynics believe that $H = \varnothing$.) □

The symbolism $\{x \in M \mid P(x)\}$ provides another example of an apparent variable. Indeed, the set $\{x \in M \mid P(x)\}$ does not depend on x in any way; so we have, for instance,

$$\{x \in M \mid P(x)\} = \{y \in M \mid P(y)\}.$$

When a set is specified in the form $\{x \in M \mid P(x)\}$, we say that it is described **implicitly**. For instance, if $\mathbb{R}$ denotes the set of all real numbers, then the implicitly described set

$$\{x \in \mathbb{R} \mid x^2 - 5x + 6 = 0\}$$

is the same as the explicitly described set $\{2, 3\}$. In elementary algebra, when you were asked to solve an equation, you were really being asked to convert the description of a certain set from implicit to explicit form.

We now come to a vexing question: Is the universe of discourse U a set? Actually, the unrestricted assumption that U is a set can lead to logical difficulties. Here's how: Suppose that U consists of all concepts that can be formulated by the human mind, and suppose that U is a set. Now, let

$$N = \{x \in U \mid x \text{ is a set and } x \notin x\}. \qquad (*)$$

Because N is a set, it is a concept; so it follows that

$$N \in U.$$

Therefore, by $(*)$,

$$N \in N \Leftrightarrow N \notin N,$$

a self-contradictory conclusion. This is one form of Russell's paradox. The paradox can be avoided by dropping the assumption that U is a set; then the notation in $(*)$ becomes meaningless, and we cannot even form the set N. Conclusion: Certain domains of discourse cannot be considered to be sets. *In the remainder of this textbook, we deal only with domains of discourse that can be regarded as being sets.* We leave the consideration of more esoteric domains of discourse (and other so-called *proper classes*) to the more formal treatises on axiomatic set theory.

Let $P(x)$ be a propositional function. Because we are assuming that U is a set, we can form the set

$$\{x \in U \mid P(x)\}.$$

This set is often abbreviated as

$$\{x \mid P(x)\}$$

and is read, in English, as *the set of all x such that* $P(x)$, it being understood that x represents an object in the domain of discourse U.

4.7 Example Let U denote the set of all real numbers. Rewrite the implicitly specified set $\{x \mid x^2 = 3\}$ in explicit form.

SOLUTION $\{x \mid x^2 = 3\} = \{\sqrt{3}, -\sqrt{3}\}$ □

If M and N are sets and if every element of M is an element of N, then we say that M is **contained** in N or that M is a **subset** of N. We write

$$M \subseteq N$$

to signify that M is contained in N. The following definition makes the idea of containment more formal.

4.8 DEFINITION

SET CONTAINMENT Let M and N be sets. Then, by definition,

$$M \subseteq N \Leftrightarrow (\forall x)(x \in M \Rightarrow x \in N)$$

and

$$M \nsubseteq N \Leftrightarrow \sim(M \subseteq N).$$

To prove that $M \subseteq N$, you must show that every member of M is also a member of N. To prove that $M \nsubseteq N$, you must show that there is at least one member of M that is not a member of N.

4.9 Example In each case, decide whether or not $M \subseteq N$:

(a) M is the set of all mammals. N is the set of all vertebrates.
(b) M is the set of all prime numbers. N is the set of all odd integers.

SOLUTION **(a)** $M \subseteq N$ is true because, indeed, every mammal is a vertebrate.
(b) $M \nsubseteq N$, because $2 \in M$ but $2 \notin N$. □

Note that $M \subseteq N$ is a proposition about the sets M and N; hence, it is either true or false, depending on M and N. Be careful: $M \subseteq N$ is not a set! Also, note that *every set M is a subset of itself*; that is,

$$M \subseteq M$$

is always a true statement. Furthermore, since all objects under consideration belong to the universe of discourse U, then *every set M is a subset of U*; that is,

$$M \subseteq U$$

is always a true statement.

4.10 Example Prove that *the null set is a subset of every set.*

SOLUTION Let M be a set. We must prove that $\varnothing \subseteq M$. According to Definition 4.8, we must prove that, for every x in U,

$$x \in \varnothing \Rightarrow x \in M.$$

But this implication is automatically true, because its hypothesis $x \in \varnothing$ is false. □

If M is any set, then both $\varnothing$ and M itself are subsets of M. We call $\varnothing$ and M the **trivial** subsets of M; all other subsets of M (if there are any) are called **nontrivial** subsets of M. If $N \subseteq M$ and $N \neq M$, then we say that N is a **proper** subset of M. Some authors indicate that N is a proper subset of M by writing

$$N \subset M.$$

If A, B, and C are sets, we write

$$A \subseteq B \subseteq C$$

to mean that $A \subseteq B$ and $B \subseteq C$. In particular, we have

$$\varnothing \subseteq M \subseteq U$$

for any set M. One of the basic properties of set inclusion is the transitivity property:

$$(A \subseteq B \wedge B \subseteq C) \Rightarrow A \subseteq C.$$

We leave the proof of this property as an exercise (Problem 15).

People who work with set theory often use diagrams to illustrate various relations among sets. These are called **Venn diagrams** in honor of the British logician John Venn (1834–1923) who devised them in 1880. First, we draw a rectangle and think of the points that lie within this rectangle as representing the objects belonging to the domain of discourse U. If we draw a circle (or any other simple closed curve) within this rectangle, then we can think of the points enclosed by the circle (or curve) as representing the elements of a set M. Two or more different sets are represented by two or more different circles. For instance, the fact that $M \subseteq N$ can be indicated as in the Venn diagram in Figure 1-2.

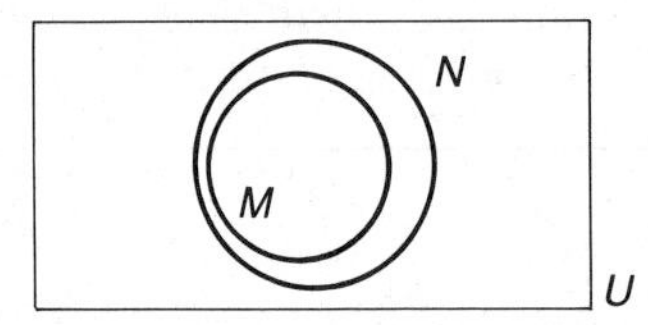

FIGURE 1-2

The following theorem expresses an important relationship between set containment and set equality.

4.11 THEOREM

EQUALITY OF SETS Let M and N be sets. Then

$$(M \subseteq N \wedge N \subseteq M) \Rightarrow M = N.$$

PROOF Assume the hypothesis, so that $M \subseteq N$ and $N \subseteq M$. Since $M \subseteq N$, every element of M is an element of N. Likewise, since $N \subseteq M$, every element of N is an element of M. Therefore, N and M have the same elements, and it follows from the axiom of extent that $M = N$. □

In words, Theorem 4.11 says that *two sets are equal if each is contained in the other.* Thus, one way to prove that two sets are equal is to prove separately that each set is a subset of the other.

Often we have to consider more sets than can be represented by the twenty-six letters of the alphabet. One way to resolve this difficulty is to use the same letter with different subscripts to represent various sets, for instance,

$$M_1, M_2, M_3, \ldots.$$

A subscript used in this way is called an **index**, and we refer to

$$M_1, M_2, M_3, \ldots$$

as an **indexed family of sets**. More generally, if I is any nonempty set and if, for each $i \in I$ we have a corresponding set M_i, then we denote this indexed family of sets by

$$(M_i)_{i \in I}.$$

For such an indexed family of sets, *there is no automatic assumption that sets with different indices are different*; that is, it is quite possible to have

$$M_i = M_j$$

even though $i \neq j$. We consider indexed families of sets in more detail in Section 1.6.

Sometimes we have to consider sets whose elements are themselves sets; that is, we have to deal with *sets of sets*. It is often convenient to use capital script letters to denote sets of sets, for instance,

$$\mathscr{E} = \{A, B, C, D\}$$

would specify that $\mathscr{E}$ is the set whose elements are the sets A, B, C, and D. Notice that an indexed family of sets

$$(M_i)_{i \in I}$$

gives rise to a set of sets

$$\mathscr{E} = \{M_i \mid i \in I\}.$$

Another example of a set of sets is provided by the set of all subsets of a given set M.

4.12 DEFINITION

POWER SET OF A SET If M is a set, then the **power set** of M, in symbols $\mathscr{P}(M)$, is the set of all subsets of M; that is,

$$\mathscr{P}(M) = \{N \mid N \subseteq M\}.$$

4.13 Example If $M = \{1, 2, 3\}$, find $\mathscr{P}(M)$.

SOLUTION $\mathscr{P}(M) = \{\varnothing, \{1\}, \{2\}, \{3\}, \{1, 2\}, \{1, 3\}, \{2, 3\}, \{1, 2, 3\}\}$. □

PROBLEM SET 1.4

In Problems 1 and 2, rewrite each set in explicit form.

1. (a) The set A of positive integers less than or equal to 8.

(b) The set B of integers whose squares are greater than 5 and less than 90.

(c) The set C consisting of all numbers that belong to both sets A and B in Parts (a) and (b).
(d) The set D consisting of all numbers that belong to set A or to set B in Parts (a) and (b).
(e) The set E of integers between 2 and 42 that are exactly divisible by 4.
(f) The set F consisting of all integers that are positive and satisfy the inequality $x + 3 < 7$, or that satisfy the equation $x^2 + 1 = 0$.

2. (a) The set P of prime numbers between 35 and 50.
(b) The set K of integers whose cubes lie between -27 and 27, inclusive.
(c) The set R of positive integers whose square root is less than 5.
(d) The set M consisting of numbers that are either odd positive integers less than 10 or are even integers between 5 and 13.
(e) The set N consisting of all integers that satisfy the equation $|x + 1| = 5$ and also satisfy the inequality $|x - 1| < 4$.
(f) The set S of real numbers that are solutions of the equation $12x^2 - 17x + 6$.

In Problems 3–8, let the universe of discourse U be the set of all real numbers. Rewrite each implicitly described set in explicit form.

3. $\{x \mid x^2 = 4\}$

4. $\{x \mid x^2 - x - 2 = 0\}$

5. $\{x \mid x^2 + x + 1 = 0\}$

6. $\{x \mid |2x + 7| = 11\}$

7. $\{x \mid (3x - 1)^{1/3} = 2\}$

8. $\{x \mid (|x| < 1) \wedge (x^2 > 4)\}$

9. Which of the following sets are equal to the empty set?
(a) The set of all women who have been President of the United States.
(b) The set of all months of the year whose names in English start with the letter O.
(c) The set of all real numbers whose square is -1.
(d) The set of all signers of the Declaration of Independence who were Presidents of the United States.

10. True or false:
(a) $1 \in \{1\}$ (b) $1 \subseteq \{1\}$ (c) $\{1\} \in \{1\}$ (d) $\{1\} \subseteq \{1\}$

In Problems 11 and 12, decide whether or not $M \subseteq N$.

11. (a) M is the set of U.S. senators, N is the set of U.S. congressmen.
(b) M is the set of all squares, N is the set of all rectangles.
(c) M is the set of all triangles, N is the set of all right triangles.
(d) M is the set of all citizens of the U.S. who are eligible to be President, N is the set of all citizens of the U.S. who are at least 35 years of age.
(e) $M = \{1, 3, 5, 7\}$, $N = \{-1, 1, 3, 5, 7, 9\}$
(f) $M = \{1, 2\}$, $N = \{1, \{2\}\}$
(g) The universe of discourse U is all real numbers, $M = \{x \mid 3x - 2 \leq 4\}$, $N = \{x \mid 8 - 5x \geq 3\}$.

(h) The universe of discourse U is all real numbers,
$M = \{x \mid x^2 + x - 12 < 0\}$, $N = \{x \mid |2x + 7| < 11\}$.

12. (a) $M = \varnothing, N = \{\{\varnothing\}\}$ (b) $M = \varnothing, N = \varnothing$
(c) $M = \{\varnothing\}, N = \{\varnothing, \{\varnothing\}\}$ (d) $M = \{1\}, N = \{\{1\}\}$
(e) $M = \{2, 3\}, N = \{1, \{2, 3\}\}$ (f) $M = \{1, 2\}, N = \{1, 2, \{3\}\}$
(g) $M = \{1, 2, 3\}, N = \{1, 2, \{3\}\}$ (h) $M = \{1, \{2\}\}, N = \{1, 2, \{2\}\}$

13. Let U, the universe of discourse, consist of all real numbers. In each case determine whether the two sets A and B are equal.
(a) $A = \{x \mid x^2 + x - 2 = 0\}$, $B = \{1, -2\}$
(b) $A = \{x \mid 1 \le x \le 3\}$, $B = \{1, 3\}$
(c) $A = \{x \mid 3x - 10 \le 5 < x + 3\}$, $B = \{x \mid 2 < x \le 5\}$
(d) $A = \{x \mid 2 \sin x = 1\}$, $B = \{\pi/6, 5\pi/6\}$
(e) $A = \{x \mid x \text{ is an integer and } |2x - 1| \le 5\}$, $B = \{-2, -1, 0, 1, 2, 3\}$

14. Let U, the universe of discourse, consist of the twenty-six letters of the alphabet. Explain why the set

$\{x \mid x$ is a letter in the word *little*$\}$

is equal to the set

$\{x \mid x$ is a letter in the word *tile*$\}$.

15. Prove that, if A, B, and C are sets with $A \subseteq B$ and $B \subseteq C$, then $A \subseteq C$.

16. True or false: If A and B are sets and $A \not\subseteq B$, then $B \subseteq A$. Explain.

17. Draw a Venn diagram illustrating three sets A, B, and C such that $A \subseteq B$, $A \subseteq C$, $B \not\subseteq C$, and $C \not\subseteq B$.

18. Let $M = \{x \mid P(x)\}$ and $N = \{x \mid Q(x)\}$. Suppose that $(\forall x)(P(x) \Rightarrow Q(x))$. Prove that $M \subseteq N$.

19. Let $I = \{1, 2, 3\}$ and consider the family of sets $(M_i)_{i \in I}$ such that $M_1 = \{a, b, d\}$, $M_2 = \{b, c, d\}$, and $M_3 = \{c, d\}$. Find, in explicit form, the set M consisting of all elements that belong to *at least one* of the sets in the family $(M_i)_{i \in I}$.

20. In Problem 19, find, in explicit form, the set D consisting of all elements that belong to all of the sets in the family $(M_i)_{i \in I}$.

In Problems 21–26, find $\mathscr{P}(M)$.

21. $M = \{1\}$ **22.** $M = \varnothing$

23. $M = \{1, 2\}$ **24.** $M = \{1, 2, 3, 4\}$

25. $M = \{a, b, c\}$ **26.** $M = \{1, \{1\}\}$

27. List all of the proper subsets of $\{1, 2, 3\}$.

28. List all of the proper subsets of $\{1, 2, 3, 4\}$.

29. List all of the nontrivial subsets of $\{1, 2, 3\}$.

30. List all of the nontrivial subsets of $\{1, 2, 3, 4\}$.

31. True or false: Every proper subset of a set is a nontrivial subset of the set. Explain.

32. Is it possible to find sets A and B such that $A \neq \emptyset$, $A \subseteq B$, and $A \in B$? If so, give an example; if not, explain why not.

33. If A and B are sets and $A \subseteq B$, prove that $\mathscr{P}(A) \subseteq \mathscr{P}(B)$.

5 THE ALGEBRA OF SETS

In much the same way that propositions are combined by propositional connectives to form new propositions, sets can be combined by suitable operations to form new sets. The result is an **algebra of sets**. If M and N are sets, then the **union** of M and N, written

$$M \cup N,$$

is defined to be *the set of all objects that belong to M or to N or to both M and N*. In other words, $M \cup N$ is the set of all elements that belong to *at least one* of the sets M or N. The **intersection** of M and N, written

$$M \cap N,$$

is defined to be *the set of all objects that belong to both M and N*. The **relative complement** of N in M, written

$$M \setminus N,$$

is defined to be *the set of all objects that belong to M but not to N*. The following definition is more formal:

5.1 DEFINITION

UNION, INTERSECTION, AND RELATIVE COMPLEMENT Let M and N be sets. Then,

(i) $M \cup N = \{x \mid x \in M \vee x \in N\}$
(ii) $M \cap N = \{x \mid x \in M \wedge x \in N\}$
(iii) $M \setminus N = \{x \mid x \in M \wedge x \notin N\}$

In words:

$M \cup N$ is read *M union N*

$M \cap N$ is read *M intersection N*

$M \setminus N$ is read *M slash N*

For instance, if M is the set of all politicians and N is the set of all honest people, then $M \cup N$ is the set of all people who are politicians or honest (or both), $M \cap N$ is the set of all honest politicians, and $M \setminus N$ is the set of all dishonest politicians. Some authors write $M \setminus N$ as $M - N$ and refer to it as M *minus* N, but we prefer to use the slash because of the possible confusion of the minus sign with numerical subtraction.

5.2 Example Let $M = \{1, 3, 6, 9\}$ and $N = \{3, 4, 9\}$. Find:

(a) $M \cup N$ **(b)** $M \cap N$ **(c)** $M \setminus N$

SOLUTION **(a)** $M \cup N = \{1, 3, 4, 6, 9\}$ **(b)** $M \cap N = \{3, 9\}$
(c) $M \setminus N = \{1, 6\}$ □

In informal mathematical writing, a comma is often used as a substitute for the word *and* or for the connective $\wedge$. For instance, the expression

$$x \in M \wedge x \in N$$

may be written in the equivalent form

$$x \in M, x \in N.$$

Thus, with this notation,

$$M \cap N = \{x \mid x \in M, x \in N\},$$

and likewise,

$$M \setminus N = \{x \mid x \in M, x \notin N\}.$$

With appropriate shading, we indicate the union, intersection, and relative complements of sets M and N in Venn diagrams, as shown in Figures 1-3, 1-4, and 1-5.

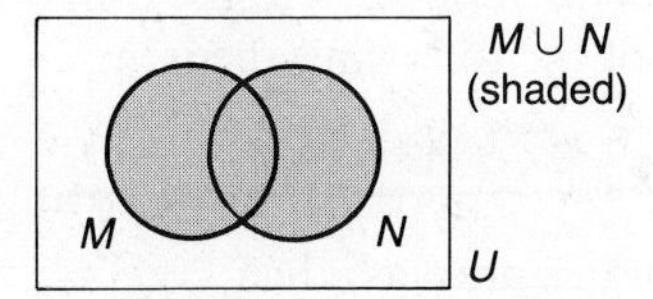

FIGURE 1-3

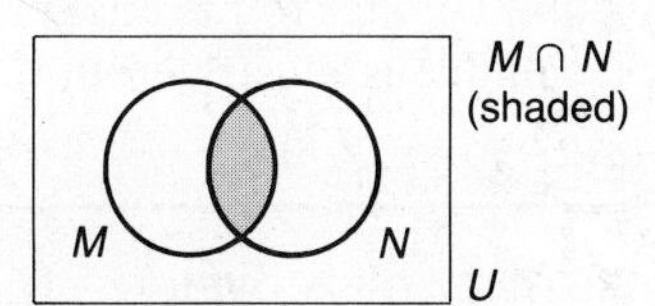

FIGURE 1-4

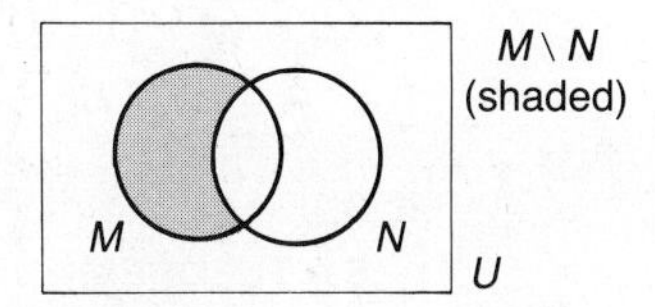

FIGURE 1-5

The set $U \setminus M$ illustrated in Figure 1-6 is called the **absolute complement** of M, or sometimes just the **complement** of M, and is often written as M'. (Other authors use the notation M^c, $\bar{M}$, or $\tilde{M}$ for the complement of M.)

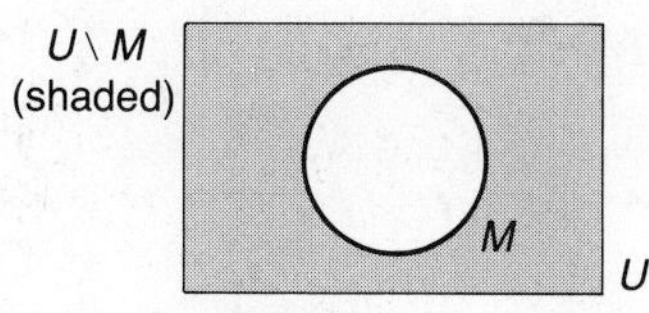

FIGURE 1-6

When you refer to the complement M' of a set M, you must be certain that U, the universe of discourse, is clearly specified or understood. In set algebra, U is usually called the **universal set**. Note that

$$U' = \varnothing \quad \text{and} \quad \varnothing' = U.$$

Also, if you complement a set twice, you obtain the original set; that is,

$$(N')' = N.$$

Because people like to use Venn diagrams to help visualize sets, the elements of a set M are often referred to as the points of M. The word *point*, used in this way, is not necessarily to be taken literally in its usual geometric sense.

Many of the theorems of set algebra state that one set, built up in a certain way from arbitrary sets by means of union, intersection, and relative (or absolute) complementation, is contained in or is equal to another such set. The **pick-a-point process** is one way to prove that a first set is contained in a second set: *Pick an arbitrary point (element) in the first set and argue that it must be in the second set.*

5.3 Example If M and N are sets, prove that $M \cap N \subseteq M$.

SOLUTION We use the pick-a-point process: Let $x \in M \cap N$. Then $x \in M$ and $x \in N$; so, in particular, $x \in M$. This shows that every element x in $M \cap N$ belongs to M, that is, that $M \cap N \subseteq M$. □

A theorem stating that two sets are equal can be proved by using the pick-a-point process twice to show that each set is contained in the other. Sometimes such a proof can be shortened by arguing directly that an arbitrary point belongs to the one set *if and only if* it belongs to the other set. This technique is illustrated in the proof of the following theorem:

5.4 THEOREM **DISTRIBUTIVITY OF ∩ OVER ∪** If L, M, and N are sets, then

$$L \cap (M \cup N) = (L \cap M) \cup (L \cap N).$$

PROOF Let x be an arbitrary point in the universal set U. Then, by Part (ii) of Definition 5.1,

$$x \in L \cap (M \cup N) \Leftrightarrow x \in L \wedge (x \in M \cup N),$$

and by Part (i) of Definition 5.1,

$$x \in M \cup N \Leftrightarrow x \in M \vee x \in N;$$

hence,

$$x \in L \cap (M \cup N) \Leftrightarrow x \in L \wedge (x \in M \vee x \in N).$$

Similarly,

$$x \in (L \cap M) \cup (L \cap N) \Leftrightarrow (x \in L \wedge x \in M) \vee (x \in L \wedge x \in N).$$

But, by tautology 8 on page 10, with P replaced by $x \in L$, Q replaced by $x \in M$, and R replaced by $x \in N$, we have

$$x \in L \wedge (x \in M \vee x \in N) \Leftrightarrow (x \in L \wedge x \in M) \vee (x \in L \wedge x \in M),$$

and it follows that

$$x \in L \cap (M \cup N) \Leftrightarrow x \in (L \cap M) \cup (L \cap N). \qquad \square$$

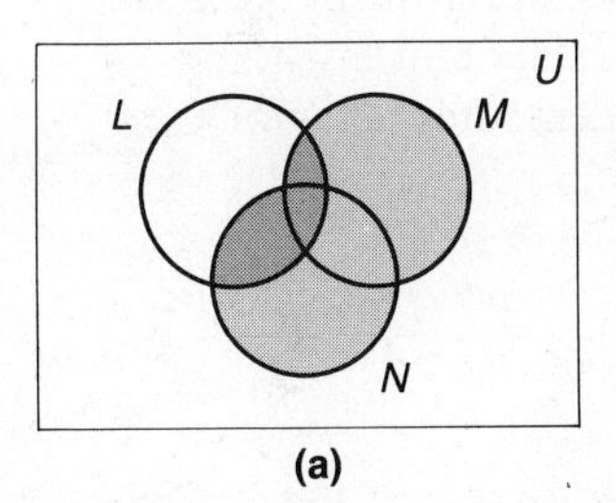

(a)

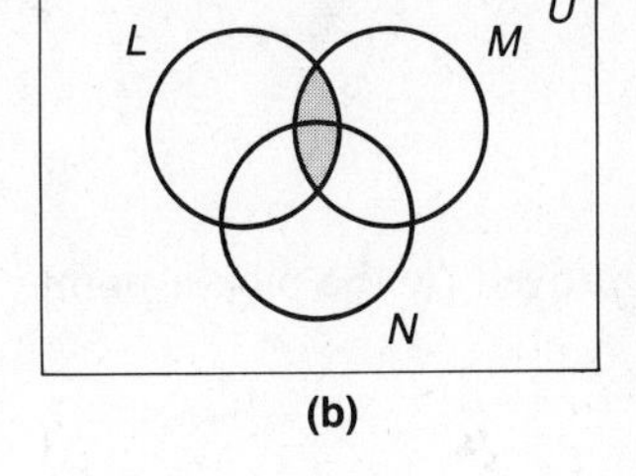

(b)

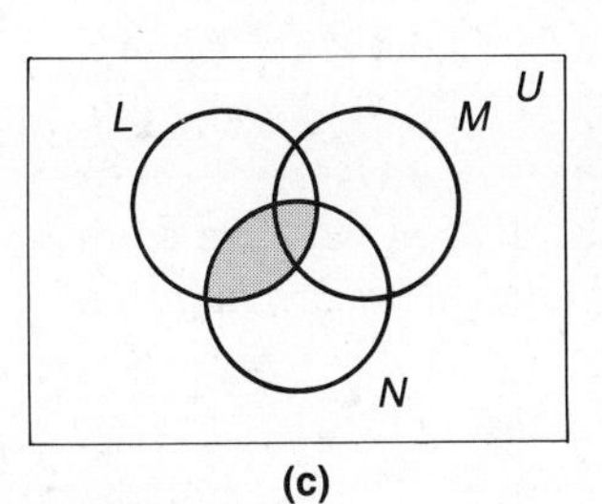

(c)

FIGURE 1-7

The Venn diagrams in Figure 1-7 illustrate the distributive law in Theorem 5.4. Part (a) shows $M \cup N$ shaded and $L \cap (M \cup N)$ with darker shading; Part (b) shows $L \cap M$ shaded, and Part (c) shows $L \cap N$ shaded. Clearly, $L \cap (M \cup N)$ in Part (a) is the union of $L \cap M$ in Part (b) and $L \cap N$ in Part (c).

Notice the analogy between the distributive law

$$L \cap (M \cup N) = (L \cap M) \cup (L \cap N)$$

for sets and the distributive law

$$P \wedge (Q \vee R) \Leftrightarrow (P \wedge Q) \vee (P \wedge R)$$

for propositions, which was, in fact, used in its proof. In a similar way, many of the tautologies of propositional calculus give rise to corresponding theorems for sets, and this creates a far-reaching analogy between propositional calculus and the algebra of sets. There is a further analogy between the distributive law above for sets and the familiar distributive law

$$x(y + z) = xy + xz$$

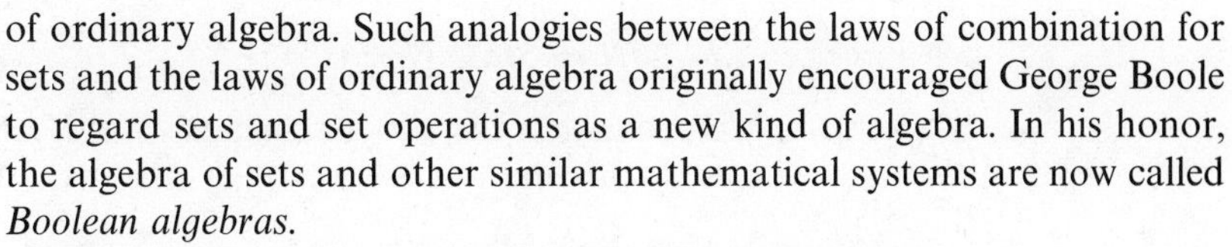

of ordinary algebra. Such analogies between the laws of combination for sets and the laws of ordinary algebra originally encouraged George Boole to regard sets and set operations as a new kind of algebra. In his honor, the algebra of sets and other similar mathematical systems are now called *Boolean algebras.*

In the following list of some of the basic theorems of set algebra, M, N, and L denote sets and U denotes the universal set.

1. $M = (M')'$	Double complementation
2. $M \cap N = N \cap M$	Commutative law for $\cap$
3. $M \cup N = N \cup M$	Commutative law for $\cup$
4. $(M \cap N) \cap L = M \cap (N \cap L)$	Associative law for $\cap$
5. $(M \cup N) \cup L = M \cup (N \cup L)$	Associative law for $\cup$
6. $M \cup M' = U$	First complementation law
7. $M \cap M' = \varnothing$	Second complementation law

8. $M \cap (N \cup L) = (M \cap N) \cup (M \cap L)$	Distributive law for $\cap$ over $\cup$
9. $M \cup (N \cap L) = (M \cup N) \cap (M \cup L)$	Distributive law for $\cup$ over $\cap$
10. $(M \cap N)' = M' \cup N'$	De Morgan's law for $\cap$
11. $(M \cup N)' = M' \cap N'$	De Morgan's law for $\cup$
12. $M \setminus N = M \cap N'$	Relative complement law
13. $M \setminus (N \cap L) = (M \setminus N) \cup (M \setminus L)$	De Morgan's law for relative complements
14. $M \setminus (N \cup L) = (M \setminus N) \cap (M \setminus L)$	De Morgan's law for relative complements
15. $(M \cup N) \cap N = N$	First absorption law
16. $(M \cap N) \cup N = N$	Second absorption law
17. $M \cap M = M$	Idempotent law for $\cap$
18. $M \cup M = M$	Idempotent law for $\cup$
19. $M \cap N \subseteq M$	
20. $M \subseteq M \cup N$	
21. $\varnothing \subseteq M \subseteq U$	
22. $M \subseteq N \Leftrightarrow N' \subseteq M'$	
23. $M \subseteq N \Leftrightarrow M \cap N = M$	
24. $M \subseteq N \Leftrightarrow M \cup N = N$	
25. $M \subseteq N \Leftrightarrow M \cap N' = \varnothing$	
26. $M \subseteq N \Leftrightarrow M' \cup N = U$	
27. $M = N \Leftrightarrow ((M \subseteq N) \wedge (N \subseteq M))$	
28. $M \cup N \subseteq L \Leftrightarrow ((M \subseteq L) \wedge (N \subseteq L))$	
29. $L \subseteq M \cap N \Leftrightarrow ((L \subseteq M) \wedge (L \subseteq N))$	

All of the theorems in the list above can be proved by the pick-a-point process.

5.5 Example Prove theorem number 26 in the list above,

$$M \subseteq N \Leftrightarrow M' \cup N = U.$$

SOLUTION We prove that the two statements $M \subseteq N$ and $M' \cup N = U$ are equivalent by proving that each implies the other. We begin by proving that

$$M \subseteq N \Rightarrow M' \cup N = U.$$

Assume that

$$M \subseteq N.$$

On the basis of this assumption, we must show that

$$M' \cup N = U.$$

Because every set is contained in U, we automatically have

$$M' \cup N \subseteq U,$$

so it remains to argue that

$$U \subseteq M' \cup N.$$

Thus, let $x \in U$. We must show $x \in M' \cup N$. If $x \in M'$, then $x \in M' \cup N$, so we only need to consider the case in which $x \notin M'$; that is, we can suppose that $x \in M$. But then, by the assumption that $M \subseteq N$, we have $x \in N$, and it follows that $x \in M' \cup N$. This completes the proof that

$$M \subseteq N \Rightarrow M' \cup N = U.$$

Now, to prove the converse implication,

$$M' \cup N = U \Rightarrow M \subseteq N,$$

we begin by assuming that

$$M' \cup N = U.$$

On the basis of this assumption, we must show that

$$M \subseteq N.$$

To this end, let $x \in M$. Our proof will be complete if we can argue that $x \in N$. Since

$$x \in U \quad \text{and} \quad U = M' \cup N,$$

it follows that

$$x \in M' \cup N;$$

hence, $x \in M'$ or $x \in N$. Because $x \in M$, we cannot have $x \in M'$, and the only alternative is that $x \in N$. □

Although the following theorem is very interesting, its proof is too difficult to include here.

5.6 THEOREM

DEDUCTIVE DEVELOPMENT OF SET ALGEBRA Every identity of the algebra of sets is a consequence of the commutative laws 2 and 3, the associative laws 4 and 5, the complementation laws 6 and 7, the distributive laws 8 and 9, the relative complementation law 12, the absorption laws 15 and 16, and the containment laws 21 and 23.

Thus, the thirteen laws mentioned in this theorem could be taken as postulates, and the algebra of sets could be derived deductively from them. (Note, however, that the algebra of sets is only a portion of the theory of sets.) By definition, an abstract mathematical system satisfying these thirteen postulates is called a **Boolean algebra**. Actually, more economical systems of postulates for a Boolean algebra have been found, and one can get along with fewer than thirteen postulates. Also, some of the thirteen laws mentioned above are redundant in the sense that they can be derived deductively from the remaining ones. For instance, the second distributive law, 9, can be derived from the remaining twelve laws.

The following example illustrates how an identity of Boolean algebra can be derived deductively from the thirteen laws mentioned in Theorem 5.6 (as opposed to a pick-a-point argument).

5.7 Example Give a deductive proof of the idempotent law 18, for union: $M \cup M = M$.

SOLUTION The first absorption law, 15, holds for all sets M and N, so we can replace the symbol N in 15 by the symbol M to obtain

$$(M \cup M) \cap M = M. \tag{1}$$

Next, we replace the symbol M in the second absorption law, 16, by $M \cup M$ to obtain

$$((M \cup M) \cap N) \cup N = N. \tag{2}$$

Now, replacing N by M in (2), we have

$$((M \cup M) \cap M) \cup M = M. \tag{3}$$

Finally, substituting (1) into (3), we have

$$M \cup M = M. \tag{4}$$

□

It is often possible to count the elements of a set M. The result of such a counting process is a nonnegative integer m called the **number** of elements in M. Counting the elements of a set M in two different ways always results in the same number m (unless a mistake in counting has been made). We introduce the notation $\# M$ for the number of elements in M. Thus, the equation

$$\# M = m$$

means that the result of counting the elements of the set M is the nonnegative integer m. For instance,

$$\#\{3, 5, 7, 9\} = 4.$$

Note that

$$\#\varnothing = 0.$$

In set theory, $\# M$ is called the **cardinal number** of the set M. Other authors use different notation for the cardinal number of a set; one popular alternative is $|M|$.

5.8 Example If U is the set of all real numbers, find

$$\#\{x \mid x^2 - 4x + 3 = 0\}.$$

SOLUTION Because

$$\{x \mid x^2 - 4x + 3 = 0\} = \{1, 3\},$$

we have

$$\#\{x \mid x^2 - 4x + 3 = 0\} = 2.$$

□

At this point in our development, it is not possible to make a rigorous formal definition of $\#M$; however, since we all know how to count, for the time being we are going to make free use of the idea of the number of elements in a set M, at least when M is a finite set. (Later, in Sections 4.6 and 4.7, we give a more formal development of the idea of the cardinal number of a set.) For our present purposes, a set M is finite if the process of counting its elements eventually terminates and results in a nonnegative integer m. We assume that finite sets have the following properties:

1. If M is a finite set, then $\#M$ is a nonnegative integer.
2. If $\#M$ is a nonnegative integer, then M is a finite set.
3. A subset of a finite set is a finite set.
4. If $M = \{a\}$ is a singleton set, then $\#M = 1$.
5. If M and N are finite sets, then $M \cup N$ is a finite set.

Of course, we cannot give rigorous proofs of theorems involving the concept of the number of elements in a set until we have a formal definition of $\#M$. In the meantime, we propose to settle for informal intuitive proofs of such theorems based on our usual understanding of the counting process.

5.9 THEOREM **THE NUMBER OF ELEMENTS IN A UNION OF TWO SETS** Let M and N be finite sets. Then,

$$\#(M \cup N) = \#M + \#N - \#(M \cap N).$$

INFORMAL PROOF If you count the elements of M and N separately, you will have counted the elements in $M \cap N$ twice: once when you count the elements of M and again when you count the elements of N. Thus, you can obtain the number of elements in the union of M and N by adding the number of elements in M to the number of elements in N and then compensating for the elements counted twice by subtracting the number of elements in $M \cap N$.

□

5.10 DEFINITION

DISJOINT SETS Two sets M and N are said to be **disjoint** if $M \cap N = \varnothing$.† The union of two disjoint sets M and N is sometimes written as

$$M \uplus N$$

and called the **disjoint union** of M and N.

5.11 THEOREM

CARDINAL NUMBER OF A DISJOINT UNION If M and N are disjoint finite sets, then

$$\#(M \uplus N) = \#M + \#N.$$

PROOF Since M and N are disjoint,

$$\#(M \cap N) = \#\varnothing = 0.$$

Therefore, by Theorem 5.9,

$$\begin{aligned}\#(M \uplus N) &= \#M + \#N - \#(M \cap N)\\ &= \#M + \#N - 0\\ &= \#M + \#N.\end{aligned}$$

□

5.12 THEOREM

THE NUMBER OF ELEMENTS IN A RELATIVE COMPLEMENT If M and N are sets and M is finite, then

$$\#(M \setminus N) = \#M - \#(M \cap N).$$

PROOF Note that $M \setminus N$ and $M \cap N$ are disjoint sets (see Problem 15a). In what follows, we use the laws of set algebra on pages 59–60. By the relative complementation law 12, we have

$$(M \cap N) \uplus (M \setminus N) = (M \cap N) \cup (M \cap N'), \tag{5}$$

and by the distributive law for $\cap$ over $\cup$, 8, we have

$$(M \cap N) \cup (M \cap N') = M \cap (N \cup N'). \tag{6}$$

Combining Equations (5) and (6), we find that

$$(M \cap N) \uplus (M \setminus N) = M \cap (N \cup N'). \tag{7}$$

† In definitions, people often write *if* when they really mean *if and only if*. Although this dubious practice would never be followed in writing very formal mathematics, it does not seem to cause any real confusion in more informal situations.

By the first complementation law 6, $N \cup N' = U$; so we can rewrite (7) as

$$(M \cap N) \cup (M \setminus N) = M \cap U. \tag{8}$$

The containment law 21 implies that $M \subseteq U$; hence, by the containment law 23, $M \cap U = M$. Therefore, (8) can be rewritten as

$$(M \cap N) \cup (M \setminus N) = M. \tag{9}$$

By (9) and Theorem 5.11, we have

$$\#(M \cap N) + \#(M \setminus N) = \#M, \tag{10}$$

from which it follows that

$$\#(M \setminus N) = \#M - \#(M \cap N). \tag{11}$$

□

5.13 COROLLARY

If M is a finite set and $N \subseteq M$, then

$$\#(M \setminus N) = \#M - \#N.$$

The proof of Corollary 5.13 follows easily from Theorem 5.12, and we leave it for you as an exercise (Problem 27).

5.14 THEOREM

THE NUMBER OF SUBSETS OF A SET If M is a finite set and $\#M = m$, then there are exactly 2^m different subsets of M. Therefore,

$$\#\mathcal{P}(M) = 2^{\#M}.$$

INFORMAL PROOF The case in which $M = \varnothing$ is addressed in Problem 28. Thus, suppose $M \neq \varnothing$; say,

$$M = \{a_1, a_2, a_3, \ldots, a_m\}.$$

Let N denote an arbitrary subset of M, and for each

$$i = 1, 2, 3, \ldots, m,$$

let P_i represent the proposition stating that a_i is an element of N, so that

$$P_i \Leftrightarrow a_i \in N.$$

By the axiom of extent, N is determined by its elements; hence, N is determined by the truth values of the propositions

$$P_1, P_2, P_3, \ldots, P_m.$$

In this way, the different subsets N of M correspond to the different truth combinations for these m propositions. In Section 1.1 we showed that there are exactly 2^m such truth combinations, and it follows that there are exactly 2^m different subsets of M. □

5.15 Example In the ABO-system, human blood is grouped according to which of the three antigens A, B, and Rh it contains. For instance, type O Rh+ blood contains the Rh antigen but not the A or the B antigen; and type AB Rh− blood contains both the A and the B antigen but not the Rh antigen. How many different blood types are possible in the ABO-system?

SOLUTION Let $M = \{A, B, Rh\}$. Corresponding to each subset N of M is a blood type containing all the antigens in N and none of the antigens in $M \setminus N$. By Theorem 5.14, there are $2^3 = 8$ such subsets; hence, in the ABO-system there are eight different blood types. □

PROBLEM SET 1.5

1. Let $M = \{1, 2, 3, 4\}$, $N = \{3, 4, 5\}$, $Q = \{3, 4, 6, 7\}$, and $P = \{6, 8, 9\}$. Find each set (in explicit form).

(a) $M \cup N$ (b) $M \cap N$ (c) $M \cap Q$
(d) $Q \cup P$ (e) $M \setminus N$ (f) $Q \setminus P$
(g) $M \setminus Q$ (h) $Q \setminus N$ (i) $M \cup (Q \cap P)$
(j) $M \cap (Q \cup P)$ (k) $(N \cup P) \cap Q$ (l) $(Q \cap P) \cup N$
(m) $(M \cup N) \setminus (Q \cap P)$ (n) $(M \cap N) \cup (M \setminus P)$

2. In Problem 1, which pairs, if any, of the sets M, N, Q, and P are disjoint?

3. If M, N, and L denote three arbitrary sets, draw Venn diagrams and use shading to illustrate the following:

(a) $M \setminus (N \cup L)$ (b) $(M \cap N) \setminus L$ (c) $M \cap N \cap L$
(d) $M' \cap N' \cap L'$ (e) $(M \setminus N) \cup (N \setminus M)$

4. If U is the universal set and M is an arbitrary set, indicate which of the following are necessarily true and which could be false.

(a) $M \cap \varnothing = \varnothing$ (b) $\varnothing \cup U = U$ (c) $M \cap U = U$
(d) $M \setminus M = \varnothing$ (e) $M \setminus U = M$ (f) $U \setminus M = \varnothing$
(g) $M \cup U = U$ (h) $\varnothing \setminus M = M'$ (i) $M \setminus \varnothing = M$

5. Let $U = \{2, 3, 4, 5, 6\}$, $M = \{2, 3, 4\}$, and $N = \{2, 4, 5, 6\}$. Find each set (in explicit form).

(a) M' (b) N' (c) $M' \cup N'$
(d) $M' \cap N'$ (e) $(M \cap N)'$ (f) $M' \cap M$
(g) $N' \cup N$ (h) $M \setminus N$ (i) $N \setminus M$

6. If M and N are sets, prove that each of the following conditions is equivalent to the condition that $M \subseteq N$:
(a) $M \cap N' = \varnothing$ (b) $M \cup N = N$ (c) $M \cap N = M$
(d) $N' \subseteq M'$ (e) $M' \cap N' = N'$ (f) $M' \cup N' = M'$

7. Prove the distributive law for $\cup$ over $\cap$.

8. Use the pick-a-point method to prove the relative complement law, $M \setminus N = M \cap N'$.

9. Prove the two De Morgan laws:
(a) $(M \cap N)' = M' \cup N'$ (b) $(M \cup N)' = M' \cap N'$

10. Let U be the universal set, and let M and N be arbitrary sets. What conditions must be imposed on M and N to make each of the following true?
(a) $M \cup N = \varnothing$ (b) $M \cap N = U$ (c) $M \cup U = M$
(d) $M \cup \varnothing = \varnothing$ (e) $M \cap U = U$ (f) $M \cap U = M$

11. Illustrate the two De Morgan laws in Problem 9 by Venn diagrams.

12. Prove the two De Morgan laws for relative complements (13 and 14 in the list on page 60) and illustrate these laws with Venn diagrams.

13. Prove the two absorption laws (15 and 16 on page 60) by the pick-a-point method.

14. Give a deductive proof, patterned after the proof of Example 5.7, for the idempotent law for intersection: $M \cap M = M$.

15. In the proof of Theorem 5.12, a deductive proof is given of the identity $(M \cap N) \cup (M \setminus N) = M$.
(a) Verify that the sets $M \cap N$ and $M \setminus N$ are disjoint.
(b) Give a pick-a-point proof of this identity.
(c) Illustrate this identity with a Venn diagram.

16. Prove that the second distributive law 9 (in the list on page 60) can be derived deductively from the remaining twelve laws mentioned in Theorem 5.6. (You may not have to use all twelve of the remaining laws to do this.)

17. True or false: $M \cup L \subseteq N \cup L \Rightarrow M \subseteq N$. If true, prove it; if false, give a counterexample.

18. If M, N, L, and P are sets with $M \subseteq N$ and $L \subseteq P$, prove that $L \setminus N \subseteq P \setminus M$.

In Problems 19–22, the notation $M \,\Delta\, N$ is used for the ***symmetric difference*** *of the two sets M and N, defined by*

$$M \,\Delta\, N = (M \setminus N) \cup (N \setminus M).$$

19. If $M = \{1, 2, 3\}$ and $N = \{2, 3, 4, 5\}$, find each set (in explicit form).
(a) $M \,\Delta\, N$ (b) $N \,\Delta\, M$ (c) $M \,\Delta\, M$ (d) $M \,\Delta\, \varnothing$

20. Draw Venn diagrams for each of the following:
(a) $M \Delta N$ (b) $M \Delta (N \Delta L)$ (c) $(M \Delta N) \Delta L$

21. Prove that $M \Delta N = (M \cup N) \setminus (M \cap N)$.

22. If U is the universal set, and M and N are arbitrary sets, prove that symmetric difference has the following properties:
(a) $M \Delta N = N \Delta M$ (b) $M \Delta M = \varnothing$
(c) $M \Delta M' = U$ (d) $M \Delta \varnothing = M$
(e) $M \Delta U = M'$ (f) $M \Delta (N \Delta L) = (M \Delta N) \Delta L$
(g) $M \cap (N \Delta L) = (M \cap N) \Delta (M \cap L)$

23. Let the universal set U consist of all real numbers. Find:
(a) $\#\{x \mid 2x^2 - x - 1 = 0\}$ (b) $\#\{x \mid 2x^2 + x + 1 = 0\}$
(c) $\#\{x \mid 9x^2 - 12x + 4 = 0\}$ (d) $\#\{x \mid x^3 = 8\}$

24. Let M be a finite set with $\# M = m$. Prove that there are exactly $(3^m + 1)/2$ different pairs of disjoint subsets of M.

25. A farmer has 41 pigs. Every fat pig is greedy; 20 of the pigs are both fat and healthy; 8 of the pigs are greedy but not healthy; and 30 of the pigs are greedy. Also, 1 pig is neither healthy nor greedy; and 5 pigs are greedy but not fat.
(a) How many pigs are healthy but not greedy?
(b) How many pigs are greedy but neither fat nor healthy?
(c) How many pigs are healthy and greedy but not fat? [Hint: Draw a Venn diagram with U = all 41 pigs, G = all greedy pigs, F = all fat pigs, and H = all healthy pigs. Label sets in the diagram with their known cardinal numbers, and deduce from this the remaining cardinal numbers.]

26. Let M be a finite set, let $N \subseteq M$, and let $d = \#(M \setminus N)$. Prove that $\#\{L \mid N \subseteq L \subseteq M\} = 2^d$.

27. Prove Corollary 5.13.

28. The proof of Theorem 5.14 does not treat the case in which $M = \varnothing$.
(a) What is $\mathscr{P}(\varnothing)$?
(b) Show that Theorem 5.14 is true even if $M = \varnothing$.

29. As in Problems 19–22, define the symmetric difference of sets M and N by $M \Delta N = (M \setminus N) \cup (N \setminus M)$. If M and N are finite sets, prove that $M \Delta N$ is a finite set and that $\#(M \Delta N) = \#M + \#N - 2\#(M \cap N)$.

30. In the theory of permutations and combinations, the symbol C_r^n, called the *number of combinations of n things taken r at a time*, is defined for nonnegative integers n and r, with $r \leq n$, by

$$C_r^n = \frac{n!}{(n-r)!r!},$$

where $n! = 1 \cdot 2 \cdot 3 \cdots (n-1)n$ for $n > 0$, and $0! = 1$. Show that if M is a set containing n elements, then the number of different subsets of M containing exactly r elements is C_r^n.

31. Let M be a finite set with $\#M = m$.
(a) How many proper subsets does M have?
(b) How many nontrivial subsets does M have?

32. By combining Theorem 5.14 and Problem 30, show that if n is a nonnegative integer, then

$$C_0^n + C_1^n + C_2^n + \cdots + C_{n-1}^n + C_n^n = 2^n.$$

33. In drawing a Venn diagram to illustrate a theorem of set algebra, you must be careful that the circles or other figures representing the sets are "in a general position." For instance, if only two sets are involved, and if the theorem does not contain a hypothesis that the two sets are disjoint, you must make certain that the circles representing these sets overlap. Try to give a careful definition of the phrase "in a general position" as it would apply to a Venn diagram involving three sets.

34. In set theory, it is possible to prove the following theorem: *A statement asserting the equality of two sets or a containment of one set in another, where the two sets are formed from sets $M_1, M_2, M_3, \ldots, M_n$ using the operations of union, intersection, and relative or absolute complementation, can be tested by replacing the sets $M_1, M_2, M_3, \ldots, M_n$ by the null set $\varnothing$ and the universal set U in every possible way.* (This requires 2^n separate tests.) *If the statement is true in each of these 2^n special cases, then it is a theorem of set algebra.* Using this theorem as a basis, describe a tabular procedure, similar to the truth-table procedure for checking tautologies, that tests for theorems in set algebra.

35. Suppose that the square array (matrix) of sets

$$\begin{bmatrix} H & L \\ M & N \end{bmatrix}$$

has the following properties: (i) Sets appearing in the same row are disjoint; that is, $H \cap L = M \cap N = \varnothing$. (ii) The union of the sets appearing in any column is a fixed set G; that is, $H \cup M = L \cup N = G$. Prove that the union of the sets appearing in any row must be the fixed set G; that is, prove that $H \cup L = M \cup N = G$.

36. Use the tabular test suggested in Problem 34 to show that

$$M_1 \backslash (M_2 \cup M_3 \cup M_4) = (M_1 \backslash M_2) \cap (M_1 \backslash M_3) \cap (M_1 \backslash M_4)$$

is a theorem of set algebra.

37. Generalize the theorem of Problem 35 to an arbitrary n-by-n square array (matrix) of sets.

*6

GENERALIZED UNIONS AND INTERSECTIONS

This section discusses generalizations of the notions of union and intersection to indexed families of sets and to sets of sets. Before we make the appropriate definitions, it is necessary to point out that there is an important (but somewhat subtle) distinction between an indexed family of sets and a set of sets.

For an indexed family of sets

$$(M_i)_{i \in I},$$

we have a definite correspondence that associates a set M_i with each element i in the index set I, whereas for a set of sets

$$\mathscr{E} = \{A, B, C, \ldots\}$$

no such correspondence is automatically provided.

For instance, suppose that

$$I = \{1, 2, 3\}$$

and consider the indexed family of sets

$$(M_i)_{i \in I}$$

given by

$$M_1 = \{\text{Olga, Carlos}\},$$
$$M_2 = \{\text{Rose, Pedro}\},$$
$$M_3 = \{\text{Anna, Boris}\}.$$

Here we have a definite correspondence that associates

the set {Olga, Carlos} with the index $1 \in I$,
the set {Rose, Pedro} with the index $2 \in I$, and
the set {Anna, Boris} with the index $3 \in I$.

However, no such correspondence can be discerned merely from the set of sets $\mathscr{E}$ given by

$$\mathscr{E} = \{\{\text{Olga, Carlos}\}, \{\text{Rose, Pedro}\}, \{\text{Anna, Boris}\}\}.$$

(Recall that no particular order is associated with the elements of a set.)

Thus, although an indexed family of sets

$$(M_i)_{i \in I}$$

gives rise to a set of sets

$$\mathscr{E} = \{M_i \mid i \in I\},$$

it does not automatically work the other way around. However, given a set of sets $\mathscr{E}$, it is possible to form a corresponding indexed family of sets

$$(M_i)_{i \in I}$$

by choosing an appropriate index set I, setting up a suitable correspondence between the elements $i \in I$ and the sets in $\mathscr{E}$, and taking M_i to be the corresponding set in $\mathscr{E}$ for each $i \in I$. The process of choosing such an I and setting up such a correspondence is called **indexing** the sets in $\mathscr{E}$. Note, however, that there are many different ways in which a given set of sets $\mathscr{E}$ can be indexed—*indexing is not a unique process*!

6.1 Example Index the set of sets $\mathscr{E}$ given by

$$\mathscr{E} = \{\{\text{red, yellow}\}, \{\text{green, blue, purple}\}, \{\text{pink, green}\}\}.$$

SOLUTION We use $I = \{p, q, r\}$ as our indexing set and define the indexed family of sets

$$(M_i)_{i \in I}$$

by taking

$$M_p = \{\text{red, yellow}\},$$
$$M_q = \{\text{green, blue, purple}\},$$
$$M_r = \{\text{pink, green}\}.$$

□

It is often possible to ignore the distinction between an indexed family of sets and a set of sets without running into mathematical or logical difficulties, and people often do this in the interest of writing or speaking more concisely. We indulge in this practice when it is convenient and harmless. But, *be careful*, in some situations serious logical trouble results from confusing an indexed family of sets with a set of sets.

Now we are ready to give the generalized definitions of union and intersections, beginning with the generalized union.

6.2 DEFINITION

UNION OF A SET OF SETS Let $\mathscr{E}$ be a set of sets. We define the **union** of the sets in $\mathscr{E}$, denoted $\bigcup\mathscr{E}$, by

$$\bigcup\mathscr{E} = \{x \mid (\exists M)(M \in \mathscr{E} \wedge x \in M)\}.$$

In words, *the union $\bigcup\mathscr{E}$ of the set of sets $\mathscr{E}$ is the set consisting of all elements that belong to at least one of the sets M in $\mathscr{E}$*. Alternative notation for $\bigcup\mathscr{E}$ is

$$\bigcup_{M \in \mathscr{E}} M \quad \text{or} \quad \bigcup\nolimits_{M \in \mathscr{E}} M.$$

6.3 Example Let $\mathscr{E} = \{A, B, C\}$ where $A = \{2, 4, 6\}$, $B = \{1, 2, 5\}$, and $C = \{2, 6, 8\}$. Find $\bigcup \mathscr{E}$.

SOLUTION $\bigcup \mathscr{E}$ consists of all elements that belong to at least one of the sets A, B, or C. Therefore,

$$\bigcup \mathscr{E} = \{1, 2, 4, 5, 6, 8\}. \qquad \square$$

If $\mathscr{E}$ is a finite set of sets, say,

$$\mathscr{E} = \{M_1, M_2, M_3, \ldots, M_n\},$$

then we often write $\bigcup \mathscr{E}$ in the alternative form

$$\bigcup \mathscr{E} = M_1 \cup M_2 \cup M_3 \cup \cdots \cup M_n.$$

For instance, in Example 6.3,

$$\bigcup \mathscr{E} = A \cup B \cup C = \{1, 2, 4, 5, 6, 8\}.$$

Also, if $(M_i)_{i \in I}$ is an indexed family of sets, we define

$$\bigcup_{i \in I} M_i = \bigcup\nolimits_{i \in I} M_i = \bigcup \{M_i \mid i \in I\},$$

so that

$$\bigcup_{i \in I} M_i = \{x \mid (\exists i)(i \in I \wedge x \in M_i)\}.$$

In words, *the union $\bigcup_{i \in I} M_i$ of an indexed family of sets is the set consisting of all elements that belong to at least one of the sets M_i, for some $i \in I$.*

6.4 Example Find the union $\bigcup_{i \in I} M_i$ for the indexed family of sets $(M_i)_{i \in I}$ in Example 6.1.

SOLUTION In Example 6.1, we have $I = \{p, q, r\}$, $M_p = \{\text{red, yellow}\}$, $M_q = \{\text{green, blue, purple}\}$, and $M_r = \{\text{pink, green}\}$. Therefore,

$$\begin{aligned} \bigcup\nolimits_{i \in I} M_i &= M_p \cup M_q \cup M_r \\ &= \{\text{red, yellow, green, blue, purple, pink}\}. \end{aligned} \qquad \square$$

The integers from 1 to n inclusive, or the set of all positive integers, is often used as the index set I for a family of sets $(M_i)_{i \in I}$. If $I = \{1, 2, 3, \ldots, n\}$, then the union of the family of sets $(M_i)_{i \in I}$ can be written in the alternative form

$$\bigcup\nolimits_{i=1}^{n} M_i \quad \text{or} \quad \bigcup_{i=1}^{n} M_i.$$

Likewise, if $I = \{1, 2, 3, \ldots\}$, the notation

$$\bigcup\nolimits_{i=1}^{\infty} M_i \quad \text{or} \quad \bigcup_{i=1}^{\infty} M_i$$

may be used for the union of the family of sets $(M_i)_{i \in I}$.

6.5 Example Let the universal set U consist of all real numbers, and for each positive integer i, let

$$M_i = \{x | i - 1 \leq x \leq i\}.$$

Thus, with the usual interval notation used in calculus, M_i is the closed interval $[i - 1, i]$. Find $\bigcup_{i=1}^{\infty} M_i$.

SOLUTION Since each nonnegative real number belongs to at least one of the intervals $M_i = [i - 1, i]$, we have

$$\bigcup_{i=1}^{\infty} M_i = \{x | 0 \leq x\} = [0, \infty).$$

□

6.6 THEOREM **GENERALIZED DISTRIBUTIVE LAW FOR ∩ OVER ∪** Let N be a set and let $(M_i)_{i \in I}$ be an indexed family of sets. Then,

$$N \cap \left(\bigcup_{i \in I} M_i \right) = \bigcup_{i \in I} (N \cap M_i).$$

PROOF We use the pick-a-point procedure. First, suppose that x is a point in the set on the left side of the equation. Then,

$$x \in N \quad \text{and} \quad x \in \bigcup_{i \in I} M_i.$$

Because

$$x \in \bigcup_{i \in I} M_i,$$

it follows that there exists an index $j \in I$ such that

$$x \in M_j.$$

Therefore,

$$x \in N \quad \text{and} \quad x \in M_j;$$

that is,

$$x \in N \cap M_j.$$

Consequently,

$$x \in \bigcup_{i \in I} (N \cap M_i);$$

so x belongs to the set on the right side of the equation.

Now, suppose that x is a point in the set on the right side of the equation. Then there exists an index $j \in I$ such that

$$x \in N \cap M_j;$$

that is,

$$x \in N \quad \text{and} \quad x \in M_j.$$

Because $x \in M_j$, it follows that

$$x \in \bigcup_{i \in I} M_i.$$

Therefore,

$$x \in N \quad \text{and} \quad x \in \bigcup_{i \in I} M_i,$$

and it follows that x belongs to the set on the left side of the equation. □

The generalized intersection is defined as follows:

6.7 DEFINITION

INTERSECTION OF A SET OF SETS Let $\mathscr{E}$ be a set of sets. We define the **intersection** of the sets in $\mathscr{E}$, denoted $\bigcap \mathscr{E}$, by

$$\bigcap \mathscr{E} = \{x \mid (\forall M)(M \in \mathscr{E} \Rightarrow x \in M)\}.$$

In words, *the intersection* $\bigcap \mathscr{E}$ *of the set of sets* $\mathscr{E}$ *is the set consisting of all elements that belong to every one of the sets M in* $\mathscr{E}$. Alternative notation for $\bigcap \mathscr{E}$ is

$$\bigcap_{M \in \mathscr{E}} M \quad \text{or} \quad \bigcap\nolimits_{M \in \mathscr{E}} M.$$

6.8 Example Let $\mathscr{E} = \{A, B, C\}$ where $A = \{2, 4, 6\}$, $B = \{1, 2, 5\}$, and $C = \{2, 6, 8\}$ as in Example 6.3. Find $\bigcap \mathscr{E}$.

SOLUTION $\bigcap \mathscr{E}$ consists of all elements that belong to *all three* of the sets A, B, and C. Therefore,

$$\bigcap \mathscr{E} = \{2\}.$$

□

If $\mathscr{E}$ is a finite set of sets, say,

$$\mathscr{E} = \{M_1, M_2, M_3, \ldots, M_n\},$$

then we often write $\bigcap \mathscr{E}$ in the alternative form,

$$\bigcap \mathscr{E} = M_1 \cap M_2 \cap M_3 \cap \cdots \cap M_n.$$

For instance, in Example 6.8,

$$\bigcap \mathscr{E} = A \cap B \cap C = \{2\}.$$

Also, if $(M_i)_{i \in I}$ is an indexed family of sets, we define

$$\bigcap_{i \in I} M_i = \bigcap_{i \in I} M_i = \bigcap\{M_i \mid i \in I\},$$

so that

$$\bigcap_{i \in I} M_i = \{x \mid (\forall i)(i \in I \Rightarrow x \in M_i\}.$$

In words, *the intersection* $\bigcap_{i \in I} M_i$ *of an indexed family of sets is the set consisting of all elements that belong to every one of the sets* M_i, *for* $i \in I$.

6.9 Example As in Example 6.1, let $I = \{p, q, r\}$ with

$M_p = \{\text{red, yellow}\}$,
$M_q = \{\text{green, blue, purple}\}$,
$M_r = \{\text{pink, green}\}$.

Find $\bigcap_{i \in I} M_i$.

SOLUTION There are no elements that belong to all three of the sets M_p, M_q, and M_r; hence,

$$\bigcap_{i \in I} M_i = \varnothing.$$

□

6.10 DEFINITION

PAIRWISE DISJOINT SETS OF SETS A set of sets $\mathscr{E}$ is said to be **pairwise disjoint** if, for all sets M and N in $\mathscr{E}$,

$$M \neq N \Rightarrow M \cap N = \varnothing.$$

Likewise, an indexed family of sets $(M_i)_{i \in I}$ is said to be pairwise disjoint if, for every pair of indices i, j in I,

$$i \neq j \Rightarrow M_i \cap M_j = \varnothing.$$

Notice that *the intersection of a pairwise disjoint set (or indexed family) of two or more sets is empty* (Problem 23). However, consider the following:

6.11 Example Is the indexed family of sets $(M_i)_{i \in I}$ in Example 6.9 pairwise disjoint?

SOLUTION No! Although the intersection of this family of sets is empty, it is not pairwise disjoint because

$$M_q \cap M_r = \{\text{green}\} \neq \varnothing.$$

□

We leave the proof of the following generalized distributive law as an exercise for you (Problem 18).

6.12 THEOREM

GENERALIZED DISTRIBUTIVE LAW FOR ∪ OVER ∩ Let N be a set and let $(M_i)_{i \in I}$ be an indexed family of sets. Then,

$$N \cup \left(\bigcap_{i \in I} M_i \right) = \bigcap_{i \in I} (N \cup M_i).$$

The De Morgan laws also generalize (Problem 24).

6.13 THEOREM

GENERALIZED DE MORGAN LAWS Let $(M_i)_{i \in I}$ be an indexed family of sets. Then,

(i) $(\bigcup_{i \in I} M_i)' = \bigcap_{i \in I} M_i'$.
(ii) $(\bigcap_{i \in I} M_i)' = \bigcup_{i \in I} M_i'$.

If $I = \{1, 2, 3, \ldots, n\}$, then the intersection of the family of sets $(M_i)_{i \in I}$ is often written in the alternative form

$$\bigcap_{i=1}^{n} M_i \quad \text{or} \quad \bigcap_{i=1}^{n} M_i.$$

Likewise, if $I = \{1, 2, 3, \ldots\}$, the notation

$$\bigcap_{i=1}^{\infty} M_i \quad \text{or} \quad \bigcap_{i=1}^{\infty} M_i$$

may be used for the intersection of the family of sets $(M_i)_{i \in I}$.

6.14 Example Let the universal set U consist of all real numbers, and, for each positive integer i, let

$$M_i = \left\{ x \,\middle|\, 1 - \frac{1}{i} < x < 1 + \frac{1}{i} \right\}.$$

Thus, with the usual interval notation used in calculus, M_i is the open interval $(1 - (1/i), 1 + (1/i))$. Find $\bigcap_{i=1}^{\infty} M_i$.

SOLUTION A real number x belongs to M_i if and only if $|x - 1| < 1/i$. Since $1/i$ becomes arbitrarily small as i gets larger and larger, x belongs to *every* M_i if and only if $x = 1$. Therefore,

$$\bigcap_{i=1}^{\infty} M_i = \{1\}.$$ □

The empty set $\varnothing$ can be regarded as a set of sets—the empty set of sets. As such, the statement $M \in \varnothing$ *is false for every set* M. Of course, we have

$$\bigcup \varnothing = \varnothing;$$

in words, *the union of the empty set of sets is the empty set* (Problem 27). For the intersection of the empty set of sets, there is a bit of a surprise: *The intersection of the empty set of sets is the universal set* U; in symbols,

$$\bigcap \varnothing = U.$$

6.15 Example Prove that $\bigcap \varnothing = U$.

SOLUTION Since every set is a subset of U, we have only to prove that $U \subseteq \bigcap \varnothing$. We use the pick-a-point process. Let $x \in U$. To prove that $x \in \bigcap \varnothing$, we must prove that, for every set M,

$$M \in \varnothing \Rightarrow x \in M.$$

But, the hypothesis $M \in \varnothing$ of this implication is *false*; hence, the implication is *true*. □

PROBLEM SET 1.6

In Problems 1–6, index the set of sets $\mathscr{E}$.

1. $\mathscr{E} = \{\{\text{Alabama, Alaska}\}, \{\text{Michigan, Ohio}\}, \{\text{Maine, Texas}\}\}$

2. $\mathscr{E} = \{\{\text{Washington, Roosevelt}\}, \{\text{Kennedy, Carter, Reagan}\}\}$

3. $\mathscr{E} = \{\{1, 4, 5, 9\}, \{1, 3, 4, 5\}, \{4, 5, 7, 8\}, \{1, 4, 7, 9\}, \{4, 5, 8, 9\}\}$

4. $\mathscr{E} = \mathscr{P}(\{a, b, c\})$

5. $\mathscr{E}$ = the set of all sets of the form

$$\{x \mid x \text{ is an integer multiple of } n\},$$

where n is a positive integer.

6. $\mathscr{E} = \{\{-3, n\} \mid n \text{ is a positive integer}\}$

7. Let $I = \{1, 2, 3, 4\}$, and for $i \in I$ let $M_i = \{i, i + 2, 3i\}$. Find each set in explicit form:
(a) M_1 (b) M_2 (c) M_4 (d) $M_1 \cup M_3$
(e) $M_1 \cap M_3$ (f) $\bigcup_{i=1}^{4} M_i$ (g) $\bigcap_{i=1}^{4} M_i$

8. Let $(M_i)_{i \in I}$ be the indexed family of sets in Problem 7 and let $\mathscr{E} = \{M_i \mid i \in I\}$. Reindex the set of sets $\mathscr{E}$ using $J = \{p, q, r, s\}$ as the new indexing set.

In Problems 9–12, let $\mathscr{E} = \{A, B, C\}$. *Find (a)* $\bigcup \mathscr{E}$ *and (b)* $\bigcap \mathscr{E}$ *in explicit form.*

9. $A = \{1, 2, 3\}$, $B = \{2, 3, 5\}$, $C = \{2, 4, 7\}$

10. $A = \{\text{tall, short, fat}\}$, $B = \{\text{slim, tall, short}\}$, $C = \{\text{tall}\}$

11. $A = \{a, b, c, d\}$, $B = \{b, c, d, e\}$, $C = \{c, d, e, f\}$

12. $A = \{-1, 0, 1, 3\}$, $B = \{-1, 0, 2, 4\}$, $C = \{2, 3, 5, 7\}$

In Problems 13–16, find (a) $\bigcup_{i\in I} M_i$ and (b) $\bigcap_{i\in I} M_i$ in explicit form.

13. $I = \{1, 2, 3\}$
$M_1 = \{\text{red, yellow, green, blue}\}$
$M_2 = \{\text{yellow, green, blue, violet}\}$
$M_3 = \{\text{green, blue, pink, orange}\}$

14. $I = \{a, b, c, d\}$
$M_a = \{1, 2, 3\}$
$M_b = \{2, 4, 5\}$
$M_c = \{2, 5, 6\}$
$M_d = \{2, 7, 8\}$

15. $I = \{1, 2, 3, 4, 5\}$
$M_1 = \{\text{right, left}\}$
$M_2 = \{\text{right, left, up}\}$
$M_3 = \{\text{right, left, up, down}\}$
$M_4 = \{\text{right, left, up, down, front}\}$
$M_5 = \{\text{right, left, up, down, front, back}\}$

16. $I = \{\text{dog, cat, lion, tiger, canary}\}$
$M_{\text{dog}} = \{\text{domestic, quadruped, } \textit{Canidae}\}$
$M_{\text{cat}} = \{\text{domestic, quadruped, } \textit{Felidae}\}$
$M_{\text{lion}} = \{\text{wild, quadruped, } \textit{Panthera}\}$
$M_{\text{tiger}} = \{\text{wild, quadruped, } \textit{Panthera}\}$
$M_{\text{canary}} = \{\text{domestic, biped, } \textit{Serinus}\}$

17. (a) Which of the sets of sets in Problems 1–6 are pairwise disjoint?
(b) Which of the indexed families of sets in Problems 13–16 are pairwise disjoint?

18. Prove Theorem 6.12.

In Problems 19–22, we use the usual notation for intervals of real numbers: (a, b) for the open interval $\{x \mid a < x < b\}$, $(a, b]$ for the half-open interval $\{x \mid a < x \le b\}$, $[a, b]$ for the closed interval $\{x \mid a \le x \le b\}$, $[a, \infty)$ for the unbounded interval $\{x \mid a \le x\}$, and so on. The answer to each question should be an interval, a singleton set, or the empty set.

19. Find: (a) $\bigcup_{i=1}^{\infty}[i-1, i)$ (b) $\bigcap_{i=1}^{\infty}[i-1, i)$

20. Find: (a) $\bigcup_{n=1}^{\infty}(-n, n)$ (b) $\bigcap_{n=1}^{\infty}(-n, n)$

21. Find: (a) $\bigcup_{j=1}^{\infty}[0, 1/j]$ (b) $\bigcap_{j=1}^{\infty}[0, 1/j]$

22. Find: (a) $\bigcup_{k=1}^{\infty}(1/k, \infty)$ (b) $\bigcap_{k=1}^{\infty}(1/k, \infty)$

23. Suppose that I contains at least two distinct elements and that the indexed family of sets $(M_i)_{i\in I}$ is pairwise disjoint. Prove that $\bigcap_{i\in I} M_i = \varnothing$.

24. Prove Theorem 6.13.

25. If $\mathscr{E}$ is a set of sets and $\varnothing \in \mathscr{E}$, prove that $\bigcap \mathscr{E} = \varnothing$.

26. If $(M_i)_{i\in I}$ and $(N_j)_{j\in J}$ are two indexed families of sets, express the set $(\bigcup_{i\in I} M_i) \cap (\bigcup_{j\in J} N_j)$ as a union of intersections. [Use Theorem 6.6.]

27. If $\varnothing$ is regarded as a set of sets, prove that $\bigcup \varnothing = \varnothing$.

28. If $(M_i)_{i\in I}$ and $(N_j)_{j\in J}$ are two indexed families of sets, express the set $(\bigcap_{i\in I} M_i) \cup (\bigcap_{j\in J} N_j)$ as an intersection of unions. [Hint: Use Theorem 6.12.]

29. Suppose $(M_i)_{i \in I}$ is an indexed family of sets and that A and B are sets such that $A \subseteq M_i \subseteq B$ for every $i \in I$.
(a) Prove that $A \subseteq \bigcap_{i \in I} M_i$. (b) Prove that $\bigcup_{i \in I} M_i \subseteq B$.

30. If $\mathscr{E}$ is a set of sets, is it always true that $\bigcap \mathscr{E} \subseteq \bigcup \mathscr{E}$? Why or why not?

31. If $\mathbb{R}$ is the set of all real numbers, then a subset G of $\mathbb{R}$ is said to be open if, for every $x \in G$, there exists a real number $\varepsilon > 0$ such that the open interval $(x - \varepsilon, x + \varepsilon)$ is contained in G. If $\mathscr{T}$ is the set of all open subsets of $\mathbb{R}$, prove that $\mathscr{T}$ has the following four properties:
(i) $\varnothing \in \mathscr{T}$ (ii) $\mathbb{R} \in \mathscr{T}$
(iii) If $G_1, G_2 \in \mathscr{T}$, then $G_1 \cap G_2 \in \mathscr{T}$. (iv) If $\mathscr{E} \subseteq \mathscr{T}$, then $\bigcup \mathscr{E} \in \mathscr{T}$.

32. Let X be a nonempty set. Show that $\mathscr{E}$ is a set of subsets of X if and only if $\mathscr{E} \in \mathscr{P}(\mathscr{P}(X))$.

33. With the notation and terminology of Problem 31, if $M \subseteq \mathbb{R}$, then $x \in \mathbb{R}$ is called an **accumulation point**, or **limit point**, of M if, for every open set G with $x \in G$, $(G \setminus \{x\}) \cap M \neq \varnothing$. The set $M \subseteq \mathbb{R}$ is said to be **closed** if every accumulation point of M belongs to M. Prove: M is closed if and only if $\mathbb{R} \setminus M$ is open.

34. If X is a nonempty set, then a set $\mathscr{C}$ of subsets of X is called a **closure system** for X if it has the following properties: $X \in \mathscr{C}$ and $\varnothing \neq \mathscr{E} \subseteq \mathscr{C} \Rightarrow \bigcap \mathscr{E} \in \mathscr{C}$. Let $\mathscr{C}$ be a closure system for X. If $M \subseteq X$, define the **closure** of M, in symbols $\bar{M}$, by $\bar{M} = \bigcap \{C \in \mathscr{C} \mid M \subseteq C\}$. If $M, N \subseteq X$, prove:
(a) $M \subseteq \bar{M}$ (b) $M \subseteq N \Rightarrow \bar{M} \subseteq \bar{N}$ (c) $N = \bar{M} \Rightarrow N = \bar{N}$

35. With the notation and terminology of Problems 31 and 33, if $M \subseteq \mathbb{R}$, then $x \in \mathbb{R}$ is called a **boundary point** of M if, for every open set G with $x \in G$, $G \cap M \neq \varnothing$ and $G \setminus M \neq \varnothing$. Let $M \subseteq \mathbb{R}$. Prove:
(a) M is open if and only if no boundary point of M belongs to M.
(b) M is closed if and only if every boundary point of M belongs to M.

36. With the notation and terminology of Problems 31, 33, and 34, let $\mathscr{C}$ be the set of all closed subsets of $\mathbb{R}$. Prove:
(a) $\mathscr{C}$ is a closure system.
(b) $M \in \mathscr{C}$ if and only if M is a closed subset of $\mathbb{R}$.
(c) If $M \subseteq \mathbb{R}$ and A is the set of all accumulation points of M, then $\bar{M} = M \cup A$.
(d) If $M, N \subseteq \mathbb{R}$, then $\overline{M \cup N} = \bar{M} \cup \bar{N}$.

HISTORICAL NOTES

Many of the ideas in this book were developed in response to difficulties in the higher realms of calculus. The book from which you recently studied calculus is a descendant of books written in the eighteenth century by the Swiss mathematician Leonhard Euler (1707–1783). Euler had learned his calculus from another Swiss, Johann Bernoulli (1667–1748), who had learned it from Gottfried Wilhelm von Leibniz (1646–1716), cofounder with Sir Isaac Newton (1642–1727) of the entire subject. Although Leibniz apparently had a glimmer of the modern concept of limit, the other early workers in the field relied more on Newton's physical and geometric notions. These intuitive notions served well enough at first, but, as the development proceeded, mathematicians became aware of serious difficulties, particularly in connection with the interchange of two limit processes.

Until the 1800s, mathematics remained the activity of a small group of extremely gifted individuals, and the difficulties alluded to above were merely annoying. A really bright person could separate the sense from the nonsense. This all changed about 1820 as a consequence of the investigations into the theory of heat by the Frenchman, Jean Baptiste Joseph Fourier (1768–1830). After Fourier, intuition was not enough—a secure methodology was necessary. The first phase of modern rigor was instituted by another Frenchman, Augustin-Louis Cauchy (1789–1857), who formulated the ε-δ definition of limit and gave one of the first rigorous definitions of the integral. Building on Cauchy's work, the German mathematician Karl Weierstrass (1815–1897) opened the second phase, known as the "arithmetization of analysis," in which a rigorous development of the system $\mathbb{R}$ of real numbers forms the basis for analysis (another word for calculus).

It is important to realize that the transition from an intuitive and geometric foundation to a more abstract and rigorous approach to calculus was not the result of philosophical pressure, but was a response to the urgent needs of working mathematicians. For example, the investigations of Georg Cantor (1845–1918) on Fourier series required the definition of a real number and careful consideration of certain sets of real numbers, for instance, the set of real numbers for which a series converges. Cantor, having become interested in the theory of sets for its own sake, published a series of papers (1875–1905) in which he sketched out the theory, identified the fundamental concepts, and demonstrated the usefulness of the set concept to mathematics in general.

Although Cantor is regarded as the originator of set theory, a fairly complete exposition of the algebra of sets had been given earlier by George Boole (1815–1864), an Englishman who was Professor of Mathematics at the University of Corcaigh in Ireland. However, Boole's ideas, and other work in the same tradition by John Venn (1834–1923) and E. Schroeder (1841–1902), had little immediate effect on the main body of mathematics. Boolean algebra infiltrated mathematics through its use by Cantor and the Italian mathematician and logician Giuseppe Peano (1858–1932); it is now recognized as an important subject in its own right.

Cantor's most fundamental contribution to set theory was his concept of transfinite numbers—numbers used to count infinite sets. (See Sections 4.6 and 4.7.) Previously, no one had suspected that infinite sets might come in different sizes. However, Cantor's methodology required that infinite sets be allowed in mathematics on the same footing as finite sets, and this aroused considerable opposition, particularly from the German mathematician Leopold Kronecker (1823–1891). Partly because of mental health problems, Cantor was inclined to view this opposition as persecution. The fact is that Cantor's revolutionary work was assimilated into standard mathematics in a remarkably short time, indeed, within Cantor's own lifetime, and, in 1900, David Hilbert (1862–1943) vowed that, "From the paradise created for us by Cantor, no one will drive us forth."

Unfortunately, certain paradoxes appeared within Cantor's set theory. Ernst Zermelo (1871–1953) had the inspired thought that these paradoxes might be eliminated by subjecting set theory itself to the ancient tradition of axiomatic development and, in 1908, he published a system of axioms that, with minor modifications, is used today as a basis for rigorous set theory.

Together with ideas of Gottlob Frege (1848–1925) and John von Neumann (1903–1957), Zermelo's set theory can be used to build the system $\mathbb{N}$ of natural numbers (Section 4.2); from $\mathbb{N}$, in turn, the system $\mathbb{Z}$ of integers (Section 4.3), and the system $\mathbb{Q}$ of rational numbers (Section 4.4) can be built. Finally, using Cantor's or Julius Richard Dedekind's (1831–1916) methods, the system $\mathbb{R}$ of real numbers can be constructed (Section 4.8). With the real number system $\mathbb{R}$, Cauchy's ε-δ definition of limit, and the formal definition of the definite integral given by Georg Bernhard Riemann (1826–1866), it seemed that calculus was at last established on a firm foundation.

The increasing complexity of the logical processes used in the development of analysis made it desirable to take a more formal approach to logic. Fortunately, the necessary discipline, symbolic logic, had become available just about the time it was really needed. Symbolic logic (without the symbols) began with Philôn of Megara (340 B.C.), who asserted that the implication $P \Rightarrow Q$ is the same as $\sim P \vee Q$. This clearly indicates that implication does not rest on some connection between P and Q, but only on their truth values and, in turn, it suggests that the form, not the subject, of an argument determines its logical validity. Once this is realized, it is only a matter of time (2000 years!) before an algebraic notation evolves for logical deliberation.

Though Leibniz expressed the opinion that an algebraic approach to logic was possible, and even made some progress in the development, his work also seems to have had little effect. The modern theory starts with George Boole's important work, *An Investigation of the Laws of Thought.* Augustus De Morgan (1806–1871) also made significant contributions to symbolic logic that often cannot be clearly distinguished from Boole's, as the two men were close friends.

Although the propositional calculus is useful, it is not, in itself, sufficiently powerful to express serious mathematical ideas. With the addition of quantification theory, it becomes possible to write mathematics in a totally symbolic language. In 1879, Frege introduced a system of quantification of the same nature (although typographically quite different) as the one used in this book. Frege's basic impulse was to derive mathematics from logical truths, and much of his work in this direction is now embedded in our modern developments. He was an early user of truth tables and the first to axiomatize the propositional calculus.

In contrast to Frege, whose interests were largely philosophical and whose notation was bizarre, Peano (in 1894) made an attempt to formulate mathematics, up through calculus, in a symbolic system. Peano's work clearly demonstrated that the project was practical and, coupled with Frege's, provided the impetus for subsequent work in the development of symbolic logic.

ANNOTATED BIBLIOGRAPHY

Halmos, Paul R. *Naive Set Theory* (New York: Springer-Verlag), 1974.

This is a classic text in set theory, widely used and admired, written by one of the great expositors of mathematics in this century.

Hamilton, A. G. *Logic for Mathematicians* (Cambridge, England: Cambridge University Press), 1978.

A widely used elementary treatment of mathematical logic.

Kline, Morris. *Mathematics: The Loss of Certainty* (New York: Oxford University Press), 1980.

A book for a general audience, by an often controversial author, on the limitations of logic and mathematics in providing sure answers. Sometimes disturbing, but at the same time stimulating.

Lakatos, Imre. *Proofs and Refutations: The Logic of Mathematical Discovery* (Cambridge, England: Cambridge University Press), 1976.

An intellectual tour de force showing the development of a mathematical theorem (Euler's formula for polyhedra) through various false starts, misstatements, and subsequent modifications. Written in the form of a Socratic dialogue.

Newsom, Carroll V. *Mathematical Discourses: The Heart of Mathematical Science* (Englewood Cliffs, N.J.: Prentice-Hall), 1960.

An interesting account for the general reader of the nature of mathematical discourse and logical argument.

Quine, Willard Van Orman. *Set Theory and Its Logic* (Cambridge, Mass.: Belknap), 1969.

This is a fairly sophisticated but readable treatment of the subject, written by one of the giants in the field.

Renz, Peter. "Mathematical Proof: What It Is and What It Ought to Be," *Two-Year College Mathematics Journal* **12** (1981), 83–103.

A somewhat philosophical discourse on what constitutes a proof in mathematics. Interesting examples.

2
RELATIONS AND FUNCTIONS

Contemporary mathematics is written in terms of sets, relations, and functions. It is widely believed that, ultimately, every mathematical notion can be so expressed. As we indicate in this chapter, it is possible to develop the theory of relations and functions on a set-theoretic basis; hence, it is conceivable that all of mathematics can be founded on logic and set theory. Thus, in the authoritative Bourbaki,† *Éléments de Mathematique*, we read that:

> Nowadays it is known to be possible, logically speaking, to derive practically the whole of known mathematics from a single source, The Theory of Sets.

The theory of relations and functions is developed on a set-theoretic basis by using the idea of an ordered pair (a, b) of elements of the universe of discourse U. Therefore, to carry out the Bourbaki program of deriving mathematics from a single set-theoretic source it is necessary to define an ordered pair (a, b) purely in terms of sets. The usual definition is

$$(a, b) = \{a, \{a, b\}\},$$

but we feel that this is a bit too sophisticated for our present purposes. Hence, in what follows, we treat ordered pairs somewhat more informally.

1 ORDERED PAIRS

If a and b are elements of the universe of discourse U, then, since the elements of a set are not supposed to occur in any particular order, we have

$$\{a, b\} = \{b, a\}.$$

† A collective pseudonym for a prestigious group of contemporary French mathematicians. See Paul Halmos, "Nicolas Bourbaki," *Scientific American* (May 1957).

For this reason, the set $\{a, b\}$ is sometimes called the *unordered* pair a, b. But suppose we want to form an **ordered** pair consisting of a **first** element a and a **second** element b. The notation

$$(a, b)$$

is used for such an ordered pair. Thus, in contrast to the unordered pair $\{a, b\}$, we understand that for $a \neq b$,

$$(a, b) \neq (b, a).$$

Our ordered pairs will be governed by just two axioms: an *existence* axiom and an *equality* axiom.

1.1 AXIOM

EXISTENCE OF ORDERED PAIRS If a and b are elements in the universe of discourse U, then the ordered pair (a, b) exists.

Axiom 1.1 is supposed to hold even if $a = b$; so an ordered pair

$$(a, a)$$

consisting of the same first and second elements is perfectly legitimate.

1.2 AXIOM

EQUALITY OF ORDERED PAIRS If a, b, c, and d are elements of U, then

$$(a, b) = (c, d) \Leftrightarrow a = c \wedge b = d.$$

In words, Axiom 1.2 says that *two ordered pairs are equal if and only if they have the same first element and the same second element.* As a consequence of Axiom 1.2, we have

$$(a, b) = (b, a) \Leftrightarrow a = b,$$

which you will prove in Problem 4.

In the seventeenth century, the French mathematician and philosopher René Descartes (1596–1650) systematically developed the idea that, by means of a coordinate system, geometry could be translated into algebra and vice versa. In the **Cartesian coordinate system** (named after Descartes), a geometric point P in the plane is assigned x and y coordinates with respect to a fixed pair of coordinate axes. The ordered pair (x, y) of these coordinates provides a numerical "address" for the point P. The resulting correspondence

$$P \leftrightarrow (x, y)$$

between the geometric point P and the ordered pair (x, y) of its coordinates is so compelling that, in some cases, it is convenient to identify P with (x, y) and write

$$P = (x, y).$$

In this way, the set of all points in the plane is identified with the set of all ordered pairs (x, y) of real numbers.

In mathematical writing, the symbol $\mathbb{R}$ is often used to represent the set of all real numbers, and the set of all ordered pairs of real numbers is denoted by

$$\mathbb{R}^2 = \{(x, y) \mid x \in \mathbb{R} \wedge y \in \mathbb{R}\}.$$

Because each $(x, y) \in \mathbb{R}^2$ can be regarded as a point in the plane, the set $\mathbb{R}^2$ is called the **Cartesian plane.**

The idea of a Cartesian plane can be generalized by considering the set of all ordered pairs whose first and second elements belong to specified sets—not necessarily the set $\mathbb{R}$ of real numbers. Thus, if M and N are sets, we define the **Cartesian product** of M and N, in symbols

$$M \times N,$$

to be the set consisting of *all ordered pairs whose first element belongs to M and whose second element belongs to N.* More formally:

1.3 DEFINITION

CARTESIAN PRODUCT OF SETS If M and N are sets, we define

$$M \times N = \{(x, y) \mid x \in M \wedge y \in N\}.$$

1.4 Example Let $M = \{1, 2, 3\}$, $N = \{a, b\}$, and $S = \{s\}$. Find:

(a) $M \times N$ **(b)** $N \times M$ **(c)** $S \times N$ **(d)** $S \times S$

SOLUTION

(a) $M \times N = \{(1, a), (1, b), (2, a), (2, b), (3, a), (3, b)\}$
(b) $N \times M = \{(a, 1), (a, 2), (a, 3), (b, 1), (b, 2), (b, 3)\}$
(c) $S \times N = \{(s, a), (s, b)\}$
(d) $S \times S = \{(s, s)\}$ □

In Parts (a) and (b) of Example 1.4, notice that $M \times N \neq N \times M$.

Just as we use Venn diagrams to illustrate unions, intersections, and complements of sets, we can visualize Cartesian products with suitable diagrams suggested by analogy with the Cartesian plane. Figure 2-1 shows such a diagram illustrating the Cartesian product $M \times N$. In this figure, the points on the lower edge of the rectangle represent the elements of M, the points on the left edge of the rectangle represent the elements of N,

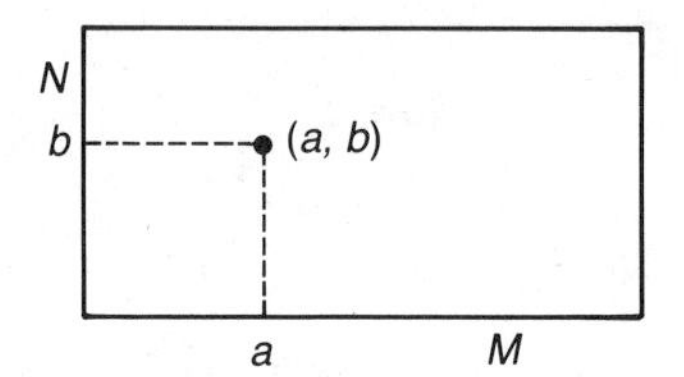

FIGURE 2-1

and the point in the rectangle with "coordinates" $a \in M$ and $b \in N$ represents the ordered pair $(a, b) \in M \times N$. In this way, we can think of the Cartesian product $M \times N$ as the set of all points in the rectangle.

If M and N actually are the sets of points on the lower and side edges of a rectangle R, then we can regard R as being represented by $M \times N$ as in Figure 2-1 and write

$$R = M \times N.$$

In this sense, *the Cartesian product of line segments is a rectangle.* In a similar way, and with a little imagination, you can visualize Cartesian products of simple geometric figures other than line segments.

1.5 Example Give a geometric interpretation of the Cartesian product of a circle and a line segment.

SOLUTION We can visualize the Cartesian product $C \times L$ of a circle C and a line segment L as forming the surface S of a right circular cylinder. Think of C as forming the circular base of S and of L as being one of the line segments, perpendicular to the base and lying on the surface S. The location of a point P on S can be specified by "coordinates" $x \in C$ and $y \in L$, determined as follows: x is the point at the foot of the perpendicular dropped from P to C; y is the point where a circle on S, parallel to C and containing P, intersects L (Figure 2-2). □

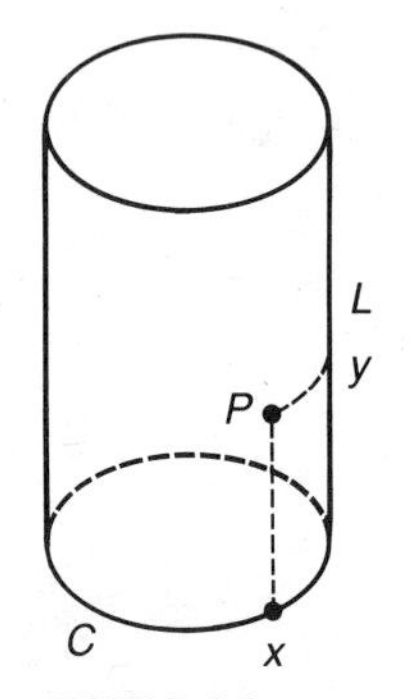

FIGURE 2-2

Example 1.5 should remind you of the "cylindrical coordinates" used in calculus.

1.6 THEOREM **THE CARDINAL NUMBER OF A CARTESIAN PRODUCT** If M and N are finite sets, then $M \times N$ is a finite set and

$$\#(M \times N) = \#M \cdot \#N.$$

INFORMAL PROOF Let $m = \#M$ and $n = \#N$. In forming an ordered pair $(x, y) \in M \times N$, we have m choices for the first element $x \in M$. For each such choice, we have n choices for the second element $y \in N$; hence, altogether, we have exactly mn such choices, and it follows that there are mn different elements $(x, y) \in M \times N$. □

1.7 Example How many different outcomes are possible if a pair of dice are tossed?

SOLUTION The answer to this question depends on whether we regard, say, a 2 on the first die and a 5 on the second die as being the same as a 5 on the first die and a 2 on the second die. Certainly, a dice player would regard these as equivalent outcomes. However, for mathematical purposes (specifically, for a proper application of the theory of probability to dice tossing), it is necessary to regard them as two distinct outcomes.

The faces of a single die correspond to the elements of the set

$$F = \{1, 2, 3, 4, 5, 6\},$$

so the possible outcomes of a toss of two dice are represented by the Cartesian product $F \times F$. For instance, if the first die falls with the face 2 upward and the second with the face 5 upward, the outcome of the toss would be represented by the ordered pair (2, 5). Because $\#F = 6$, it follows from Theorem 1.6 that

$$\#(F \times F) = \#F \cdot \#F = 6 \cdot 6 = 36;$$

so there are 36 possible outcomes of a toss of the two dice. □

Theorems concerning the equality or containment of sets built up from combinations of unions, intersections, complements, and Cartesian products can be proved by the pick-a-point procedure.

1.8 Example If A, B, and C are sets, prove that

$$(A \cup B) \times C \subseteq (A \times C) \cup (B \times C).$$

SOLUTION We begin by picking an arbitrary point in the set on the left of the inclusion. Such a point has the form (x, y) where

$$x \in (A \cup B) \quad \text{and} \quad y \in C.$$

Since $x \in (A \cup B)$, it follows that

$$x \in A \quad \text{or} \quad x \in B.$$

If $x \in A$, then we have

$$x \in A \quad \text{and} \quad y \in C;$$

hence, in this case,

$$(x, y) \in A \times C.$$

Likewise, if $x \in B$, then we have

$$x \in B \quad \text{and} \quad y \in C,$$

so that

$$(x, y) \in B \times C.$$

This shows that at least one of the two statements,

$$(x, y) \in A \times C \quad \text{or} \quad (x, y) \in B \times C,$$

is true, and therefore (x, y) belongs to the set on the right of the inclusion.

□

In Problem 29, we ask you to show that the inclusion in Example 1.8 can be strengthened to an equality.

PROBLEM SET 2.1

1. If x and y are real numbers and the ordered pairs $(x + 2y, 5)$ and $(3, 7x - 2y)$ are equal, find x and y.
2. True or false: $(1, (2, 3)) = ((1, 2), 3)$. Explain.
3. Find the ordered pairs corresponding to the points P_1, P_2, P_3, and P_4 in the diagram of $\{1, 2, 3, 4\} \times \{1, 2, 3, 4\}$ in Figure 2-3.

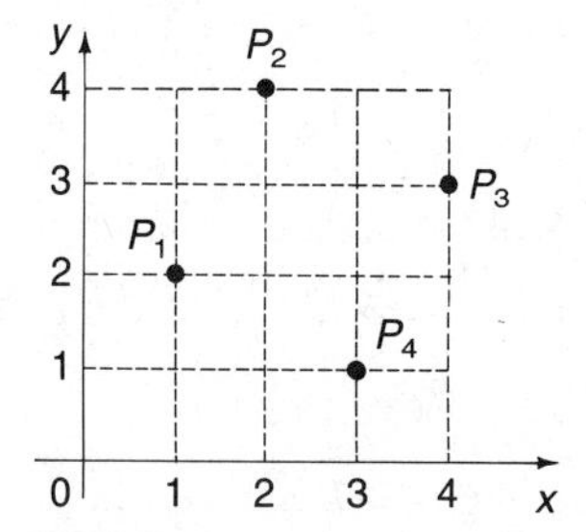

FIGURE 2-3

4. Prove that $(a, b) = (b, a)$ if and only if $a = b$.

In Problems 5–12, let $M = \{2, 4, 5\}$ and $N = \{1, 2\}$. Write out each set explicitly.

5. $M \times N$
6. $M \times M$
7. $N \times M$
8. $N \times N$
9. $\varnothing \times M$
10. $N \times \varnothing$
11. $(M \times N) \cup (N \times M)$
12. $(M \times N) \cap (N \times M)$

In Problems 13–16, let $M = \{a, b\}$, $N = \{c, d\}$, and $R = \{e, f\}$. Write out each set explicitly.

13. $M \times (N \cup R)$
14. $M \times (N \cap R)$
15. $(M \times N) \cup (M \times R)$
16. $(M \times N) \cap (M \times R)$
17. Give a geometric interpretation of the Cartesian product of a triangle and a line segment.

18. Give a geometric interpretation of the Cartesian product of two circles. [Hint: It is not a sphere.]

19. Give a geometric interpretation of the Cartesian product of a circular disk and a line segment.

20. Prove: $A \times B = \emptyset \Rightarrow A = \emptyset \vee B = \emptyset$.

In Problems 21–24, find $\#(M \times N)$.

21. $M = \{1, 2, 3\}$, $N = \{2, 7\}$

22. $M = N =$ the set of major planets in the solar system.

23. $M = \{1, 2, 3\}$, $N =$ the set of letters in the alphabet.

24. $M = N =$ the set of cards in a poker deck (excluding the joker).

25. How many different outcomes are possible if three dice are tossed?

26. If $A \neq \emptyset$ and $B \neq \emptyset$, prove that $A \times B = B \times A \Rightarrow A = B$.

27. If $A \subseteq X$ and $B \subseteq Y$, prove that $A \times B \subseteq X \times Y$.

28. Prove that $(A \cap B) \times C = (A \times C) \cap (B \times C)$.

29. Prove that $(A \cup B) \times C = (A \times C) \cup (B \times C)$.

30. True or false: $(A \setminus B) \times C = (A \times C) \setminus (B \times C)$. If true, prove it; if false, give a counterexample.

31. Prove that $(A \cap B) \times (C \cap D) = (A \times C) \cap (B \times D)$.

32. Generalize Problem 28, replacing $A \cap B$ by an intersection of an indexed family of sets.

33. Illustrate Problem 29 with diagrams.

34. True or false: $(A \cup B) \times (C \cup D) = (A \times C) \cup (B \times D)$. If true, prove it; if false, give a counterexample.

35. If M is a finite set with $\#M = m$, what is $\#(\mathscr{P}(M) \times \mathscr{P}(M))$?

36. If we define $(a, b) = \{a, \{a, b\}\}$, show that Axiom 1.2 becomes a theorem.

2 RELATIONS

The word *relation* is used in mathematics in much the same way as in English: A certain relation holds between two objects when there is a particular type of connection between them. For instance, in conversation we speak of the relation of parenthood between a mother or father and their children, and in mathematics we speak about the relation of similarity between two triangles.

Suppose we have in mind a definite mathematical relation, say, the relation of similarity between triangles or, perhaps, the relation of exact divisibility between whole numbers. Some triangles are similar to others, and some are not; some whole numbers divide others exactly, and some do not. Now, consider the propositional function

$$P(x, y)$$

of two variables x and y in U which states that *the relation in question holds for x and y.* For instance, if U is the set of all triangles and the relation under consideration is similarity, then $P(x, y)$ would be the statement

x is similar to y.

Likewise, if U is the set of all whole numbers and the relation in question is exact divisibility, then $P(x, y)$ would be the statement

x divides y.

According to the discussion above, *a mathematical relation can be represented by a propositional function $P(x, y)$ of two variables.* Conversely, *a propositional function $P(x, y)$ of two variables can be regarded as stating that a certain definite relation holds between x and y.* What relation? The relation that holds between x and y when $P(x, y)$ is true. For this reason, logicians refer to a propositional function of two variables as an **intensional binary relation**. The word *intensional* is a technical term used in logic to indicate that all of the qualities or properties inherent in the relation are expressed by the propositional function. The word *binary* simply means that *two* variables are involved.

If $P(x, y)$ is an intensional binary relation (that is, a propositional function of two variables), then the set

$$\{(x, y) \mid P(x, y)\},$$

consisting of all ordered pairs (x, y) for which $P(x, y)$ is true, is called the **graph** of the relation. Logicians refer to this graph as the binary relation in **extension**.

2.1 Example If U, the universe of discourse, consists of all the integers, find the graph R of

$$(2 < x < 7) \wedge (y = 3x \vee x = 2y).$$

SOLUTION The only possible values of x are 3, 4, 5, and 6. For these values, $y = 3x$ has corresponding values 9, 12, 15, and 18; so the ordered pairs (3, 9), (4, 12), (5, 15), and (6, 18) belong to the graph R. Furthermore, if x is 4 or 6, we have $x = 2y$ when y is 2 or 3, respectively; so the ordered pairs (4, 2) and (6, 3) also belong to R. Since these are the only pairs in R, it follows that

$$R = \{(3, 9), (4, 2), (4, 12), (5, 15), (6, 3), (6, 18)\}.$$ □

2.2 Example Show that every set R of ordered pairs is the graph of some propositional function $P(x, y)$.

SOLUTION Let R be a set of ordered pairs. Define the propositional function $P(x, y)$ of two variables to be the statement asserting that $(x, y) \in R$. Then the graph of $P(x, y)$ is given by

$$\{(x, y) | P(x, y)\} = \{(x, y) | (x, y) \in R\} = R. \qquad \square$$

An intensional binary relation $P(x, y)$ uniquely determines its graph,

$$R = \{(x, y) | P(x, y)\},$$

but it is not clear whether the original statement $P(x, y)$ can be recaptured from the graph R. You might stare at the set

$$R = \{(3, 9), (4, 2), (4, 12), (5, 15), (6, 3), (6, 18)\}.$$

obtained in Example 2.1 for a long time without realizing that it is the graph of

$$(2 < x < 7) \wedge (y = 3x \vee x = 2y).$$

Indeed, you might never hit on this fact. This should make it plain that there is a difference between binary relations in *intension* (propositional functions of two variables) and binary relations in *extension* (sets of ordered pairs). However, when speaking or writing informally, mathematicians usually do not bother to distinguish between relations in these two senses.

In more formal mathematical definitions, theorems, and proofs, it is usually convenient to deal with binary relations in extension (sets of ordered pairs); hence, we make the following definition:

2.3 DEFINITION

RELATION A **binary relation** or, in what follows, simply a **relation**, is a set of ordered pairs of elements of the domain of discourse U.

Thus, in all of our definitions and proofs, a relation R is a subset of $U \times U$. If x and y are in U, we say that *x is related to y by R* if and only if

$$(x, y) \in R.$$

The statement $(x, y) \in R$ is often abbreviated by writing

$$xRy.$$

Whether a relation R is expressed in words or specified in mathematical symbols, it is always understood that R is the *set of all ordered pairs that satisfy the condition or conditions expressed by the words or symbols.* For instance, if the domain of discourse U is the set of all real numbers and if the relation R is defined by

$$xRy \Leftrightarrow x = y^2,$$

then,

$$R = \{(x, y) | x = y^2\}.$$

2.4 DEFINITION

DOMAIN AND CODOMAIN Let R be a relation.

(i) The **domain** of R, written dom(R) is the set given by

$$\text{dom}(R) = \{x | (\exists y)(xRy)\}.$$

(ii) The **codomain** of R, written codom(R) is the set given by

$$\text{codom}(R) = \{y | (\exists x)(xRy)\}.$$

(iii) If $\text{dom}(R) \subseteq X$ and $\text{codom}(R) \subseteq X$, then we say that R is a relation **on** the set X.

In words, *the domain of R is the set of all first elements of ordered pairs in R, and the codomain of R is the set of all second elements of ordered pairs in R.* Note that R is a relation on the set X if and only if $R \subseteq X \times X$ (Problem 3).

2.5 Example If $R = \{(\pi, 3.14), (e, 2.17), (\sqrt{2}, 1.41)\}$, find:

(a) dom(R) **(b)** codom(R)

SOLUTION **(a)** $\text{dom}(R) = \{\pi, e, \sqrt{2}\}$
(b) $\text{codom}(R) = \{3.14, 2.17, 1.41\}$ □

Other authors use other words for the codomain of a relation. Some people call it the *range*, and others call it the *image* of the relation.

2.6 DEFINITION

CONVERSE, OR INVERSE If R is a relation, then the **converse**, or **inverse**, of R, written R^{-1}, is the relation defined by

$$R^{-1} = \{(y, x) | (x, y) \in R\}.$$

Note that *the converse, or inverse, of R is obtained by reversing all of the ordered pairs in R.* Thus,

$$yR^{-1}x \Leftrightarrow xRy.$$

For instance, let U be the set of all human beings, and let M be the relation defined by xMy if and only if x is the biological mother of y. Then $yM^{-1}x$ holds if and only if x is a female and y is a son or daughter of x.

2.7 Example If R is the relation of Example 2.5, find R^{-1}.

SOLUTION Since

$$R = \{(\pi, 3.14), (e, 2.17), (\sqrt{2}, 1.41)\},$$

it follows that

$$R^{-1} = \{(3.14, \pi), (2.17, e), (1.41, \sqrt{2})\}.$$ □

Note that *the domain of the converse of a relation R is the codomain of R*. This accounts for the terminology *codomain*, which stands for *converse domain.*

2.8 DEFINITION **DIAGONAL AND UNIVERSAL RELATION** Let X be a set. Then:

(i) The **diagonal relation** on X, denoted by Δ_X or simply by Δ if X is understood, is the relation defined by

$$\Delta_X = \{(x, x) | x \in X\}.$$

(ii) The **universal relation** on X is the relation $X \times X$.

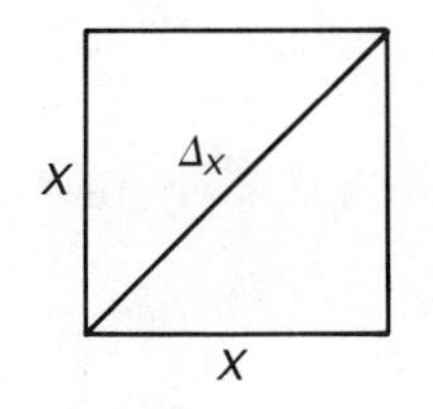

FIGURE 2-4

Note that $x(\Delta_X)y$ holds if and only if x and y are elements of X and $x = y$. Graphically, the relation Δ_X can be visualized as shown in Figure 2-4, which explains why it is called the *diagonal* relation on X. The universal relation $X \times X$ holds between *any* two elements x and y in X, and so it can be visualized as *all* points in the square shown in the figure.

2.9 DEFINITION **REFLEXIVE AND TRANSITIVE RELATIONS** Let R be a relation, and let X be a set.

(i) R is said to be **reflexive on** X if xRx holds for all $x \in X$.
(ii) R is said to be **transitive** if, for all x, y, and z,

$$xRy \wedge yRz \Rightarrow xRz.$$

Note that both Δ_X and $X \times X$ are reflexive on X (Problem 9a). Also, both Δ_X and $X \times X$ are transitive (Problem 9c).

2.10 Example Let $\mathbb{R}$ denote the set of all real numbers, and let L be the relation on $\mathbb{R}$ defined by $xLy \Leftrightarrow x < y$.

(a) Is L reflexive on $\mathbb{R}$? **(b)** Is L transitive?

SOLUTION **(a)** L is *not* reflexive on $\mathbb{R}$; indeed, no real number is less than itself; so xLx never holds.

(b) L is transitive because, if $x < y$ and $y < z$, then it follows that $x < z$. □

2.11 DEFINITION

SYMMETRIC AND ANTISYMMETRIC RELATIONS Let R be a relation.

(i) R is said to be **symmetric** if, for every x and every y,

$$xRy \Rightarrow yRx.$$

(ii) R is said to be **antisymmetric** if, for every x and every y,

$$xRy \wedge yRx \Rightarrow x = y.$$

Note that both Δ_X and $X \times X$ are symmetric relations (Problem 9b). Furthermore, Δ_X is antisymmetric (Problem 8a), but $X \times X$ is antisymmetric if and only if X is either a singleton set or the empty set (Problem 8b).

2.12 Example Let $\mathbb{R}$ denote the set of all real numbers, and let G be the relation on $\mathbb{R}$ defined by $xGy \Leftrightarrow x \geq y$.

(a) Is G symmetric? **(b)** Is G antisymmetric?

SOLUTION **(a)** G is *not* symmetric because, in general, $x \geq y$ *does not imply* that $y \geq x$. For instance, $3 \geq 2$, but $2 \not\geq 3$.

(b) G is antisymmetric because, if both $x \geq y$ and $y \geq x$ hold, then it follows that $x = y$. □

Relations are often used to order, rank, or otherwise organize the elements of a set. Among the relations that are used in this way, perhaps the most general are the preorder relations, defined as follows:

2.13 DEFINITION

PREORDER Let X be a set. By a **preorder** relation on X we mean a relation R on X such that

(i) R is reflexive on X, and
(ii) R is transitive.

Note that both Δ_X and $X \times X$ are preorder relations on X (Problem 13a). Also, the relation G in Example 2.12 is a preorder on the set $\mathbb{R}$ of all real numbers (Problem 13b), but the relation L in Example 2.10 is not (because it is not reflexive on $\mathbb{R}$).

2.14 Example Let T denote the set of all triangles in the plane, and let R be the relation defined on T by the condition that xRy if and only if the area of triangle y is at least as large as the area of triangle x. Is R a preorder relation on T?

SOLUTION Because the area of a triangle is at least as large as the area of that same triangle, R is reflexive on T. To see that R is transitive, suppose that

$$xRy \quad \text{and} \quad yRz.$$

Let A_x, A_y, and A_z denote the areas of triangles x, y, and z, respectively. Since xRy and yRz, we have

$$A_x \leq A_y \quad \text{and} \quad A_y \leq A_z,$$

and it follows that

$$A_x \leq A_z,$$

that is,

$$xRz.$$

Therefore, R is both reflexive on T and transitive; so it is a preorder on T. □

2.15 DEFINITION

PARTIAL ORDER Let X be a set. By a **partial order** relation on X we mean a relation R on X such that

(i) R is reflexive on X,
(ii) R is antisymmetric, and
(iii) R is transitive.

If you compare Definitions 2.13 and 2.15, you will see that *a partial order relation is the same thing as a preorder relation that is antisymmetric.* A memory device for partial order relations is RAT (Reflexive, Antisymmetric, and Transitive). Note that the diagonal relation Δ_X is a partial order relation on X; but, unless X is either empty or a singleton set, the universal relation $X \times X$ on X is not (Problem 14).

2.16 Example Let $\mathbb{R}$ denote the set of all real numbers, and let R be the relation defined on $\mathbb{R}$ by

$$xRy \Leftrightarrow x \leq y.$$

Is R a partial order relation on $\mathbb{R}$?

SOLUTION Yes. Because

$$x \leq x$$

holds for all $x \in \mathbb{R}$, it follows that x is reflexive on $\mathbb{R}$. Also,

$$x \leq y \wedge y \leq x \Rightarrow x = y,$$

and so R is antisymmetric. Finally,

$$x \leq y \wedge y \leq z \Rightarrow x \leq z,$$

which shows that R is transitive. □

2.17 Example Is the relation R in Example 2.14 a partial order relation on the set T of all triangles in the plane?

SOLUTION No, because in Example 2.14 the relation R is *not* antisymmetric. Indeed, if x and y are triangles in the plane and if both xRy and yRx hold, then x and y are triangles with the same area; but this does not imply that they are the same triangle. □

The relation $\leq$ in Example 2.16 on the set $\mathbb{R}$ of all real numbers is the prototype for all partial order relations; and for this reason, an arbitrary partial order relation on a set X is frequently denoted by $\leq$ rather than by R. Thus, if $\leq$ is a partial order relation on X, we read the statement $x \leq y$ as x *is less than or equal to* y. This has the advantage of making general partial order relations seem less abstract, but it sometimes has the disadvantage of suggesting *too much.* For instance, if $\leq$ is a partial order relation on a set X and if the statement $x \leq y$ is false, then you might be tempted to conclude that $y \leq x$ must be true (by analogy with the usual order relation for the real numbers). However, as the following example shows, this conclusion is not necessarily correct for partial order relations in general.

2.18 Example Let X denote the set of all positive integers and define the relation $\leq$ on X by

$$x \leq y \Leftrightarrow y \textit{ is an integer multiple of } x.$$

(Thus, in this example, $\leq$ does not have its usual meaning.) Show that it is possible for *both* of the statements $x \leq y$ and $y \leq x$ to be false.

SOLUTION Let $x = 2$, and let $y = 3$. Then, neither x nor y is an integer multiple of the other. (In Problem 15, we ask you to show that $\leq$, as defined here, is a partial order relation on X.) □

Example 2.18 should convince you of the need to be careful when dealing with general partial order relations denoted by $\leq$. With this firmly in mind, we make the following definition:

2.19 DEFINITION

PARTIALLY ORDERED SET A **partially ordered set**, or **poset** for short, is a pair $(X, \leq)$ consisting of a nonempty set X and a partial order relation $\leq$ on X.

Thus, $(X, \leq)$ is a poset if and only if $\leq$ is a relation, defined on the nonempty set X and satisfying the following three conditions for all elements x, y, and z in X:

(i) $x \leq x$ (reflexivity)

(ii) $x \leq y$ and $y \leq x \Rightarrow x = y$ (antisymmetry)

(iii) $x \leq y$ and $y \leq z \Rightarrow x \leq z$ (transitivity)

Notice that a partially ordered set, or poset, is more than just a set; it is a set *together with* a specified partial order relation on that set. Nevertheless, people often speak (incorrectly) of "the poset X," it being understood that a partial order relation $\leq$ is specified. In the interest of avoiding wordiness, mathematicans occasionally indulge in such abuses of language.

2.20 DEFINITION

UPPER AND LOWER BOUNDS Let $(X, \leq)$ be a poset, suppose that $M \subseteq X$, and let $a, b \in X$.

(i) We say that a is an **upper bound** for the set M in X if the relation

$$m \leq a$$

holds for every element $m \in M$.

(ii) We say that b is a **lower bound** for the set M in X if the relation

$$b \leq m$$

holds for every element $m \in M$.

2.21 Example In the poset $(\mathbb{R}, \leq)$ of real numbers, ordered in the usual way by $\leq$, find, if possible:

(a) A lower bound for the set P of all positive real numbers
(b) An upper bound for the set P of all positive real numbers

SOLUTION **(a)** Because $0 \leq p$ holds for every $p \in P$, it follows that 0 is a lower bound for P. Of course, there are many other lower bounds for P; for instance, any negative number is such a lower bound.

(b) There is no real number that is greater than or equal to every positive number; so P has no upper bound in $\mathbb{R}$. □

A set, such as P in Example 2.21, which has a lower bound in a poset is said to be **bounded below**. Likewise, a set that has an upper bound is said to be **bounded above**. The set P in Example 2.21 is bounded below, but unbounded above. Notice that 0 is a lower bound for P, but it does not belong to P. In general, an upper or lower bound for a set might or might not belong to the set. This leads to a definition.

2.22 DEFINITION

GREATEST AND LEAST ELEMENTS Let $(X, \leq)$ be a poset, and let $M \subseteq X$.

(i) If $a \in M$, and a is an upper bound for M, then we say that a is the **greatest element** of M.
(ii) If $b \in M$, and b is a lower bound for M, then we say that b is the **least element** of M.

If M has a least element b or a greatest element a, then this least or greatest element is unique (Problem 37). For instance, in Example 2.21, if M is the set of all positive integers, then 1 is the (unique) least element of M; but M has no upper bound and hence has no greatest element.

In a poset $(X, \leq)$, we write $x \geq y$ (by definition) to mean $y \leq x$. Statements of the form $x \geq y$ or $y \leq x$ are called **inequalities**. By definition, $x < y$ means that

$$x \leq y \wedge x \neq y.$$

The relation $x < y$, which is called a *strict* inequality, can also be written in the alternative form $y > x$. If x, y, and z are elements of X, we write the *compound* inequality

$$x \leq y \leq z$$

to mean that $x \leq y$ *and* $y \leq z$. This can also be written in the alternative form

$$z \geq y \geq x.$$

Both strict and nonstrict inequalities can be combined in compound inequalities; for instance, $x < y \leq z$.

If $(X, \leq)$ is a poset and M is a subset of X, then, by the **least upper bound** of M we mean the least element (if it exists) of the set A of all upper bounds of M in X. (Note that A could be empty.) Likewise, by the **greatest lower bound** of M, we mean the greatest element (if it exists) of the set B of all lower bounds of M in X. The least upper and greatest lower bounds of M (if they exist) are written as

$$\text{LUB}(M) \quad \text{and} \quad \text{GLB}(M),$$

respectively. The least upper bound of M is also called the **supremum** of M, and the greatest lower bound of M is also called the **infimum** of M. We denote the supremum and infimum of M by

$$\sup(M) \quad \text{and} \quad \inf(M),$$

respectively.

2.23 Example In the poset $(\mathbb{R}, \leq)$ of Example 2.21, let

$$I = \{x \mid 0 < x \leq 1\}.$$

Find $\text{LUB}(I) = \sup(I)$ and $\text{GLB}(I) = \inf(I)$.

SOLUTION The set of all upper bounds of I is the set

$$A = \{a \mid 1 \leq a\},$$

and the set of all lower bounds of I is the set

$$B = \{b \mid b \leq 0\}.$$

The least element of A is 1, and the greatest element of B is 0; hence,

$$\text{LUB}(I) = \sup(I) = 1$$

and

$$\text{GLB}(I) = \inf(I) = 0.$$

□

In Example 2.23, notice that $0 = \inf(I)$ does not belong to I, but $1 = \sup(I)$ does belong to I. In general, practically anything can happen: A subset M of a poset X might have a supremum and an infimum, it might have one but not the other, or it might have neither. Even if M has a supremum or an infimum, this supremum or infimum might or might not belong to the set M.

If x and y are elements of a poset X, we say that x and y are **comparable** if at least one of the two conditions $x \leq y$ or $y \leq x$ holds. In the poset $(\mathbb{R}, \leq)$ of Examples 2.21 and 2.23, any two elements are comparable, but Example 2.18 shows that this is not true for posets in general. This leads us to our final definition in this section:

2.24 DEFINITION **TOTALLY ORDERED OR LINEARLY ORDERED SET** A poset $(X, \leq)$ is said to be **totally ordered**, or **linearly ordered**, if, for every pair of elements x, y in X,

$$x \leq y \vee y \leq x.$$

In words, *a totally ordered or linearly ordered set is a poset in which any two elements are comparable.* Thus, the real numbers $\mathbb{R}$, ordered in the usual way by $\leq$, form a totally ordered set; however, the positive integers, ordered by the relation given in Example 2.18, do not. If $(X, \leq)$ is a totally ordered set, we refer to $\leq$ as a **total order relation.**

PROBLEM SET 2.2

In Problems 1 and 2, let U, the universe of discourse, consist of all integers. Find the graph R of each relation.

1. (a) $(1 < x < 4) \wedge (y = x^2)$ (b) $(1 < x < 4) \wedge (x = 5 - y)$
(c) $(1 < x < 4) \wedge (y = x^2 \vee x = 5 - y)$

2. (a) $(3 < x < 10) \wedge (x = y^2 \vee x = 5 - y)$
(b) $(3 < x < 10) \wedge (y = \sin(\pi x/2) \vee x = 3y)$

3. Prove that R is a relation on the set X if and only if $R \subseteq X \times X$.

4. If $X = \{2, 3, 4, 5, 6\}$, write the relation

$$R = \{(x, y) \in X \times X \,\big|\, |x - y| \text{ is exactly divisible by } 3\}$$

explicitly as a set of ordered pairs.

5. For each relation R, find dom(R), codom(R), and R^{-1}.
(a) $R = \{(1, 5), (2, 5), (1, 4), (2, 6), (3, 7), (7, 6)\}$
(b) $R = \{(c, b), (b, g), (c, e), (b, b), (b, e), (a, e), (a, c)\}$
(c) $\mathbb{R}$ is the set of all real numbers, and
$R = \{(x, y) \in \mathbb{R} \times \mathbb{R} \,|\, 4x^2 + 9y^2 = 36\}$
(d) $T = \{1, 2, 3, 4, 5\}$ and $R \subseteq T \times T$ is the set of points displayed in Figure 2-5.

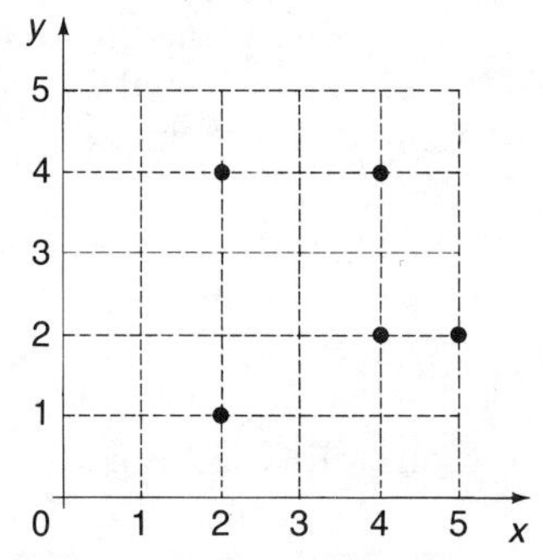

FIGURE 2-5

6. If X is a set, find:
(a) dom(Δ_X) (b) codom(Δ_X) (c) $(\Delta_X)^{-1}$
(d) dom($X \times X$) (e) codom($X \times X$) (f) $(X \times X)^{-1}$

7. Let $R = \{(1, 2), (3, 4), (2, 2), (3, 3), (2, 1)\}$.
(a) Is R symmetric? (b) Is R antisymmetric?
Explain.

8. If X is a set, prove:
(a) Δ_X is antisymmetric.
(b) $X \times X$ is antisymmetric if and only if X is either a singleton set or the empty set.

9. If X is a set, prove that both of the relations Δ_X and $X \times X$ are:
(a) Reflexive (b) Symmetric (c) Transitive

10. True or false: If a relation R is both symmetric and antisymmetric, then there exists a set X such that $R = \Delta_X$. If true, prove it; if false, give a counterexample.

11. Is the relation $R = \{(1, 2), (2, 2)\}$ transitive? Explain.

12. Let R be a relation. Prove:
(a) If R is transitive, so is R^{-1}. (b) If R is symmetric, so is R^{-1}.

13. (a) If X is a set, prove that both Δ_X and $X \times X$ are preorder relations on X.
(b) Prove that the relation G in Example 2.12 is a preorder relation on the set $\mathbb{R}$ of all real numbers.

14. If X is a set, prove that Δ_X is a partial order relation on X but that, unless X is either empty or a singleton set, $X \times X$ is not.

15. In Example 2.18, show that the relation $\leq$ is a partial order relation on the set X of all positive integers.

16. Let X and Y be sets and let $R \subseteq X \times Y$. Then, picturing R as in Figure 2-6, explain why dom(R) can be pictured as the vertical projection of R down onto X and codom(R) can be pictured as the horizontal projection of R over onto Y.

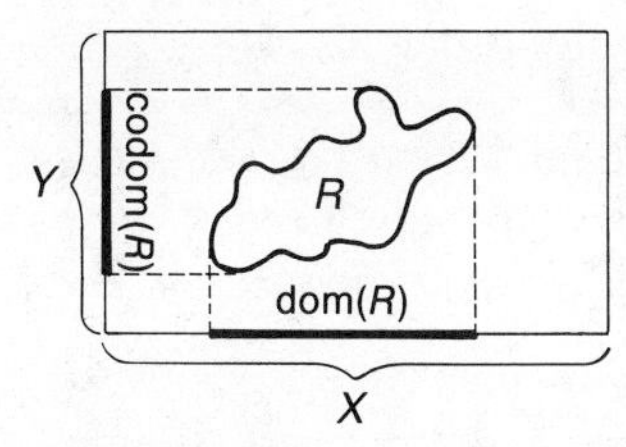

FIGURE 2-6

17. Decide which of the given relations on the indicated set X are (i) reflexive, (ii) symmetric, (iii) antisymmetric, (iv) transitive, (v) preorder relations, (vi) partial order relations, and/or (vii) total order relations.
(a) X is any set, and xRy means $x \neq y$.
(b) X is the set of humans, and xRy means that $x = y$ or x is a sibling of y.

(c) X is the set of humans, and xRy means that $x = y$ or x is an ancestor of y.
(d) X is the set of all triangles, and xRy means that x is congruent to y
(e) X is the set of all circles in the plane, and xRy means that x and y are concentric and the radius of x does not exceed the radius of y.
(f) X is the set of all real numbers, and xRy means $x^3 = y^3$.
(g) X is the set of all integers, n is a fixed positive integer, and xRy means that $x - y$ is exactly divisible by n.
(h) X is the set of all line segments in the plane, and xRy means that x is at least as long as y.

18. Let X be a set, and let R be a relation on X. Picturing R as a subset of $X \times X$, give a pictorial description of each of the following:
(a) R^{-1}
(b) The condition that R is symmetric
(c) The condition that R is reflexive

19. Let X be a set, and let R be a relation on X. Prove:
(a) R is reflexive on X if and only if $\Delta_X \subseteq R$.
(b) R is symmetric if and only if $R^{-1} = R$.

20. Let X be a finite set with $\#X = n$.
(a) Prove that there are exactly 2^{n^2} different relations on X.
(b) Of the 2^{n^2} relations in Part (a), prove that there are exactly $(2^n - 1)^n$ relations whose domain is X.
(c) Of the $(2^n - 1)^n$ relations in Part (b), prove that there are exactly $2^{n^2 - n}$ relations that are reflexive on X.
(d) Of the 2^{n^2} relations in Part (a), prove that there are exactly $2^{(n^2 + n)/2}$ symmetric relations.
(e) Of the $2^{(n^2 + n)/2}$ relations in Part (d), show that there are exactly $2^{(n^2 - n)/2}$ relations that are both symmetric and reflexive on X.
(f) Of the $2^{n^2 - n}$ relations in Part (c), prove that there are exactly $n!$ relations that totally order X.

In Problems 21–32, let X be a set, and let R and S be relations on X. Use the fact that relations are sets, so that you can form unions, intersections, and complements of relations just as you can for any sets.

21. Prove that $R \cup \Delta_X$ is reflexive on X.

22. If R and S are reflexive on X, prove that $R \cap S$ is reflexive on X.

23. If R is reflexive on X, prove that $R \cup S$ is reflexive on X.

24. If R and S are transitive, prove that $R \cap S$ is transitive.

25. Prove:
(a) $(R \cap S)^{-1} = R^{-1} \cap S^{-1}$
(b) $(R \cup S)^{-1} = R^{-1} \cup S^{-1}$

26. Prove:
(a) $R \cap R^{-1}$ is a symmetric relation.
(b) $R \cup R^{-1}$ is a symmetric relation.

27. If R is symmetric, prove that $(X \times X) \setminus R$ is symmetric.

28. If R is a partial order relation on X, prove that $R \cap R^{-1} = \Delta_X$.

29. If R is a partial order relation on X, prove that R^{-1} is also a partial order relation on X.

30. If R is a partial order relation on X, prove that R is a total order relation on X if and only if $R \cup R^{-1} = X \times X$.

31. If R and S are partial order relations on X, prove that $R \cap S$ is a partial order relation on X.

32. True or false: If R and S are partial order relations on X, then $R \cup S$ is a partial order relation on X. If true, prove it; if false, give a counterexample.

33. Let $\mathbb{R}$ denote the set of all real numbers, and let $\leq$ be the usual order relation on $\mathbb{R}$. If $I = \{x \in \mathbb{R} \mid 0 < x \leq 1\}$, find:
(a) $\text{lub}(I)$ (b) $\text{glb}(I)$
(c) The greatest element of I, if it exists.
(d) The least element of I, if it exists.

34. Let $\mathbb{Q}$ denote the set of all rational numbers (ratios of integers with nonzero denominators) ordered in the usual way by the relation $\leq$. Let $M = \{x \in \mathbb{Q} \mid x^2 < 3\}$. Find, if it exists:
(a) An upper bound for M. (b) A lower bound for M.
(c) $\sup(M)$ (d) $\inf(M)$

35. Let X be a nonempty set.
(a) Prove that $(\mathscr{P}(X), \subseteq)$ is a poset.
(b) If $M, N \in \mathscr{P}(X)$, find $\sup\{M, N\}$ and $\inf\{M, N\}$ in the poset $(\mathscr{P}(X), \subseteq)$.

36. Prove that a poset X is totally ordered if and only if every nonempty finite subset of X contains a least element.

37. Let X be a poset, and let $M \subseteq X$.
(a) If M has a greatest element a, prove that this greatest element is unique.
(b) If M has a least element b, prove that this least element is unique.

3 PARTITIONS AND EQUIVALENCE RELATIONS

As we showed in Section 2.2, the elements of a set can be ranked or arranged by means of a preorder, partial order, or total order relation. Another useful procedure for organizing the elements of a set is to classify them according to certain features or properties that some may possess and others may not. This is accomplished by subdividing the set in such a way that elements possessing common features or properties are grouped together. For instance, the set of all real numbers can be classified according to algebraic sign by subdividing it into three parts: the positive numbers, the negative numbers, and zero.

In general, the result of a classification of the elements of a set X is a collection $\mathscr{D}$ of pairwise disjoint subsets of X called **equivalence classes**. Elements of X that possess certain features or properties in common are placed in the same equivalence class; otherwise, they are placed in different equivalence classes. Thus, $\mathscr{D}$ is a set of nonempty subsets of X such that every element of X belongs to one and only one of the sets in $\mathscr{D}$. This suggests the following definition:

3.1 DEFINITION

PARTITION OF A SET By a **partition** of a set X, we mean a set of sets $\mathscr{D}$ such that

(i) every set $C \in \mathscr{D}$ is a nonempty subset of X;
(ii) if $C \in \mathscr{D}$ and $D \in \mathscr{D}$ with $C \neq D$, then $C \cap D = \varnothing$; and
(iii) if $x \in X$, then there exists $C \in \mathscr{D}$ with $x \in C$.

In other words, *a partition of X is a decomposition, or "chop up," of X into nonempty, pairwise disjoint pieces called equivalence classes.* Condition (ii) in Definition 3.1, that these equivalence classes are pairwise disjoint (so that no two different classes have an element in common), is sometimes expressed by saying that they are **mutually exclusive**. Condition (iii), that every element of X belongs to at least one equivalence class (so that X is the union of these classes), is sometimes expressed by saying that they are **exhaustive**.

A partition $\mathscr{D}$ of X into mutually exclusive and exhaustive equivalence classes can be visualized as shown in Figure 2-7, where the set X is partitioned into 21 equivalence classes,

$$C_1, C_2, C_3, \ldots, C_{21}$$

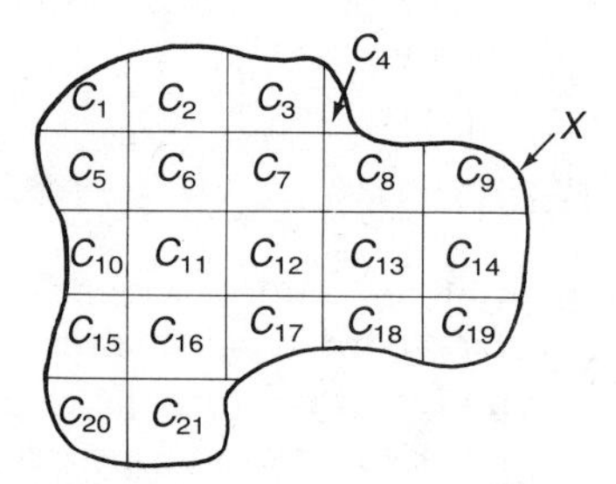

FIGURE 2-7

and

$$\mathscr{D} = \{C_1, C_2, C_3, \ldots, C_{21}\}.$$

As suggested by the appearance of Figure 2-7, the equivalence classes in a partition are sometimes called the *cells* of the partition.

Here are several examples of partitions $\mathscr{D}$:

1. The registered voters in a certain district are classified according to political party; $\mathscr{D} = \{R, D, I\}$, *where* R is the set of Republicans, D the set of Democrats, and I the set of Independents or members of other parties.
2. The integers are classified according to *parity* (even or odd); $\mathscr{D} = \{E, O\}$, where E is the set of even integers and O the set of odd integers.
3. The points in the Cartesian plane are classified according to the quadrant in which they lie; $\mathscr{D} = \{Q_1, Q_2, Q_3, Q_4, A\}$, where Q_1, Q_2, Q_3, and Q_4 are the points lying in quadrants I, II, III, and IV, respectively, and A is the set of points lying on the coordinate axes.
4. The line segments in three-dimensional space are classified according to their lengths; $\mathscr{D} = \{S_x | x \in \mathbb{R} \wedge x > 0\}$, where $\mathbb{R}$ is the set of real numbers, and for each positive real number x, S_x is the set of all line segments of length x.

If $\mathscr{D}$ is a partition of the set X, we can define a relation E on X by specifying that xEy holds if and only if x and y *belong to the same cell of the partition.* For instance, if E is defined in this way for the last example above, then xEy holds if and only if the line segments x and y *have the same length.* This idea is sufficiently important to warrant a more formal definition.

3.2 DEFINITION

EQUIVALENCE RELATION DETERMINED BY $\mathscr{D}$ Let $\mathscr{D}$ be a partition of the set X. Define the relation E on X as follows: If x and y are elements of X, then

$$xEy \Leftrightarrow (\exists C)(C \in \mathscr{D} \wedge x \in C \wedge y \in C).$$

The relation E is called the **equivalence relation determined by the partition $\mathscr{D}$.**

The condition in Definition 3.2, that both x and y belong to C, is often abbreviated by writing

$$x, y \in C$$

rather than

$$x \in C \wedge y \in C.$$

3.3 Example Consider the partition $\mathscr{D} = \{P, N, Z\}$ of the set $\mathbb{R}$ of all real numbers, where

$$P = \{x \mid x > 0\},$$
$$N = \{x \mid x < 0\},$$
$$Z = \{0\}.$$

Describe in words the equivalence relation E determined by this partition.

SOLUTION xEy holds if and only if either x and y have the same algebraic sign, or they are both zero. □

3.4 Example Let $\mathscr{D}$ be a partition of a set X, and let E be the equivalence relation determined by $\mathscr{D}$. Show that the relation E is:

(a) Reflexive on X **(b)** Symmetric **(c)** Transitive

SOLUTION
(a) Let $x \in X$. Then there exists a cell $C \in \mathscr{D}$ such that $x \in C$. It follows that $(\exists C)(C \in \mathscr{D} \wedge x \in C \wedge x \in C)$; hence, xEx holds.
(b) Suppose that xEy, so that there exists a cell $C \in \mathscr{D}$ with $x, y \in C$. Then we have $y, x \in C$, and it follows that yEx.
(c) Suppose that xEy and yEz. Since xEy, there exists a cell $C \in \mathscr{D}$ such that $x, y \in C$. Since yEz, there exists a cell $D \in \mathscr{D}$ such that $y, z \in D$. Because no two different cells in $\mathscr{D}$ have an element in common, the fact that $y \in C$ and $y \in D$ implies that $C = D$. From $z \in D$ and $D = C$, we conclude that $z \in C$; hence, we have $x, z \in C$, and it follows that xEz. □

Example 3.4 suggests the following definition.

3.5 DEFINITION

EQUIVALENCE RELATION ON A SET Let X be a set. By an **equivalence relation on** X, we mean a relation E that is reflexive on X, symmetric, and transitive.

Example 3.4 shows that the relation E determined by a partition $\mathscr{D}$ of a set X is, in fact, an equivalence relation on X. The diagonal relation Δ_X and the universal relation $X \times X$ are examples of equivalence relations on X (Problem 6).

Here are more examples of equivalence relations E:

1. Define the relation E on the set $\mathbb{R}$ of all real numbers by

$$E = \{(x, y) \mid x^2 = y^2\}.$$

2. Define the relation E on the set $\mathbb{R}$ of all real numbers by

$$E = \{(x, y) \mid \sin x = \sin y\}.$$

3. Define the relation E on the set X of all human beings by xEy if and only if x and y have the same biological father.
4. Let X denote the set of all ordered pairs (m, n) of integers m and n such that $n \neq 0$, and define $(m, n)E(p, q)$ to mean that

$$\frac{m}{n} = \frac{p}{q}.$$

In Problems 9–12, we ask you to show that these relations are equivalence relations on the indicated sets.

We have seen above that *every partition determines an equivalence relation.* Now we are going to show that, vice versa, *every equivalence relation determines a partition.* We begin with the following definition:

3.6 DEFINITION

***E*-EQUIVALENCE CLASSES** Let E be an equivalence relation on the set X. If $x \in X$, we define the ***E*-equivalence class determined by** x, in symbols $[x]_E$, by

$$[x]_E = \{y \in X \mid yEx\}.$$

If the equivalence relation E under consideration is understood, we sometimes write $[x]$ rather than $[x]_E$ and simply refer to it as the *equivalence class determined by x.* Because of the square brackets, $[x]$ is also called *brackets x.*

3.7 Example Let E be the equivalence relation defined on the set $\mathbb{R}$ of all real numbers by

$$E = \{(x, y) \mid \sin x = \sin y\}.$$

Find the equivalence class $[0]$ determined by 0.

SOLUTION

$$\begin{aligned}[0] = [0]_E &= \{y \mid yE0\} \\ &= \{y \mid \sin y = \sin 0\} \\ &= \{y \mid \sin y = 0\} \\ &= \{y \mid y = n\pi \text{ for some integer } n\}.\end{aligned}$$

In other words, the equivalence class $[0]$ is the set of all integer multiples of π. Although it is not in strict accordance with the usual notation for implicitly defined sets, it is permissible to write this in the slightly abbreviated form

$$[0] = \{n\pi \mid n \text{ is an integer}\}.$$ □

3.8 LEMMA

Let E be an equivalence relation on a set X, and let $x, y \in X$. Then:

(i) $y \in [x] \Leftrightarrow yEx$.
(ii) $x \in [x]$.
(iii) $[x]$ is a nonempty subset of X.

PROOF

(i) By Definition 3.6, $y \in [x] \Leftrightarrow yEx$.
(ii) That $x \in [x]$ is a consequence of Part (i) above and the fact that E is reflexive on X, so that xEx.
(iii) By Part (i) again, if $y \in [x]$, then [by Part (i) of Definition 2.4], $y \in \text{dom}(E)$. Since E is a relation on x, it follows [from Part (iii) of Definition 2.4] that $\text{dom}(E) \subseteq X$; hence, $y \in X$. This shows that $[x]$ is a subset of X. That $[x]$ is nonempty follows from Part (ii) above. □

3.9 THEOREM

PROPERTIES OF [x] Let E be an equivalence relation on a set X, and let $x, y \in X$. Then:

(i) $[x] \cap [y] \neq \varnothing \Leftrightarrow xEy$.
(ii) $[x] \cap [y] \neq \varnothing \Leftrightarrow [x] = [y]$.

PROOF

(i) Suppose that $[x] \cap [y] \neq \varnothing$, so there exists $z \in [x] \cap [y]$. Then $z \in [x]$; so by Part (i) of Lemma 3.8, zEx. Likewise, $z \in [y]$; so zEy. Because E is symmetric, it follows from zEx that xEz. Now, from xEz, zEy, and the fact that E is transitive, we have xEy. This shows that

$$[x] \cap [y] \neq \varnothing \Rightarrow xEy.$$

To prove the converse, we begin by assuming that xEy. Then, by Part (i) of Lemma 3.8 again, we have $x \in [y]$. Since $x \in [x]$ by Part (ii) of Lemma 3.8, it follows that $x \in [x] \cap [y]$ and, hence, that $[x] \cap [y] \neq \varnothing$. This proves that

$$xEy \Rightarrow [x] \cap [y] \neq \varnothing,$$

and completes the proof of Part (i).

(ii) Again, we begin by supposing that $[x] \cap [y] \neq \varnothing$. Then, by Part (i) (which has been proved), xEy. Now we must prove that $[x] = [y]$. We do this by the pick-a-point process. Let $z \in [x]$, so that zEx by Part (i) of Lemma 3.8. From zEx, xEy, and the transitivity of E, it follows that zEy; and therefore, by Part (i) of Lemma 3.8 again, $z \in [y]$. We have shown that, if $z \in [x]$, then $z \in [y]$, and it follows that $[x] \subseteq [y]$. The inclusion $[y] \subseteq [x]$ follows from a similar argument, which we leave as an exercise for you (Problem 20). Thus, we have

$$[x] \cap [y] \neq \varnothing \Rightarrow [x] = [y].$$

To prove the converse, we begin by assuming that $[x] = [y]$. By Part (ii) of Lemma 3.8, $x \in [x]$, and it follows that $x \in [y]$. Therefore, $x \in [x] \cap [y]$, and so $[x] \cap [y] \neq \varnothing$. This shows that

$$[x] = [y] \Rightarrow [x] \cap [y] \neq \varnothing,$$

and the proof is complete. □

If E is an equivalence relation on a set X, then the set of all E-equivalence classes is denoted by

$$X/E.$$

This set of sets is called X **modulo** E, or X **mod** E for short. (The Latin word *modulo* is used in mathematics to mean *with respect to*.) More formally, we make the following definition:

3.10 DEFINITION

X* MODULO *E Let E be an equivalence relation on a set X. Then

$$X/E = \{[x]_E \mid x \in X\}.$$

If $[x]$ is understood to mean $[x]_E$, we can simply write

$$X/E = \{[x] \mid x \in X\}.$$

3.11 Example Let $X = \{\text{Matilda, Carlos, Emily}\}$, and let E be the relation defined on X by xEy if and only if x and y are of the same sex. Find X/E.

SOLUTION Here we have

$$[\text{Matilda}] = \{\text{Matilda, Emily}\},$$
$$[\text{Emily}] = \{\text{Matilda, Emily}\},$$

and

$$[\text{Carlos}] = \{\text{Carlos}\}.$$

Therefore,

$$X/E = \{\{\text{Matilda, Emily}\}, \{\text{Carlos}\}\}.$$ □

Notice that X/E in Example 3.11 is a partition of X. That this is true for any equivalence relation is the content of the next theorem.

3.12 THEOREM

THE PARTITION DETERMINED BY *E* If E is an equivalence relation on a set X, then X/E is a partition of X.

PROOF By Part (iii) of Lemma 3.8, X/E is a set of nonempty subsets of X. By Part (ii) of Theorem 3.9, the sets in X/E are mutually exclusive; and by Part (ii) of Lemma 3.8, they are exhaustive. □

3.13 Example Let E be the equivalence relation of Example 3.11, so that $X/E = \{\{\text{Matilda, Emily}\}, \{\text{Carlos}\}\}$. Find the equivalence relation determined by the partition X/E.

SOLUTION Let F be the equivalence relation determined by the given partition X/E. Then, xFy holds if and only if x and y belong to the same cell of the partition, that is, if and only if x and y are of the same sex. Therefore, we have $F = E$. □

The following theorem, whose proof we leave to you as an exercise (Problem 21), shows that the phenomenon illustrated in Example 3.13 is general.

3.14 THEOREM

RECAPTURING *E* FROM *X/E* If you start with an equivalence relation E on X and form the partition $\mathscr{D} = X/E$, then the equivalence relation determined by $\mathscr{D}$ is the original equivalence relation E.

3.15 THEOREM

***E* AS A UNION OF CARTESIAN PRODUCTS** Let $\mathscr{D}$ be a partition of X into a finite number of cells,

$$\mathscr{D} = \{C_1, C_2, C_3, \ldots, C_n\}.$$

Then, the equivalence relation E determined by $\mathscr{D}$ is given by

$$E = (C_1 \times C_1) \cup (C_2 \times C_2) \cup (C_3 \times C_3) \cup \cdots \cup (C_n \times C_n).$$

PROOF By definition, xEy holds if and only if both x and y belong to the same cell, say, C_i of the partition $\mathscr{D}$. But the condition $x, y \in C_i$ is equivalent to $(x, y) \in C_i \times C_i$; hence, E is the union of the sets $C_i \times C_i$ as i runs from 1 to n. □

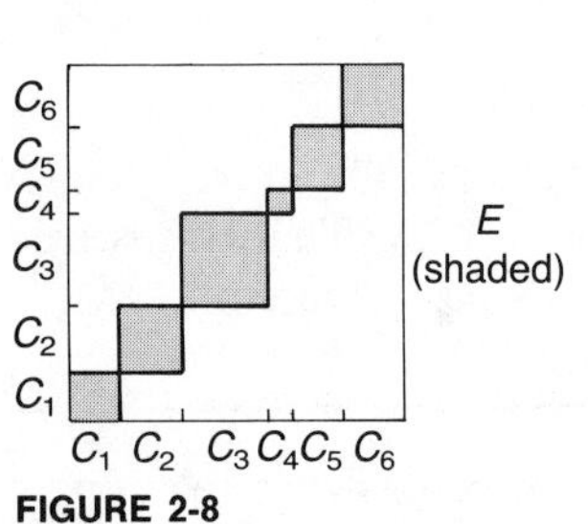

FIGURE 2-8

As a matter of fact, Theorem 3.15 is true without any finiteness restriction on the partition $\mathscr{D}$, but then E is the union of an *infinite* set of Cartesian products. (See Section 1.6.) The proof of the more general theorem is similar to the one given above (Problem 22). In view of this theorem, the equivalence relation E determined by a partition $\mathscr{D}$ can be visualized as shown in Figure 2-8 (in which, for purposes of illustration, we have taken $\mathscr{D}$ to consist of six cells).

3.16 Example Let $X = \{a, b, c, d, e\}$, and consider the partition

$$\mathscr{D} = \{\{a, b\}, \{c, d\}, \{e\}\}$$

of X. Find **(a)** the equivalence relation E on X determined by $\mathscr{D}$ and **(b)** the partition X/E.

SOLUTION **(a)** By Theorem 3.15,

$$\begin{aligned} E &= (\{a, b\} \times \{a, b\}) \cup (\{c, d\} \times \{c, d\}) \cup (\{e\} \times \{e\}) \\ &= \{(a, a), (a, b), (b, a), (b, b)\} \cup \{(c, c), (c, d), (d, c), (d, d)\} \cup \{(e, e)\} \\ &= \{(a, a), (a, b), (b, a), (b, b), (c, c), (c, d), (d, c), (d, d), (e, e)\}. \end{aligned}$$

(b) From Part (a), we have

$$[a] = [b] = \{a, b\},$$
$$[c] = [d] = \{c, d\},$$

and

$$[e] = \{e\}.$$

Hence,

$$X/E = \{\{a, b\}, \{c, d\}, \{e\}\} = \mathscr{D}. \qquad \square$$

In Part (b) of Example 3.16, we found that X/E coincides with the original partition $\mathscr{D}$. Again, this phenomenon is general, and we have the following theorem, whose proof we leave as an exercise (Problem 23).

3.17 THEOREM

RECAPTURING $\mathscr{D}$ FROM E If you start with a partition $\mathscr{D}$ of a set X and form the equivalence relation E determined by $\mathscr{D}$, then the partition X/E coincides with the original partition $\mathscr{D}$.

In view of Theorems 3.14 and 3.17, partitions and equivalence relations are intimately connected; they are "two sides of the same coin." This one-to-one correspondence between partitions and equivalence relations is exploited throughout modern mathematics. Sometimes it is convenient to translate a problem about equivalence relations into a problem about partitions; and, vice versa, it sometimes helps to translate a problem about partitions into a problem about equivalence relations. There is also an interesting connection between preorder relations (Definition 2.13) and equivalence relations.

3.18 THEOREM

THE EQUIVALENCE RELATION DETERMINED BY A PREORDER Let R be a preorder relation on a set X. Then the relation E defined on X by

$$xEy \Leftrightarrow xRy \wedge yRx$$

is an equivalence relation on X.

PROOF We must show that E is reflexive on X, symmetric, and transitive. Because the preorder relation R is reflexive on X, we have xRx for every element x in X, and it follows from the definition of E that xEx holds as well. Thus, E is reflexive on X. Since

$$xRy \wedge yRx \Leftrightarrow yRx \wedge xRy,$$

it is clear that E is symmetric. Finally, to prove that E is transitive, suppose that

$$xEy \wedge yEz.$$

Then, from the definition of E we have

$$xRy \wedge yRx \wedge yRz \wedge zRy.$$

From xRy, yRz, and the fact that the preorder R is transitive, we infer that xRz. Likewise, from zRy and yRx, we infer that zRx. Thus, we have

$$xRz \wedge zRx,$$

and it follows that xEz. □

PROBLEM SET 2.3

In Problems 1–4, determine whether or not each set of sets is a partition of the given set X.

1. $\{C_1, C_2, C_3, C_4\}$, where $C_1 = \{1, 3\}$, $C_2 = \{7, 8, 10\}$, $C_3 = \{2, 5, 6\}$, $C_4 = \{4, 9\}$, and $X = \{1, 2, 3, 4, 5, 6, 7, 8, 9, 10\}$
2. $\{A, B, C\}$, where $A = \{a, c, e\}$, $B = \{b\}$, $C = \{d, g\}$, and $X = \{a, b, c, d, e, f, g\}$
3. $\{\{1, 3, 5\}, \{2, 6, 10\}, \{4, 8, 9\}\}$, where $X = \{1, 2, 3, 4, 5, 6, 7, 8, 9, 10\}$
4. $\{\{a, b, c, d, e, f, g\}\}$, where $X = \{a, b, c, d, e, f, g\}$
5. Decide which of the relations given in Parts (a) through (h) of Problem 17 in Problem Set 2.2 are equivalence relations.

6. If X is a set, show that the diagonal relation Δ_X and the universal relation $X \times X$ are equivalence relations on X.

7. Let $X = \{$Los Angeles, Miami, Dallas, San Francisco, Fort Lauderdale, San Antonio, Daytona Beach, Chicago$\}$. Define a relation E on X by xEy if and only if x and y are in the same state.
(a) Show that E is an equivalence relation on X.
(b) Find the equivalence classes in the partition X/E determined by E.

8. In Problem 6, find the partition X/Δ_X and the partition $X/(X \times X)$.

In Problems 9–12, show that the relation E is an equivalence relation on the indicated set.

9. $\mathbb{R}$ is the set of all real numbers, and E is defined on $\mathbb{R}$ by $E = \{(x, y) | x^2 = y^2\}$.

10. $\mathbb{R}$ is the set of all real numbers, and E is defined on $\mathbb{R}$ by $E = \{(x, y) | \sin x = \sin y\}$.

11. X is the set of all humans, and E is defined on X by xEy if and only if x and y have the same biological father.

12. X is the set of all ordered pairs (m, n) if integers m and n are such that $n \neq 0$, and $(m, n)E(p, q)$ if and only if $m/n = p/q$.

13. Let E be the equivalence relation defined on the set $\mathbb{R}$ of all real numbers by $E = \{(x, y) | \cos x = \cos y\}$. Find the equivalence class $[0]_E$.

14. (a) In Problem 9, find $\mathbb{R}/E$.
(b) In Problem 12, find X/E.

15. Describe the partitions X/R determined by those relations in Problem 5 that are equivalence relations.

16. In Problem 10, show that, for $x \in \mathbb{R}$,

$$[x]_E = \{x + 2n\pi | n \in \mathbb{Z}\} \cup \{(2n + 1)\pi - x | n \in \mathbb{Z}\},$$

where $\mathbb{Z}$ denotes the set of all integers.
Hint: Use the trigonometric identity

$$\sin x - \sin y = 2 \cos\left(\frac{x + y}{2}\right) \sin\left(\frac{x - y}{2}\right).$$

In Problems 17 and 18, (a) find, as an explicit set of ordered pairs, the equivalence relation E on the set X determined by the partition $\mathscr{D}$, and (b) find the partition of X determined by E.

17. $X = \{a, b, c, d, e\}$ and $\mathscr{D} = \{\{a, d\}, \{b, e\}, \{c\}\}$

18. $X = \{1, 2, 3, 4, 5, 6, 7, 8, 9, 10\}$ and $\mathscr{D} = \{\{2, 5, 8\}, \{3, 6, 9\}, \{1, 4, 7, 10\}\}$

19. Find, as an explicit set of ordered pairs, the equivalence relation on X determined by the partition found in Problem 7.

20. Complete the proof of Part (ii) of Theorem 3.9 by proving that $[x] \cap [y] \neq \emptyset \Rightarrow [y] \subseteq [x]$.
21. Prove Theorem 3.14.
22. State and prove a general version of Theorem 3.15 in which no assumption is made that $\#\mathscr{D}$ is finite.
23. Prove Theorem 3.17.
24. True or false: Every equivalence relation is a preorder relation. If true, prove it; if false, give a counterexample.
25. Prove that, if R is a symmetric and transitive relation and if $X = \text{dom}(R)$, then R is an equivalence relation on X. Does this mean that the reflexivity condition is not needed in Definition 3.5?
26. Let R be a reflexive and transitive relation on X. Prove that $E = R \cap R^{-1}$ is an equivalence relation on X.
27. Suppose that R is both a partial order relation on X and an equivalence relation on X. Prove that $R = \Delta_X$.
28. Let R be a symmetric relation that is reflexive on the set X. Define a relation E on X as follows: xEy holds if and only if there exists a finite sequence of elements $x_1, x_2, \ldots, x_n$ in X such that x_jRx_{j+1} holds for $j = 1, 2, \ldots, n-1$. Prove that E is an equivalence relation on X.
29. Let E and F be equivalence relations on the set X. Since E and F are sets (of ordered pairs), we can form the intersection $G = E \cap F$.
 (a) Prove that G is an equivalence relation on X.
 (b) If $x \in X$, find $[x]_G$ in terms of $[x]_E$ and $[x]_F$.
30. Let E and F be equivalence relations on the set X. Give necessary and sufficient conditions on the partitions X/E and X/F so that (as sets of ordered pairs) $E \subseteq F$.
31. Let X be a nonempty set.
 (a) Prove that $\{X\}$ is a partition of X.
 (b) Find the equivalence relation determined by the partition $\{X\}$ of X.
32. If X is a finite set and E is an equivalence relation on X, prove that $\#E$ is a sum of squares of positive integers.

In Problems 33–36, let R be a preorder relation on the set X, and let E be the corresponding equivalence relation as in Theorem 3.18. If $x \in X$, denote the equivalence class $[x]_E$ by $[x]$, and let $\mathscr{D} = X/E$.

33. If $a, b, c, d \in X$ with $[a] = [c]$ and $[b] = [d]$, show that $aRb \Rightarrow cRd$.
34. We propose to define a relation $\leq$ on $\mathscr{D}$ by $[a] \leq [b]$ if and only if aRb. However, there could be some difficulty with this definition because $[a]$ does not determine the element $a \in X$ uniquely. Using the result of Problem 33, show that there is no difficulty with the definition.

35. If R is a partial order relation on X, show that $E = \Delta_X$.

36. Show that the relation $\leq$ defined in Problem 34 is a partial order relation on $\mathscr{D}$.

37. Let X be a set, and let $\mathscr{E}$ be the set of all equivalence relations on X. If $E, F \in \mathscr{E}$, define $E \leq F$ to mean that (as sets of ordered pairs) $E \subseteq F$. Prove that $\leq$ is a partial order on the set $\mathscr{E}$.

38. In Problem 37, if X has at least three members, prove that $\leq$ is not a total order on $\mathscr{E}$.

39. Let X be a finite nonempty set, let m be a positive integer, and let $\mathscr{D}$ be a partition of X such that, for every $C \in \mathscr{D}$, $\#C = m$.
(a) Prove that m divides $\#X$ exactly.
(b) Prove that $\#\mathscr{D}$ divides $\#X$ exactly.

40. If $\mathbb{R}$ denotes the set of all real numbers and $f(x)$ is a real-valued function defined on all of $\mathbb{R}$, show that the relation

$$E = \{(x, y) \mid f(x) = f(y)\}$$

is an equivalence relation on $\mathbb{R}$.

41. If X is a set with $\#X = n$, let B_n denote the number of different equivalence relations on X. Note that B_n is also the number of different partitions of X. For convenience, put $B_0 = 1$. By listing all possibilities, show that $B_2 = 2$ and that $B_3 = 5$. It can be shown that

$$B_{n+1} = \sum_{k=0}^{n} C_k^n \cdot B_k,$$

where C_k^n denotes the number of combinations of n things taken k at a time. [See Problem 30 in Problem Set 1.5.] Using this formula, compute B_4, B_5, B_6, B_7, and B_8.

42. Let X be a set with $\#X = 8$. A relation on X is selected at random. What is the probability that it is an equivalence relation? [Hint: See Problem 41.]

43. With the notation of Problem 41, explain why B_n is the number of different rhyme schemes for a poem with n lines.

44. It can be shown that the number B_n defined in Problem 41 can be obtained as the sum of an infinite series,

$$B_n = \frac{1}{e} \sum_{k=1}^{\infty} \frac{k^n}{k!},$$

where $e = 2.718281828459\ldots$ is the base of the natural logarithm. Try this out, using a calculator or a computer and forming the partial sums of the first 20 or 30 terms of the series, for $n = 2, 3, 4, 5, 6, 7$, and 8.

4

FUNCTIONS

In elementary algebra and calculus textbooks, a function is usually defined to be a *rule of correspondence f that assigns to each element x in a set X a unique element $f(x)$ in a set Y*. Although this definition conveys the intuitive idea of a function, it is not sufficiently precise for the purposes of more advanced mathematics. What, exactly, is a rule of correspondence? A more formal definition of a function avoids imprecision but is considerably more abstract:

4.1 DEFINITION

FUNCTION A **function** is a relation R such that, for every x in dom(R) and every y and z in codom(R),

$$xRy \wedge xRz \Rightarrow y = z.$$

Thus, a *function is a special kind of relation; it is a relation such that no object in the domain is related to two different objects in the codomain.* At first glance, this definition seems to have little to do with the more intuitive definition of a function as a rule of correspondence. In the end, however, you will see that it is merely a logically clean statement of the same basic idea.

4.2 Example Which of the following sets of ordered pairs are functions?

(a) $R = \{(a, 2), (a, 3), (b, 4), (c, 1)\}$
(b) $S = \{(a, 4), (b, 4), (c, 4), (d, 2)\}$

SOLUTION **(a)** R is not a function because it contains two ordered pairs $(a, 2)$ and $(a, 3)$ with the same first element but different second elements.
(b) S is a function because no two ordered pairs in S have the same first element and different second elements. □

Although relations in general are usually symbolized by capital letters, relations that are functions are often denoted by small letters. Also, if f is a function, we usually write $(x, y) \in f$ in preference to xfy, reserving the latter notation for the study of general relations. With this notation, Definition 4.1 can be rewritten as follows:

*A **function** is a relation f such that, for every $x \in$ dom(f), there exists a unique $y \in$ codom(f) such that $(x, y) \in f$.*

4.3 DEFINITION

IMAGE OF AN ELEMENT UNDER A FUNCTION If f is a function and x is an element in the domain of f, then the unique element y in the codomain of f such that $(x, y) \in f$ is denoted by $f(x)$ and called the **image of** x **under** f, or the **value of** f **at** x.

Thus, if f is a function and $x \in \text{dom}(f)$, we have

$$y = f(x) \Leftrightarrow (x, y) \in f.$$

The expression $f(x)$ is read in English as *f of x*.

4.4 Example For the function $f = \{(a, 2), (b, 1), (c, 2)\}$ find $f(a)$, $f(b)$, and $f(c)$.

SOLUTION $f(a) = 2$, $f(b) = 1$, and $f(c) = 2$. □

We leave it as a simple exercise (Problem 2) for you to show that, if f is a function, then

$$\text{codom}(f) = \{y \mid (\exists x)(x \in \text{dom}(f) \wedge y = f(x))\},$$

or, with a slight abuse of notation,

$$\text{codom}(f) = \{f(x) \mid x \in \text{dom}(f)\}.$$

For functions, the word *range* is usually used in preference to *codomain*. Let's make this official:

4.5 DEFINITION

RANGE OF A FUNCTION If f is a function, then the **range** of f, in symbols *range*(f), is defined by

$$\text{range}(f) = \{f(x) \mid x \in \text{dom}(f)\}.$$

Thus, *the range of f is the set of all images under f of elements in the domain of f.*

4.6 Example Find the domain and range of the function

$$f = \{(a, 2), (b, 1), (c, 2)\}$$

in Example 4.4.

SOLUTION

$$\text{dom}(f) = \{a, b, c\}$$

and

$$\text{range}(f) = \{f(a), f(b), f(c)\} = \{2, 1\}.$$ □

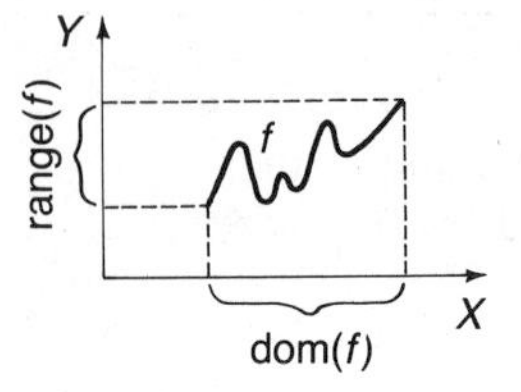

FIGURE 2-9

If $\mathbb{R}$ denotes the set of all real numbers, then a function f with $\text{dom}(f) \subseteq \mathbb{R}$ and $\text{range}(f) \subseteq \mathbb{R}$ is called a **real-valued** function of a **real variable**. In your calculus courses you learned to draw the graphs of such functions in a Cartesian coordinate system. The same idea can be applied to provide schematic diagrams (still called graphs) to aid your intuition when dealing with more general functions. Thus, if X and Y are sets and f is a function such that $\text{dom}(f) \subseteq X$ and $\text{range}(f) \subseteq Y$, you can visualize the function f as the set of ordered pairs

$$(x, f(x))$$

such that $x \in \text{dom}(f)$ (Figure 2-9).

In the figure, notice that the domain of f is the vertical projection of f down onto the X axis, and the range of f is the horizontal projection of f over onto the Y axis.

4.7 DEFINITION

MAPPING NOTATION The symbolism

$$f : X \to Y,$$

called **mapping notation**, indicates that

(i) f is a function,
(ii) $\text{dom}(f) = X$, and
(iii) $\text{range}(f) \subseteq Y$.

Note carefully that, in Definition 4.7, we do *not* require that Y is the range of f, only that it *contains* the range of f.

If

$$f : X \to Y,$$

we say that f is a function, or mapping, **from** X **into** Y. The alternative notation

$$X \xrightarrow{f} Y$$

is also used to indicate that f is a mapping from X into Y. When mapping notation is used, the statement that

$$y = f(x)$$

is often written in the alternative form

$$x \overset{f}{\mapsto} y,$$

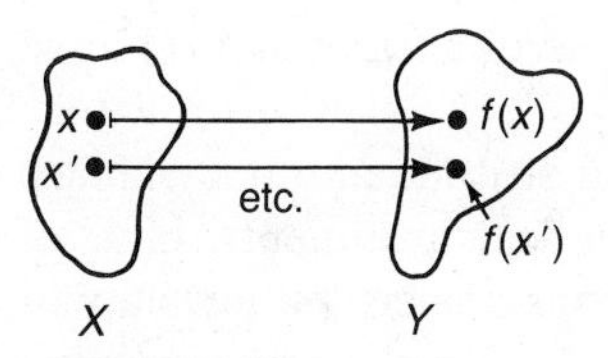

FIGURE 2-10

or, if f is understood, simply as

$$x \mapsto y.$$

If $x \mapsto y$, we say that x is **mapped**, or **transformed**, into y. (Notice that a "tail" is placed on the arrow when mapping notation is used in this fashion.)

The idea of a function as a mapping, or transformation,

$$f: X \to Y,$$

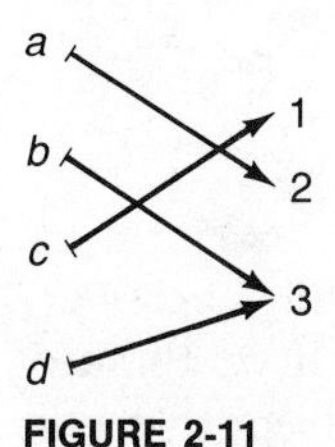

FIGURE 2-11

is nicely depicted by a **transformation diagram** in which arrows are drawn connecting each element $x \in X$ with the element $f(x) \in Y$ into which it is mapped, or transformed, by f (Figure 2-10). In fact, if the sets X and Y are *finite*, a mapping $f: X \to Y$ can be specified completely by such a diagram (Figure 2-11).

4.8 Example Determine as an explicit set of ordered pairs the function f depicted in the transformation diagram in Figure 2-11.

SOLUTION $f = \{(a, 2), (b, 3), (c, 1), (d, 3)\}$. □

An interesting variation of the idea of a mapping, or transformation, is the picture of a function as a machine that takes objects x in the domain X and converts them into objects $f(x)$ in the set Y. The function f can be depicted as a "black box" with some mysterious inner workings that can accept any element x in X as an **input**, process it internally, and eject $f(x)$ as an **output** (Figure 2-12). This machine picture is useful in the study of computing devices.

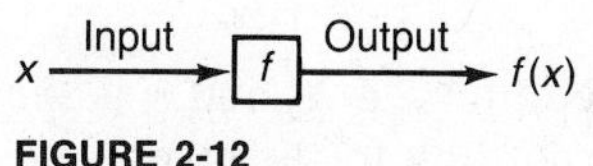

FIGURE 2-12

Functions and variables are closely related. In the old days, when mathematicians spoke of a *variable*, they meant a quantity that is changing, or at least capable of changing. Although it is easy to think about a variable as a quantity subject to change, this idea is somewhat imprecise by modern standards of mathematical rigor. Contemporary mathematicians often regard a variable as a "blank spot" that can be "filled" by any particular object from the set of objects under consideration. From this point of view, an equation such as

$$(x + 1)^2 = x^2 + 2x + 1$$

can be thought of as having the form

$$(\square + 1)^2 = \square^2 + 2\square + 1,$$

where any particular real number can be inserted into the three blank spots marked by the empty boxes. When a number is placed in the boxes, the result is a definite proposition about that number. The advantage of this viewpoint is that the philosophically elusive ideas of time and change are not required; the disadvantage is that variables become rather dull, more or less static entities, rather than dynamic quantities that change even as we speak about them.

We recommend that you adopt the following attitude about variables: When you use variables in a formal mathematical argument, think of them as blank spots that can be filled by particular objects, but, when you are trying to gain an intuitive understanding of some concept involving variables, think of them as quantities that change in time by ranging over the set of objects involved.

If f is a function and

$$y = f(x),$$

we call x the **argument** of f, or the **independent variable**; and, since y depends on x, we often refer to y as the **dependent variable**. If we think of x and y as variable quantities, then, as x changes, y is subject to a corresponding change, and we obtain a compelling dynamic picture of the function f. This dynamic picture is especially useful in calculus, where the derivative dy/dx can be interpreted as the *rate of change of* y *with respect to* x.

Now let us consider the idea, mentioned at the beginning of this section, that a function is a rule of correspondence. Suppose we have two sets, X and Y, and a specific *rule*, *formula*, *correspondence*, or *prescription* that associates with each object x in X a unique object y in Y. This determines a definite function, or mapping,

$$f:X \to Y,$$

where f is the set of all ordered pairs (x, y) such that $x \in X$, and y is the unique object in Y that is associated with x according to the given rule, formula, correspondence, or prescription. If you want to define a particular function, or mapping, in this way, here is what you should do:

Step 1. Specify the set X that is to be the domain of f.

Step 2. Specify the set Y that is to contain the range of f.

Step 3. Be certain that the rule, formula, correspondence, or prescription unambiguously assigns to each object x in X *one and only one* object y in Y.

In carrying out Step 3, you are making sure that the function, or mapping, is **well defined**.

In less formal mathematical writing, especially in calculus textbooks, Steps 1 and 2 are often ignored. Step 1 is ignored because it is understood that the domain X is the set of all elements in the domain of discourse for which the rule, formula, or prescription is meaningful. Step 2 is ignored because it is understood that Y comprises all elements y that correspond to elements x in X according to the rule, formula, or prescription.

4.9 Example In a calculus book, we find the words "Consider the function $y = \sqrt{1-x}$." Translate these words into more precise mathematical language.

SOLUTION Here it is understood that the universe of discourse is the set $\mathbb{R}$ of all real numbers. We carry out the three steps suggested above:

Step 1. Since the domain is not given, we assume that it consists of all real numbers for which the expression $\sqrt{1-x}$ is defined; that is, all real numbers x for which $x \leq 1$, so that $0 \leq 1 - x$. Therefore, the domain of the function is the set

$$X = \{x \mid x \leq 1\},$$

or, using interval notation,

$$X = (-\infty, 1].$$

Step 2. We can take Y to be any set containing all of the numbers $y = \sqrt{1-x}$ as x runs through X. The simplest choice seems to be $Y = \mathbb{R}$ (although any subset of $\mathbb{R}$ containing all nonnegative numbers would do as well).

Step 3. The formula $y = \sqrt{1-x}$ assigns one and only one value of y to each value of x in X. (Recall that the square root symbol is always understood to represent the *principal* square root, that is, the non-negative square root.) Therefore, the function

$$f: X \to Y$$

under consideration is given by

$$f = \{(x, y) \mid x \in \mathbb{R} \wedge y \in \mathbb{R} \wedge x \leq 1 \wedge y = \sqrt{1-x}\}. \qquad \square$$

Summarizing our discussion so far, a function, or mapping,

$$f: X \to Y$$

can be thought of or visualized in (at least) six different ways:

1. The *mathematical definition* (Definition 4.1) as a set of ordered pairs in which no two different pairs have the same first element.
2. The *graphical picture*, either as a graph (Figure 2-9) or as a transformation diagram (Figures 2-10 and 2-11).
3. The *mapping*, or *transformation*, point of view in which f is thought of as transforming each element $x \in X$ into a corresponding element $y \in Y$, denoted by $x \mapsto y$.
4. The *machine* picture as a *processing device* that converts each input object $x \in X$ into a corresponding output object $f(x) \in Y$.
5. The *dynamic view*, as a relation between variable quantities x and y.
6. The *rule*, *formula*, *correspondence*, or *prescription* view.

In a formal mathematical proof, only the official definition (Definition 4.1) of a function is acceptable; the other five interpretations are useful for informal proofs and for building your intuition about functions.

Note that there is a technical distinction between a function f and a mapping $f:X \to Y$. A mapping $f:X \to Y$ is more than the function f because it involves the specification of a set Y that contains, but is not necessarily equal to, range(f). In practice, however, mathematicians often disregard this distinction and speak of functions and mappings more or less interchangeably.

Although Definition 4.1 requires that no two different ordered pairs in f have the same *first* element, it is noncommittal about whether two different ordered pairs can have the same *second* element. Thus, as far as this definition is concerned, it is quite possible to have ordered pairs (say)

$$(a, c) \in f \quad \text{and} \quad (b, c) \in f \qquad \text{with } a \neq b.$$

This is the same as having

$$f(a) = f(b) \qquad \text{with } a \neq b.$$

Functions for which this does not happen are the subject of the following definition.

4.10 DEFINITION

ONE-TO-ONE, OR INJECTIVE, FUNCTIONS A function, or mapping, $f:X \to Y$ is said to be **one-to-one**, or **injective**, if, for all elements a and b in X,

$$f(a) = f(b) \Rightarrow a = b.$$

The condition that f is one-to-one, or injective, can also be expressed by the contrapositive

$$a \neq b \Rightarrow f(a) \neq f(b)$$

of the implication in Definition 4.10. Thus, *a function, or mapping, is one-to-one, or injective, if and only if distinct elements of its domain have distinct images under the function.*

The terminology *one-to-one* is used in most calculus textbooks, but the word *injective* is often used in more advanced mathematical writing. An injective function, or mapping, is sometimes referred to as an **injection.** The so-called **horizontal-line test** determines whether a real-valued function f of a real variable is injective: *f is injective if and only if no straight line parallel to the x axis intersects the graph of f more than once* (Problem 12). You can also tell from its transformation diagram whether a mapping is injective: *$f: X \to Y$ is injective if and only if no two arrows start at different elements of X and point to the same element of Y* (Problem 15).

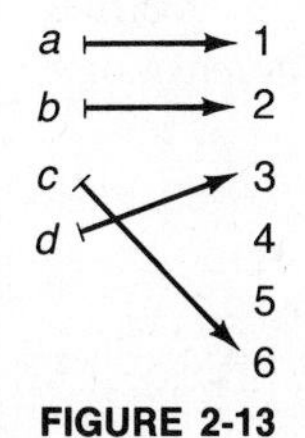

FIGURE 2-13

4.11 Example Is the mapping $f: X \to Y$ whose transformation diagram is shown in Figure 2-13 an injection?

SOLUTION Since no two arrows point to the same element in Y, it follows that f is an injection. □

4.12 Example Determine which of the following real-valued functions of a real variable is an injection:

(a) $f(x) = x^2$
(b) $g(x) = 2x - 5$

SOLUTION **(a)** f is not an injection because, for instance,

$$f(-1) = f(1), \quad \text{but} \quad -1 \neq 1.$$

(b) g is an injection because, if $g(a) = g(b)$, then

$$2a - 5 = 2b - 5,$$

so

$$2a = 2b,$$

and therefore

$$a = b.$$ □

In using the mapping notation

$f: X \to Y$

you have considerable latitude in your choice of the set Y; the only requirement is that it *contain* the range of f. The "smallest" possible Y would be range(f) itself. This leads us to the following definition:

4.13 DEFINITION

ONTO, OR SURJECTIVE, MAPPING A mapping $f: X \to Y$ is said to be **onto**, or **surjective**, if, for every element $y \in Y$, there exists at least one element $x \in X$ such that

$$y = f(x).$$

Thus, *$f: X \to Y$ is onto, or surjective, if and only if*

$$Y = \text{range}(f).$$

A surjective mapping, $f:X \to Y$ is also called a **surjection**. You can tell from its transformation diagram whether a mapping is surjective: *$f:X \to Y$ is surjective if and only if every element in Y has at least one arrow pointing to it.*

4.14 Example Is the mapping in Figure 2-13 a surjection?

SOLUTION No, because there are elements in Y—namely, 4 and 5—with no arrows pointing to them. □

4.15 Example If $\mathbb{R}$ denotes the set of all real numbers, determine which of the following mappings is a surjection:

(a) $f:\mathbb{R} \to \mathbb{R}$ given by $f(x) = x^2$
(b) $g:\mathbb{R} \to \mathbb{R}$ given by $g(x) = 2x - 5$

SOLUTION **(a)** $f:\mathbb{R} \to \mathbb{R}$ is not a surjection because the range of f is the set of all nonnegative real numbers, a *proper* subset of $\mathbb{R}$.
(b) $g:\mathbb{R} \to \mathbb{R}$ is a surjection because every real number y can be written in the form

$$y = g(x) = 2x - 5$$

for a suitable choice of x, namely,

$$x = \frac{y+5}{2}.$$

(This value of x was obtained by solving the equation $y = 2x - 5$ for x in terms of y.) □

There are mappings that are neither injective nor surjective, mappings that are one but not the other, and mappings that are both injective and surjective. The last possibility suggests the following definition:

4.16 DEFINITION

ONE-TO-ONE CORRESPONDENCE, OR BIJECTION A mapping $f:X \to Y$ is said to be a **one-to-one correspondence**, or a **bijective** mapping, if it is both injective (one-to-one) and surjective (onto).

A bijective mapping is also called a **bijection**.

4.17 Example Show that the mapping $g:\mathbb{R} \to \mathbb{R}$ defined by

$$g(x) = 2x - 5$$

is a bijection.

SOLUTION By Example 4.12b, $g:\mathbb{R} \to \mathbb{R}$ is an injection, and by Example 4.15b, it is a surjection. □

Now, suppose

$$f:X \to Y,$$

so that f is a function, $\text{dom}(f) = X$, and $\text{range}(f) \subseteq Y$. Then f is a relation (a set of ordered pairs), and

$$f \subseteq X \times Y.$$

It follows that the converse, or inverse, f^{-1} of f is a relation and that

$$f^{-1} \subseteq Y \times X;$$

however, in general, it would be incorrect to write

$$f^{-1}:Y \to X.$$

In the first place, unless $f:X \to Y$ is surjective, so that $\text{range}(f) = Y$, *the domain of f^{-1} need not be all of Y.* In the second place, unless $f:X \to Y$ is injective, *f^{-1} need not be a function.* However, we do have the following theorem, whose proof we leave to you as an exercise (Problem 32):

4.18 THEOREM

THE INVERSE OF A BIJECTION Let $f:X \to Y$. Then:

(i) f is injective (one-to-one) if and only if f^{-1} is a function.
(ii) If $f:X \to Y$ is a bijection (one-to-one correspondence), then so is $f^{-1}:Y \to X$.

Some of the concepts that were introduced in earlier sections can be tied in quite nicely with the idea of a function or mapping. For instance, an indexed family $(M_i)_{i \in I}$ of subsets of a set X is just a mapping

$$i \mapsto M_i$$

from the indexing set I into the set $\mathcal{P}(X)$ of all subsets of X. Also, a predicate, or propositional function, $P(x)$ really is a function whose domain is the universe of discourse U and which maps each object $x \in U$ into a corresponding proposition $P(x)$,

$$x \mapsto P(x).$$

Furthermore, the idea of a function, or mapping, can be used to clarify ideas that you may have encountered in your study of algebra or calculus. For instance, a **permutation** of a set X is the same thing as a bijection

$f: X \to X$. Also, a **sequence of real numbers** is just a mapping

$$s: \mathbb{N} \to \mathbb{R}$$

from the set $\mathbb{N} = \{1, 2, 3, \ldots\}$ of positive integers into the set $\mathbb{R}$ of real numbers. If s is such a sequence and $n \in \mathbb{N}$, we usually write $s_n = s(n)$, and refer to s_n as the ***n*th term** of the sequence.

PROBLEM SET 2.4

1. Determine which sets of ordered pairs are functions. [$\mathbb{R}$ denotes the set of all real numbers.]
   (a) $\{(1, 1), (2, 1), (1, 2)\}$
   (b) $\{(1, 1), (2, 1), (3, 1)\}$
   (c) $\{(x, y) | x$ is a triangle in the plane and $y =$ area of $x\}$
   (d) $\{(x, y) | x, y \in \mathbb{R}, x^2 = y\}$
   (e) $\{(x, y) | x, y \in \mathbb{R}, y^2 = x\}$
   (f) $\{(x, y) | x, y \in \mathbb{R}, x = \sin y\}$
   (g) $\{(x, y) | x, y \in \mathbb{R}, x \leq y\}$
   (h) $\{(x, y) | x, y \in \mathbb{R}, x^2 = y^2\}$
2. If f is a function, prove that $\text{codom}(f) = \{y | (\exists x)(x \in \text{dom}(f) \wedge y = f(x))\}$.
3. For the function $f = \{(1, c), (2, d), (3, a), (5, b), (7, c)\}$, find:
   (a) $\text{dom}(f)$ (b) $\text{range}(f)$ (c) $f(1)$
   (d) $f(2)$ (e) $f(5)$ (f) $f(7)$
4. In calculus, we often refer to $f(x)$ as a function; however, this is really an abuse of language. Why?
5. Find the domain and range of each function. [$\mathbb{R}$ denotes the set of all real numbers.]
   (a) $f = \{(1, a), (2, a), (4, c), (6, b)\}$
   (b) $g = \{(1, b), (2, b), (3, b), (5, b)\}$
   (c) $h = \{(x, y) | x, y \in \mathbb{R}, y = \sin x\}$
   (d) $F = \{(x, y) | x, y \in \mathbb{R}, y = x^2 + 4\}$
   (e) $G = \{(x, y) | x, y \in \mathbb{R}, y = \sec x\}$
   (f) $H = \{(x, y) | x, y \in \mathbb{R}, y = 1/(x + 2)\}$
6. (a) Is the null set a function?
   (b) If Y is a set, does there exist a mapping $f: \varnothing \to Y$? If so, how many such mappings are there?
7. Determine, as an explicit set of ordered pairs, the function f depicted in the transformation diagram in Figure 2-14.
8. A function f (or a mapping $f: X \to Y$) is called a **constant function** (or a **constant mapping**) if $f(a) = f(b)$ holds for every $a, b \in \text{dom}(f)$.

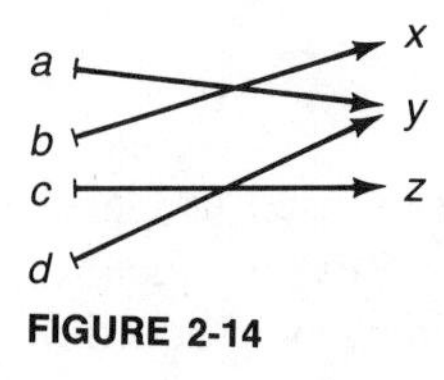

FIGURE 2-14

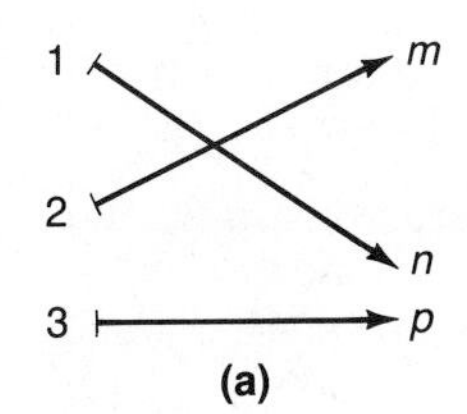

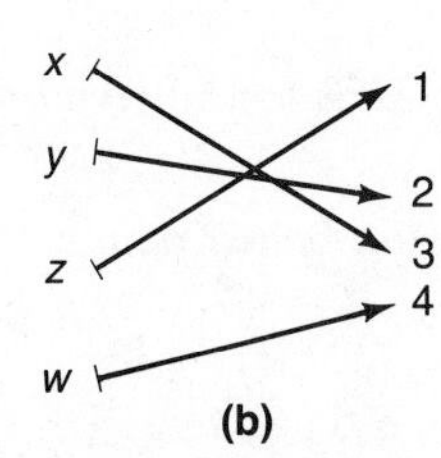

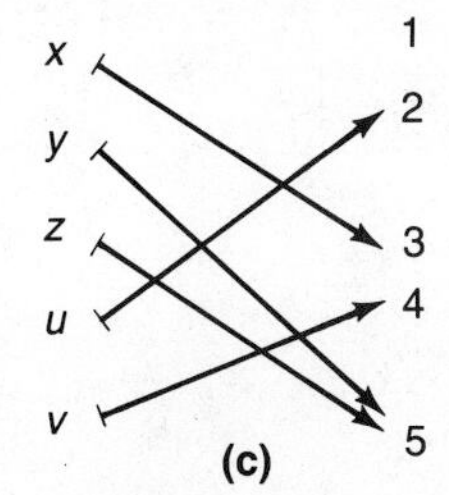

FIGURE 2-15

(a) Describe the graph of a constant function.
(b) If X is a nonempty set and Y is a finite set with $\# Y = m$, how many different constant mappings $f:X \to Y$ are there?

9. In a calculus textbook, we find the words "Consider the function $y = 1/\sqrt{x + 2}$." Translate these words into more precise mathematical language.

10. In a calculus textbook, we find the words "Consider the function $f(x) = (x + 1)/(x - 1)$." Translate these words into more precise mathematical language.

11. Which of the mappings whose transformation diagrams are shown in Figure 2-15 are injections?

12. If $\mathbb{R}$ denotes the set of all real numbers and f is a function with $f \subseteq \mathbb{R} \times \mathbb{R}$, show that f is injective if and only if no horizontal straight line in the Cartesian plane $\mathbb{R} \times \mathbb{R}$ intersects f more than once.

13. Determine which of the following real-valued functions of a real variable are injections:
(a) $f(x) = x$ (b) $g(x) = x^3 - 1$ (c) $h(x) = \sin x$

14. If f is a function and A is a subset of dom(f), then the **image** of A under f, written $f(A)$ is defined by

$$f(A) = \{f(x) | x \in A\}.$$

Prove that f is an injection if and only if $f(A \cap B) = f(A) \cap f(B)$ holds for every pair of subsets A, B of dom(f).

15. Show that a mapping $f:X \to Y$ is injective if and only if, in its transformation diagram, no two arrows start at different elements of X and point to the same element of Y.

16. There are eight different mappings from $X = \{1, 2, 3\}$ into $Y = \{a, b\}$.
(a) Draw transformation diagrams for each of these mappings.
(b) How many of these mappings are injective?
(c) How many of these mappings are surjective?
(d) How many of these mappings are bijective?

17. Let $X = \{1, 2, 3, 4, 5\}$, $Y = \{a, b, c, d, e\}$, and

$$f = \{(1, d), (2, c), (3, d), (4, c), (5, d)\}.$$

(a) Show that $f:X \to Y$ is a mapping.
(b) Find range(f).
(c) Is $f:X \to Y$ injective?
(d) Is $f:X \to Y$ surjective?

18. Suppose that f is a function. Does it make any sense to ask whether f is surjective? Explain.

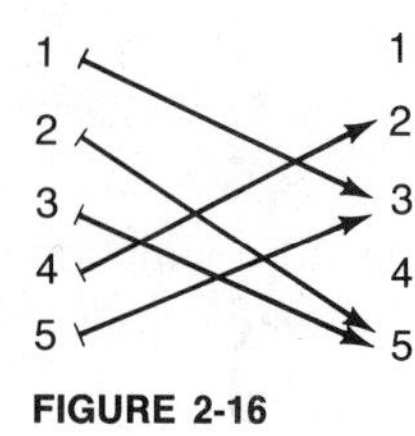

FIGURE 2-16

19. Let $X = \{1, 2, 3, 4, 5\}$, and let the mapping $f: X \to X$ be defined by the diagram in Figure 2-16. Is f a surjection? Why or why not?

20. If X is a nonempty set and $M \subseteq X$, we define the **characteristic function** of M, in mapping notation $\chi_M: X \to \{0, 1\}$, by

$$\chi_M(x) = \begin{Bmatrix} 1, & \text{if } x \in M \\ 0, & \text{if } x \notin M \end{Bmatrix} \quad \text{for all } x \in X.$$

Show that $\chi_M: X \to \{0, 1\}$ is a surjection if and only if M is a nontrivial subset of X.

21. If $\mathbb{R}$ denotes the set of all real numbers, then each formula below defines a mapping $f: \mathbb{R} \to \mathbb{R}$. In each case, determine whether $f: \mathbb{R} \to \mathbb{R}$ is injective and whether it is surjective.
(a) $f(x) = x$ (b) $f(x) = \sin x$ (c) $f(x) = 3x + 2$
(d) $f(x) = \text{Arctan } x$ (e) $f(x) = 2$ for all $x \in \mathbb{R}$

22. Let $\mathbb{R}$ denote the set of real numbers, and define $f: \mathbb{R} \to \mathbb{R}$ by

$$f(x) = \begin{Bmatrix} 2 - x, & \text{if } x \leq 1 \\ 1/x, & \text{if } x > 1 \end{Bmatrix} \quad \text{for all } x \in \mathbb{R}.$$

(a) Is $f: \mathbb{R} \to \mathbb{R}$ injective?
(b) Is $f: \mathbb{R} \to \mathbb{R}$ surjective?

23. If $X = \{1, 2, 3\}$, $Y = \{a, b, c\}$, and $f = \{(1, a), (2, c), (3, b)\}$, show that $f: X \to Y$ is a bijection.

24. If $\mathbb{R}$ denotes the set of all real numbers, then show that the mapping $T: \mathbb{R} \times \mathbb{R} \to \mathbb{R} \times \mathbb{R}$ defined by $T(x, y) = (x + y, x - y)$ is a bijection.

25. In the set $\mathbb{R}$ if all real numbers, let $(0, 1)$ be the open interval from 0 to 1, and let $(0, \infty)$ be the unbounded open interval of all positive numbers. Prove that the mapping $f: (0, 1) \to (0, \infty)$ defined by $f(x) = x/(1 - x)$ for all $x \in (0, 1)$ is a bijection.

26. Let X be a finite set. Explain why a mapping $f: X \to X$ is an injection if and only if it is a bijection. Show by a suitable example that this is not true for an infinite set X.

27. If $\mathbb{R}$ denotes the set of all real numbers, give examples of mappings $f: \mathbb{R} \to \mathbb{R}$ that are
(a) injective but not surjective;
(b) surjective but not injective;
(c) neither injective nor surjective; and
(d) bijective.

28. In calculus, the natural logarithm function and the exponential function are inverses of each other. Are the sine and the Arcsine functions inverses of each other? Explain.

29. Each of the formulas given below defines a real-valued function of a real variable. (Assume, as usual, that the domain of each function is the set of all real numbers for which the formula makes sense.) In each case, write a formula for the inverse of the function.
(a) $f(x) = -7x + 2$ (b) $g(x) = 1/x$
(c) $h(x) = (2x - 3)/(3x - 2)$ (d) $F(x) = 1 + \sqrt{x}$
(e) $G(x) = (e^x - e^{-x})/2$

30. Define a real-valued function f of a real variable by

$$f(x) = \frac{1}{2} \ln \frac{1 + x}{1 - x}, \quad \text{for } |x| < 1.$$

Prove that f is injective and find a formula for f^{-1}.

31. Two sets X and Y are said to be **equinumerous**, or to have the same **cardinal number**, if there exists a bijection $f: X \to Y$.
(a) Explain, in your own words, the rationale for this definition.
(b) Show that the set X of all positive integers and the set Y of all even positive integers are equinumerous.

32. Prove Theorem 4.18.

In Problems 33–36, let X and Y be finite sets with $\#X = m$ and $\#Y = n$.

33. Prove that there are n^m different mappings $f: X \to Y$.

34. Prove that there are $(n + 1)^m$ different functions g with $g \subseteq X \times Y$. [Hint: Let u be an element that does not belong to Y and let $Y^* = Y \cup \{u\}$. Note that $\#Y^* = n + 1$. If g is a function with $g \subseteq X \times Y$, define $g^*: X \to Y^*$ by $g^*(x) = g(x)$ for $x \notin \text{dom}(g)$ and $g^*(x) = u$ for $x \in \text{dom}(g)$. Show that $g \mapsto g^*$ produces a one-to-one correspondence between the functions $g \subseteq X \times Y$ and the mappings $g^*: X \to Y^*$ and then use the result of Problem 33.]

35. Of the n^m different mappings $f: X \to Y$, prove that, if $m > n$, none of them are injections, whereas if $m \leq n$, precisely $n!/(n - m)!$ of them are injections.

36. If $m \geq n > 0$, it can be shown that, of the n^m different mappings $f: X \to Y$, exactly

$$\sum_{k=1}^{n} (-1)^{n-k} C_k^n \cdot k^m$$

are surjections, where C_k^n denotes the number of combinations of n things taken k at a time. [See Problem 30 in Problem Set 1.5.] Use this formula to compute the number of surjections $f: X \to Y$ if $X = \{1, 2, 3\}$ and $Y = \{a, b\}$ and compare the result with your answer to Part (c) of Problem 16.

5

COMPOSITION OF FUNCTIONS

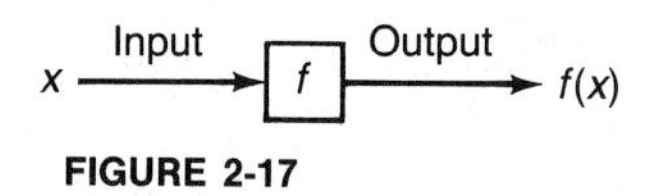

FIGURE 2-17

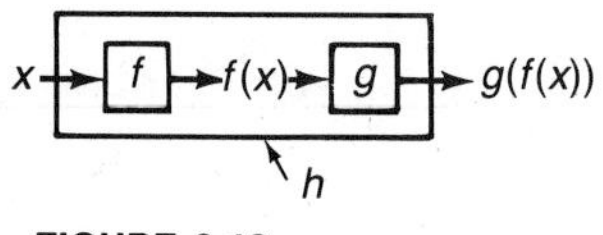

FIGURE 2-18

The idea of a function as a processing device or machine, and the fact that machines can be operated in tandem suggests the concept of *function composition.* In the machine picture, a function f is regarded as a black box that can accept as an input any object x in the domain X, convert it into a new object $f(x)$, and eject $f(x)$ as an output (Figure 2-17).

If f and g are two functions, regarded as machines, we can imagine hooking up the f-machine and the g-machine in tandem, feeding the output of the f-machine directly into the input of the g-machine. If we enclose the resulting assembly in a larger black box called h, we obtain the diagram shown in Figure 2-18.

The composite h-machine can be regarded as a function that transforms x into

$$h(x) = g(f(x)).$$

The *composite function* h is usually written as

$$h = g \circ f,$$

read g *composed with* f. Thus, we have

$$(g \circ f)(x) = g(f(x)).$$

Because of the last equation in the paragraph above, the notation

$$g \circ f$$

for the composition of g and f seems reasonable; however, from the point of view of Figure 2-18, it seems inappropriate because the f-machine is *first* and the g-machine *second* in the diagram. This notational inversion of order is annoying—and can be confusing—but the symbolism $g \circ f$ is entrenched in the mathematical literature. Thus, when translating from diagrams to the symbolism of function composition, you must pay careful attention to the resulting inversion of order.

Let us give a formal definition of function composition:

5.1 DEFINITION

COMPOSITION OF FUNCTIONS Let f and g be functions. Then, the relation $g \circ f$ defined by

$$g \circ f = \{(x, z) \mid (\exists y)((x, y) \in f \wedge (y, z) \in g)\}$$

is called the **composition** of g and f.

Our first task is to prove that $g \circ f$ is actually a function.

5.2 THEOREM **THE COMPOSITION OF FUNCTIONS IS A FUNCTION** If f and g are functions, then $g \circ f$ is a function.

PROOF We have to show that, if two ordered pairs in the relation $g \circ f$ have the same first element, then they have the same second element. Thus, suppose that

$$(x, z) \in g \circ f \quad \text{and} \quad (x, w) \in g \circ f.$$

We must prove that $z = w$. Because $(x, z) \in g \circ f$, there exists an element y with

$$(x, y) \in f \quad \text{and} \quad (y, z) \in g.$$

Therefore, since f and g are functions, we can write

$$y = f(x) \quad \text{and} \quad z = g(y).$$

Because $(x, w) \in g \circ f$, there exists an element u with

$$(x, u) \in f \quad \text{and} \quad (u, w) \in g.$$

Therefore, since f and g are functions, we can write

$$u = f(x) \quad \text{and} \quad w = g(u).$$

Because $y = f(x)$ and $u = f(x)$, we have $y = u$; hence,

$$w = g(u) = g(y) = z.$$ □

In the next theorem, we show that Definition 5.1 produces the desired formula $(g \circ f)(x) = g(f(x))$ for the composite function.

5.3 THEOREM **THE IMAGE OF AN ELEMENT UNDER A COMPOSITE FUNCTION** Let f and g be functions, let $x \in \text{dom}(f)$, and suppose that $f(x) \in \text{dom}(g)$. Then,

$$(g \circ f)(x) = g(f(x)).$$

PROOF Let $x \in \text{dom}(f)$. Then, there exists an element y such that $(x, y) \in f$. Since f is a function, we can write $y = f(x)$. Because

$$y = f(x) \in \text{dom}(g),$$

there exists an element z such that $(y, z) \in g$, that is,

$$z = g(y).$$

Since we have

$$(x, y) \in f \wedge (y, z) \in g,$$

it follows from Definition 5.1 that

$$(x, z) \in g \circ f.$$

Because $g \circ f$ is a function, we can write

$$z = (g \circ f)(x).$$

Therefore,

$$(g \circ f)(x) = z = g(y) = g(f(x)).$$

□

5.4 Example Let f and g be the real-valued functions of a real variable defined by

$$f(x) = 2x - 1 \quad \text{and} \quad g(x) = 3x + 5.$$

Find **(a)** $(g \circ f)(x)$ and **(b)** $(f \circ g)(x)$.

SOLUTION **(a)** $(g \circ f)(x) = g(f(x)) = g(2x - 1) = 3(2x - 1) + 5 = 6x + 2$
(b) $(f \circ g)(x) = f(g(x)) = f(3x + 5) = 2(3x + 5) - 1 = 6x + 9$ □

As this example shows, when you use function composition, you must be careful about the order in which you write the functions. In general, $g \circ f$ is not the same as $f \circ g$. In other words, *function composition is not a commutative operation.*

The *chain rule*, one of the most important theorems of differential calculus, is actually a rule for computing the derivative of the composition of two functions. According to the chain rule, *if g is differentiable at x, and if f is differentiable at $g(x)$, then $f \circ g$ is differentiable at x and*

$$(f \circ g)'(x) = f'(g(x))g'(x).$$

In employing the chain rule and in dealing with function composition in general, *you must be careful about the domains and ranges of the functions involved.*

5.5 THEOREM **DOMAIN AND RANGE OF A COMPOSITION OF FUNCTIONS** Let f and g be functions. Then:

(i) $\text{dom}(f \circ g) = \{x \mid x \in \text{dom}(g) \wedge g(x) \in \text{dom}(f)\}$
(ii) $\text{range}(f \circ g) = \{f(y) \mid y \in \text{range}(g)\}$

We leave the proof of Theorem 5.5 to you as an exercise (Problem 16).

5.6 Example In a calculus textbook, we are asked to consider the function

$$y = \sin^{-1} \sqrt{x} = \text{Arcsin} \sqrt{x}.$$

What is the domain of this function?

SOLUTION Here we are working with a composition $f \circ g$, where

$$g(x) = \sqrt{x}$$

and

$$f(x) = \sin^{-1} x = \text{Arcsin } x.$$

The domain of g is the unbounded interval $[0, \infty)$ consisting of all nonnegative real numbers, and the domain of f is the closed interval $[-1, 1]$ consisting of all real numbers between -1 and 1. By Part (i) of Theorem 5.5, the domain of $f \circ g$ is the set of all numbers x in $[0, \infty)$ such that $g(x) \in [-1, 1]$; that is, all real numbers $x \geq 0$ such that

$$-1 \leq \sqrt{x} \leq 1.$$

Thus, the domain of $f \circ g$ is the closed interval $[0, 1]$. □

Although function composition is not, in general, commutative, it turns out that it is *associative.*

5.7 THEOREM

ASSOCIATIVE LAW FOR FUNCTION COMPOSITION If f, g, and h are any three functions, then,

$$(f \circ g) \circ h = f \circ (g \circ h).$$

PROOF We are going to prove that, as sets of ordered pairs,

$$(f \circ g) \circ h = f \circ (g \circ h).$$

We do this by the pick-a-point process. To begin with, suppose

$$(x, z) \in (f \circ g) \circ h,$$

so that, by Definition 5.1, there exists an element y such that

$$(x, y) \in h \quad \text{and} \quad (y, z) \in f \circ g.$$

Because $(y, z) \in f \circ g$, it follows from Definition 5.1 that there exists an element w such that

$$(y, w) \in g \quad \text{and} \quad (w, z) \in f.$$

Now, we have

$$(x, y) \in h, \quad (y, w) \in g, \quad \text{and} \quad (w, z) \in f.$$

Because $(x, y) \in h$ and $(y, w) \in g$, it follows from Definition 5.1 that

$$(x, w) \in g \circ h.$$

Now, we have

$$(x, w) \in g \circ h \quad \text{and} \quad (w, z) \in f;$$

so a final application of Definition 5.1 shows that

$$(x, z) \in f \circ (g \circ h).$$

This proves that

$$(f \circ g) \circ h \subseteq f \circ (g \circ h).$$

The proof of the opposite inclusion is much the same and is left as an exercise (Problem 18). □

Because of Theorem 5.7, you can write compositions of three or more functions without the aid of parentheses. For instance,

$$f \circ g \circ h$$

is completely unambiguous; it makes no difference whether it is interpreted as $(f \circ g) \circ h$ or as $f \circ (g \circ h)$. In particular, the functions

$$f, f \circ f, f \circ f \circ f, f \circ f \circ f \circ f, \ldots,$$

which are called the **iterates** of f, can be written without parentheses.

5.8 Example If f is the real-valued function of a real variable defined by the formula

$$f(x) = 2x - x^2,$$

find a formula for the third iterate $f \circ f \circ f$ of f.

SOLUTION

$$\begin{aligned}(f \circ f)(x) &= f(f(x)) \\ &= f(2x - x^2) \\ &= 2(2x - x^2) - (2x - x^2)^2 \\ &= 4x - 2x^2 - (4x^2 - 4x^3 + x^4) \\ &= -x^4 + 4x^3 - 6x^2 + 4x.\end{aligned}$$

Therefore,

$$\begin{aligned}(f \circ f \circ f)(x) &= f((f \circ f)(x)) \\ &= f(-x^4 + 4x^3 - 6x^2 + 4x) \\ &= 2(-x^4 + 4x^3 - 6x^2 + 4x) - (-x^4 + 4x^3 - 6x^2 + 4x)^2 \\ &= -x^8 + 8x^7 - 28x^6 + 56x^5 - 70x^4 + 56x^3 - 28x^2 + 8x.\end{aligned}$$

□

The mapping notation makes it easier to deal with function composition by automatically taking care of questions involving the domains and ranges of the functions involved. This is accomplished by adopting the following convention: We speak about the composition of the mappings

$$f:X \to Y \quad \text{and} \quad g:W \to Z$$

only when $Y = W$. Then we have

$$f:X \to Y \quad \text{and} \quad g:Y \to Z,$$

in which case,

$$\text{range}(f) \subseteq Y = \text{dom}(g),$$

and it follows that

$$\text{dom}(g \circ f) = X$$

(Problem 21a) and

$$\text{range}(g \circ f) \subseteq Z$$

(Problem 21b), so that

$$(g \circ f):X \to Z.$$

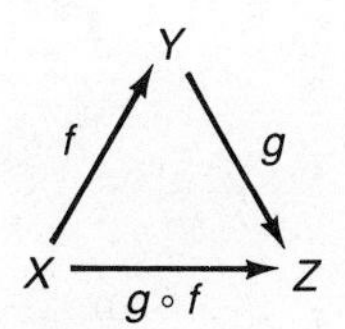

FIGURE 2-19

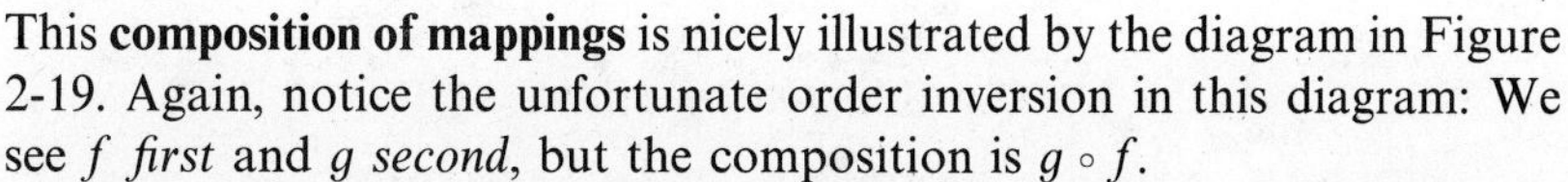

This **composition of mappings** is nicely illustrated by the diagram in Figure 2-19. Again, notice the unfortunate order inversion in this diagram: We see f *first* and g *second*, but the composition is $g \circ f$.

There is an interesting interplay between composition of mappings and the conditions of injectivity, surjectivity, and bijectivity. For instance, as the following theorem shows, injective mappings are the mappings that "cancel on the left" under composition.

5.9 THEOREM

INJECTIVITY AND COMPOSITION A mapping $f:X \to Y$ is injective if and only if the following condition holds: If $g:Z \to X$ and $h:Z \to X$ are mappings, then

$$f \circ g = f \circ h \Rightarrow g = h.$$

PROOF First, suppose that $f:X \to Y$ is injective and that

$$f \circ g = f \circ h.$$

Then, for every $z \in Z$,

$$(f \circ g)(z) = (f \circ h)(z),$$

that is,

$$f(g(z)) = f(h(z)).$$

Since f is injective, it follows that

$$g(z) = h(z).$$

Because z is an arbitrary element of the common domain Z of g and h, we conclude that $g = h$.

Conversely, suppose that the condition

$$f \circ g = f \circ h \Rightarrow g = h$$

holds for all mappings g and h from a common domain Z into X. We are going to make an indirect proof to show that f is injective. Thus, suppose that f is *not* injective, so that there exist elements a and b in the domain X of f with

$$f(a) = f(b) \quad \text{and} \quad a \neq b.$$

Let $Z = X = \{a, b\}$, and define mappings g and h from Z into X by

$$g = \{(a, a), (b, b)\} \quad \text{and} \quad h = \{(a, b), (b, a)\}.$$

Then,

$$(f \circ g)(a) = f(g(a)) = f(a)$$

and

$$(f \circ g)(b) = f(g(b)) = f(b),$$

so that

$$f \circ g = \{(a, f(a)), (b, f(b))\}.$$

A similar calculation shows that

$$f \circ h = \{(a, f(b)), (b, f(a))\}.$$

However, since $f(a) = f(b)$, it follows that

$$f \circ g = f \circ h;$$

hence, by hypothesis, $g = h$. Therefore,

$$a = g(a) = h(a) = b,$$

contradicting the fact that $a \neq b$ and completing the proof. □

As the next theorem shows, surjective mappings are those that "cancel on the right" under composition.

5.10 THEOREM

SURJECTIVITY AND COMPOSITION A mapping $f: X \to Y$ is surjective if and only if the following condition holds: If $g: Y \to Z$ and $h: Y \to Z$ are mappings, then,

$$g \circ f = h \circ f \Rightarrow g = h.$$

PROOF First, suppose that $f:X \to Y$ is surjective and that

$$g \circ f = h \circ f.$$

We must prove that $g = h$; that is, we must prove that $g(y) = h(y)$ for every element y in the common domain Y of g and h. Thus, let y be any element in Y. Because $f:X \to Y$ is surjective, there exists $x \in X$ such that

$$f(x) = y.$$

Therefore, using the fact that $g \circ f = h \circ f$, we find that $g(y) = g(f(x)) = (g \circ f)(x) = (h \circ f)(x) = h(f(x)) = h(y)$.

Conversely, suppose that the condition

$$g \circ f = h \circ f \Rightarrow g = h$$

holds for all mappings g and h from Y into Z. We are going to make an indirect proof to show that f is surjective. We consider the case in which $X \neq \varnothing$, leaving the case when $X = \varnothing$ as an exercise (Problem 30). Thus, suppose that $f:X \to Y$ is *not* surjective, so that there exists an element $b \in Y$ such that $b \notin \text{range}(f)$. Choose and fix any element $a \in X$. (This is where we use the hypothesis that X is nonempty.) Let $Z = Y$, and define

$$g:Y \to Y \quad \text{and} \quad h:Y \to Y$$

by the rules

$$g(y) = y$$

and

$$h(y) = \begin{cases} y, & \text{if } y \neq b \\ f(a), & \text{if } y = b \end{cases}$$

for all $y \in Y$. Then, for every $x \in X$,

$$(g \circ f)(x) = g(f(x)) = f(x).$$

Because $b \notin \text{range}(f)$, we have $f(x) \neq b$ for every $x \in X$, and, therefore,

$$(h \circ f)(x) = h(f(x)) = f(x)$$

holds for every $x \in X$. Consequently,

$$g \circ f = h \circ f$$

and so, by hypothesis,

$$g = h.$$

Therefore,

$$b = g(b) = h(b) = f(a),$$

contradicting the supposition that $b \notin \text{range}(f)$. □

5.11 COROLLARY

(i) The composition of two injective mappings is again injective.
(ii) The composition of two surjective mappings is again surjective.
(iii) The composition of two bijective mappings is again bijective.

PROOF (i) Suppose that $p:X \to W$ and $q:W \to Y$ are injections. We are going to prove that $(q \circ p):X \to Y$ is an injection by using Theorem 5.9. Thus, suppose that

$$g:Z \to X \quad \text{and} \quad h:Z \to X$$

are mappings such that

$$(q \circ p) \circ g = (q \circ p) \circ h.$$

We must prove that $g = h$. By the associative law for composition (Theorem 5.7), we can rewrite the last equation as

$$q \circ (p \circ g) = q \circ (p \circ h).$$

Since q is injective, we can cancel it on the left to obtain

$$p \circ g = p \circ h.$$

Likewise, since p is injective, we can cancel it on the left and conclude that

$$g = h,$$

finishing the proof of Part (i).

(ii) The proof of Part (ii), using Theorem 5.10, is quite similar to the proof of Part (i) and is left as an exercise (Problem 27a).

(iii) Part (iii) is a direct consequence of Parts (i) and (ii). □

5.12 COROLLARY

Let $f:X \to Y$ and $g:Y \to Z$.

(i) If $(g \circ f):X \to Z$ is injective, then $f:X \to Y$ is injective.
(ii) If $(g \circ f):X \to Z$ is surjective, then $g:Y \to Z$ is surjective.
(iii) If $(g \circ f):X \to Z$ is bijective, then $f:X \to Y$ is injective and $g:Y \to Z$ is surjective.

PROOF Evidently, (i) and (ii) together imply (iii). We prove Part (i) and leave the analogous proof of Part (ii) as an exercise (Problem 27b). Suppose

$$(g \circ f):X \to Z$$

is injective. To prove that f is injective, we are going to use Theorem 5.9. Thus, let

$$p:W \to X \quad \text{and} \quad q:W \to X$$

and suppose

$$f \circ p = f \circ q.$$

We must prove that $p = q$. Because $f \circ p = f \circ q$, we have

$$g \circ (f \circ p) = g \circ (f \circ q),$$

or

$$(g \circ f) \circ p = (g \circ f) \circ q.$$

Since $g \circ f$ is injective, the last equation implies that $p = q$, and the proof of Part (i) is complete. □

Actually, the results in Corollaries 5.11 and 5.12 are easy to prove directly, without using Theorems 5.9 and 5.10 at all (Problems 29 and 31). However, the technique of proof shown above generalizes to situations that are frequently encountered in advanced mathematics, whereas the more direct proofs do not.

The following definition for composition of relations is obtained simply by replacing the functions f and g in Definition 5.1 by relations R and S:

5.13 DEFINITION

COMPOSITION OF RELATIONS Let R and S be relations. Then, the relation $R \circ S$ defined by

$$R \circ S = \{(x, z) | (\exists y)((x, y) \in S \land (y, z) \in R)\}$$

is called the **composition** of R and S.

Some authors use a different definition for relation composition, interchanging S and R on the right of the equation in Definition 5.13. We prefer our definition because it makes function composition a special case of relation composition. In Problems 35–39, we ask you to prove a few simple facts about relation composition.

5.14 Example If U, the universe of discourse, consists of all humans, if xMy means that x is the mother of y, and if yFz means that y is the father of z, give an interpretation of the composite relation $F \circ M$.

SOLUTION Suppose $(x, z) \in F \circ M$. Then, by Definition 5.13, there exists y such that $(x, y) \in M$ and $(y, z) \in F$. This means that x is the mother of y and that y is the father of z. Hence, x is the paternal grandmother of z. Thus, the composite relation $F \circ M$ is the relation *paternal grandmother of*. □

PROBLEM SET 2.5

In Problems 1–10, let f and g be the real-valued functions of a real variable defined by the indicated formulas. In each case, assume that the domains of f and g consist of all real numbers x for which the formula makes sense. Find: (a) $\text{dom}(g \circ f)$, *(b)* $(g \circ f)(x)$ *for* $x \in \text{dom}(g \circ f)$, *(c)* $\text{dom}(f \circ g)$, *and (d)* $(f \circ g)(x)$ *for* $x \in \text{dom}(f \circ g)$.

1. $f(x) = 7x + 2,\ g(x) = -3x$

2. $f(x) = 2x,\ g(x) = 1/x$

3. $f(x) = 3x,\ g(x) = \sqrt{x + 1}$

4. $f(x) = 1 - 3x,\ g(x) = (x + 7)^{1/3}$

5. $f(x) = \dfrac{x + 1}{x + 3},\ g(x) = x^2 + 2$

6. $f(x) = x^3 + 1,\ g(x) = \sqrt{x - 1}$

7. $f(x) = \sin x,\ g(x) = 3x^2 + 2$

8. $f(x) = \tan x,\ g(x) = 4x - 3$

9. $f(x) = |5x + 1|,\ g(x) = 3$

10. $f(x) = |x|,\ g(x) = \dfrac{1}{2x - 3}$

In Problems 11 and 12, find (a) $f \circ g$ *and (b)* $g \circ f$ *as an explicit set of ordered pairs.*

11. $f = \{(1, b), (2, c), (3, a)\},\ g = \{(a, 1), (b, 2), (c, 2)\}$

12. $f = \{(1, 1), (-1, 5), (2, 1), (-3, 9)\},\ g = \{(1, 3), (5, 2), (9, 4), (3, -3)\}$

13. In a calculus textbook, we are asked to consider the real-valued functions of a real variable defined by the indicated formulas. In each case, write the function as a composition $f \circ g$ of real-valued functions f, g of real variables, and find the domain of the function.

(a) $y = \sqrt{\dfrac{x + 1}{x - 1}}$ (b) $y = \cos^{-1} \sqrt{x} = \text{Arccos} \sqrt{x}$

(c) $y = \ln(x^2 - x - 6)$

14. Some mathematicians (mostly algebraists) write $(x)f$ rather than $f(x)$ for the image of x under the function f to obviate the "order inversion" difficulty mentioned in the text. Explain how this maneuver removes the difficulty.

15. Let a, b, c, and d be fixed real numbers and let real-valued functions f and g of real variables be given by $f(x) = ax + b$ and $g(x) = cx + d$ for all real numbers x.

(a) Find necessary and sufficient conditions on a, b, c, and d so that $f \circ g = g \circ f$.

(b) If $a \neq 1$, $c \neq 1$, and $f \circ g = g \circ f$, show that there exists a unique real number x such that $f(x) = g(x) = x$.

16. Prove Theorem 5.5.

17. Let f be the real-valued function of a real variable defined by the indicated formula. Find a formula for the fourth iterate $f \circ f \circ f \circ f$ of f.

(a) $f(x) = 2x - 3$ (b) $f(x) = 1 - 4x$

18. Finish the proof of Theorem 5.7 by proving that $f \circ (g \circ h) \subseteq (f \circ g) \circ h$.

19. Let $\mathbb{R}$ denote the set of all real numbers and suppose that $f:\mathbb{R} \to \mathbb{R}$ and $g:\mathbb{R} \to \mathbb{R}$. If $x \in \mathbb{R}$, show that $(f \circ g)(x)$ can be obtained graphically as follows: Start at the point $(x, 0)$ on the x axis, move vertically to the graph of g, then horizontally to the graph of $y = x$, then vertically to the graph of f, and finally horizontally to the point $(0, y)$ on the y axis (Figure 2-20). Conclude that $y = (f \circ g)(x)$.

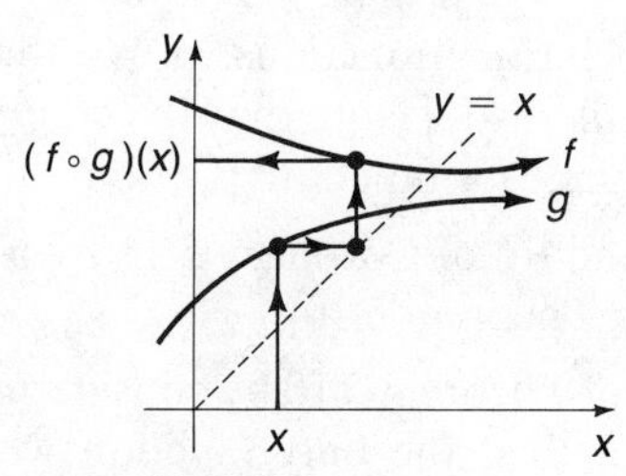

FIGURE 2-20

20. Show how to modify the procedure in Problem 19 to find the values $f(x)$, $(f \circ f)(x)$, $(f \circ f \circ f)(x)$, and so on, of the successive iterates of f at x. The set consisting of x and these values is called the **orbit** of x under the iterates of the function f.

21. Let $f:X \to Y$ and $g:Y \to Z$. Prove:
(a) $\text{dom}(g \circ f) = X$ (b) $\text{range}(g \circ f) \subseteq Z$

22. (a) Let $f = \{(1, a), (2, a)\}$ and $g = \{(a, 1)\}$. Show that g is injective, but $g \circ f$ is not injective.
(b) Give an example of a pair of functions f and g such that $g \circ f$ is injective, but g is not injective.

23. Let the mappings $f:X \to Y$ and $g:Y \to Z$ be defined by the diagram in Figure 2-21.
(a) Draw a diagram for $(g \circ f):X \to Z$.
(b) Find $\text{range}(g \circ f)$.

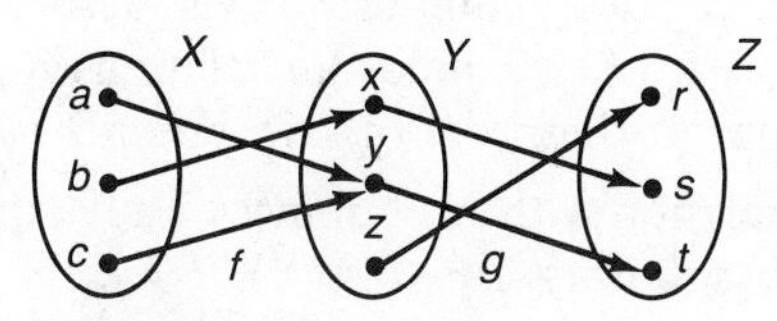

FIGURE 2-21

24. Give an example of a pair of mappings $f:X \to Y$ and $g:Y \to Z$ such that $(g \circ f):X \to Z$ is surjective, but $f:X \to Y$ is not surjective.

25. A mapping $g:Y \to X$ is called a **cross section** of the mapping $f:X \to Y$ if $f \circ g = \Delta_Y$.

(a) If $f:X \to Y$ has a cross section, show that $f:X \to Y$ is surjective.
(b) If $g:Y \to X$ is a cross section of $f:X \to Y$, show that $g:Y \to X$ is injective.

26. Give an example of a pair of mappings $f:X \to Y$ and $g:Y \to Z$ such that $(g \circ f):X \to Z$ is bijective, but neither $f:X \to Y$ nor $g:Y \to Z$ is bijective.

27. Using Theorem 5.10:
(a) Prove Part (ii) of Corollary 5.11.
(b) Prove Part (ii) of Corollary 5.12.

28. If f and g are arbitrary functions, prove that $g \circ f = \varnothing$ if and only if $\text{range}(f) \cap \text{dom}(g) = \varnothing$. [Note that the empty set, regarded as an empty set of ordered pairs, is a function!]

29. Prove Parts (i) and (ii) of Corollary 5.11 by a direct argument (not using Theorem 5.9 nor Theorem 5.10).

30. Finish the proof of Theorem 5.10 by considering the case in which $X = \varnothing$. [Hint: If $X = \varnothing$, then the only function with domain X is the empty function, that is, the empty set $\varnothing$.]

31. Prove Parts (i) and (ii) of Corollary 5.12 by a direct argument (not using Theorem 5.9 nor Theorem 5.10).

32. Let $X = \{1, 2, 3\}$ and let $f:X \to X$ be given by the rules $f(1) = 2, f(2) = 3$, and $f(3) = 1$. There are 27 different mappings $g:X \to X$. Find all such mappings $g:X \to X$ such that $f \circ g = g \circ f$.

33. Let X be a nonempty set. By a **permutation** of X, we mean a bijection $f:X \to X$. The **active set** of a permutation $f:X \to X$ is defined to be the set $\{x \in X \mid f(x) \neq x\}$. Prove that, if two permutations of X have disjoint active sets, then they commute under composition.

34. In Problem 33, give an example of two permutations that commute under composition but whose active sets are not disjoint.

35. If R and S are relations, prove that $(R \circ S)^{-1} = S^{-1} \circ R^{-1}$.

36. Prove that a relation R is transitive if and only if, as sets of ordered pairs, $R \circ R \subseteq R$.

37. Prove that relation composition is associative; that is, prove that $R \circ (S \circ T) = (R \circ S) \circ T$ holds for relations R, S, and T. [Hint: The proof is very similar to the proof of Theorem 5.7.]

38. Let E be a relation with $X = \text{dom}(E)$. Prove that E is an equivalence relation on X if and only if $E = E^{-1} = E \circ E$.

39. If f is a function, then f^{-1} is a relation but not necessarily a function. Let $E = f^{-1} \circ f$. Prove that E is an equivalence relation on the domain of f.

40. If U, the universe of discourse, consists of all humans, if xPy means that y is a parent of x, and if ySz means that z is a sibling of y, give an interpretation of the composite relation $S \circ P$. Is it true that $P \circ S = P$? Explain.

6 BINARY OPERATIONS

A *binary operation*, or *binary composition*, on a set S is a way of combining or composing pairs of elements x and y in S to obtain a third element $x * y$ in S, in symbols,

$$(x, y) \mapsto x * y.$$

For instance, take S to be the set $\mathbb{R}$ of all real numbers, and let $*$ be addition,

$$(x, y) \mapsto x + y,$$

or take S to be $\mathbb{R}$, and let $*$ be multiplication,

$$(x, y) \mapsto x \cdot y,$$

or take S to be all positive real numbers, and let $*$ be division,

$$(x, y) \mapsto x \div y.$$

More formally, we have the following definition:

6.1 DEFINITION

BINARY OPERATION, OR COMPOSITION If S is a nonempty set, then by a **binary operation**, or **binary composition**, on S, we mean a mapping

$$*: S \times S \to S.$$

If $*: S \times S \to S$ is a binary operation on S, we usually write

$$x * y$$

rather than

$$*((x, y))$$

for the image under $*$ of the ordered pair $(x, y) \in S \times S$. The abbreviation $x * y$ for $*((x, y))$, which is called **infix notation**, greatly simplifies calculations with binary operations. Imagine how complicated arithmetic would look if we wrote

$$+((9, 16)) = 25$$

rather than using the usual infix notation

$$9 + 16 = 25.$$

However, there are situations in more advanced mathematics in which it is useful to regard $+$ as what it really is, a mapping

$$+: \mathbb{R} \times \mathbb{R} \to \mathbb{R}.$$

If $*$ is a binary operation on a set $S = \{a, b, c, \ldots\}$ with a finite number of elements, $\#S = n$, then we can give a complete description of $*$ by writing its **operation table**. To make such a table, we label n horizontal rows and n vertical columns by the symbols $a, b, c, \ldots$, and we fill in the table by placing the value of $x * y$ at the intersection of the row labeled by x and the column labeled by y. (Remember the times table for multiplication from your grammar school days?) The resulting operation table has the following form:

$*$	a	b	c	$\cdots$
a	$a * a$	$a * b$	$a * c$	$\cdots$
b	$b * a$	$b * b$	$b * c$	$\cdots$
c	$c * a$	$c * b$	$c * c$	$\cdots$
$\vdots$	$\vdots$	$\vdots$	$\vdots$	

Notice that in the body of the table itself there are n^2 positions to be filled. These positions can be filled in any way whatsoever with the n elements of S, depending on the binary operation $*$ being described; hence, *on a set S with n elements there are exactly*

$$n^{n^2}$$

different binary operations (Problem 8).

6.2 DEFINITION

ALGEBRAIC SYSTEM WITH ONE BINARY OPERATION An **algebraic system with a single binary operation** is an ordered pair $(S, *)$ whose first element is a nonempty set S and whose second element is a binary operation $*$ on S.

It is customary to speak of an *algebraic system S with a single binary operation $*$*, although, technically, this is an abuse of language. After all, S is just a nonempty set; the algebraic system is really the ordered pair $(S, *)$. However, provided there is a clear understanding of what is really meant, no harm seems to come from this custom.

Later in this textbook we study certain algebraic systems (such as the *real number system* $\mathbb{R}$) with more than one binary operation (such as addition and multiplication); however, in the present section we confine our attention to algebraic systems S with only one binary operation $*$. Therefore, in this section, when we speak of an algebraic system, we always mean an algebraic system with one binary operation.

In the examples of algebraic systems $(S, *)$ that follow,

$\mathbb{R}$ is the set of all real numbers,

$\mathbb{R}^3$ is the set of all vectors in three-dimensional space,

$\mathbf{x} \times \mathbf{y}$ denotes the vector cross product of $\mathbf{x}$ and $\mathbf{y}$ in $\mathbb{R}^3$,

X is any nonempty set,

$\mathscr{P}(X)$ is the set of all subsets of X,

$\mathscr{F}(X)$ is the set of all functions $f: X \to X$, and

$\mathscr{B}(X)$ is the set of all bijections $f: X \to X$.

For easy reference, we denote the algebraic systems by AS1, AS2, and so forth.

AS1 $S = \mathbb{R}$ $x * y = x + y$

AS2 $S = \mathbb{R}$ $x * y = x \cdot y$

AS3 $S = \mathbb{R}$ $x * y = x - y$

AS4 $S = \mathbb{R}$ $x * y = x + y - xy$

AS5 $S = \mathbb{R} \setminus \{0\}$ $x * y = x \cdot y$

AS6 $S = \mathbb{R} \setminus \{0\}$ $x * y = x/y$

AS7 $S = \mathbb{R}^3$ $\mathbf{x} * \mathbf{y} = \mathbf{x} \times \mathbf{y}$

AS8 $S = \mathscr{F}(X)$ $f * g = f \circ g$

AS9 $S = \mathscr{B}(X)$ $f * g = f \circ g$

AS10 $S = \mathscr{P}(X)$ $M * N = M \cup N$

AS11 $S = \mathscr{P}(X)$ $M * N = M \cap N$

AS12 $S = \mathscr{P}(X)$ $M * N = (M \cup N) \setminus (M \cap N)$

AS13 $S = \{a, b, c, e\}$

*	a	b	c	e
a	e	c	b	a
b	c	e	a	b
c	b	a	e	c
e	a	b	c	e

AS14 $S = \{e, a, b, c, d, f\}$

*	e	a	b	c	d	f
e	e	a	b	c	d	f
a	a	b	e	d	f	c
b	b	e	a	f	c	d
c	c	f	d	e	b	a
d	d	c	f	a	e	b
f	f	d	c	b	a	e

AS15 $S = \{a, b, c, d, e\}$

*	a	b	c	d	e
a	b	c	d	e	a
b	e	d	a	c	b
c	d	a	e	b	c
d	c	e	b	a	d
e	a	b	c	d	e

6.3 Example In the algebraic system AS15, find:

(a) $a * a$ **(b)** $a * b$ **(c)** $b * a$ **(d)** $a * (a * b)$

SOLUTION From the table above:

(a) $a * a = b$ **(b)** $a * b = c$

(c) $b * a = e$ **(d)** $a * (a * b) = a * c = d$ □

6.4 DEFINITION

ASSOCIATIVE BINARY OPERATION A binary operation $*:S \times S \to S$ is said to be **associative** if

$$x * (y * z) = (x * y) * z$$

holds for every x, y, and z in S.

If $*$ is associative, we say that $(S, *)$ is an associative algebraic system. Of the fifteen examples given, all are associative except for AS3, AS6, AS7, and AS15.

6.5 Example Show that the algebraic system $(\mathbb{R}, *)$,

$$x * y = x - y,$$

given in example AS3, is not associative.

SOLUTION We must find real numbers x, y, and z such that

$$x * (y * z) \neq (x * y) * z,$$

that is,

$$x - (y - z) \neq (x - y) - z.$$

Let us investigate the question of exactly when

$$x - (y - z) = (x - y) - z.$$

The latter equation is equivalent to

$$x - y + z = x - y - z,$$

or

$$z = -z.$$

The only real number z for which $z = -z$ is $z = 0$; hence, unless $z = 0$, we have

$$x - (y - z) \neq (x - y) - z.$$

In particular, for (say) $x = y = z = 1$, we have

$$1 - (1 - 1) \neq (1 - 1) - 1.$$ □

If S is an algebraic system and $\#S = n$ is finite, then you can write n^3 different equations of the form

$$x * (y * z) = (x * y) * z$$

with x, y, and z in S. Therefore, to check associativity directly from the definition, you would have to verify n^3 equations. Needless to say, if n is larger than 2 or 3, this could require considerable effort. For instance, to verify by direct computation that algebraic system AS13 is associative would require verifying $4^3 = 64$ equations.

6.6 Example For the algebraic system AS13, verify by direct calculation that

$$x * (y * z) = (x * y) * z$$

for the following cases:

(a) $x = a, y = b, z = c$
(b) $x = b, y = c, z = c$

SOLUTION **(a)** From the operation table for AS13,

$$a * (b * c) = a * a = e \quad \text{and} \quad (a * b) * c = c * c = e.$$

(b) $b * (c * c) = b * e = b$ and $(b * c) * c = a * c = b$. □

To complete the verification (by direct calculation) that AS13 is associative would require checking 62 additional cases.

By bringing some theory to bear, it is often possible to shorten the labor of checking whether an algebraic system is associative. For instance, because function composition is associative (Theorem 5.7), it follows that both AS8 and AS9 are associative algebraic systems. Also, it is not difficult to write a computer program to check the associativity of algebraic systems.

6.7 DEFINITION **COMMUTATIVE BINARY OPERATION** A binary operation $*: S \times S \to S$ is said to be **commutative** if

$$x * y = y * x$$

holds for every x and y in S.

If $*$ is commutative, we say that $(S, *)$ is a commutative algebraic system. Of the fifteen examples given above, all are commutative except for AS3, AS6, AS7, AS8, AS9, AS14, and AS15. It is not difficult to verify that an algebraic system is commutative by examining its operation table: You merely check that the table is symmetric about the diagonal going from the upper left corner to the lower right corner. For instance, AS13 is commutative but AS14 is not.

6.8 DEFINITION **NEUTRAL ELEMENT** An element $e \in S$ is said to be a **neutral element** for the algebraic system $(S, *)$ provided that, for every $x \in S$,

$$e * x = x \quad \text{and} \quad x * e = x.$$

A neutral element is also called a **unity element**, a **unit element**, or an **identity element**. Of the fifteen examples given above, all have neutral elements except for AS3, AS6, and AS7. It is not difficult to check for the existence of a neutral element by examining the operation table for an algebraic system. In fact, e is a neutral element for S if and only if the entries in the column labeled e reproduce the row labels and the elements in the row labeled e reproduce the column labels, as shown below.

$*$	a	b	$\cdots$	e	$\cdots$	y	z
a				a			
b				b			
$\vdots$				$\vdots$			
e	a	b	$\cdots$	e	$\cdots$	y	z
$\vdots$				$\vdots$			
y				y			
z				z			

For instance, in AS13, AS14, and AS15, the symbol e denotes a neutral element. In AS1 and AS4, the real number 0 serves as a neutral element. In AS2 and AS5, the real number 1 is a neutral element. In AS8 and AS9, the function Δ_X is effective as a neutral element. The empty set $\varnothing$ is a neutral element in AS10 and AS12, and the set X is a neutral element in AS11.

According to the following theorem, if an algebraic system has a neutral element, it has exactly one such; and, therefore, we are entitled to speak of *the* neutral element for the system.

6.9 THEOREM

UNIQUENESS OF THE NEUTRAL ELEMENT If an algebraic system $(S, *)$ has a neutral element, then this neutral element is unique.

PROOF Suppose that e and f are two (possibly different) neutral elements for S. Then, since e is a neutral element for S, and since $f \in S$, we have

$$e * f = f.$$

But, since f is a neutral element for S, and since $e \in S$, we also have

$$e * f = e.$$

Therefore, $e = f$. □

6.10 DEFINITION

SEMIGROUP AND MONOID An associative algebraic system $(S, *)$ is called a **semigroup**. A semigroup with a neutral element is called a **monoid**.

Of the fifteen examples given above, all except AS3, AS6, AS7, and AS15 are monoids. We leave it as an exercise for you to give an example of a semigroup that has no neutral element and therefore fails to be a monoid (Problem 28).

If S is a semigroup, then the identity

$$x * (y * z) = (x * y) * z$$

holds for all choices of x, y, and z in S; that is, *the result of a composition of three elements of a semigroup is independent of the way in which they are grouped.* This leads us to define

$$x * y * z = (x * y) * z,$$

or, what is the same thing,

$$x * y * z = x * (y * z).$$

Similarly, if x, y, z, and w are elements of S, we define

$$x * y * z * w = (x * y * z) * w.$$

There are five different ways in which the four elements x, y, z, and w can be grouped under the binary composition $*$ (Problem 24); for instance,

$$(x * y) * (z * w) \quad \text{and} \quad (x * (y * z)) * w$$

are two of the ways. It is not difficult to show (Problem 26) that all five of these different groupings produce the same result—namely, $x * y * z * w$. If $x_1, x_2, x_3, \ldots, x_n$ are elements of S, then, proceeding by induction, we define the ***n*-fold composition** $x_1 * x_2 * x_3 * \cdots * x_n$ by

$$x_1 * x_2 * x_3 * \cdots * x_n = (x_1 * x_2 * x_3 * \cdots * x_{n-1}) * x_n.$$

Using the associative law over and over again, you can see that *in a semigroup S, the result of a composition of any finite number of elements is independent of the way in which they are grouped.* Note, however, that, unless S is commutative, the result of a composition

$$x_1 * x_2 * x_3 * \cdots * x_n$$

might very well depend on the *order* in which the elements appear.

6.11 Example Let S be a semigroup, and let x, y, and z be elements in S. If S is commutative, show that

(a) $x * y * z = y * x * z$
(b) $x * y * z = z * y * x$

SOLUTION **(a)** $x * y * z = (x * y) * z = (y * x) * z = y * x * z$
(b) $x * y * z = (x * y) * z = z * (x * y) = z * (y * x)$
$= (z * y) * x = z * y * x$ □

Arguing as in Example 6.11, and using the associative and commutative laws over and over again, you can see that *in a commutative semigroup, the value of an n-fold composition* $x_1 * x_2 * x_3 * \cdots * x_n$ *is unaffected by any rearrangement of the order in which the elements appear.*

Consider the following two algebraic systems $(S, *)$ and $(T, \star)$:

AS16 $S = \{a, b, e\}$

$*$	a	b	e
a	b	e	a
b	e	a	b
e	a	b	e

AS17 $T = \{\alpha, \beta, \varepsilon\}$

$\star$	α	β	ε
α	β	ε	α
β	ε	α	β
ε	α	β	ε

Although the elements that form AS16 are not the same as the elements that form AS17, there is an obvious sense in which these two algebraic systems have the same structure. Indeed, we can set up a bijection

$a \mapsto \alpha$

$b \mapsto \beta$

$e \mapsto \varepsilon$

between S and T in such a way that *corresponding elements compose in corresponding ways.* If we call this bijection

$$\Phi: S \to T,$$

so that

$$\Phi(a) = \alpha, \quad \Phi(b) = \beta, \quad \text{and} \quad \Phi(e) = \varepsilon,$$

then the fact that corresponding elements compose in corresponding ways is expressed by the equation

$$\Phi(x * y) = \Phi(x) \star \Phi(y),$$

which holds for all elements x and y in S.

More generally, we have the following definition:

6.12 DEFINITION

ISOMORPHISM If $(S, *)$ and $(T, \star)$ are algebraic systems, then a bijection

$$\Phi: S \to T$$

is called an **isomorphism** of S onto T if

$$\Phi(x * y) = \Phi(x) \star \Phi(y)$$

holds for all elements x and y in S.

If there exists an isomorphism of S onto T, then we say that the two algebraic systems S and T are **isomorphic**. The word *isomorphic* comes from the Greek *iso*, meaning *equal* or *alike*, and *morphe*, meaning *form* or *structure*.

6.13 Example Let $(\mathbb{R}, +)$ be the algebraic system of all real numbers under the operation $+$ of ordinary addition, and let $(P, \cdot)$ be the algebraic system of all positive real numbers under the operation $\cdot$ of ordinary multiplication. Show that these two systems are isomorphic.

SOLUTION The exponential function

$$x \mapsto e^x$$

is a bijection from $\mathbb{R}$ onto P and

$$e^{x+y} = e^x \cdot e^y$$

holds for all x and y in $\mathbb{R}$. Therefore, the exponential function is an isomorphism from $\mathbb{R}$ onto P. □

Example 6.13 shows that, as *abstract algebraic systems*, the real numbers under addition and the positive real numbers under multiplication *have the same algebraic structure.*

Another example of an isomorphism is obtained as follows: In AS9, take $X = \{1, 2, 3\}$, so that $S = \mathscr{B}(X)$ denotes the set of all bijections from X to X, that is, all *permutations* of the set X. There are $3! = 6$ such permutations, namely,

$$\varepsilon, \alpha, \beta, \gamma, \delta, \phi,$$

where, as sets of ordered pairs,

$$\varepsilon = \{(1, 1), (2, 2), (3, 3)\} \qquad \alpha = \{(1, 2), (2, 3), (3, 1)\}$$
$$\beta = \{(1, 3), (2, 1), (3, 2)\} \qquad \gamma = \{(1, 1), (2, 3), (3, 2)\}$$
$$\delta = \{(1, 2), (2, 1), (3, 3)\} \qquad \phi = \{(1, 3), (2, 2), (3, 1)\}$$

The identity permutation ε serves as a neutral element for $\mathscr{B}(X)$; and, since function composition is associative, $\mathscr{B}(X)$ is a monoid. By direct calculation (Problem 22), you can verify the following operation table for the algebraic system $\mathscr{B}(X)$:

$*$	ε	α	β	γ	δ	ϕ
ε	ε	α	β	γ	δ	ϕ
α	α	β	ε	δ	ϕ	γ
β	β	ε	α	ϕ	γ	δ
γ	γ	ϕ	δ	ε	β	α
δ	δ	γ	ϕ	α	ε	β
λ	ϕ	δ	γ	β	α	ε

A glance at the operation table for AS14 shows that the mapping

$$\Phi:\mathscr{B}(X) \to S$$

given by

$$\varepsilon \mapsto e$$
$$\alpha \mapsto a$$
$$\beta \mapsto b$$
$$\gamma \mapsto c$$
$$\delta \mapsto d$$
$$\phi \mapsto f$$

establishes an isomorphism from the monoid $\mathscr{B}(X)$ onto the algebraic system AS14.

If two algebraic systems are isomorphic, then they share the same algebraic properties. For instance, if one of the systems is associative, so is the other (Problem 33); if one is commutative, so is the other (Problem 34); if one has a neutral element, so does the other (Problem 35); and so forth. In particular, since the monoid $\mathscr{B}(X)$ discussed above is isomorphic to the algebraic system AS14, it follows that AS14 is also a monoid. This is one of the things we had in mind when we said that by "bringing some theory to bear" it is possible to shorten the labor of proving that an algebraic system is associative.

6.14 THEOREM

THE INVERSE OF AN ISOMORPHISM Let $(S, *)$ and $(T, \star)$ be two algebraic systems and let

$$\Phi:S \to T$$

be an isomorphism of S onto T. Then

$$\Phi^{-1}:T \to S$$

is an isomorphism of T onto S.

PROOF By Part (ii) of Theorem 4.18, the inverse of a bijection is a bijection; hence,

$$\Phi^{-1}:T \to S$$

is a bijection. Let a and b be any two elements of T. We must prove that

$$\Phi^{-1}(a \star b) = \Phi^{-1}(a) * \Phi^{-1}(b).$$

Let

$$a' = \Phi^{-1}(a) \quad \text{and} \quad b' = \Phi^{-1}(b).$$

Then,

$$a = \Phi(a') \quad \text{and} \quad b = \Phi(b').$$

Because Φ is an isomorphism, we have

$$\Phi(a' * b') = \Phi(a') \star \Phi(a') = a \star b,$$

that is,

$$\Phi^{-1}(a \star b) = a' * b' = \Phi^{-1}(a) * \Phi^{-1}(b).$$

□

PROBLEM SET 2.6

1. In the algebraic system AS1, find:
(a) $1 * 3$ (b) $3 * 1$ (c) $2 * (-2)$ (d) $(-1) * 0$
(e) $5 * (5 * 5)$ (f) $a * (a * b)$ (g) $(a * a) * b$ (h) $0 * a$

2. In the algebraic system AS4, find:
(a) $1 * 3$ (b) $3 * 1$ (c) $5 * (-5)$ (d) $4 * 4$
(e) $3 * (3 * 4)$ (f) $(3 * 3) * 4$ (g) $0 * a$ (h) $1 * a$

3. In the algebraic system AS13, find:
(a) $a * a$ (b) $a * b$ (c) $b * a$
(d) $e * c$ (e) $c * (b * e)$ (f) $(c * b) * e$

4. Let $S = \{a, b, c\}$, and define a binary operation $*$ on S by

$$x * y = \begin{cases} z \in S, & \text{such that } x \neq z \text{ and } y \neq z \text{ if } x \neq y \\ x, & \text{if } x = y \end{cases}$$

for all $x, y \in S$.
(a) Write out the operation table for $(S, *)$.
(b) Is $(S, *)$ an associative algebraic system?
(c) Is $(S, *)$ a commutative algebraic system?
(d) Does $(S, *)$ have a neutral element? If so, what is it?

5. In the algebraic system AS14, find:
(a) $a * b$ (b) $b * a$ (c) $b * f$
(d) $d * d$ (e) $a * (a * c)$ (f) $(a * a) * c$

6. Let $S = [0, 1]$ be the closed unit interval in $\mathbb{R}$, and define

$$x * y = (x + y + |x - y|)/2$$

for $x, y \in S$.
(a) Prove that $(S, *)$ is an algebraic system.
(b) Prove that $(S, *)$ is associative.
(c) Prove that $(S, *)$ is commutative.
(d) Does $(S, *)$ have a neutral element? If so, what is it?

7. In each case, determine whether or not $*$, as given by the indicated formula for $x, y \in S$, is a binary operation on the set S. Give reasons for your answers.
(a) $S =$ all the negative integers, $x * y = 4x - y^2$
(b) $S =$ all positive integers, $x * y = 15x + y$
(c) $S =$ all positive integers, $x * y = \sqrt{x^2 + y^2}$
(d) $S =$ all real numbers, $x * y = (x + y)/(1 + x^2 + y^2)$
(e) $S =$ all real numbers, $x * y = y^x$

8. If S is a finite set with $\# S = n$, show that there are precisely n^{n^2} different binary operations $*$ on S.

9. Show that the algebraic system AS6 is not associative.

10. Show that the algebraic system AS7 is not associative. [Hint: Use the vector-triple-product identities from vector algebra.]

11. Show that the algebraic system AS10 is associative.

12. Show that the algebraic system AS12 is associative. [Hint: See Problem 21 and Part (f) of Problem 22 in Problem Set 1.5.]

13. Prove that the algebraic system AS4 is associative by calculating both $x * (y * z)$ and $(x * y) * z$ and showing that they are equal for all real numbers x, y, and z.

14. Show that the algebraic system AS15 is not associative by finding three elements x, y, and z such that $x * (y * z) \neq (x * y) * z$.

15. For the algebraic system AS14, verify by direct calculation that $x * (y * z) = (x * y) * z$ for the following cases:
(a) $x = a, y = b, z = c$ (b) $x = f, y = f, z = b$

16. Let $(S, *)$ be an algebraic system with a neutral element e. If $x, y, z \in S$ and if at least one of the elements x, y, z is equal to e, show that $x * (y * z) = (x * y) * z$.

17. Let $S = \{1, 2, 3, 4\}$ and define $*$ by $x * y = (x + y - |x - y|)/2$ for $x, y \in S$.
(a) Show that $*$ is a binary operation on S.
(b) Write out the operation table for $(S, *)$.
(c) Show that $(S, *)$ is a commutative algebraic system.
(d) Show that $(S, *)$ has a neutral element.

18. If $(S, *)$ is an algebraic system with a neutral element e, and if S is finite with $\# S = n$, use the result of Problem 16 to show that it is necessary only to check $(n - 1)^3$ equations of the form $x * (y * z) = (x * y) * z$ to verify that $(S, *)$ is associative.

19. In Problem 17, prove that the algebraic system $(S, *)$ is associative.

20. Give an example of an algebraic system $(S, *)$ containing an element a such that $a * (a * a) \neq (a * a) * a$.

21. If $X = \{a, b\}$ and $\mathscr{F}(X)$ denotes the set of all functions $f: X \to X$, then $\#\mathscr{F}(X) = 4$. Denote the four elements of $\mathscr{F}(X)$ by e, f, g, and h, with $e = \Delta_X$, and write out the operation table for the monoid $(\mathscr{F}(X), \circ)$, where $\circ$ denotes function composition.

22. By direct calculation, verify the operation table for the algebraic system $\mathscr{B}(X)$ on page 151.

23. If $(S, *)$ is an algebraic system and S is a finite set, then $\#S$ is called the **order** of $(S, *)$. If X is a finite set with $\#X = n$, find:
(a) The order of AS8 (b) The order of AS9
(c) The order of AS10, AS11, and AS12

24. On page 149, we showed two ways in which the four elements x, y, z, and w can be grouped under a binary composition $*$ without changing the order in which they occur. Find the remaining three ways in which these four elements can be grouped under the binary composition $*$ without changing this order.

25. Let the binary operation $*$ be defined on the set $\mathbb{R}$ of all real numbers by $x * y = x + y + xy$ for $x, y \in \mathbb{R}$. Prove that $(\mathbb{R}, *)$ is a commutative monoid.

26. If $*$ is associative in Problem 24, prove that all five of the groupings produce the same result, namely, $x * y * z * w$.

27. Let S be a nonempty set, and define a binary operation $*$ on S by $x * y = x$ for all $x, y \in S$.
(a) Prove that $(S, *)$ is a semigroup.
(b) Is $(S, *)$ a monoid? Why or why not?

28. Give an example of a semigroup that is not a monoid.

29. Let S be a nonempty set, and let a be a fixed element of S. Define a binary operation $*$ on S by $x * y = a$ for all $x, y \in S$.
(a) Prove that $(S, *)$ is a semigroup.
(b) Is $(S, *)$ a monoid? Why or why not?

30. Let A, B, C, D, E, and K be six fixed real numbers. Define a binary composition $*$ on the set $\mathbb{R}$ of all real numbers by

$$x * y = Ax^2 + Bxy + Cy^2 + Dx + Ey + K$$

for all $x, y \in \mathbb{R}$. Find all possible values of A, B, C, D, E, and K for which the system $(\mathbb{R}, *)$ is a semigroup.

31. In Example 6.13, we showed that the exponential function $x \mapsto e^x$ provides an isomorphism of the algebraic system $(\mathbb{R}, +)$ of real numbers under addition onto the algebraic system $(P, \cdot)$ of positive real numbers under multiplication. Find the inverse of this isomorphism.

32. Is the isomorphism $x \mapsto e^x$ of Example 6.13 the *only* isomorphism of $(\mathbb{R}, +)$ onto $(P, \cdot)$?

In Problems 33–36, let $(S, *)$ and $(T, \star)$ be two algebraic systems, and suppose that $\Phi: S \to T$ is an isomorphism of S onto T.

33. If $(S, *)$ is associative, prove that $(T, \star)$ is associative.

34. If $(S, *)$ is commutative, prove that $(T, \star)$ is commutative.

35. If $(S, *)$ has a neutral element e, prove that $\Phi(e)$ is a neutral element for $(T, \star)$.

36. If $(S, *)$ is a monoid, prove that $(T, \star)$ is a monoid.

37. In the algebraic system AS12, let $X = \{1, 2\}$. Show that the resulting algebraic system is isomorphic to the algebraic system AS13.

*7 MORE ABOUT FUNCTIONS AND MAPPINGS

We devote this section to a brief indication of the interplay among functions, sets, and equivalence relations.

7.1 DEFINITION **IMAGE AND INVERSE IMAGE OF A SET** Let f be a function, and let M be a set. We define the **image of M under f**, in symbols $f(M)$, and the **inverse image of M under f**, in symbols $f^{-1}(M)$, as follows:

(i) $f(M) = \{y \in \text{range}(f) \mid (\exists x)(x \in M \cap \text{dom}(f) \wedge y = f(x))\}$.
(ii) $f^{-1}(M) = \{x \in \text{dom}(f) \mid f(x) \in M\}$.

If $M \subseteq \text{dom}(f)$, then the definition of $f(M)$ is often abbreviated as

$$f(M) = \{f(x) \mid x \in M\},$$

so that *$f(M)$ is the set of all elements of the form $f(x)$ as x runs through the set M.*

7.2 Example If f is a function and if M and N are sets, show that

$$f(M \cap N) \subseteq f(M) \cap f(N).$$

SOLUTION We use the pick-a-point procedure. First, suppose that $y \in f(M \cap N)$. By Part (i) of Definition 7.1, there exists

$$x \in M \cap N \cap \text{dom}(f)$$

such that

$$y = f(x).$$

Because $x \in M \cap N \cap \operatorname{dom}(f)$ and $y = f(x)$, it follows that

$$x \in M \cap \operatorname{dom}(f) \wedge y = f(x);$$

hence, $y \in f(M)$. Likewise,

$$x \in N \cap \operatorname{dom}(f) \wedge y = f(x);$$

hence, $y \in f(N)$. Therefore,

$$y \in f(M) \cap f(N).$$ □

The following example shows that the inclusion obtained in Example 7.2 cannot, in general, be strengthened to an equality.

7.3 Example Give an example of a function f and two sets M and N, contained in the domain of f, such that

$$f(M) \cap f(N) \not\subseteq f(M \cap N).$$

SOLUTION Let $f = \{(1, 1), (2, 1)\}$, $M = \{1\}$, and $N = \{2\}$. Then,

$$f(M) = f(\{1\}) = \{f(1)\} = \{1\}$$

and

$$f(N) = f(\{2\}) = \{f(2)\} = \{1\},$$

so that

$$f(M) \cap f(N) = \{1\} \cap \{1\} = \{1\}.$$

However,

$$f(M \cap N) = f(\{1\} \cap \{2\}) = f(\varnothing) = \varnothing$$

(see Problem 3c); hence, in this case,

$$f(M) \cap f(N) \not\subseteq f(M \cap N)$$

because

$$\{1\} \not\subseteq \varnothing.$$ □

Additional features of images and inverse images of sets under functions are developed in Problems 5–17, Problem 21, and Problem 22.

7.4 DEFINITION

IMPLICIT FUNCTION Let R be a relation. We say that a function f is **implicit** in R if, as sets of ordered pairs,

$$f \subseteq R.$$

In other words, a function f is implicit in the relation R if and only if, for every $x \in \text{dom}(f)$,

$$(x, f(x)) \in R.$$

7.5 Example Find a function f that is implicit in the relation

$$x^2 + y^2 = 1.$$

SOLUTION To begin with, notice that $x^2 + y^2 = 1$ is an *equation*, not a *set of ordered pairs R*. The relation being referred to is obviously

$$R = \{(x, y) \mid x^2 + y^2 = 1\}.$$

The function f with domain

$$X = \{x \mid -1 \leq x \leq 1\}$$

and given by

$$f(x) = \sqrt{1 - x^2}$$

is implicit in R because, for every $x \in X$,

$$x^2 + (f(x))^2 = 1. \qquad \square$$

For a given relation, there may be many different implicit functions. For instance, the function g given by

$$g(x) = \begin{cases} -\sqrt{1 - x^2}, & \text{if } -1 \leq x < 0 \\ \sqrt{1 - x^2}, & \text{if } 0 \leq x \leq 1 \end{cases}$$

is also implicit in the relation $x^2 + y^2 = 1$ of Example 7.5. Usually, one is interested in finding implicit functions that are "well behaved" in some appropriate sense, such as being continuous or differentiable. Note that f in Example 7.5 is continuous, but g, given above, is not. Recall from calculus the technique of *implicit differentiation* for finding the derivative of an implicit function (provided, of course, that the implicit function is differentiable).

Implicit functions often arise in connection with the problem of inversion. Thus, if f is a function, then f^{-1} is a relation but not necessarily a function. However, f^{-1} might contain an interesting implicit function. For instance, consider the function

$$f(x) = \sin x.$$

Then f^{-1} is a relation but not a function (why?). However, the function g with domain

$$X = \{x \mid -1 \leq x \leq 1\}$$

given by the rule

$$y = g(x) \Leftrightarrow x = \sin y \wedge (-\pi/2 \leq y \leq \pi/2)$$

is a continuous function implicit in the relation f^{-1}. Notice that g is just the familiar Arcsin function from calculus.

Implicit functions for equivalence relations are of special interest, particularly when they satisfy the additional condition in the following definition.

7.6 DEFINITION

CANONICAL FORM Let E be an equivalence relation on the set X. A function f with $\text{dom}(f) = X$ is called a **canonical form** for E if the following two conditions are satisfied:

(i) $f \subseteq E$.
(ii) $xEy \Rightarrow f(x) = f(y)$, for all $x, y \in X$.

A canonical form f for an equivalence relation E on X has the effect of selecting a unique representative element $f(x)$ from each equivalence class $[x]$ of the partition X/E determined by E. Thus, for $x, y \in X$, you can check whether or not xEy holds by checking whether or not $f(x) = f(y)$.

7.7 Example Let $\mathbb{R}$ denote the set of all real numbers, and let E be the equivalence relation defined on $\mathbb{R}$ by

$$E = \{(x, y) \mid x, y \in \mathbb{R} \text{ and } x - y \text{ is an integer}\}.$$

Find a canonical form for E.

SOLUTION For each real number x, let $[\![x]\!]$ denote the greatest integer that is less than or equal to x. Define

$$f:\mathbb{R} \to \mathbb{R}$$

by

$$f(x) = x - [\![x]\!].$$

To establish Condition (i) of Definition 7.6, we must show that, for $x \in \mathbb{R}$,

$$xEf(x),$$

that is, $x - f(x)$ is an integer. But, $x - f(x) = [\![x]\!]$, which is indeed an integer.

To verify Condition (ii), suppose that xEy holds, so that

$$x - y = n,$$

where n is an integer. We must prove that $f(x) = f(y)$. We have

$$x = y + n.$$

Let $m = [\![y]\!]$, so that m is an integer and

$$m \le y < m + 1.$$

Then,

$$m + n \le y + n < m + n + 1,$$

that is,

$$m + n \le x < (m + n) + 1.$$

Because $m + n$ is an integer, it follows from the last set of inequalities that $m + n$ is the greatest integer less than or equal to x; so

$$[\![x]\!] = m + n.$$

Consequently,

$$\begin{aligned} f(x) &= x - [\![x]\!] = x - (m + n) = x - n - m \\ &= x - n - [\![y]\!] = y - [\![y]\!] \\ &= f(y). \end{aligned}$$

Therefore, Condition (ii) holds, and f is a canonical form for the equivalence relation E. (Also see Problem 28.) □

Condition (ii) of Definition 7.6 suggests the following:

7.8 DEFINITION

INVARIANT FOR AN EQUIVALENCE RELATION Let E be an equivalence relation on the set X. A function f is said to be an **invariant** for E if $\text{dom}(f) = X$ and, for all $x, y \in X$,

$$xEy \Rightarrow f(x) = f(y).$$

This definition can be paraphrased as follows: *An invariant for an equivalence relation is a function that does not discriminate between equivalent elements.*

7.9 Example Find an invariant (other than the canonical form already found) for the equivalence relation E in Example 7.7.

SOLUTION The function g with domain $\mathbb{R}$ defined by

$$g(x) = \sin(2\pi x)$$

is an invariant for E. Indeed, suppose that xEy, so that

$$x - y = n,$$

where n is an integer. Then,

$$g(x) = \sin(2\pi x) = \sin(2\pi x - 2\pi n) = \sin(2\pi(x - n)) = \sin(2\pi y)$$
$$= g(y).$$ □

Invariants for equivalence relations help us to understand and deal with these relations. In this regard, complete sets of invariants, as defined below, are especially useful.

7.10 DEFINITION

COMPLETE SET OF INVARIANTS Let E be an equivalence relation on X, and let $\mathscr{F}$ be a nonempty set of functions, each of which is an invariant for E. We say that $\mathscr{F}$ is a **complete set of invariants** for E if the condition that

$$f(x) = f(y)$$

for every $x, y \in X$ and every $f \in \mathscr{F}$ implies that

$$xEy.$$

If $\mathscr{F}$ is a complete set of invariants for the equivalence relation E on the set X, then, for $x, y \in X$, you can use $\mathscr{F}$ to tell whether or not xEy holds: Indeed, xEy holds if and only if $f(x) = f(y)$ holds for every function $f \in \mathscr{F}$.

7.11 Example Let E be the equivalence relation of Example 7.7. Find a complete set $\mathscr{F}$ of invariants for E that contains the function g already found in Example 7.9.

SOLUTION We already know that the function g defined on $\mathbb{R}$ by

$$g(x) = \sin(2\pi x)$$

is an invariant for E. So is the function h defined on $\mathbb{R}$ by

$$h(x) = \cos(2\pi x),$$

and, taken together, we claim that these two functions form a complete set

$$\mathscr{F} = \{g, h\}$$

of invariants for E. Indeed, if $x, y \in \mathbb{R}$ and

$$\sin(2\pi x) = \sin(2\pi y) \quad \text{and} \quad \cos(2\pi x) = \cos(2\pi y),$$

then x and y must differ by an integer. □

Now we are going to show that any function f gives rise to an equivalence relation for which $\{f\}$ is a complete set of invariants. The appropriate relation is defined as follows:

7.12 DEFINITION

EQUIVALENCE RELATION DETERMINED BY *f* Let f be any function, and let $X = \text{dom}(f)$. Define the relation E_f on X by

$$E_f = \{(x, y) \mid x, y \in X \land f(x) = f(y)\}.$$

7.13 Example Let f be a function, and let $X = \text{dom}(f)$. Show that the relation E_f is an equivalence relation on X and that $\{f\}$ is a complete set of invariants for E_f.

SOLUTION For $x \in X$, we have $f(x) = f(x)$, and so E_f is reflexive on X. Suppose that $(x, y) \in E_f$. Then $x, y \in X$ and $f(x) = f(y)$, so that $y, x \in X$ and $f(y) = f(x)$; that is, $(y, x) \in E_f$. Therefore, E_f is symmetric. The transitivity of E_f follows from the fact that, if $x, y, z \in X$ with $f(x) = f(y)$ and $f(y) = f(z)$, then $f(x) = f(z)$. Consequently, E_f is an equivalence relation on X. Also, for $x, y \in X$,

$$(x, y) \in E_f \Leftrightarrow f(x) = f(y),$$

and so $\{f\}$ is a complete set of invariants for E_f. □

Given an equivalence relation E on a set X, can we always find a function f with $\text{dom}(f) = X$ such that $\{f\}$ is a complete set of invariants for E? The answer is yes, as is shown by the following definition and example.

7.14 DEFINITION

CANONICAL SURJECTION Let E be an equivalence relation on the set X. The mapping

$$f: X \to X/E$$

defined by

$$f(x) = [x]$$

for all $x \in X$ is called the **canonical surjection induced by** E.

The canonical surjection $f: X \to X/E$ induced by E simply maps each element of X onto the unique equivalence class in X/E that contains that element. Notice that $f: X \to X/E$ really is surjective.

7.15 Example Let E be an equivalence relation on the set X, and let $f:X \to X/E$ be the canonical surjection induced by E. Show that $\{f\}$ is a complete set of invariants for E.

SOLUTION For $x, y \in X$ we must show that

$$xEy \Leftrightarrow f(x) = f(y),$$

that is,

$$xEy \Leftrightarrow [x] = [y].$$

But this is an immediate consequence of Theorem 3.9 on page 108. □

PROBLEM SET 2.7

1. Let f be the function given by

$$f = \{(1, a), (2, b), (3, a), (4, c), (5, b), (6, d)\}.$$

Find:
(a) $f(\{1, 3\})$ (b) $f(\{2, 5, 6\})$ (c) $f(\{1, 2, 4, 6\})$ (d) $f^{-1}(\{a\})$
(e) $f^{-1}(\{b\})$ (f) $f^{-1}(\{b, c\})$ (g) $f^{-1}(\{c, d\})$

2. Let $f:\mathbb{R} \to \mathbb{R}$ be given by $f(x) = \sin x$. Find:
(a) $f(\{x | 0 \le x \le \pi/2\})$ (b) $f(\{x | 0 \le x \le \pi\})$
(c) $f^{-1}(\{0\})$ (d) $f^{-1}(\{-1, 1\})$

3. Let $f:X \to Y$. Show that:
(a) $f(X) = \text{range}(f)$ (b) $f^{-1}(Y) = X$
(c) $f(\varnothing) = \varnothing$ (d) $f^{-1}(\varnothing) = \varnothing$

4. Give an example of a function f and two sets M and N with $M \subseteq N \subseteq \text{dom}(f)$ and $f(N \setminus M) \nsubseteq f(N) \setminus f(M)$.

In Problems 5–17, f and g denote functions, and M and N denote sets. Show that each statement is true.

5. $f(M) = f(M \cap \text{dom}(f))$

6. $f^{-1}(N) = f^{-1}(N \cap \text{range}(f))$

7. $M \subseteq N \Rightarrow f(M) \subseteq f(N)$

8. $M \subseteq N \Rightarrow f^{-1}(M) \subseteq f^{-1}(N)$

9. $M \subseteq \text{dom}(f) \Rightarrow M \subseteq f^{-1}(f(M))$

10. $N \subseteq \text{range}(f) \Rightarrow N = f(f^{-1}(N))$

11. $f(M \cup N) = f(M) \cup f(N)$

12. $f^{-1}(M \cup N) = f^{-1}(M) \cup f^{-1}(N)$

13. $f^{-1}(M \cap N) = f^{-1}(M) \cap f^{-1}(N)$ [Cf. Examples 7.2 and 7.3.]

14. $f(M) \setminus f(N) \subseteq f(M \setminus N)$

15. $f^{-1}(M) \setminus f^{-1}(N) = f^{-1}(M \setminus N)$

16. $(f \circ g)(M) = f(g(M))$

17. $(f \circ g)^{-1}(N) = g^{-1}(f^{-1}(N))$

18. Let $f: X \to Y$.
(a) Give an example to show that it is possible to have $M \subseteq X$ and $M \neq f^{-1}(f(M))$.
(b) Show that $f: X \to Y$ is injective if and only if $M = f^{-1}(f(M))$ holds for all sets $M \subseteq X$.
(c) Give an example to show that it is possible to have $N \subseteq Y$ and $N \neq f(f^{-1}(N))$.
(d) Show that $f: X \to Y$ is surjective if and only if $N = f(f^{-1}(N))$ holds for all sets $N \subseteq Y$.

19. Suppose that $f(M) = g(M)$ holds for every set M. Show that $f = g$.

20. Suppose that $f^{-1}(N) = g^{-1}(N)$ holds for every set N. Show that $f = g$.

21. Let $f: X \to Y$, and suppose that $M \subseteq X$. Show that

$$f^{-1}(Y \setminus f(X \setminus M)) \subseteq M.$$

22. Generalize Example 7.2 and Problems 11, 12, and 13 for images and inverse images under f of indexed families of sets.

23. Let X denote the set of all nonnegative real numbers. Find three different implicit functions, each with domain X, contained in the relation $y^2 = x$.

24. Find three different implicit functions, each with domain $\mathbb{R}$ = the set of all real numbers, contained in the relation $x < y$.

25. If $g(x) = \cos x$, find a continuous function f that is implicit in g^{-1}.

26. Find two different implicit functions, each with domain $X = \{x \mid x \geq -37/4\}$, contained in the relation $7 + x + 3y = y^2$.

27. Let E be the equivalence relation defined on $\mathbb{R}$ (the set of all real numbers) by $xEy \Leftrightarrow x$ and y have the same algebraic sign. Find a canonical form f for E.

28. Verify that the relation E in Example 7.7 is an equivalence relation.

29. If E is an equivalence relation on the set X, and f is a canonical form for E, show that $f: X \to X$.

30. Explain the distinctions between a canonical form and the canonical surjection for an equivalence relation E on X.

31. Let X denote the set of all triangles in the plane, and let C be the relation defined on X by $(T_1, T_2) \in C$ if and only if triangle T_1 is congruent to triangle T_2. If $T \in X$, let $P(T)$ denote the perimeter of T, let $v(T)$ denote the smallest vertex angle of T, and let $V(T)$ denote the largest vertex angle of T.
(a) Show that P, v, and V are invariants for C.

(b) Is $\{v, V\}$ a complete set of invariants for C? Why or why not?
(c) Is $\{P, v, V\}$ a complete set of invariants for C? Why or why not?
(d) Find other invariants for C that are studied in elementary geometry.

32. In Problem 31, replace the relation C of congruence by the relation S of similarity.
(a) Show that v and V are invariants for S but that P is not.
(b) Is $\{v, V\}$ a complete set of invariants for S? Why or why not?

33. Let $\mathbb{Z}$ denote the set of integers, and define the relation E on $\mathbb{Z}$ by $xEy \Leftrightarrow x - y$ is an even integer.
(a) Show that E is an equivalence relation on $\mathbb{Z}$.
(b) Find a canonical form for E.

34. Let $\mathbb{Z}$ denote the set of integers, and let n be a fixed integer with $n > 1$. Define the relation $\equiv_n$ on $\mathbb{Z}$ by

$$p \equiv_n q \Leftrightarrow (\exists k)(k \in \mathbb{Z} \wedge p - q = kn).$$

(The relation $\equiv_n$ is called **congruence modulo *n*.**)
(a) Show that $\equiv_n$ is an equivalence relation on $\mathbb{Z}$.
(b) Show that, for each integer $p \in \mathbb{Z}$, there exists a unique integer $q \in \mathbb{Z}$ such that $p \equiv_n q$ and $0 \leq q \leq n - 1$. (The integer q is called the **reduced residue of *p* modulo *n*.**)
(c) Define $f:\mathbb{Z} \to \mathbb{Z}$ by $f(p) = q$, where q is the reduced residue of p modulo n, for each $p \in \mathbb{Z}$. Show that f is a canonical form for $\equiv_n$.

35. In Problem 33, describe in words the set $\mathbb{Z}/E$, and explain in words just how the canonical surjection $f:\mathbb{Z} \to \mathbb{Z}/E$ works.

36. In Problem 34, the set $\mathbb{Z}/\equiv_n$ is called the set of **integers modulo *n*.** For $x \in \mathbb{Z}$, denote by $[x]_n$ the equivalence class in $\mathbb{Z}/\equiv_n$ that contains x. Notice that it is quite possible to have $[x]_n = [y]_n$ with $x \neq y$. We propose to define a mapping $\phi:\mathbb{Z}/\equiv_n \to \mathbb{Z}/\equiv_n$ by

$$\phi([x]_n) = [n - x]_n$$

for all $x \in \mathbb{Z}$. Is ϕ well-defined? Why or why not?

37. Let $f:X \to Y$. Prove that f is an injection if and only if $E_f = \Delta_X$.

38. Let E be an equivalence relation on X, and let $\mathscr{F}$ be a set of invariants for E. Prove that $\mathscr{F}$ is a complete set of invariants for E if and only if

$$\bigcap_{f \in \mathscr{F}} E_f = E.$$

39. Let E be an equivalence relation on X. Prove that there exists a function f such that $E = E_f$.

40. Let X be a finite set, and let E be an equivalence relation on X. Suppose that X/E consists of k distinct equivalence classes $C_1, C_2, \ldots, C_k$, and let $m_i = \# C_i$ for $i = 1, 2, \ldots, k$. Prove that there are precisely $m_1 m_2 \cdots m_k$ different canonical forms for E.

41. Let E be an equivalence relation on X, and let f be a function with $\text{dom}(f) = X$. Prove that f is an invariant for E if and only if $E \subseteq E_f$.

42. Let $f:X \to Y$, and let $g:Y \to Z$. Define $\phi:(X \times Y) \times Z \to X \times Z$ and $\theta:X \times (Y \times Z) \to (X \times Y) \times Z$ by $\phi((x, y), z) = (x, z)$ and $\theta(x, (y, z)) = ((x, y), z)$ for $x \in X$, $y \in Y$, and $z \in Z$. Prove that

$$g \circ f = \phi((f \times Z) \cap \theta(X \times g)).$$

43. Let $f:X \to Y$, and define $\pi_1:X \times Y \to X$ and $\pi_2:X \times Y \to Y$ by $\pi_1((x, y)) = x$ and $\pi_2((x, y)) = y$ for all $x \in X$ and all $y \in Y$. Let $M \subseteq X$, and let $N \subseteq Y$. Prove:
(a) $f(M) = \pi_2(f \cap \pi_1^{-1}(M))$ (b) $f^{-1}(N) = \pi_1(f \cap \pi_2^{-1}(N))$

44. Let P denote the set of all polynomials

$$P(x, y) = Ax^2 + Bxy + Cy^2 + Dx + Ey + F$$

in the two variables x and y, where the coefficients A, B, C, D, E, and F are constant real numbers. Define

$$d:P \to \mathbb{R}, \qquad t:P \to \mathbb{R}, \quad \text{and} \quad f:P \to \mathbb{R}$$

by $d(p(x, y)) = B^2 - 4AC$, $t(p(x, y)) = A + C$, and $f(p(x, y)) = F$. Each polynomial $p(x, y) \in P$ determines a corresponding conic section

$$\{(x, y) \mid p(x, y) = 0\}$$

in the plane. Define a relation R on P by the condition that $p(x, y)Rq(x, y)$ if and only if the conic section corresponding to $p(x, y)$ can be brought into coincidence with the conic section corresponding to $q(x, y)$ by a rigid motion of the plane. (A **rigid motion** is obtained by first performing a translation and then a rotation.)
(a) By using the equations for translations and rotations of the xy plane, show that R is an equivalence relation.
(b) Prove that d, t, and f are invariants for R.
(c) Is $\{d, t, f\}$ a complete set of invariants for R?
(d) Find a canonical form for R. [Hint: Recall the standard forms for the equations of the conic sections.]

45. Let X denote the set of all ordered pairs (m, n) of integers such that $n \neq 0$. Let $\mathbb{Q}$ denote the set of all rational numbers, so that

$$\mathbb{Q} = \{m/n \mid (m, n) \in X\}.$$

Define $f:X \to \mathbb{Q}$ by $f(m, n) = m/n$ for all $(m, n) \in X$. Consider the equivalence relation E_f determined by f.
(a) Show that $(m, n)E_f(p, q) \Leftrightarrow mq = np$.
(b) Explain why the process of converting a fraction to "lowest terms" can be regarded as a canonical form for E_f.

*8

MORE ABOUT COMPOSITION: COMMUTATIVE DIAGRAMS

The following theorem shows that a mapping

$$f:X \to Y$$

can always be "factored" under composition as

$$f = i \circ g \circ p,$$

where p is a surjection, g is a bijection, and i is an injection.

8.1 THEOREM

CANONICAL FACTORIZATION OF A MAPPING Let

$$f:X \to Y$$

be a mapping, and let E_f be the equivalence relation determined by f (Definition 7.12). Denote by

$$p:X \to X/E_f$$

the canonical surjection (Definition 7.14). Let

$$i:\text{range}(f) \to Y$$

be the injective mapping defined by

$$i(y) = y$$

for all $y \in \text{range}(f)$. Then there exists a bijection

$$g:X/E_f \to \text{range}(f)$$

such that

$$f = i \circ g \circ p.$$

PROOF Each element $x \in X$ determines a unique equivalence class $[x]$ in X/E_f, and we have

$$p(x) = [x].$$

We propose to define

$$g([x]) = f(x),$$

but we must show that g is well-defined. The problem is to show that the proposed definition of $g([x])$ does not depend on the choice of the representative element x in the equivalence class $[x]$. Thus, suppose that

$x, x' \in X$ with $[x] = [x']$. We must prove that $f(x) = f(x')$. But, if $[x] = [x']$, then xE_fx'; so, by the definition of E_f, it follows that $f(x) = f(x')$. To show that

$$g: X/E_f \to \text{range}(f)$$

is surjective, let $y \in \text{range}(f)$. Then, there exists $x \in X$ such that $y = f(x)$; so

$$g([x]) = f(x) = y,$$

and it follows that g is surjective. To show that g is injective, suppose that $x, x' \in X$ with

$$g([x]) = g([x']).$$

Then, by the definition of g, $f(x) = f(x')$; so

$$[x] = [x'],$$

and, therefore, g is injective. Finally, if $x \in X$, we have

$$f(x) = g([x]) = g(p(x)) = i(g(p(x))) = (i \circ g \circ p)(x),$$

and it follows that

$$f = i \circ g \circ p. \qquad \square$$

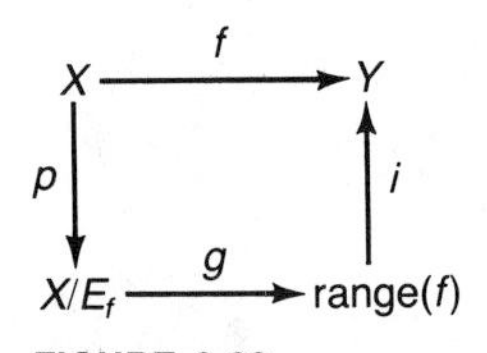

FIGURE 2-22

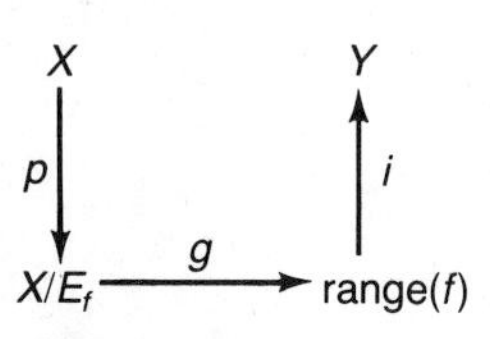

FIGURE 2-23

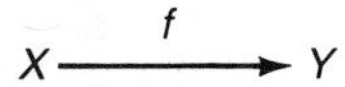

X —f→ Y

FIGURE 2-24

The canonical factorization theorem can be illustrated by the diagram in Figure 2-22. Mapping diagrams, frequently much more complicated than this, are used extensively in contemporary mathematics. By a **path** in a mapping diagram, we mean a finite sequence of "arrows" that join in a "head-to-tail" fashion. For instance, Figure 2-23 shows a path in the mapping diagram of Figure 2-22.

Notice that a "singleton path," as illustrated in Figure 2-24, is perfectly acceptable.

8.2 DEFINITION

COMMUTATIVE MAPPING DIAGRAM A mapping diagram is said to be **commutative** if, when two paths in the diagram start at the same place and end at the same place, then the composition of the mappings involved gives the same result, either way.

For instance, *the diagram in Figure 2-22 is commutative*, and the fact that it is commutative nicely summarizes the content of Theorem 8.1.

The mapping $i:\text{range}(f) \to Y$ in Figure 2-22 is called the *inclusion mapping* from range(f) into Y. More generally:

8.3 DEFINITION

INCLUSION MAPPING If X and Y are sets with $X \subseteq Y$, then the mapping

$$i:X \to Y$$

defined for every $x \in X$ by

$$i(x) = x$$

is called the **inclusion mapping** from X into Y.

Notice that the *inclusion mapping* $i:X \to Y$ *is an injection, but it is a surjection if and only if* $X = Y$ (Problem 1). Although it might seem that inclusion mappings are somewhat trivial, they turn out to be useful. We have already seen the role played by an inclusion mapping in formulating the canonical factorization of a mapping (Theorem 8.1). The following definition illustrates another use for inclusion mappings.

8.4 DEFINITION

EXTENSION OF A MAPPING Let X and Z be sets with $X \subseteq Z$, and let $i:X \to Z$ be the inclusion mapping. If

$$f:X \to Y \quad \text{and} \quad g:Z \to Y$$

are mappings, and if

$$g \circ i = f,$$

then we say that g is an **extension** of f.

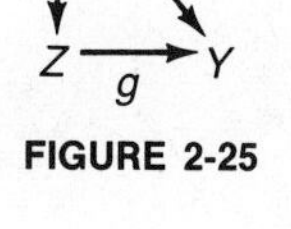

FIGURE 2-25

The condition that g be an extension of f is equivalent to the condition that the mapping diagram in Figure 2-25 is commutative. If g is an extension of f as in the figure, then, for every $x \in X$,

$$f(x) = g(i(x)),$$

that is,

$$f(x) = g(x).$$

In other words, f and g are the "same" on the domain X of f, but g is defined on the (possibly) larger domain Z. This is also expressed by saying that f is a *restriction* of g. More generally:

8.5 DEFINITION

RESTRICTION OF A FUNCTION If g is a function, and X is a set, then the **restriction** of g to the set X, in symbols $g|_X$, is defined by

$$g|_X = \{(x, y) | (x, y) \in g \wedge x \in X\}.$$

Note that the restriction $g|_X$ of a function is again a function (Problem 5).

8.6 Example By definition, a **complex number** has the form $x + iy$, where $x, y \in \mathbb{R}$ (the set of all real numbers), and i is a so-called imaginary number with the property that $i^2 = -1$. The set of all complex numbers is denoted by $\mathbb{C}$. The **complex exponential function**, denoted by **exp**, is defined for $x + iy \in \mathbb{C}$ by

$$\exp(x + iy) = e^x(\cos y + i(\sin y)).$$

Find $\exp|_{\mathbb{R}}$.

SOLUTION For $x \in \mathbb{R}$,

$$\exp|_{\mathbb{R}}(x) = \exp(x + i(0)) = e^x(\cos 0 + i(\sin 0)) = e^x.$$

In other words, the restriction of exp to $\mathbb{R}$ is just the ordinary exponential function of elementary calculus. Equivalently, the complex exponential function is an extension to $\mathbb{C}$ of the real exponential function $x \mapsto e^x$. □

If we let $h = g \circ p$ in Theorem 8.1, then we have

$$h:X \to \text{range}(f)$$

and

$$f = i \circ h.$$

Because

$$g:X/E_f \to \text{range}(f)$$

is a bijection and hence, in particular, a surjection; and because

$$p:X \to X/E_f$$

is a surjection, it follows from Part (ii) of Corollary 5.11 that

$$h:X \to X/E_f$$

is a surjection. Therefore, an arbitrary mapping

$$f:X \to Y$$

can be "factored" under composition as

$$f = i \circ h,$$

where i is an injection and h is a surjection. Can f also be factored into the form

$$f = q \circ j$$

where j is an injection and q is a surjection? The answer is given by the following theorem, the proof of which we leave as an exercise (Problem 9).

8.7 THEOREM

ALTERNATIVE FACTORIZATION OF A MAPPING Let X be a non-empty set, and let $f:X \to Y$. Define

$$j:X \to X \times Y$$

by $j(x) = (x, f(x))$ for all $x \in X$. Define

$$q:X \times Y \to Y$$

by $q((x, y)) = y$. Then j is an injection, q is a surjection, and

$$f = q \circ j,$$

so that the diagram in Figure 2-26 is commutative.

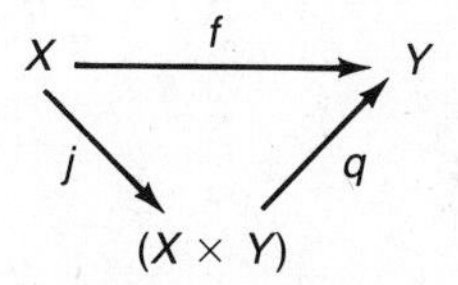

FIGURE 2-26

8.8 DEFINITION

RIGHT- AND LEFT-INVERSES Let $f:X \to Y$. A mapping $g:Y \to X$ is called a **right-inverse** for $f:X \to Y$ if

$$f \circ g = \Delta_Y.$$

A mapping $h:Y \to X$ is called a **left-inverse** for $f:X \to Y$ if

$$h \circ f = \Delta_X.$$

Mapping diagrams for a right-inverse g and for a left-inverse h of f are shown in Figure 2-27a and Figure 2-27b, respectively.

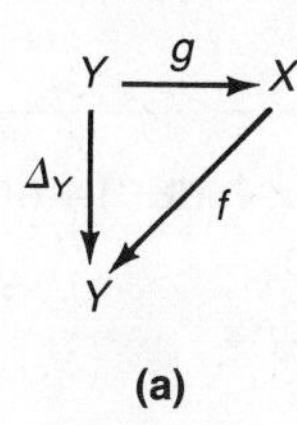

(a)

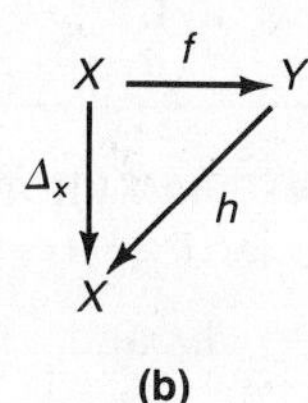

(b)

FIGURE 2-27

8.9 THEOREM

EXISTENCE OF RIGHT- AND LEFT-INVERSES

(i) $f:X \to Y$ has a right-inverse if and only if it is surjective and the equivalence relation E_f on X has a canonical form.
(ii) $f:X \to Y$ has a left-inverse if and only if it is injective.

PROOF The proof is left as an exercise (Problem 11). □

8.10 DEFINITION

DIRECT PRODUCT OF SETS Let A and B be two sets. By a **direct product** of A and B, we mean a set P and two mappings $p:P \to A$ and $q:P \to B$ with the following property: Given any set M and two mappings $f:M \to A$ and $g:M \to B$, there exists a unique mapping $h:M \to P$ such that the diagram in Figure 2-28a is commutative.

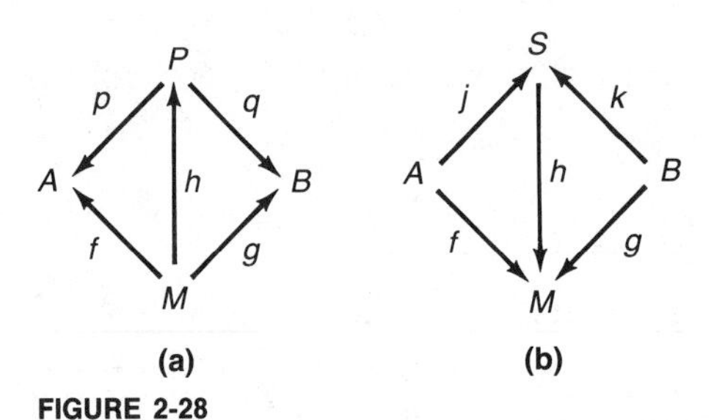

FIGURE 2-28

By reversing all the arrows in Figure 2-28a, we obtain the "dual" diagram in Figure 2-28b, which corresponds to the following definition:

8.11 DEFINITION

DIRECT SUM OF SETS Let A and B be two sets. By a **direct sum** of A and B, we mean a set S and two mappings $j:A \to S$ and $k:B \to S$ with the following property: Given any set M and two mappings $f:A \to M$ and $g:B \to M$, there exists a unique mapping $h:S \to M$ such that the diagram in Figure 2-28b is commutative.

8.12 THEOREM

THE CARTESIAN PRODUCT IS A DIRECT PRODUCT Let A and B be sets, and let $P = A \times B$. Define

$$\pi_1:P \to A \quad \text{and} \quad \pi_2:P \to B$$

by

$$\pi_1((x, y)) = x \quad \text{and} \quad \pi_2((x, y)) = y$$

for all $(x, y) \in P$. Then, the set P and the two mappings π_1 and π_2 constitute a direct product of the sets A and B.

PROOF The proof is left as an exercise (Problem 12). □

The mappings π_1 and π_2 in Theorem 8.12 are called the **canonical projection mappings** from the Cartesian product $P = A \times B$ to the "factors" A and B. Thus, Theorem 8.12 says that *the Cartesian product of two sets, together with the two canonical projection mappings, constitutes a direct product of the sets.* The following theorem can be regarded as a converse of this result:

8.13 THEOREM

DIRECT PRODUCTS AND CARTESIAN PRODUCTS Let P and the two mappings $p:P \to A$ and $q:P \to B$ constitute a direct product of the sets A and B. Let $\pi_1:A \times B \to A$ and $\pi_2:A \times B \to B$ be the canonical projection mappings. Then, there exists a unique bijection

$$k:P \to A \times B$$

such that $p = \pi_1 \circ k$ and $q = \pi_2 \circ k$.

PROOF The proof is left as an exercise (Problem 13). □

Theorems 8.12 and 8.13 show that the direct product of two sets can be realized or represented by the Cartesian product of the sets. A similar realization for the direct sum of two sets is provided by the following definition and theorems:

8.14 DEFINITION

DISJOINT UNION Let A and B be two sets. By the **disjoint union** of A and B, we mean the set $A \uplus B$ defined by

$$A \uplus B = \{(a, 1) | a \in A\} \cup \{(b, 2) | b \in B\}.$$

Define the **canonical injection mappings**

$$i_1:A \to A \uplus B \quad \text{and} \quad i_2:B \to A \uplus B$$

by

$$i_1(a) = (a, 1) \quad \text{and} \quad i_2(b) = (b, 2)$$

for all $a \in A$ and all $b \in B$.

8.15 THEOREM

DISJOINT UNION Let $A \uplus B$ be the disjoint union of the sets A and B. Then, the canonical injection mappings

$$i_1:A \to A \uplus B \quad \text{and} \quad i_2:B \to A \uplus B$$

are, in fact, injections. Moreover,

$$\text{range}(i_1) \cap \text{range}(i_2) = \varnothing$$

and

$$\text{range}(i_1) \cup \text{range}(i_2) = A \uplus B.$$

PROOF The proof is left as an exercise (Problem 15a). □

8.16 THEOREM

THE DISJOINT UNION IS A DIRECT SUM Let A and B be sets, and let $S = A \uplus B$. Then S, together with the two canonical injection mappings $i_1 : A \to S$ and $i_2 : B \to S$, constitute a direct sum of A and B.

PROOF The proof is left as an exercise (Problem 15b). □

8.17 THEOREM

DIRECT SUMS AND DISJOINT UNIONS Let S and the two mappings $j : A \to S$ and $k : B \to S$ constitute a direct sum of the sets A and B. Let $i_1 : A \to A \uplus B$ and $i_2 : B \to A \uplus B$ be the canonical injection mappings. Then, there exists a unique bijection

$$s : A \uplus B \to S$$

such that $j = s \circ i_1$ and $k = s \circ i_2$.

PROOF The proof is left as an exercise (Problem 16). □

PROBLEM SET 2.8

1. Let X and Y be sets with $X \subseteq Y$, and let $i : X \to Y$ be the inclusion mapping.
   (a) Prove that $i : X \to Y$ is an injection.
   (b) Prove that $i : X \to Y$ is a surjection if and only if $X = Y$.
2. Let $\mathbb{R}$ and $\mathbb{C}$ denote the real and complex numbers, respectively, and let $f : \mathbb{R} \to \mathbb{C}$ be defined by $f(x) = \sin x$ for all $x \in \mathbb{R}$. Find an extension $g : \mathbb{C} \to \mathbb{C}$ of f that is "natural" in the sense that it retains the important properties of the sine function in elementary calculus.
3. Let $\mathbb{C}$ denote the set of all complex numbers, and let $f : \mathbb{C} \to \mathbb{C}$ be defined by

$$f(x + iy) = \frac{e^{-y}(\cos x + i \sin x) + e^{y}(\cos x - i \sin x)}{2}$$

   for all $x, y \in \mathbb{R}$. Let $g = f|_{\mathbb{R}}$, where $\mathbb{R}$ denotes the set of all real numbers. Identify g as one of the functions studied in elementary calculus.
4. Repeat Problem 2, but start with the function $f(x) = \sinh x$.
5. If g is a function and X is a set, prove that $g|_X$ is a function.
6. Show that $f : X \to Y$ has a unique right-inverse if and only if it is a bijection.
7. Show that $f : X \to Y$ has a unique left-inverse if and only if it is a bijection.

8. A mapping $r:X \to M$ is called a **retraction of** X **onto** M if it is a surjection and $r(r(x)) = x$ holds for all $x \in X$. Show that $r:X \to M$ is a retraction of X onto M if and only if $M \subseteq X$ and the inclusion mapping $i:M \to X$ is a right-inverse for $r:X \to M$.

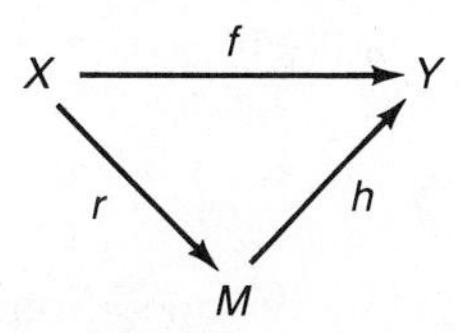

FIGURE 2-29

9. Prove Theorem 8.7.

10. Let $f:X \to Y$ be a mapping. Prove that the equivalence relation E_f has a canonical form if and only if there exists a retraction $r:X \to M$ of X onto M (see Problem 8) and an injection $h:M \to Y$ such that the diagram in Figure 2-29 is commutative.

11. Prove Theorem 8.9.

12. Prove Theorem 8.12.

13. Prove Theorem 8.13.

14. Suppose that the set P and the mappings $p:P \to A$ and $q:P \to B$ constitute a direct product of the sets A and B. Prove that $p:P \to A$ and $q:P \to B$ are surjections.

15. Prove: (a) Theorem 8.15 (b) Theorem 8.16

16. Prove Theorem 8.17.

17. Let A and B be finite sets with $\#A = m$ and $\#B = n$. Let P together with the mappings $p:P \to A$ and $q:P \to B$ constitute a direct product of A and B. Prove that $\#P = mn$.

18. Suppose that the set S and the mappings $i:A \to S$ and $j:B \to S$ constitute a direct sum of the sets A and B. Prove that $i:A \to S$ and $j:B \to S$ are injections.

19. Let A and B be finite sets with $\#A = m$ and $\#B = n$. Let S together with the mappings $i:A \to S$ and $j:B \to S$ constitute a direct sum of A and B. Prove that $\#S = m + n$.

20. Let P be a set, and let $p:P \to A$ and $q:P \to B$ be mappings such that the following two conditions hold:
 (i) $\{p, q\}$ is a complete set of invariants for the equivalence relation Δ_P.
 (ii) For every $a \in A$ and every $b \in B$, there exists $z \in P$ such that $p(z) = a$ and $q(z) = b$.

 Prove that the set P together with the mappings $p:P \to A$ and $q:P \to B$ constitute a direct product of A and B.

21. Let A and B be two sets with $A \cap B = \emptyset$. Let $S = A \cup B$, and let $i:A \to S$ and $j:B \to S$ be the inclusion mappings. Show that S together with the mappings $i:A \to S$ and $j:B \to S$ constitute a direct sum of A and B.

22. Let A_1, A_2, B_1, and B_2 be sets. Let $A = A_1 \times A_2$, and let $B = B_1 \times B_2$. Define mappings

$$p_1:A \to A_1,\ p_2:A \to A_2,\ q_1:B \to B_1, \text{ and } q_2:B \to B_2$$

by

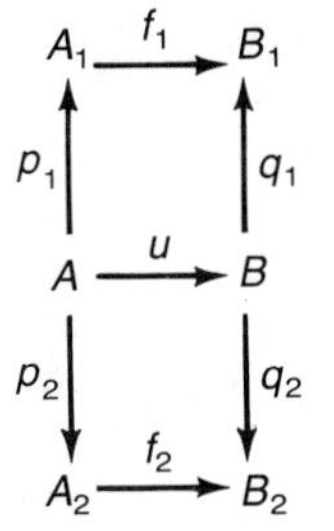

$$p_1(a_1, a_2) = a_1, \quad p_2(a_1, a_2) = a_2,$$
$$q_1(b_1, b_2) = b_1, \quad \text{and} \quad q_2(b_1, b_2) = b_2$$

for all $(a_1, a_2) \in A$ and all $(b_1, b_2) \in B$. Let $f_1: A_1 \to B_1$ and $f_2: A_2 \to B_2$ be arbitrary mappings. Prove that there exists a unique mapping $u: A \to B$ such that the diagram in Figure 2-30 is commutative. The mapping $u: A \to B$ is called the **direct product** of the mappings $f_1: A_1 \to B_1$ and $f_2: A_2 \to B_2$.

FIGURE 2-30

23. Let $(S, *)$ be an •algebraic system with a single binary operation $*: S \times S \to S$, and define $\sigma: S \times S \to S \times S$ by $\sigma((x, y)) = (y, x)$ for all $(x, y) \in S \times S$.
(a) Show that $(S, *)$ is commutative if and only if $* = * \circ \sigma$.
(b) Illustrate the condition in Part (a) by means of a commuting diagram.

24. Let $(S, *)$ be an algebraic system. Define a bijection

$$\alpha: S \times (S \times S) \to (S \times S) \times S$$

by $\alpha((a, (b, c))) = ((a, b), c)$ for all $a, b, c \in S$. Also define mappings

$$f: S \times (S \times S) \to S \times S, \quad \text{and} \quad g: (S \times S) \times S \to S \times S$$

by $f((a, (b, c))) = (a, b * c)$ and $g(((a, b), c)) = (a * b, c)$ for all $a, b, c \in S$. Prove that $(S, *)$ is associative if and only if the diagram in Figure 2-31 is commutative.

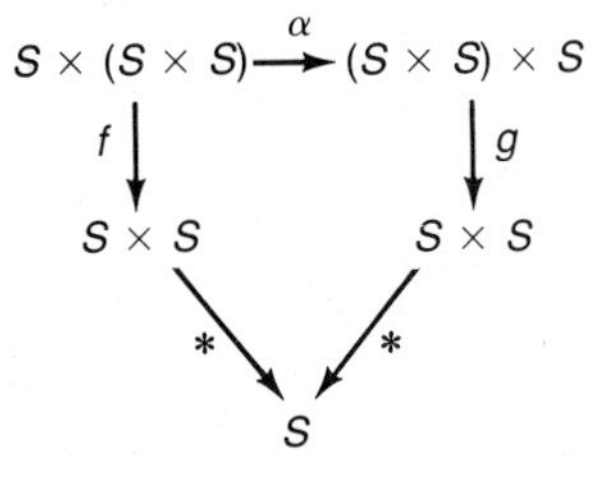

FIGURE 2-31

25. Let $(S, *)$ and $(T, \star)$ be algebraic systems, and let $\Phi: S \to T$ be a bijection. Define $h: S \times S \to T \times T$ by

$$h((x, y)) = (\Phi(x), \Phi(y))$$

for all $(x, y) \in S \times S$. Prove that Φ is an isomorphism if and only if the diagram in Figure 2-32 is commutative.

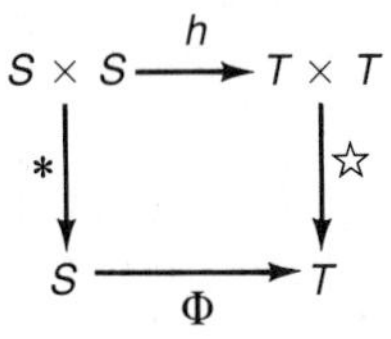

FIGURE 2-32

HISTORICAL NOTES

The algebra of relations was originated by Augustus De Morgan (1806–1871) and further elaborated by Charles S. Peirce (1839–1914) and E. Schroeder (1841–1902). By the late 1800s, the algebraic development was fairly complete, but response from the general mathematical world was small. There was a long philosophical debate over whether a relation—for example, a binary relation—should be identified with a set of ordered pairs. Many philosophers and some mathematicians feared that the set

$$\ldots, (1, 2), (1, 3), (1, 4), (2, 3), (2, 4), (3, 4), \ldots$$

could not capture the true meaning of the "less than" relation on the positive integers. As late as 1903, Bertrand Russell (1872–1970) was still discussing this question, but in *Principia Mathematica* (1910–1913), Russell and Alfred North Whitehead (1861–1947) decisively opted for this explanation of relations, which became the mathematical standard.

The explicit use of equivalence relations begins with the *Disquisitiones Arithmeticae* of Carl Friedrich Gauss (1777–1855), where he introduced the notion of congruence for integers. However, the basic idea of equivalence relations is found in ancient Greek geometry in the notions of similarity and congruence.

The notion of function in the sense of one quantity depending upon another is implicitly present even in ancient mathematics. The notion begins to take on a visible role with the scholastic philosophers (1300s) with their discussions of the various ways (linear, quadratic, reciprocal) quantities may depend on one another. With René Descartes's (1596–1650) invention of analytic geometry (1637) it became necessary to associate some analytic object with the geometric curve, and this object is, of course, the function. Descartes himself would allow only polynomials and considered other kinds of curves to belong to mechanics, not mathematics.

By 1755, Leonhard Euler had given a definition of function coinciding with ours (at least for numerical functions), but the concept was too general for the times and no use was made of it. Nikolai Lobachevski (in 1834) and Gustav Lejeune Dirichlet (in 1837) gave our definition (with the additional requirement of continuity), and Hermann Hankel (in 1870) gave our definition in full generality. The notion of function as a set of ordered pairs dates from Whitehead and Russell's *Principia Mathematica.* They write $f'x$ for $f(x)$, and this notation is still occasionally used by philosophers.

ANNOTATED BIBLIOGRAPHY

Hillman, A. P., et al. *Discrete and Combinatorial Mathematics* (San Francisco: Dellen), 1987.

A recent, fairly comprehensive text on many topics from discrete mathematics, including relations, functions, and binary operations.

Pfleeger, Shari Lawrence, and David W. Straight. *Introduction to Discrete Structures* (New York: Wiley), 1985.

A detailed treatment of sets, relations, functions, Boolean algebras, lattices, and other topics in discrete mathematics.

Ross, Kenneth A., and Charles R. B. Wright. *Discrete Mathematics* (Englewood Cliffs, N.J.: Prentice-Hall), 1985.

A fairly detailed treatment of relations and functions, along with related material, all with applications in computer science in mind. A text by well-respected mathematicians.

3

GROUPS, RINGS, FIELDS, AND VECTOR SPACES

In this chapter we continue the study of algebraic systems begun in Section 2.6. In Section 1, we consider monoids in which every element is "invertible." Such a monoid is called a *group*, and the theory of groups is one of the most important and most highly developed branches of contemporary algebra. In Section 2, we begin our study of algebraic systems with two binary operations called "addition" and "multiplication," and subject to certain postulates suggested by the familiar binary operations of addition and multiplication in ordinary arithmetic. This study continues in Section 3, where we introduce the idea of a field. Ordered fields are studied in Section 4, and the chapter concludes with Sections 5 and 6 on vector spaces, subspaces, and linear transformations.

1 GROUPS

Recall from Section 2.6 that a *monoid* is an algebraic system $(S, *)$ with the following two properties:

(i) $(S, *)$ is *associative*, that is, for all $x, y, z \in S$,

$$x * (y * z) = (x * y) * z.$$

(ii) $(S, *)$ has a *neutral element*, that is, there is an element $e \in S$ such that, for all $x \in S$,

$$e * x = x = x * e.$$

In Theorem 6.9 of Section 2.6, we proved that such a neutral element is unique; so we can speak of *the* neutral element e of the monoid $(S, *)$.

1.1 DEFINITION

LEFT- AND RIGHT-INVERSES Let $(S, *)$ be a monoid with neutral element e, and let $x \in S$.

(i) An element $a \in S$ is called a **left-inverse** for x if

$$a * x = e.$$

(ii) An element $b \in S$ is called a **right-inverse** for x if

$$x * b = e.$$

(iii) An element $c \in S$ is called an **inverse** for x if

$$c * x = e = x * c.$$

Thus, an inverse for x is an element that is both a right-inverse and a left-inverse for x. Because

$$e * e = e,$$

it follows that *the neutral element e is its own inverse.*

1.2 THEOREM

EQUALITY OF LEFT- AND RIGHT-INVERSES Let $(S, *)$ be a monoid, and let $x \in S$. If $a, b \in S$, a is a left-inverse for x, and b is a right-inverse for x, then $a = b$.

PROOF If e is the neutral element of S, then we have

$$a * x = e \quad \text{and} \quad x * b = e.$$

Therefore,

$$a = a * e = a * (x * b) = (a * x) * b = e * b = b. \qquad \square$$

1.3 COROLLARY

UNIQUENESS OF INVERSES Let $(S, *)$ be a monoid, and let $x \in S$. If x has an inverse in S, then this inverse is unique.

PROOF Suppose that both a and b are inverses of x in S. Then, in particular, a is a left-inverse for x in S and b is a right-inverse for x in S. Consequently, by Theorem 1.2, $a = b$. $\square$

1.4 DEFINITION

INVERTIBLE ELEMENT OF A MONOID Let $(S, *)$ be a monoid, and let $x \in S$. If x has an inverse in S, we say that x is an **invertible** element of S.

Rather than saying that x is an invertible element of S, we often just say that x is **invertible in** S. If x is invertible in S, then, by Corollary 1.3, it has a unique inverse in S.

1.5 DEFINITION

NOTATION FOR THE INVERSE OF AN ELEMENT Let $(S, *)$ be a monoid, and let x be an invertible element of S. In this section we shall use the notation x^{-} for the unique inverse of x in S.

If $(S, *)$ is a monoid and x is invertible in S, the element x^{-} is called *x-inverse.* If e is the neutral element of $(S, *)$, we have

$$(x^{-}) * x = x * (x^{-}) = e.$$

As we remarked above, the neutral element e is its own inverse, so that

$$e^{-} = e.$$

Other authors use different symbols for the inverse of x, one popular alternative being x^{-1}. However, we want to reserve the symbol x^{-1} for the inverse of x in a so-called *multiplicatively written* algebraic system, and so we have chosen to use the symbol x^{-} for the general concept of the inverse of x in $(S, *)$.

Recall that, of the fifteen algebraic systems AS1–AS15 introduced in Section 2.6 (pages 144–145), all but AS3, AS6, AS7, and AS15 are monoids.

1.6 Example In each of the algebraic systems AS1 to AS14 that are monoids, identify the neutral element and the invertible elements. Specify the inverses of the invertible elements.

SOLUTION

AS1 The neutral element is the real number 0, every element is invertible, and $x^{-} = -x$ for every element x.

AS2 The neutral element is the real number 1, and an element x is invertible if and only if $x \neq 0$. If x is invertible, then $x^{-} = x^{-1}$, the reciprocal of x.

AS4 The neutral element is the real number 0, and an element x is invertible if and only if $x \neq 1$. If x is invertible, then $x^{-} = x/(x - 1)$.

AS5 The neutral element is the real number 1, every element x is invertible, and $x^{-} = x^{-1}$.

AS8 The neutral element is Δ_X, and an element f is invertible if and only if $f: X \to X$ is a bijection. If f is invertible, then $f^{-} = f^{-1}$.

AS9 The neutral element is Δ_X, and every element f is invertible with $f^{-} = f^{-1}$.

AS10 The neutral element is the empty set $\varnothing$, and $\varnothing$ is the only invertible element. Of course, $\varnothing^{-} = \varnothing$.

AS11 The neutral element is the set X, and X is the only invertible element. Of course, $X^- = X$.

AS12 The neutral element is the empty set $\varnothing$, and every element is invertible. If $M \subseteq X$, then $M^- = M$.

AS13 The neutral element is e, and every element is invertible. We have $e^- = e$, $a^- = a$, $b^- = b$, and $c^- = c$.

AS14 The neutral element is e, and every element is invertible. We have $e^- = e$, $a^- = b$, $b^- = a$, $c^- = c$, $d^- = d$, and $f^- = f$. □

The following theorem establishes two of the most important properties of inverses.

1.7 THEOREM

PROPERTIES OF INVERSES Let $(S, *)$ be a monoid and let $x, y \in S$. Then:

(i) If x is invertible in S, so is x^-, and

$$(x^-)^- = x.$$

(ii) If x and y are invertible in S, so is $x * y$, and

$$(x * y)^- = (y^-) * (x^-).$$

PROOF (i) If x is invertible, then we have

$$(x^-) * x = x * (x^-) = e,$$

where e is the neutral element of S. This shows that x^- is invertible and that its inverse is x, that is,

$$(x^-)^- = x.$$

(ii) Suppose that x and y are invertible in S. To show that $x * y$ is invertible in S, we must find an element $z \in S$ such that

$$(x * y) * z = z * (x * y) = e,$$

and, if such an element z can be found, then $z = (x * y)^-$. Thus, let $z = (y^-) * (x^-)$. Then, since $*$ is associative,

$$\begin{aligned}(x * y) * z &= x * y * z = x * y * (y^-) * (x^-) \\ &= x * [y * (y^-)] * (x^-) = x * e * (x^-) \\ &= x * (x^-) = e.\end{aligned}$$

A similar calculation (Problem 7) shows that $z * (x * y) = e$, and completes the proof. □

1.8 Example Interpret Theorem 1.7 for the monoid $(\mathbb{R}, \cdot)$ of real numbers under the operation of multiplication.

SOLUTION The neutral element is the real number 1, and $x \in \mathbb{R}$ is invertible if and only if $x \neq 0$, in which case, $x^- = 1/x$. Thus, in this example, Part (i) of Theorem 1.7 is just the familiar fact that

$$x \neq 0 \Rightarrow 1/x \neq 0 \quad \text{and} \quad 1/(1/x) = x.$$

Part (ii) is the equally familiar fact that

$$x \neq 0 \wedge y \neq 0 \Rightarrow xy \neq 0 \wedge 1/(xy) = (1/y)(1/x).$$

□

Because we are so used to working in commutative algebraic systems such as $(\mathbb{R}, \cdot)$, the "order inversion" in Part (ii) of Theorem 1.7 may come as a bit of a surprise. In a commutative monoid, it is irrelevant whether we write $(y^-) * (x^-)$ or $(x^-) * (y^-)$; however, the order inversion is quite essential in a noncommutative monoid. For instance, in the noncommutative monoid AS14, we have

$$(a * d)^- = f^- = f = d * b = (d^-) * (a^-)$$

in conformity with Part (ii) of Theorem 1.7. However,

$$(a^-) * (d^-) = b * d = c \neq (a * d)^-.$$

In Example 1.6, notice that each of the algebraic systems AS1, AS5, AS9, AS12, AS13, and AS14 has the property that *all of their elements are invertible*. This leads us to the following fundamental definition:

1.9 DEFINITION

GROUP A **group** is a monoid in which every element is invertible. A **commutative** group is a commutative monoid that is a group.

Commutative groups are often referred to as **abelian** groups in honor of the Norwegian mathematician Niels Henrik Abel (1802–1829), a leader in the development of several important branches of modern mathematics. Thus, AS1, AS5, AS12, and AS13 are commutative, or abelian, groups; and AS14 is a noncommutative group. Unless $\#X \leq 2$, AS9 is a noncommutative group (Problem 10). If $(S, *)$ is a group, then $\#S$ is called the **order** of the group. If $\#S$ is finite, then $(S, *)$ is called a **finite group**; otherwise, it is called an **infinite group**. Thus, AS13 and AS14 are finite groups, AS9 and AS12 are finite groups if and only if $\#X$ is finite, and both AS1 and AS5 are infinite groups.

1.10 Example Let $(S, *)$ be a monoid, and let G be the subset of S consisting of all of the invertible elements in S. If $x, y \in G$, then $x * y \in G$ by Part (ii) of

Theorem 1.7. This permits us to define a binary operation

$$☆: G \times G \to G$$

on G by

$$x ☆ y = x * y$$

for all $x, y \in G$. Show that $(G, ☆)$ is a group.

SOLUTION Notice that the binary operation $☆$ is the same as $*$, except that it is restricted to elements of G; hence, $(G, ☆)$ inherits the associative law from $(S, *)$. Because the neutral element $e \in S$ is invertible (with $e^- = e$), it follows that $e \in G$. Evidently, e is a neutral element for $(G, ☆)$. If $x \in G$, then $x^- \in G$ by Part (i) of Theorem 1.7, and

$$(x^-) ☆ x = (x^-) * x = e = x * (x^-) = x ☆ (x^-);$$

so x is invertible in G. Thus, $(G, ☆)$ is a monoid in which every element is invertible. By Definition 1.9, it is a group. □

The group $(G, ☆)$ in Example 1.10 is called the **group of invertible elements in the monoid** $(S, *)$. In practice, people use the same symbol $*$ for the binary operation in S and for its restriction to G. (This slight abuse of notation usually causes no difficulty.) We shall follow this custom and speak of "the group $(G, *)$ of invertible elements in the monoid $(S, *)$."

1.11 Example Find the group of invertible elements in the monoid AS2.

SOLUTION As we observed in Example 1.6, the invertible elements in AS2 are the nonzero real numbers; hence, the group of invertible elements in AS2 is AS5. □

1.12 DEFINITION

CANCELLATION LAWS Let $(S, *)$ be an algebraic system.

(i) $(S, *)$ is said to satisfy the **left-cancellation law** if, for all elements $a, x, y \in S$,

$$a * x = a * y \Rightarrow x = y.$$

(ii) $(S, *)$ is said to satisfy the **right-cancellation law** if, for all elements $a, x, y \in S$,

$$x * a = y * a \Rightarrow x = y.$$

Note that $(S, *)$ satisfies the left-cancellation law if and only if no element appears twice in any horizontal row of its operation table (Problem 19). Likewise, $(S, *)$ satisfies the right-cancellation law if and only if no element appears twice in any vertical column of its operation table. If an algebraic system $(S, *)$ satisfies both the left- and the right-cancellation laws, we simply say that it satisfies the **cancellation laws**.

1.13 THEOREM **CANCELLATION LAWS IN A GROUP** A group satisfies the cancellation laws.

PROOF Let $(S, *)$ be a group with neutral element e, and suppose that $a, x, y \in S$ with

$$a * x = a * y.$$

Because a is invertible in S, a^- exists in S, and we have

$$(a^-) * (a * x) = (a^-) * (a * y).$$

Therefore, since $*$ is associative,

$$[(a^-) * a] * x = [(a^-) * a] * y,$$

that is,

$$e * x = e * y,$$

so that

$$x = y.$$

This proves the left-cancellation law. The proof of the right-cancellation law is similar and is left as an exercise (Problem 21). □

Because AS1, AS5, AS9, AS12, AS13, and AS14 are groups, they satisfy the cancellation laws by Theorem 1.13. Notice that AS3, AS6, and AS15 also satisfy the cancellation laws, even though they are not groups. In the remaining examples AS2, AS4, AS7, AS8, AS10, and AS11, the cancellation laws fail (Problem 20). For instance, in AS2, we have $0 \cdot 1 = 0 \cdot 2$, but $1 \neq 2$.

As a consequence of Theorem 1.13, no element appears more than once in any horizontal row or in any vertical column of the operation table for a group $(S, *)$. Moreover, each element appears *at least once* in each such row or column. Indeed, to say that the element $b \in S$ appears at least once in the row labeled by the element $a \in S$ is to say that the equation

$$a * x = b$$

can be solved for $x \in S$ (Problem 23), and this equation has the unique solution

$$x = (a^-) * b$$

(Problem 24). Likewise, to say that the element b appears at least once in the column labeled by the element a is to say that the equation

$$y * a = b$$

can be solved for $y \in S$ (Problem 23), and this equation has the unique solution

$$y = b * (a^{-})$$

(Problem 24). Therefore:

> *In the operation table for a group, each element appears once and only once in each row and in each column.*

Notice, however, that AS15 has this property but is *not* a group.

The following theorem shows that the solvability of the equations $a * x = b$ and $y * a = b$ in a semigroup implies that the semigroup is a group. (Recall that a semigroup is an associative algebraic system.)

1.14 THEOREM **SOLVING EQUATIONS IN A SEMIGROUP** Let $(S, *)$ be a semigroup for which the following two conditions hold:

(i) For all $a, b \in S$, the equation $a * x = b$ has a solution $x \in S$.
(ii) For all $a, b \in S$, the equation $y * a = b$ has a solution $y \in S$.

Then $(S, *)$ is a group.

PROOF Choose and fix an element $c \in S$. By Part (ii), with $a = c$ and $b = c$, there is an element $e \in S$ such that

$$e * c = c.$$

In what follows, let s denote an arbitrary element of S. By Part (i), with $a = c$ and $b = s$, there is an element $x \in S$ such that

$$c * x = s.$$

Therefore,

$$e * s = e * (c * x) = (e * c) * x = c * x = s.$$

Thus, we have an element $e \in S$ such that

$$e * s = s$$

holds for every $s \in S$. In particular, we have

$$e * e = e.$$

By (ii) again, this time with a replaced by s and with b replaced by e, there is an element $d \in S$ with

$$d * s = e.$$

Applying (ii) once more, but this time with $a = d$ and $b = e$, we conclude that there is an element $y \in S$ such that

$$y * d = e.$$

Now,

$$s = e * s = (y * d) * s = y * (d * s) = y * e.$$

Using the fact that $s = y * e$ twice, we find that

$$s * e = (y * e) * e = y * (e * e) = y * e = s.$$

Thus, for every $s \in S$, we have

$$e * s = s = s * e,$$

and it follows that e is a neutral element for S. Finally, applying (i) and (ii) with $a = s$ and $b = e$, we see that s has both a left- and a right-inverse in S; hence, every element $s \in S$ is invertible, and so $(S, *)$ is a group. □

1.15 COROLLARY A semigroup $(S, *)$ is a group if and only if each element of S appears exactly once in every row and in every column of its operation table.

PROOF We have already noted that the solvability of the equations $a * x = b$ and $y * a = b$ in S is equivalent to the condition that each element of S appears at least once in each row and in each column of the operation table for S. On the other hand, we have noted that the cancellation laws, which hold in any group, are equivalent to the condition that each element of S appears at most once in each row and in each column of the operation table. □

1.16 COROLLARY A finite semigroup in which the cancellation laws hold is a group.

PROOF Let $(S, *)$ be a semigroup in which the cancellation laws hold, and suppose that $\#S$ is finite. Let $n = \#S$. A row in the operation table for S contains n entries, and no element of S is repeated in this row; hence, all of the elements of S must be used to fill the row. Thus, each element of S appears once and only once in each row. Likewise for columns, so $(S, *)$ is a group by Corollary 1.15. □

PROBLEM SET 3.1

1. In AS4 (page 145), show that:
(a) 0 is the neutral element.
(b) An element x is invertible if and only if $x \neq 1$.
(c) If $x \neq 1$, then $x^{-} = x/(x - 1)$.

2. Define a binary operation $*$ on the set $\mathbb{R}$ of all real numbers by $x * y = x + y + xy$.
(a) Show that $(\mathbb{R}, *)$ is a commutative monoid.
(b) Identify the neutral element and the invertible elements.
(c) If x is an invertible element in $(\mathbb{R}, *)$, find a formula for x^{-}.

3. Let $\mathbb{N} = \{1, 2, 3, \ldots\}$ denote the set of all natural numbers (positive integers), and let S be the set of all mappings $f: \mathbb{N} \to \mathbb{N}$. Consider the monoid $(S, \circ)$, where $\circ$ denotes function composition. Let $s \in S$ be given by $s(n) = n + 1$ for all $n \in \mathbb{N}$.
(a) Show that s has a left-inverse in $(S, \circ)$.
(b) Show that s has no right-inverse in $(S, \circ)$.

4. In Problem 3, show that the element s has infinitely many different left-inverses in $(S, \circ)$.

5. An element a in an algebraic system $(S, *)$ is called an **idempotent** if $a * a = a$. If $(S, *)$ is a monoid and if y is a left-inverse for x in S, show that $x * y$ is an idempotent in S.

6. Show that the only idempotent element in a group is the neutral element. (See Problem 5.)

7. Complete the proof of Theorem 1.7 by showing that, if $z = (y^{-}) * (x^{-})$, then $z * (x * y) = e$.

8. Interpret Theorem 1.7 for the group AS9 (page 145).

9. Interpret Theorem 1.7 for the abelian group $(\mathbb{R}, +)$ of real numbers under the operation of addition.

10. If $\#X > 2$, show that AS9 is a nonabelian (that is, noncommutative) group.

11. Does there exist a group of order 1? Explain.

12. Let i be the complex number with $i^2 = -1$. Show that $V = \{1, i, -1, -i\}$ forms a group $(V, \cdot)$ of order 4 under multiplication.

13. Write an operation table for a group $(G, *)$ of order 2, denoting the elements by $G = \{e, a\}$, where e is the neutral element. Explain why there is only one way to form such an operation table.

14. Show that the group $(V, \cdot)$ of Problem 12 is isomorphic to the group AS13 (page 145).

15. Write an operation table for a group $(G, *)$ of order 3, denoting the elements by $G = \{e, a, b\}$, where e is the neutral element. Explain why there is only one way to form such an operation table.

16. Write two essentially different operation tables for a group $(G, *)$ of order 4, denoting the elements by $G = \{e, p, q, r\}$, where e is the neutral element. Explain why the resulting two groups are *not* isomorphic.

17. Find the group of invertible elements in the monoid AS8 (page 145).

18. Let $\mathcal{M}_2(\mathbb{R})$ denote the set of all two-by-two matrices with real number entries. Multiply two such matrices as follows:

$$\begin{bmatrix} a & b \\ c & d \end{bmatrix} \cdot \begin{bmatrix} x & y \\ z & w \end{bmatrix} = \begin{bmatrix} ax + bz & ay + bw \\ cx + dz & cy + dw \end{bmatrix}.$$

(a) Show that $(\mathcal{M}_2(\mathbb{R}), \cdot)$ is a monoid and identify its neutral element.
(b) Find necessary and sufficient conditions on the real numbers a, b, c, and d so that the matrix on the left is invertible.

19. Show that an algebraic system $(S, *)$ satisfies the left- (respectively, the right-) cancellation law if and only if no element appears twice in any horizontal row (respectively, vertical column) of its operation table.

20. Show that the cancellation laws fail in AS4, AS7, AS8, AS10, and AS11.

21. Complete the proof of Theorem 1.13 by showing that the right-cancellation law holds in a group.

22. Show that the cancellation laws fail in the monoid $(\mathcal{M}_2(\mathbb{R}), \cdot)$ of Problem 18.

23. Let $(S, *)$ be an algebraic system, and let $a, b \in S$. Show that the equation $a * x = b$ (respectively, $y * a = b$) can be solved for $x \in S$ (respectively, for $y \in S$) if and only if b appears at least once in the row (respectively, column) labeled by a in the operation table for $(S, *)$.

24. Let $(S, *)$ be a group, and let $a, b \in S$. Show that the equations $a * x = b$ and $y * a = b$ have unique solutions $x, y \in S$.

25. Prove that if an algebraic system $(S, *)$ is isomorphic to a group $(G, \star)$, then $(S, *)$ is a group.

26. By definition, a **quasigroup** is an algebraic system $(S, *)$ such that, for every $a, b \in S$, the equations $a * x = b$ and $y * a = b$ have unique solutions $x, y \in S$.
(a) Give an example of a quasigroup that is not a group.
(b) Explain why a group is the same thing as an associative quasigroup.

27. Suppose that $(S, *)$ is a semigroup and that there is an element $e \in S$ with the following two properties:
(i) $e * x = x$ for every $x \in S$.
(ii) For every $x \in S$, there exists $y \in S$ such that $y * x = e$.
Prove that $(S, *)$ is a group with e as its neutral element.

28. Let $(S, *)$ be a commutative quasigroup (see Problem 26), and assume that $\#S$ is finite and odd. Prove that, for every element $a \in S$, there exists a unique element $b \in S$ such that $b * b = a$ (that is, every element has a unique "square root").

29. Give an example of an infinite commutative semigroup $(S, *)$ in which the cancellation laws hold but that is not a group.

30. Let $(S, *)$ be a quasigroup (see Problem 26), and suppose that $\#S$ is finite. Let $A, B \subseteq S$, and suppose that there exists an element $g \in S$ that cannot be expressed in the form $g = a * b$ with $a \in A$ and $b \in B$. Prove that $\#A + \#B \leq \#S$.

2 RINGS

Until now, we have studied algebraic systems $(S, *)$ with a single binary operation $*$. A *ring* is an algebraic system with two binary operations, called "addition" and "multiplication," governed by certain postulates suggested by the behavior of ordinary addition and multiplication in the real number system $\mathbb{R}$.

2.1 DEFINITION

RING A **ring** is an ordered triple $(R, +, \cdot)$ consisting of a nonempty set R and two binary operations $+$ and $\cdot$, called **addition** and **multiplication** on R, that satisfy the following conditions:

(i) $(R, +)$ is a commutative group.
(ii) $(R, \cdot)$ is a semigroup.
(iii) The **distributive laws** hold, so that, for all $x, y, z \in R$,

$$x \cdot (y + z) = (x \cdot y) + (x \cdot z)$$

and

$$(x + y) \cdot z = (x \cdot z) + (y \cdot z).$$

The first equation in Part (iii) of the definition above expresses the **left-distributive law**, and the second equation expresses the **right-distributive law**. If $(R, +, \cdot)$ is a ring, then the neutral element of the group $(R, +)$ is written as 0 and called the **zero element** of the ring. If $x \in R$, then the inverse of x in the group $(R, +)$ is written as $-x$ rather than as x^- and is called the **additive inverse** of x, or the **negative** of x in the ring. Of course, the real number system $(\mathbb{R}, +, \cdot)$ with the usual operations of addition and multiplication provides an example of a ring. However, it must not be supposed that the addition and multiplication operations in a ring $(R, +, \cdot)$ have anything to do with ordinary numerical addition and multiplication, *except for the fact that they obey some of the same algebraic laws.*

By abuse of language, we often speak of "the ring R," when what we really mean is the ring $(R, +, \cdot)$. Also, we frequently omit the symbol $\cdot$ for the ring multiplication and write xy rather than $x \cdot y$ for the "product" of the two elements x and y in R. We recall that, because of the associative laws, parentheses are not needed for iterated additions or multiplications; so expressions such as

$$x + y + z, \quad x + y + z + w, \quad xyz, \quad xyzw,$$

and so forth make perfectly good sense in a ring. Also, because $(R, +)$ is a commutative, or abelian, group, we are permitted to "shuffle" the terms in an iterated sum; for instance,

$$x + y + z + w = z + y + w + x.$$

However, in a general ring, there is no assumption that multiplication is commutative; thus, the order of the factors in a product might be of significance.

Note that, in a ring, an expression such as

$$x + yz$$

could be ambiguous. Does it mean

$$x + (yz)$$

or does it mean

$$(x + y)z?$$

We resolve this ambiguity for a general ring, just as we do in ordinary algebra, by making the following notational agreement:

> *In writing expressions involving combinations of addition and multiplication in a ring, multiplication takes precedence over addition.*

In other words, you must carry out indicated multiplications *first* before you carry out the additions. Thus, we understand that $x + yz$ means $x + (yz)$. With this notational agreement in mind, we can write the distributive laws as

$$x(y + z) = xy + xz \quad \text{and} \quad (x + y)z = xz + yz.$$

The following theorem is almost a straight transcription of Definition 2.1 (Problem 4).

2.2 THEOREM

POSTULATES FOR A RING Let R be a set containing a special object 0 and equipped with binary operations $(x, y) \mapsto x + y$ and $(x, y) \mapsto x \cdot y = xy$. Then, $(R, +, \cdot)$ is a ring if and only if the following conditions hold for all $x, y, z \in R$:

(i) $x + (y + z) = (x + y) + z$
(ii) $x + y = y + x$
(iii) $x + 0 = x$
(iv) For every $x \in R$, there exists $-x \in R$ with $x + (-x) = 0$.
(v) $x(yz) = (xy)z$
(vi) $x(y + z) = xy + xz$ and $(x + y)z = xz + yz$

We call $(R, +, \cdot)$ a **finite ring** if $\#R$ is finite. For such a finite ring, the binary operations $+$ and $\cdot$ can be specified by operation tables. For instance, here are the operation tables for a ring called the **integers modulo 4** (see Problem 27) and denoted by $(\mathbb{Z}_4, +, \cdot)$, or simply by $\mathbb{Z}_4$, where $\mathbb{Z}_4 = \{0, 1, 2, 3\}$:

$+$	0	1	2	3
0	0	1	2	3
1	1	2	3	0
2	2	3	0	1
3	3	0	1	2

$\cdot$	0	1	2	3
0	0	0	0	0
1	0	1	2	3
2	0	2	0	2
3	0	3	2	1

In Problem 5, we ask you to verify that $\mathbb{Z}_4$ really is a ring. Of course, 0, 1, 2, and 3 do not have their usual numerical meanings in $\mathbb{Z}_4$—they are used merely as convenient symbols for the four elements of this ring! Nevertheless, portions of the two tables reproduce familiar numerical facts; for instance, in the multiplication table, $0 \cdot x = x \cdot 0 = 0$ holds for all elements $x \in \mathbb{Z}_4$. The following theorem shows that this is no accident:

2.3 THEOREM

MULTIPLICATION BY ZERO IN A RING If R is a ring and $x \in R$, then $0 \cdot x = x \cdot 0 = 0$.

PROOF We prove that $0 \cdot x = 0$ and leave the analogous proof that $x \cdot 0 = 0$ as an exercise (Problem 11). By the right-distributive law, we have

$$(0 + 0) \cdot x = 0 \cdot x + 0 \cdot x.$$

Because 0 is the neutral element in $(R, +)$,

$$0 + 0 = 0,$$

and so we can rewrite the first equation as

$$0 \cdot x = 0 \cdot x + 0 \cdot x.$$

Again, because 0 is the additive neutral element,

$$0 \cdot x + 0 = 0 \cdot x.$$

Combining the last two equations, we have

$$0 \cdot x + 0 = 0 \cdot x + 0 \cdot x.$$

Because $(R, +)$ is a group, it satisfies the left-cancellation law (Theorem 1.13), and, therefore, from the last equation we infer that

$$0 = 0 \cdot x.$$ □

A number of additional properties of a ring R follow easily from the fact that $(R, +)$ is a commutative group with 0 as its neutral element. For instance, since the neutral element in a group is its own inverse, we have

$$-0 = 0.$$

Three more properties, all of which seem quite natural, appear in the following theorem:

2.4 THEOREM

PROPERTIES OF NEGATION IN A RING Let R be a ring, and let $x, y \in R$. Then:

(i) $x + y = 0 \Rightarrow x = -y$
(ii) $-(-x) = x$
(iii) $-(x + y) = (-x) + (-y)$

PROOF

(i) If $x + y = 0$, then x is a left-inverse for y in $(R, +)$. Since $-y$ is a right-inverse for y in $(R, +)$, the fact that $x = -y$ follows from Theorem 1.2.
(ii) This follows directly from Part (i) of Theorem 1.7.
(iii) This follows from Part (ii) of Theorem 1.7 and the fact that $(R, +)$ is commutative. □

The next theorem establishes some additional properties of negation in a ring, indicating how it interacts with multiplication.

2.5 THEOREM

PROPERTIES OF NEGATION AND MULTIPLICATION Let R be a ring, and let $x, y \in R$. Then:

(i) $(-x)y = -(xy)$
(ii) $x(-y) = -(xy)$
(iii) $(-x)(-y) = xy$

PROOF (i) By the right-distributive law,

$$(-x)y + xy = ((-x) + x)y.$$

Because $(-x) + x = 0$, we can rewrite the last equation as

$$(-x)y + xy = 0 \cdot y.$$

By Theorem 2.3, $0 \cdot y = 0$, and so we have

$$(-x)y + xy = 0.$$

Therefore, by Part (i) of Theorem 2.4,

$$(-x)y = -(xy).$$

(ii) The proof is analogous to the proof of Part (i) and is left as an exercise (Problem 13a).

(iii) Using Parts (i) and (ii) above, we have

$$(-x)(-y) = -(x(-y)) = -(-(xy)).$$

By Part (ii) of Theorem 2.4, we have

$$-(-(xy)) = xy,$$

and it follows that

$$(-x)(-y) = xy.$$ □

Notice that Theorem 2.5 generalizes the well-known "rules of signs" in elementary algebra. However, here are some important words of warning: In a ring, the element $-x$ is often called "minus x" or "the negative of x," but *you must not make the mistake of supposing that* $-x$ *is necessarily negative*! In the first place, we are dealing with abstract rings here, and the idea of being positive or negative may not even make sense. For instance, in the ring $\mathbb{Z}_4$ discussed above, which elements are positive and which are negative? (Answer: None of them—positive and negative make no sense in $\mathbb{Z}_4$.) In the second place, any element y in a ring R can be written in the form $y = -x$ for some $x \in R$. (Just take $x = -y$ and use Part (ii) of Theorem 2.4.) Thus, even in the ring $(\mathbb{R}, +, \cdot)$ of real numbers, where it makes sense to talk about positive and negative elements, $-x$ *need not be negative*!

By combining addition and negation, we can introduce the idea of *subtraction* in a ring.

2.6 DEFINITION

SUBTRACTION IN A RING Let R be a ring, and let $x, y \in R$. We define

$$x - y = x + (-y).$$

The element $x - y$ is read in English as "x minus y," and the binary operation $(x, y) \mapsto x - y$ is called **subtraction**. We leave the proof of the following theorem as an exercise (Problem 13b).

2.7 LEMMA

TRANSPOSITION IN A RING Let R be a ring, and let $x, y, z \in R$. Then,

$$x + y = z \Leftrightarrow x = z - y.$$

Lemma 2.7 justifies the familiar procedure (called *transposition* in elementary algebra) of moving a term from one side of an equation to another at the expense of changing the algebraic sign of the term.

The ring $\mathbb{R}$ of real numbers and the ring $\mathbb{Z}_4$ both contain an element 1 that is effective as a multiplicative neutral element. This observation leads us to the next definition.

2.8 DEFINITION

RING WITH UNITY Let $(R, +, \cdot)$ be a ring. If $(R, \cdot)$ has a neutral element different from the additive neutral element 0, then this neutral element of $(R, \cdot)$ is called the **unity element** of the ring, and R is called a **ring with unity**.

In a ring R with unity, the unity element is ordinarily denoted by 1. Thus, 1 is the unique element in R such that

$$1 \cdot x = x \cdot 1 = x$$

holds for all elements $x \in R$. The condition that $1 \neq 0$, which is part of Definition 2.8, just rules out the trivial case in which R contains only one element, namely, 0.

2.9 Example Give an example of a ring $R \neq \{0\}$ that has no unity element.

SOLUTION Let R be the set of all *even* integers,

$$R = \{0, \pm 2, \pm 4, \pm 6, \ldots\}.$$

If $+$ and $\cdot$ denote the ordinary arithmetic operations of addition and multiplication, then $(R, +, \cdot)$ is a ring with no unity element. □

If $(R, +, \cdot)$ is a ring with unity 1, then $(R, \cdot)$ is a monoid, and we can speak of the *invertible* elements in this monoid. Some authors use the word *unit* to mean an invertible element in the monoid $(R, \cdot)$, but we prefer to avoid this word here because of its possible confusion with the word *unity*.

2.10 DEFINITION

INVERTIBLE ELEMENTS IN A RING Let $(R, +, \cdot)$ be a ring with unity. If $x \in R$, and x is an invertible element of the monoid $(R, \cdot)$, then we call x a **multiplicatively invertible** element of the ring R, and we write its inverse in $(R, \cdot)$ as x^{-1}.

It is customary to refer to the multiplicatively invertible elements of a ring simply as the *invertible* elements of the ring, and we propose to follow this practice. Because all of the elements of a ring $(R, +, \cdot)$ are automatically additively invertible, that is, invertible in the group $(R, +)$, this abbreviation is unlikely to lead to any confusion.

In the ring $\mathbb{R}$ of real numbers, the invertible elements are the nonzero elements. As a consequence of Example 1.10 on page 182, *the invertible elements in a ring with unity always form a group under the ring multiplication.* This group is called the **group of invertible elements of the ring**. (Some authors would call it the **group of units** of the ring.) For instance, the group of invertible elements of the ring $\mathbb{R}$ is the group AS5 (page 145).

2.11 Example Find the group of invertible elements in the ring $\mathbb{Z}_4$ of integers modulo 4.

SOLUTION Examining the table for multiplication in $\mathbb{Z}_4$, we find that 1 and 3 are the only invertible elements. Notice that $1^{-1} = 1$, and $3^{-1} = 3$. The group G of invertible elements of $\mathbb{Z}_4$ is the group $G = \{1, 3\}$ of order 2 with the following operation table:

$\cdot$	1	3
1	1	3
3	3	1

□

In a ring, the operation of addition is, by definition, commutative; however, there is no such requirement for the operation of multiplication.

2.12 DEFINITION

COMMUTATIVE RING If $(R, +, \cdot)$ is a ring and if $(R, \cdot)$ is a commutative semigroup, then we say that R is a **commutative** ring.

All of the rings that we have considered up to now have been commutative. An example of a noncommutative ring is the ring $\mathcal{M}_2(\mathbb{R})$ of all two-by-two matrices with real number entries. Two such matrices are added by adding their corresponding entries:

$$\begin{bmatrix} a & b \\ c & d \end{bmatrix} + \begin{bmatrix} x & y \\ z & w \end{bmatrix} = \begin{bmatrix} a+x & b+y \\ c+z & d+w \end{bmatrix}.$$

They are multiplied by the row-by-column rule:

$$\begin{bmatrix} a & b \\ c & d \end{bmatrix} \cdot \begin{bmatrix} x & y \\ z & w \end{bmatrix} = \begin{bmatrix} ax + bz & ay + bw \\ cx + dz & cy + dw \end{bmatrix}.$$

In Problem 23, we ask you to show that $\mathscr{M}_2(\mathbb{R})$ forms a ring with unity under these operations. The invertible elements of this ring are the **nonsingular** two-by-two matrices—those whose determinant is not equal to zero.

2.13 Example Show that $\mathscr{M}_2(\mathbb{R})$ is a noncommutative ring.

SOLUTION There are lots of pairs of two-by-two matrices that do not commute under multiplication. For example,

$$\begin{bmatrix} 1 & 0 \\ 0 & 0 \end{bmatrix} \cdot \begin{bmatrix} 1 & 1 \\ 0 & 0 \end{bmatrix} = \begin{bmatrix} 1 & 1 \\ 0 & 0 \end{bmatrix},$$

but

$$\begin{bmatrix} 1 & 1 \\ 0 & 0 \end{bmatrix} \cdot \begin{bmatrix} 1 & 0 \\ 0 & 0 \end{bmatrix} = \begin{bmatrix} 1 & 0 \\ 0 & 0 \end{bmatrix}.$$ □

In the ring $\mathbb{R}$ of real numbers with the usual algebraic operations of addition and multiplication, the product of nonzero elements is nonzero; however, glancing back at the multiplication table on page 191, we notice that $2 \cdot 2 = 0$ in the ring $\mathbb{Z}_4$! There are also nonzero elements in $\mathscr{M}_2(\mathbb{R})$ whose product is zero (Problem 25). This leads us to the next definition.

2.14 DEFINITION

ZERO-DIVISORS Let $(R, +, \cdot)$ be a ring. A *nonzero* element $x \in R$ is called a **left zero-divisor** (respectively, a **right zero-divisor**) in R if there exists a *nonzero* element $y \in R$ (respectively, a *nonzero* element $z \in R$) such that $xy = 0$ (respectively, $zx = 0$).

In a commutative ring, a left (or, what is the same thing, a right) zero-divisor is called a **zero-divisor**.

2.15 DEFINITION

INTEGRAL DOMAIN A commutative ring with unity in which there are no zero-divisors is called an **integral domain**.

As you might guess from the terminology, the integers

$\mathbb{Z} = \{0, \pm 1, \pm 2, \pm 3, \ldots\},$

with the usual numerical operations of addition and multiplication, form an integral domain. Of course, the real numbers $\mathbb{R}$ do, too; however, $\mathbb{Z}_4$, the integers modulo 4, do not form an integral domain.

PROBLEM SET 3.2

1. If $R = \{0\}$, show that $(R, +, \cdot)$ is a ring if we define $0 + 0 = 0$ and $0 \cdot 0 = 0$.
2. Let $(G, +)$ be any commutative group with 0 as its neutral element. Define a binary operation $(x, y) \mapsto x \star y$ on G by $x \star y = 0$ for all $x, y \in G$. Is $(G, +, \star)$ a ring? Why or why not?
3. The usual rules for parity (that is, evenness or oddness) of integers can be written symbolically as

$$\text{even} + \text{even} = \text{odd} + \text{odd} = \text{even},$$
$$\text{even} + \text{odd} = \text{odd} + \text{even} = \text{odd},$$
$$\text{even} \cdot \text{even} = \text{even} \cdot \text{odd} = \text{odd} \cdot \text{even} = \text{even},$$

and

$$\text{odd} \cdot \text{odd} = \text{odd}.$$

Taking $R = \{\text{even}, \text{odd}\}$ and using the rules above for addition and multiplication, show that $(R, +, \cdot)$ is a commutative ring.
4. Prove Theorem 2.2.
5. Verify that $\mathbb{Z}_4$ is a ring.
6. Suppose that $(R, +, \cdot)$ is a system such that:
(i) $(R, +)$ is a group with neutral element 0.
(ii) $(R, \cdot)$ is a monoid with neutral element 1.
(iii) The distributive laws hold.
Prove that $(R, +, \cdot)$ is a ring with unity. [Hint: The only thing that is missing is the commutative law for addition. Prove that $x + y = y + x$ by beginning with $(1 + 1) \cdot (x + y)$ and expanding it in two ways using the distributive laws.]
7. Prove that the identity

$$(a + b)(c + d) = ac + ad + bc + bd$$

holds in a ring.
8. If $\mathbb{Z}$ denotes the ring of integers, show that the set of all real numbers of the form $m + n\sqrt{2}$, where $m, n \in \mathbb{Z}$, forms a commutative ring under the usual operations of addition and multiplication.
9. True or false: In a ring, $(a + b) \cdot (a - b) = a \cdot a - b \cdot b$. Explain.
10. An element x in a ring R is said to be an **idempotent** if $x \cdot x = x$. (For instance, 0 and, if it exists, 1 are idempotents.) An **idempotent ring** is a ring in which *every* element is an idempotent. Prove that, in an idempotent

ring, the identity $x + x = 0$ holds, so that every element is its own negative. [Hint: Use the fact that $(x + x) \cdot (x + x) = x + x$.]

11. Complete the proof of Theorem 2.3 by showing that $x \cdot 0 = 0$ in a ring.

12. Prove that an idempotent ring is commutative (see Problem 10). [Hint: Expand $(x + y) \cdot (x + y)$.]

13. (a) Prove Part (ii) of Theorem 2.5.
(b) Prove Lemma 2.7.

14. Let X be a nonempty set, and let $\mathscr{P}(X)$ denote the set of all subsets of X. For $M, N \in \mathscr{P}(X)$, define $M + N$, $M \cdot N \in \mathscr{P}(X)$ by the equation $M + N = (M \cup N) \setminus (M \cap N)$ and $M \cdot N = M \cap N$. Note that $+$ is the binary operation $*$ in AS12 on page 145. Show that $(\mathscr{P}(X), +, \cdot)$ is an idempotent ring (see Problem 10). Show that this ring has a unity.

15. If $(R, +, \cdot)$ is a ring such that 0 is a neutral element for the monoid $(R, \cdot)$, show that $R = \{0\}$ (see Problem 1).

16. Find the group of invertible elements in the ring $(\mathscr{P}(X), +, \cdot)$ of Problem 14.

17. Find the group of invertible elements in the ring of integers $\mathbb{Z}$.

18. Let $(R, +, \cdot)$ be a commutative ring, and let E denote the set of all idempotents in R (see Problem 10). For $a, b \in E$ define $a \oplus b = a + b - ab$ and $a \otimes b = a \cdot b$.
(a) If $a, b \in E$, show that $a \oplus b \in E$ and $a \otimes b \in E$.
(b) Show that $(E, \oplus, \otimes)$ is an idempotent ring.

19. Let i denote the "imaginary" number with $i^2 = -1$, and let $\mathbb{C}$ be the **complex numbers**, $\mathbb{C} = \{x + yi \mid x, y \in \mathbb{R}\}$. Add and multiply complex numbers as follows:

$$(x + yi) + (u + vi) = (x + u) + (y + v)i,$$
$$(x + yi) \cdot (u + vi) = (xu - yv) + (xv + yu)i.$$

Show that $(\mathbb{C}, +, \cdot)$ is a commutative ring with unity.

20. Find the invertible elements in the ring $(\mathbb{C}, +, \cdot)$ of complex numbers (see Problem 19).

21. Let $\mathbb{Z}$ denote the ring of integers. A complex number of the form $m + ni$, where $m, n \in \mathbb{Z}$, is called a **Gaussian integer**, in honor of the German mathematician Carl Friedrich Gauss (1777–1855). Show that the Gaussian integers form a commutative ring with unity under operations of addition and multiplication defined as in Problem 19.

22. Find the group of invertible elements in the ring of Gaussian integers (see Problem 21).

23. Show that the set $\mathscr{M}_2(\mathbb{R})$ of all two-by-two matrices with real number entries forms a ring with unity under the operations of entrywise addition and row-by-column multiplication.

24. Show that the two-by-two matrices having the form

$$\begin{bmatrix} \cos^2 \phi & \cos \phi \sin \phi \\ \cos \phi \sin \phi & \sin^2 \phi \end{bmatrix}$$

are the idempotent elements in the ring $\mathcal{M}_2(\mathbb{R})$ (see Problem 10).

25. Find two nonzero elements of $\mathcal{M}_2(\mathbb{R})$ whose product is the zero element of $\mathcal{M}_2(\mathbb{R})$.

26. An element x of a ring R is said to be a **nilpotent of order 2** if $x \cdot x = 0$. Are there any nonzero nilpotents of order 2 in the ring $\mathcal{M}_2(\mathbb{R})$?

27. Let $\mathbb{Z}$ denote the ring of integers, and let n be a fixed positive integer. Define a relation $\equiv_n$ on $\mathbb{Z}$ by

$$a \equiv_n b \Leftrightarrow b - a \text{ is an integer multiple of } n$$

for $a, b \in \mathbb{Z}$. The relation $\equiv_n$ is called **congruence modulo** n.

(a) Prove that $\equiv_n$ is an equivalence relation on $\mathbb{Z}$.

(b) If $a \in \mathbb{Z}$, denote by $[a]_n$ the equivalence class modulo $\equiv_n$ determined by a, and denote by $\mathbb{Z}_n$ the set of all such equivalence classes. We propose to define a binary operation $+$ on $\mathbb{Z}_n$ by

$$[a]_n + [b]_n = [a + b]_n$$

for all $a, b \in \mathbb{Z}$. Prove that $+$ is well defined on $\mathbb{Z}_n$.

(c) We propose to define a binary operation $\cdot$ on $\mathbb{Z}_n$ by

$$[a]_n \cdot [b]_n = [ab]_n$$

for all $a, b \in \mathbb{Z}$. Prove that $\cdot$ is well defined on $\mathbb{Z}_n$.

(d) Prove that $(\mathbb{Z}_n, +, \cdot)$ forms a commutative ring with unity. The ring $(\mathbb{Z}_n, +, \cdot)$ is called the **ring of integers modulo** n.

28. If p is a prime positive integer (that is, $p > 1$, and p has no integer factors other than ± 1 and $\pm p$), show that the ring $\mathbb{Z}_p$ of integers modulo p (see Problem 27) contains no zero-divisors.

29. Let $(R, +, \cdot)$ and $(S, +, \cdot)$ be two rings. A bijection

$$\phi: R \to S$$

is called an **isomorphism** of $(R, +, \cdot)$ onto $(S, +, \cdot)$ if it is at the same time an isomorphism of $(R, +)$ onto $(S, +)$ and an isomorphism of $(R, \cdot)$ onto $(S, \cdot)$. If such an isomorphism exists, then we say that the two rings are **isomorphic**. Show that the ring $\mathbb{Z}_4$ defined by the operation tables on page 191 is isomorphic to the ring of integers modulo 4 as defined in Problem 27.

30. Show that the ring $\mathbb{Z}_2$ of integers modulo 2 (see Problem 27) is isomorphic to the ring $R = \{\text{even, odd}\}$ in Problem 3 (see Problem 29).

31. A **polynomial function** $p:\mathbb{R} \to \mathbb{R}$ has the form

$$p(x) = a_n x^n + a_{n-1}x^{n-1} + \cdots + a_1 x + a_0,$$

where $x \in \mathbb{R}$ and the **coefficients** $a_n, a_{n-1}, \ldots, a_1, a_0$ are fixed real numbers. Let P denote the set of all polynomial functions $p:\mathbb{R} \to \mathbb{R}$. Show that P is a commutative ring with unity under the usual operations of addition and multiplication of functions as defined in calculus. Is P an integral domain? Explain.

32. Let P be the ring of polynomial functions defined in Problem 31. Find the group of invertible elements in P.

3 FIELDS

In the commutative ring $(\mathbb{R}, +, \cdot)$ of real numbers, every nonzero element is invertible. Thus, the real numbers form a *field* in the sense of the following definition:

3.1 DEFINITION

DIVISION RING, FIELD A **division ring** is a ring with unity in which every nonzero element is multiplicatively invertible. A **field** is a commutative division ring.

In Problem 36, we give an example of a noncommutative division ring. For the rest of this section, we focus our attention on fields. We begin with the following theorem, which is almost a straight transcription of the definition of a field (Problem 2).

3.2 THEOREM

POSTULATES FOR A FIELD Let F be a set containing two special objects 0 and 1 and equipped with two binary operations $(x, y) \mapsto x + y$ and $(x, y) \mapsto x \cdot y = xy$. Then, $(F, +, \cdot)$ is a field if and only if the following conditions hold for all $x, y, z \in F$:

(i) Associative Laws

$$x + (y + z) = (x + y) + z$$

and

$$x(yz) = (xy)z$$

(ii) Commutative Laws

$$x + y = y + x$$

and

$$xy = yx$$

(iii) Neutral Elements

$$x + 0 = x,$$
$$x \cdot 1 = x,$$

and

$$0 \neq 1$$

(iv) Inverses

$$\forall x \in F, \quad \exists(-x) \in F, \quad x + (-x) = 0$$

and

$$\forall x \in F, \quad x \neq 0 \Rightarrow \exists x^{-1} \in F, \quad x \cdot x^{-1} = 1$$

(v) Distributive Law

$$x(y + z) = xy + xz$$

Just as we did for rings, we often speak of "the field F," when what we really mean is the field $(F, +, \cdot)$. Here are some examples of fields:

1. $\mathbb{R}$, the field of *real numbers.*
2. $\mathbb{Q}$, the field of *rational numbers,* that is, all real numbers of the form m/n such that m and n are integers and $n \neq 0$.
3. $\mathbb{C}$, the field of *complex numbers,* that is, all numbers of the form $x + yi$, where $x, y \in \mathbb{R}$ and $i^2 = -1$.
4. $\mathbb{Q}(\sqrt{2})$, the field obtained by *adjoining* $\sqrt{2}$ *to the rationals,* that is, all real numbers of the form $q + r\sqrt{2}$, where $q, r \in \mathbb{Q}$.
5. $\mathbb{R}(X)$, the field of *rational functions in X with coefficients in* $\mathbb{R}$, that is, the field consisting of all expressions of the form $p(X)/q(X)$, where $p(X)$ and $q(X)$ are polynomials in X with coefficients in $\mathbb{R}$ and $q(X)$ is not the zero polynomial.
6. GF(p), the *Galois field of order p,* where p is a prime number (that is, p is a positive integer, $p \neq 1$, and p has no proper integer divisors). The elements of GF(p) are represented by the positive integers $0, 1, 2, \ldots, p - 1$, but the addition and multiplication operations are not the ordinary ones. For $x, y \in \mathrm{GF}(p)$:

   $x + y =$ the remainder upon division of the ordinary sum of x and y by p.

   $xy =$ the remainder upon division of the ordinary product of x and y by p.

Notice that the examples 1–5 are *infinite* fields, whereas the Galois fields GF(p) in example 6 are *finite* fields. In general, a finite field is called a **Galois field**, in honor of the French mathematician Evariste Galois (1811–1832), who used group theory to settle the classic problem of solvability of equations by radicals. If p is a prime, then GF(p) is isomorphic to the ring $\mathbb{Z}_p$ of integers modulo p constructed in Problem 27 of Problem Set 3.2 (Problem 12).

3.3 Example Write the addition and multiplication tables for GF(3).

SOLUTION

$$\text{GF}(3) = \{0, 1, 2\}$$

+	0	1	2
0	0	1	2
1	1	2	0
2	2	0	1

·	0	1	2
0	0	0	0
1	0	1	2
2	0	2	1

Notice that 0, 1, and 2 do not have their usual meanings in GF(3), and the addition and multiplication operations have some apparently peculiar features. For instance, to find $2 \cdot 2$ in the multiplication table, we compute the ordinary product 4, but then we write the *remainder*, 1, upon division of 4 by 3 as the product of 2 and 2 in GF(3). □

The "smallest" possible field is the field GF(2) consisting only of the two elements 0 and 1. Its addition and multiplication tables are as follows:

$$\text{GF}(2) = \{0, 1\}$$

+	0	1
0	0	1
1	1	0

·	0	1
0	0	0
1	0	1

Since a field is a ring, all of the facts about rings established in Section 3.2 hold for fields as well.

3.4 THEOREM **THE GROUP OF INVERTIBLE ELEMENTS IN A FIELD** Let $(F, +, \cdot)$ be a field. Then, the group of invertible elements in F is $F \setminus \{0\}$.

PROOF By the definition of a field, every element in $F \setminus \{0\}$ is invertible in F. Conversely, if x is an invertible element of F, then x cannot be equal to 0. Indeed, if 0 were invertible in F, we would have

$$0 \cdot 0^{-1} = 1.$$

Moreover, by Theorem 2.3, we would also have

$$0 \cdot 0^{-1} = 0.$$

Combining these two equations, we conclude that $1 = 0$, in contradiction with Part (iii) of Theorem 3.2. □

The multiplicative inverse x^{-1} of a nonzero element x in a field is often called the **reciprocal** of x. The following corollary of Theorem 3.4 is obtained by applying the facts about invertible elements in a group (Section 3.1) to the group $(F \setminus \{0\}, \cdot)$ of invertible elements of a field F. We leave its routine verification to you as an exercise (Problem 11).

3.5 COROLLARY

PROPERTIES OF RECIPROCALS IN A FIELD Let $(F, +, \cdot)$ be a field, and let $x, y \in F$.

(i) $1^{-1} = 1$
(ii) $x \neq 0 \Rightarrow x^{-1} \neq 0$ and $(x^{-1})^{-1} = x$
(iii) $x \neq 0 \wedge y \neq 0 \Rightarrow xy \neq 0 \wedge (xy)^{-1} = y^{-1}x^{-1}$
(iv) $xy = 1 \Rightarrow x, y \neq 0 \wedge y = x^{-1}$

Because multiplication in a field is commutative, Part (iii) of Corollary 3.5 can be rewritten in the form

$$x \neq 0 \wedge y \neq 0 \Rightarrow xy \neq 0 \wedge (xy)^{-1} = x^{-1}y^{-1}.$$

In particular, we have

$$x \neq 0 \wedge y \neq 0 \Rightarrow xy \neq 0,$$

from which it follows that

$$xy = 0 \Rightarrow x = 0 \vee y = 0,$$

and so *there are no divisors of zero in a field*. In other words, *a field is an integral domain* (Problem 13). This is the justification for solving equations in elementary algebra by the method of factorization (Problem 14).

By combining multiplication and the formation of reciprocals, we can introduce the idea of *division* in a field.

3.6 DEFINITION

DIVISION IN A FIELD Let F be a field. If $x, y \in F$ and $y \neq 0$, we define $x/y \in F$ by

$$x/y = xy^{-1}.$$

We read x/y in English as "x divided by y," or as "x over y," and we refer to the expression x/y as a **fraction** with **numerator** x and **denominator** y. Notice that x/y is defined only if $y \neq 0$. Thus:

A fraction with a zero denominator is meaningless.

Alternative notation for x/y is

$$x \div y \quad \text{or} \quad \frac{x}{y}.$$

As a particular case of Definition 3.6, we have, for $y \neq 0$,

$$\frac{1}{y} = 1 \cdot y^{-1} = y^{-1}.$$

Thus, the notation $1/y$ can be used rather than y^{-1} for the reciprocal of a nonzero element in a field.

3.7 Example Find 3/2 in the Galois field GF(5).

SOLUTION By Definition 3.6, $3/2 = 3 \cdot 2^{-1}$, so the first problem is to find 2^{-1} in GF(5). Now, 2^{-1} must be one of the elements 1, 2, 3, or 4, and $2 \cdot 2^{-1} = 1$ in GF(5). Trying the possibilities one at a time, we find that $2 \cdot 3 = 1$, so that $2^{-1} = 3$ in GF(5). (Recall that we compute $2 \cdot 3$ in GF(5) by taking the usual product 6, dividing it by 5, and writing the remainder 1 for the result.) Thus, in GF(5),

$$\frac{3}{2} = 3 \cdot 2^{-1} = 3 \cdot 3 = 4.$$

As strange and unfamiliar as this may seem, "arithmetic" in the Galois field GF(5) is just as logically self-consistent as ordinary arithmetic! □

The familiar process of "dividing both sides of an equation by a nonzero quantity" is justified by the following theorem, the proof of which we leave as an exercise (Problem 19).

3.8 THEOREM **DIVIDING AN EQUATION BY A NONZERO QUANTITY** Let F be a field, and suppose that $a, b \in F$ with $b \neq 0$. Then:

$$ab = c \Leftrightarrow a = \frac{c}{b}.$$

The next theorem justifies the rule that *the reciprocal of a fraction is obtained by inverting the fraction.*

3.9 THEOREM

THE RECIPROCAL OF A FRACTION Let F be a field, and suppose that $a, b \in F$ with $a \neq 0$ and $b \neq 0$. Then, $a/b \neq 0$ and

$$\left(\frac{a}{b}\right)^{-1} = \frac{b}{a}.$$

PROOF Let $x = a/b$, and let $y = b/a$. (Note that, by hypothesis, $b \neq 0$ and $a \neq 0$; so we can legitimately form these fractions.) By Definition 3.6, $x = ab^{-1}$ and $y = ba^{-1}$. Therefore,

$$xy = ab^{-1}ba^{-1} = aa^{-1}bb^{-1} = 1 \cdot 1 = 1.$$

By Part (iv) of Theorem 3.5, it follows that $a/b = x \neq 0$ and that

$$\frac{b}{a} = y = x^{-1} = \left(\frac{a}{b}\right)^{-1}.$$

□

The usual rule of arithmetic for multiplying two fractions is justified by the following theorem.

3.10 THEOREM

THE PRODUCT OF TWO FRACTIONS Let F be a field, and suppose that $a, b, c, d \in F$ with $b, d \neq 0$. Then,

$$\left(\frac{a}{b}\right)\left(\frac{c}{d}\right) = \frac{ac}{bd}.$$

PROOF By Definition 3.6, $a/b = ab^{-1}$ and $c/d = cd^{-1}$; hence,

$$\left(\frac{a}{b}\right)\left(\frac{c}{d}\right) = ab^{-1}cd^{-1} = acb^{-1}d^{-1}.$$

By Part (iii) of Theorem 3.5,

$$b^{-1}d^{-1} = (bd)^{-1},$$

and it follows that

$$\left(\frac{a}{b}\right)\left(\frac{c}{d}\right) = ac(bd)^{-1} = \frac{ac}{bd}$$

by Definition 3.6 again.

□

By a computation similar to the one in the proof of Theorem 3.10, you can prove the following theorem, which justifies the standard procedure of canceling a common factor in the numerator and denominator of a fraction (Problem 20).

3.11 THEOREM

CANCELLATION LAW FOR FRACTIONS Let F be a field and suppose that $a, b, c \in F$ with $b, c \neq 0$. Then,

$$\frac{ac}{bc} = \frac{a}{b}.$$

In a field F, the distributive law justifies the usual rule for adding fractions with a common denominator. Indeed, if $a, b, d \in F$ with $d \neq 0$, we have

$$\frac{a}{d} + \frac{b}{d} = ad^{-1} + bd^{-1} = (a + b)d^{-1} = \frac{a + b}{d}.$$

To add two fractions with different denominators, we use the cancellation law "in reverse" to obtain a common denominator and then add the fractions by the rule above.

3.12 THEOREM

THE SUM OF TWO FRACTIONS Let F be a field, and suppose that $a, b, c, d \in F$ with $b, d \neq 0$. Then,

$$\frac{a}{b} + \frac{c}{d} = \frac{ad + bc}{bd}.$$

PROOF By Theorem 3.11, we can multiply numerator and denominator of the first fraction by d, and we can multiply numerator and denominator of the second fraction by b, to obtain

$$\frac{a}{b} + \frac{c}{d} = \frac{ad}{bd} + \frac{cb}{db} = \frac{ad}{bd} + \frac{bc}{bd} = \frac{ad + bc}{bd}.$$ □

We leave it to you as an exercise to prove the analogous rule for subtraction of fractions in a field, namely,

$$\frac{a}{b} - \frac{c}{d} = \frac{ad - bc}{bd},$$

provided, of course, that $b, d \neq 0$ (Problem 23).

The following theorem justifies the familiar rules for negation of a fraction.

3.13 THEOREM **NEGATION OF A FRACTION** Let F be a field, and suppose that $a, b \in F$ with $b \neq 0$. Then,

$$-\left(\frac{a}{b}\right) = \frac{-a}{b} = \frac{a}{-b}.$$

PROOF By Part (i) of Theorem 2.5, we have

$$-\left(\frac{a}{b}\right) = -(ab^{-1}) = (-a)b^{-1} = \frac{-a}{b}.$$

If $-b$ were equal to 0, we would have $b = -(-b) = -0 = 0$, contradicting the hypothesis that $b \neq 0$. Therefore, $-b \neq 0$. Since $b \neq 0$ and $-b \neq 0$, the fact that F is an integral domain implies that $b(-b) \neq 0$. By Part (iii) of Theorem 2.5,

$$(-a)(-b) = ab;$$

hence,

$$\frac{(-a)(-b)}{b(-b)} = \frac{ab}{b(-b)} = \frac{ab}{(-b)b},$$

and, applying the cancellation law (Theorem 3.11) to the first and last fractions, we conclude that

$$\frac{-a}{b} = \frac{a}{-b}.$$ □

Integer exponents can be used in an arbitrary field F in the same way they are used in ordinary arithmetic. Thus, if n is a positive integer, and $a \in F$, we understand that

$$a^n = a \cdot a \cdot a \cdot \cdots \cdot a,$$

where there are n factors on the right. Positive integer exponents are characterized by the following two properties:

(i) $a^1 = a$
(ii) $a^{n+1} = a^n \cdot a$ for every positive integer n.

Properties (i) and (ii) are especially useful for making proofs by mathematical induction of theorems involving exponents. If n and m are positive integers, it is easy to see that

$$a^{n+m} = a^n a^m$$

just by noting that there are $n + m$ factors on each side of the equation. (If desired, a more rigorous proof can be made by holding m constant and using mathematical induction on n.)

If $a \neq 0$, we follow the convention of ordinary arithmetic and define

$$a^0 = 1.$$

Note that 0^0 *is undefined.* Finally, if $a \neq 0$ and n is a positive integer, we define

$$a^{-n} = (a^{-1})^n,$$

noting that $-n$ is a negative integer. We leave it to you to check that all the usual rules for integer exponents hold in the field F (Problems 27–32).

PROBLEM SET 3.3

1. Check that GF(2) is a field.
2. Prove Theorem 3.2.
3. Check that GF(3) (Example 3.3) is a field.
4. Check that the complex numbers $\mathbb{C}$ form a field. [Hint: If at least one of the two real numbers a and b is nonzero, so that $z = a + bi \neq 0$ in $\mathbb{C}$, then $z^{-1} = (a/d) - (b/d)i$, where $d = a^2 + b^2$.]
5. Show that, if F is a field and $a \in F$, then $a = 0 \Leftrightarrow -a = 0$.
6. Check that $\mathbb{Q}(\sqrt{2})$ is a field.
7. Write the addition and multiplication tables for GF(5).
8. Is it possible in a field F to have $1 = -1$? Explain.
9. Let $R = \{0, 1, 2, 3\}$, and introduce addition and multiplication operations into R by defining $x + y$ (respectively, xy) to be the remainder upon division of the usual arithmetic sum (respectively, product) of x and y by 4.
   (a) Write addition and multiplication tables for R.
   (b) Show that $(R, +, \cdot)$ is not a field.
10. An **idempotent** in a field F is an element $a \in F$ such that $a \cdot a = a$. Show that 0 and 1 are the only two idempotents in F.
11. Prove Corollary 3.5.
12. If the positive integer p is a prime, show that GF(p) is isomorphic to the ring $\mathbb{Z}_p$ in Problem 27 of Problem Set 3.2.
13. Prove that a field is an integral domain.
14. Explain what the fact that the field $\mathbb{R}$ is an integral domain has to do with the usual technique of solving equations by factoring.
15. Find (a) 2/3 and (b) $-(2/3)$ in GF(5).
16. If F is a field, find all solutions in F of the equation $x = x^{-1}$ for $x \neq 0$.
17. Find (a) 1/4 and (b) $-(1/4)$ in GF(5).
18. A **nilpotent** in a field F is an element $a \in F$ such that $a^n = 0$ for some positive integer n. Show that 0 is the only nilpotent in F.

19. Prove Theorem 3.8.

20. Prove Theorem 3.11.

21. Prove the rule of **cross multiplication** in a field F: If $a, b, c, d \in F$ with $b, d \neq 0$, then,

$$\frac{a}{b} = \frac{c}{d} \Leftrightarrow ad = bc.$$

22. In the field $\mathbb{R}$ of real numbers, the following condition holds: $x^2 + y^2 = 0 \Leftrightarrow x = y = 0$. Is this a special property of the real numbers, or does it hold in any field F?

23. Let F be a field, and suppose that $a, b, c, d \in F$ with $b, d \neq 0$. Prove that

$$\frac{a}{b} - \frac{c}{d} = \frac{ad - bc}{bd}.$$

24. Let $F = \{0, a, 1\}$ be a field with three elements.
(a) Show that $a^2 = 1$. [Hint: There are only three possibilities for a^2; it must be one of the elements 0, a, or 1.]
(b) Show that $1 + 1 = a$.
(c) Show that F is isomorphic to GF(3).

25. Prove the *invert-and-multiply rule*: If F is a field, and $a, b, c, d \in F$ with $b, c, d \neq 0$, then $(a/b)/(c/d) = (a/b)(d/c)$.

26. Show that, in the field $\mathbb{C}$ of complex numbers, the equation $z^2 = a$ has at least one solution for each element $a \in \mathbb{C}$.

In Problems 27–32, let F be a field, let $a, b \in F$ with $a, b \neq 0$, and let m and n be integers. Prove each statement.

27. $a^{n+m} = a^n a^m$ **28.** $(a^n)^m = a^{nm}$ **29.** $(ab)^n = a^n b^n$

30. $\left(\dfrac{a}{b}\right)^n = \dfrac{a^n}{b^n}$ **31.** $a^{n-m} = \dfrac{a^n}{a^m}$ **32.** $\dfrac{a^{-n}}{b^{-m}} = \dfrac{b^m}{a^n}$

33. Show that the quadratic equation $x^2 + x + 1 = 0$ has no solution in the field GF(2).

34. Show that there exists a field F containing four distinct elements, $F = \{0, 1, a, b\}$, in which 0 and 1 add and multiply as in GF(2) and in which both of the elements a and b are solutions of $x^2 + x + 1 = 0$ in Problem 33. Write the addition and multiplication tables for this field.

35. Assume that F_1 and F_2 are two fields such that, as sets, $F_1 \subseteq F_2$, and such that the addition and multiplication operations in F_1 are just the restrictions to F_1 of the corresponding operations in F_2. Then F_1 is said to be a **subfield** of F_2, and F_2 is said to be an **extension field** of F_1.
(a) Find two different subfields of $\mathbb{R}$ (each different from $\mathbb{R}$ itself).
(b) Find an extension field of $\mathbb{R}$ (other than $\mathbb{R}$ itself).

36. Let $\mathbb{R}^3$ denote the set of all vectors $\mathbf{v} = (x, y, z)$, $x, y, z \in \mathbb{R}$. Define the dot product $\mathbf{v} \cdot \mathbf{w}$ and the cross product $\mathbf{v} \times \mathbf{w}$ of vectors in $\mathbb{R}^3$ as usual. Let $\mathbb{H}$ be the Cartesian product $\mathbb{H} = \mathbb{R} \times \mathbb{R}^3$. Define addition and multiplication operations on $\mathbb{H}$ by

$$(a, \mathbf{v}) + (b, \mathbf{w}) = (a + b, \mathbf{v} + \mathbf{w})$$

and

$$(a, \mathbf{v})(b, \mathbf{w}) = (ab - \mathbf{v} \cdot \mathbf{w}, a\mathbf{w} + b\mathbf{v} + \mathbf{v} \times \mathbf{w}).$$

Show that $(\mathbb{H}, +, \cdot)$ is a division ring but not a field. [Hint: If $h = (a, \mathbf{v}) \neq 0$, so that $a \neq 0$ and $\mathbf{v} \neq \mathbf{0}$, then $h^{-1} = (ca, -c\mathbf{v})$, where $c = (a^2 + \mathbf{v} \cdot \mathbf{v})^{-1}$.] [$\mathbb{H}$ is isomorphic to the division ring of **real quaternions** discovered by the Irish mathematician W. R. Hamilton (1805–1865) in 1843.]

4 ORDERED FIELDS

Visualizing the real number field $\mathbb{R}$, in the usual way, as points on a line, we see that there is a natural order relation $\leq$ on $\mathbb{R}$. Indeed, for $x, y \in \mathbb{R}$, $x < y$ means that the point corresponding to x lies to the left of the point corresponding to y on the line (Figure 3-1), and $x \leq y$ means that either $x < y$ or else $x = y$. The *positive* real numbers, whose corresponding points lie to the *right* of 0 on the line, are the numbers $x \in \mathbb{R}$ such that $0 < x$. Notice that, if P denotes the set of all positive real numbers, then, for $x, y \in \mathbb{R}$,

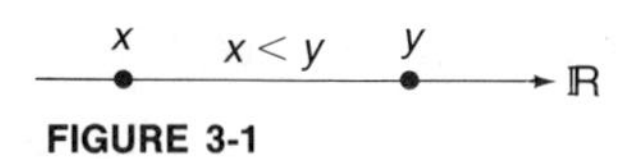

FIGURE 3-1

$$x < y \Leftrightarrow y - x \in P.$$

Thus, the order relation on $\mathbb{R}$ both determines and is determined by the set P of positive real numbers.

In the present section, we study fields F in which it is possible to introduce a notion of *order*, patterned after the order relation $\leq$ in $\mathbb{R}$. The most efficient way to do this is to begin with an analogue $P \subseteq F$ of the set of positive real numbers.

4.1 DEFINITION

ORDERED FIELD An **ordered field** is a field F together with a fixed subset $P \subseteq F$, called the set of *positive* elements of F, which satisfies the following conditions:

(i) $0 \notin P$
(ii) $x \in F, x \neq 0, x \notin P \Rightarrow -x \in P$
(iii) $x, y \in P \Rightarrow x + y \in P$
(iv) $x, y \in P \Rightarrow xy \in P$

Let F be an ordered field as in Definition 4.1. If $x \in P$, we say that x is **positive**. Condition (i) says that 0 *is not positive*. A nonzero element $x \in F$ is said to be **negative** if $-x$ is positive. Condition (ii) says that *if a nonzero element is not positive, then it is negative.* Conditions (iii) and (iv) require that the set P be *closed under addition and multiplication.* Thus, in an ordered field, *the sum and the product of positive elements are positive.*

4.2 THEOREM

LAW OF TRICHOTOMY Let F be an ordered field with P as its set of positive elements. Then, for $x \in F$, *exactly one* of the following is true:

(i) $x \in P$
(ii) $x = 0$
(iii) $-x \in P$

PROOF By condition (ii) in Definition 4.1, at least one of the three statements (i), (ii), and (iii) must be true. We must show that *no two* of these three statements can be true simultaneously. Since $0 \notin P$, we cannot have both (i) and (ii). Since $-0 = 0$ and $0 \notin P$, we cannot have both (ii) and (iii). Finally, if (i) and (iii) were both true, then, since P is closed under addition, we would again have $0 = x + (-x) \in P$, contradicting Part (i) of Definition 4.1. □

The law of trichotomy can be paraphrased by the statement that, *in an ordered field, every element is positive, zero, or negative, and no element satisfies any two of these three conditions.*

4.3 THEOREM

SQUARES OF NONZERO ELEMENTS ARE POSITIVE Let F be an ordered field with P as its set of positive elements. Let $x \in F$ with $x \neq 0$. Then $x^2 \in P$.

PROOF Since $x \neq 0$, the law of trichotomy (Theorem 4.2) implies that either $x \in P$ or else $-x \in P$. If $x \in P$, then, since P is closed under multiplication, we have

$$x^2 = x \cdot x \in P.$$

Likewise, if $-x \in P$, then, using Part (iii) of Theorem 2.5, we have

$$x^2 = x \cdot x = (-x)(-x) \in P.$$

In any case, then, provided that $x \neq 0$, we have $x^2 \in P$. □

4.4 COROLLARY

POSITIVITY OF 1 In an ordered field F with P as its set of positive elements, it is always the case that

$$1 \in P.$$

PROOF $1 \neq 0$ and $1 = 1^2$. □

4.5 COROLLARY

POSITIVE PRODUCTS AND QUOTIENTS Let F be an ordered field, and let $x, y \in F$ with $y \neq 0$. Then,

$$xy \in P \Leftrightarrow \frac{x}{y} \in P.$$

PROOF We prove the implication from left to right and leave the converse implication as an exercise (Problem 1). Suppose that $xy \in P$. By Theorem 4.3, $(1/y)^2 \in P$; hence, since the product of positive elements is positive,

$$\frac{x}{y} = xy\left(\frac{1}{y}\right)^2 \in P.$$ □

4.6 Example Show that the field $\mathbb{C}$ of complex numbers *cannot* be made into an ordered field.

SOLUTION Suppose that, somehow, $\mathbb{C}$ could be made into an ordered field with P as its set of positive elements. Then, by Theorem 4.3,

$$-1 = i^2 \in P.$$

However, by Corollary 4.4, we also have

$$1 \in P,$$

contradicting the law of trichotomy (Theorem 4.2). □

Here are some examples of ordered fields:

1. $\mathbb{R}$, the field of real numbers, is an ordered field if we take P to be the set of real numbers that are positive in the usual sense; that is, $P = \{x \in \mathbb{R} \mid \exists y \in \mathbb{R}, y \neq 0 \wedge x = y^2\}$.
2. $\mathbb{Q}$, the field of rational numbers, is an ordered field if we take P to be the set of all rational numbers that can be written in the form m/n, where $m \cdot n \in \mathbb{N} = \{1, 2, 3, \ldots\}$.
3. $\mathbb{R}(X)$, the field of all **rational functions** with real coefficients (that is, all expressions of the form $f(X) = p(X)/q(X)$, where $p(X)$ and $q(X)$ are polynomials with real coefficients, and $q(X)$ is not the zero polynomial)

can be made into an ordered field as follows (Problem 2): Call a nonzero polynomial $p(X)$ *positive* if its leading coefficient is a positive real number. Let P be the set of all rational functions $f(X) = p(X)/q(X)$ such that $p(X) \cdot q(X)$ is a positive polynomial.

4. Let $\mathbb{Z}$ be the set of all integers, and let

$$\mathbf{L} = \{\phi : \mathbb{Z} \to \mathbb{R} \mid \exists N \in \mathbb{Z}, \forall n \in \mathbb{Z}, n < N \Rightarrow \phi(n) = 0\}.$$

For $\phi, \psi \in \mathbf{L}$, define $\phi + \psi, \phi \cdot \psi \in \mathbf{L}$ by

$$(\phi + \psi)(n) = \phi(n) + \psi(n), \forall n \in \mathbb{Z}$$

and

$$(\phi \cdot \psi)(n) = \sum_{j \in \mathbb{Z}} \phi(j)\psi(n - j), \forall n \in \mathbb{Z}.$$

Let

$$P = \{\phi \in \mathbf{L} \mid \exists N \in \mathbb{Z}, \phi(N) > 0 \wedge (\forall n \in \mathbb{Z}, n < N \Rightarrow \phi(n) = 0)\}.$$

Then, with P as its set of positive elements, $\mathbf{L}$ is an ordered field (Problem 40).

The last example above is admittedly abstruse; it is included here only to show that there are ordered fields other than the ones you have already encountered in high-school algebra.

4.7 DEFINITION

THE RELATION < ON AN ORDERED FIELD Let F be an ordered field with P as its set of positive elements. We define the binary relation $<$ on F by

$$x < y \Leftrightarrow y - x \in P$$

for all $x, y \in F$.

If $x < y$ we say that x is **less than** y or, equivalently, that y is **greater than** x. The statement $x < y$ is often written in the alternative form $y > x$. An expression of the form $x < y$, or $y > x$, is called a **strict inequality**. We define the **nonstrict inequality** $x \le y$ by

$$x \le y \Leftrightarrow x < y \vee x = y.$$

If $x \le y$, we say that x is **less than or equal to** y or, equivalently, that y is **greater than or equal to** x. The statement $x \le y$ is often written in the alternative form $y \ge x$. Notice that

$$x \le x$$

holds for any element x in an ordered field F; that is, *the binary relation* $\leq$ *is reflexive on* F. (See Definition 2.9 in Chapter 2.)

We leave the proof of the following lemma as an exercise (Problem 3).

4.8 LEMMA Let F be an ordered field with P as its set of positive elements. Then, for $x, y \in F$:

(i) $x \in P \Leftrightarrow 0 < x$
(ii) $-x \in P \Leftrightarrow x < 0$
(iii) $x < y \Rightarrow x \neq y$

Using Definition 4.7, we can restate the law of trichotomy (Theorem 4.2) in the following equivalent form (Problem 5).

4.9 THEOREM **LAW OF TRICHOTOMY (ALTERNATIVE VERSION)** If F is an ordered field, then, for $x, y \in F$, *exactly one* of the following is true:

(i) $x < y$
(ii) $x = y$
(iii) $x > y$

The next theorem shows that both strict and nonstrict inequalities are transitive relations. (See Definition 2.9 in Chapter 2.)

4.10 THEOREM **TRANSITIVITY OF INEQUALITIES** Let F be an ordered field, and let $x, y, z \in F$.

(i) $x < y \wedge y < z \Rightarrow x < z$
(ii) $x \leq y \wedge y \leq z \Rightarrow x \leq z$

PROOF We prove (i) and leave (ii) as an exercise (Problem 7a). Thus, suppose that $x < y$ and $y < z$. Then, by Definition 4.7,

$$y - x \in P \quad \text{and} \quad z - y \in P.$$

Therefore, by Part (iii) of Definition 4.1,

$$z - x = (y - x) + (z - y) \in P;$$

hence, by Definition 4.7 again, $x < z$. □

We have noted that the relation $\leq$ is reflexive on an ordered field F, and Part (ii) of Theorem 4.10 shows that it is transitive. As a consequence of the law of trichotomy, $\leq$ is also *antisymmetric* (see Definition 2.11 in Chapter 2); and in fact, $(F, \leq)$ *is a totally ordered set* (see Definition 2.24 in Chapter 2). We ask you to prove these facts in Problem 9.

In an ordered field F, we write the **compound inequality**

$$x < y < z$$

as an abbreviation for the statement that

$$x < y \wedge y < z.$$

It is good mathematical practice to arrange it so that all inequalities in a compound inequality "run in the same direction." Thus, writing an expression such as

$$x < y > w$$

to mean that $x < y$ and $w < y$ would be regarded as bad form. However, it is acceptable to abbreviate the statement that

$$x < y \wedge w < y$$

as

$$x, w < y.$$

Strict and nonstrict inequalities can be combined in a compound inequality, so that, for instance,

$$x < y \leq z$$

is understood to mean that

$$x < y \wedge y \leq z.$$

Note that, for such a "mixed" compound inequality, we have

$$x < y \wedge y \leq z \Rightarrow x < z$$

(Problem 13).

4.11 THEOREM

ADDITION OF INEQUALITIES Let F be an ordered field, and let $x, y, z, w \in F$. Then:

(i) $x < y \Rightarrow x + z < y + z$
(ii) $x < y$ and $z < w \Rightarrow x + z < y + w$

PROOF (i) Suppose that $x < y$. Then, by Definition 4.7, we have $y - x \in P$. But then,

$$(y + z) - (x + z) = y - x \in P,$$

and consequently,

$$x + z < y + z.$$

(ii) Suppose that $x < y$ and $z < w$. Then, by Part (i) above,

$$x + z < y + z$$

and likewise,

$$z + y < w + y.$$

Therefore,

$$x + z < y + z = z + y < w + y,$$

and it follows from Part (i) of Theorem 4.10 (transitivity of the $<$ relation) that

$$x + z < w + y. \qquad \square$$

Suitably modified versions of Parts (i) and (ii) of Theorem 4.11 hold for nonstrict inequalities; for instance, we have

$$x \le y \Rightarrow x + z \le y + z$$

and

$$x < y \quad \text{and} \quad z \le w \Rightarrow x + z < y + w$$

(Problem 15b).

4.12 THEOREM

MULTIPLICATION OF INEQUALITIES Let F be an ordered field, and let $x, y, z, w \in F$. Then:

(i) $x < y$ and $0 < z \Rightarrow xz < yz$
(ii) $x < y$ and $z < 0 \Rightarrow yz < xz$
(iii) $0 < x < y$ and $0 < z < w \Rightarrow 0 < xz < yw$

PROOF (i) Suppose that $x < y$ and $0 < z$. Then we have

$$y - x \in P \quad \text{and} \quad z \in P,$$

and it follows from Part (iv) of Definition 4.1 that

$$(y - x)z \in P.$$

Therefore,

$$yz - xz \in P$$

and consequently,

$$xz < yz.$$

(ii) Suppose that $x < y$ and $z < 0$. Then,

$$y - x \in P \quad \text{and} \quad -z \in P,$$

and therefore,

$$(y - x)(-z) \in P.$$

Consequently,

$$xz - yz \in P$$

and so

$$yz < xz.$$

(iii) We leave the proof of Part (iii) as an exercise (Problem 7b). □

By Part (i) of Theorem 4.12, we are permitted to multiply both sides of an inequality by a positive element of F; but if we multiply by a negative element of F, we must *reverse the inequality*. Suitably modified versions of Parts (i), (ii), and (iii) of the theorem hold for nonstrict inequalities; for instance, we have

$$x \leq y \quad \text{and} \quad 0 \leq z \Rightarrow xz \leq yz,$$

$$x \leq y \quad \text{and} \quad z \leq 0 \Rightarrow yz \leq xz,$$

and

$$0 \leq x \leq y \quad \text{and} \quad 0 \leq z \leq w \Rightarrow 0 \leq xz \leq yw$$

(Problem 19).

4.13 COROLLARY Let x and y be elements in an ordered field F, and suppose that $0 < x, y$. Then,

$$x < y \Leftrightarrow x^2 < y^2.$$

PROOF If $0 < x < y$, then $x^2 = x \cdot x < y \cdot y = y^2$ follows from Part (iii) of Theorem 4.12. Conversely, suppose that $x^2 < y^2$ but that $x < y$ is false. Then, by the law of trichotomy, $y \leq x$. If $y = x$, then $y^2 = x^2$, contradicting $x^2 < y^2$. However, if $y < x$, then (by the same argument given above), $y^2 < x^2$, again contradicting $x^2 < y^2$. □

4.14 DEFINITION

SIGNUM FUNCTION Let F be an ordered field. We define a function $\operatorname{sgn}: F \to F$, called the **signum function** on F, by

$$\operatorname{sgn}(x) = \left\{ \begin{array}{rl} 1, & \text{if } x > 0 \\ 0, & \text{if } x = 0 \\ -1, & \text{if } x < 0 \end{array} \right\}.$$

Note that, by the law of trichotomy, sgn is a well-defined function on the ordered field F. The signum function has the property that

$$\operatorname{sgn}(xy) = \operatorname{sgn}(x) \cdot \operatorname{sgn}(y)$$

for all $x, y \in F$ (Problem 21). Using the signum function, we can define the idea of absolute value as follows:

4.15 DEFINITION

ABSOLUTE VALUE Let F be an ordered field. For $x \in F$, we define the **absolute value** of x, in symbols $|x|$, by

$$|x| = x \cdot \operatorname{sgn}(x).$$

Note that, if $x \geq 0$, then $|x| = x$; however, if $x < 0$, then $|x| = -x$ (Problem 23). In any case, then, $|x| \geq 0$ (Problem 24).

4.16 LEMMA

PRODUCTS AND ABSOLUTE VALUES Let F be an ordered field, and let $x, y \in F$. Then:

(i) $|xy| = |x|\,|y|$
(ii) $|x^2| = x^2$

PROOF Part (ii) is just a special case of Part (i). To prove Part (i), we use the fact that

$$\operatorname{sgn}(xy) = \operatorname{sgn}(x) \cdot \operatorname{sgn}(y).$$

Therefore, by Definition 4.15,

$$\begin{aligned} |xy| &= xy \cdot \operatorname{sgn}(xy) = xy \cdot \operatorname{sgn}(x) \cdot \operatorname{sgn}(y) = x \cdot \operatorname{sgn}(x) \cdot y \cdot \operatorname{sgn}(y) \\ &= |x|\,|y|. \end{aligned}$$

□

Some of the properties of absolute value are most easily obtained by a *case analysis*, as illustrated in the proof of the following lemma.

4.17 LEMMA Let F be an ordered field. Then, for $x \in F$,

$$-|x| \leq x \leq |x|.$$

PROOF We consider separately the case in which $0 \leq x$ and the case in which $x < 0$. (By the law of trichotomy, these cases cover all possibilities.) If $0 \leq x$, then $x = |x|$ and

$$-|x| = -x \leq 0 \leq x \leq |x|.$$

If $x < 0$, then $-x = |x|$ and

$$-|x| = x < 0 \leq |x|.$$

Thus, in either case,

$$-|x| \leq x \leq |x|.$$ □

4.18 THEOREM **TRIANGLE INEQUALITY** Let F be an ordered field. Then, for $x, y \in F$,

$$|x + y| \leq |x| + |y|.$$

PROOF By Lemma 4.17, we have

$$-|x| \leq x \leq |x| \quad \text{and} \quad -|y| \leq y \leq |y|. \tag{1}$$

Adding these inequalities, we obtain

$$x + y \leq |x| + |y| \tag{2}$$

and

$$-(|x| + |y|) \leq x + y. \tag{3}$$

Multiplying the inequality in (3) by -1, and reversing it, we find that

$$-(x + y) \leq |x| + |y|. \tag{4}$$

Now we consider first the case in which $0 \leq x + y$ and second the case in which $x + y < 0$. In the first case, we use Inequality (2) to conclude that

$$|x + y| = x + y \leq |x| + |y|. \tag{5}$$

In the second case, we use Inequality (4) to obtain

$$|x + y| = -(x + y) \leq |x| + |y|. \tag{6}$$

Thus, in either case,

$$|x + y| \leq |x| + |y|.$$ □

PROBLEM SET 3.4

1. Let F be an ordered field with P as its set of positive elements, and let $x, y \in F$ with $y \neq 0$. Complete the proof of Corollary 4.5 by showing that $x/y \in P \Rightarrow xy \in P$.

2. Show that the field $\mathbb{R}(X)$ of rational functions with real coefficients becomes an ordered field if P is defined as in example 3 on page 212.

3. Prove Lemma 4.8.

4. Let F be an ordered field with P as its set of positive elements, and let K be a subfield of F. (See Problem 35 in Problem Set 3.3.) Show that K becomes an ordered field if we take its set of positive elements to be $P \cap K$.

5. Prove Theorem 4.9.

6. The rational numbers $\mathbb{Q}$ form a subfield of the real numbers $\mathbb{R}$. If $\mathbb{Q}$ is made into an ordered field as in Problem 4, show that the positive elements in $\mathbb{Q}$ are exactly those rational numbers that can be written in the form m/n, where

$$m \cdot n \in \mathbb{N} = \{1, 2, 3, \ldots\}.$$

7. (a) Prove Part (ii) of Theorem 4.10.
(b) Prove Part (iii) of Theorem 4.12.

8. Let $\mathbb{Q}(\sqrt{2})$ be the field obtained by adjoining $\sqrt{2}$ to the rationals. (See example 4 on page 201.) Is it possible to make $\mathbb{Q}(\sqrt{2})$ into an ordered field? Explain.

9. Let F be an ordered field.
(a) Prove that the relation $\leq$ on F is antisymmetric.
(b) Prove that F is totally ordered by the relation $\leq$.

10. Prove that, in an ordered field F, it is impossible to have $1 + 1 = 0$.

11. Prove that GF(2) cannot be made into an ordered field.

12. True or false: In the ordered field $\mathbb{R}$ of real numbers, $0 \leq x < y \Rightarrow \sqrt{x} < \sqrt{y}$. If true, prove it; if false, give a counterexample.

In Problems 13–32, suppose that F is an ordered field and that $x, y, z, w \in F$. Prove each statement.

13. $x < y \wedge y \leq z \Rightarrow x < z$.

14. $0 < x \Rightarrow 0 < 1/x$.

15. (a) $x \leq y \Rightarrow x + z \leq y + z$.
(b) $x < y \wedge z \leq w \Rightarrow x + z < y + w$.
(c) $x \leq y \wedge z \leq w \Rightarrow x + z \leq y + w$.

16. $x < y \Rightarrow \exists a \in F,\ x < a < y$. [Hint: Let $a = (x + y)/(1 + 1)$.]

17. (a) $x < y \Leftrightarrow x + z < y + z$.
(b) $x \leq y \Leftrightarrow x + z \leq y + z$.

18. F contains infinitely many elements. [Hint: Use the result of Problem 16.]

19. (a) $x \le y \wedge 0 \le z \Rightarrow xz \le yz$.
(b) $x \le y \wedge z \le 0 \Rightarrow yz \le xz$.
(c) $0 \le x \le y \wedge 0 \le z \le w \Rightarrow 0 \le xz \le yw$.

20. F cannot contain a largest element nor a smallest element.

21. $\operatorname{sgn}(xy) = \operatorname{sgn}(x) \cdot \operatorname{sgn}(y)$.

22. $x = 0 \Leftrightarrow |x| = 0$.

23. (a) $x \ge 0 \Rightarrow |x| = x$. (b) $x < 0 \Rightarrow |x| = -x$.

24. $|x| \ge 0$ with equality if and only if $x = 0$.

25. $|-x| = |x|$.

26. (a) $|x| = x \Rightarrow x \ge 0$. (b) $|x| = -x \Rightarrow x \le 0$.

27. $y \ne 0 \Rightarrow |x/y| = |x|/|y|$.

28. $y \ne 0 \Rightarrow \operatorname{sgn}(y) = y/|y|$.

29. $|x - y| \le |x| + |y|$.

30. If $x \le y$ and $-x \le y$, then $|x| \le y$.

31. $|x| < y \Leftrightarrow -y < x < y$.

32. $||x| - |y|| \le |x - y|$.

33. Prove that, in an ordered field F, the formula

$$\frac{x + y + |x - y|}{1 + 1}$$

gives the larger of the two elements x and y (or gives their common value if they are equal).

34. Show that there is just one way to make the field $\mathbb{Q}$ of rational numbers into an ordered field.

35. Find a formula similar to that in Problem 33 for the smaller of two elements x and y in an ordered field F.

36. Show that there is just one way to make the field $\mathbb{R}$ of real numbers into an ordered field.

37. Show that, in an ordered field F, $|x + y| = |x| + |y|$ holds if and only if $xy \ge 0$.

38. Does there exist a field F that can be made into an ordered field in two different ways? Explain.

39. If x and y are elements in an ordered field F, find a necessary and sufficient condition, similar to that in Problem 37, for the equation $|x - y| = |x| + |y|$ to hold.

40. (a) Show that **L**, equipped with the operations defined in example 4 on page 213, is a field.
(b) Show that **L** is an ordered field.

*5 VECTOR SPACES

In elementary mathematics, a *vector* is defined as *a quantity that has a magnitude and a direction.* Such a quantity is represented pictorially by an arrow, pointing in the appropriate direction, with the length of the arrow equal to the magnitude of the vector. When working with vectors, the real numbers are often called *scalars,* since they are visualized as points on a number scale. Vectors are often printed in boldface type (for instance, $\mathbf{v}$, $\mathbf{w}$) to distinguish them from scalars.

A special *zero vector* $\mathbf{0}$ with magnitude 0 is introduced to play the same role for vectors that the real number 0 plays in ordinary arithmetic. Two vectors $\mathbf{v}$ and $\mathbf{w}$ are added by the *parallelogram rule* to produce the vector $\mathbf{v} + \mathbf{w}$ (Figure 3-2). To multiply a vector by a positive real number s, keep it in the same direction and multiply its length by s. To multiply it by a negative real number s, reverse its direction and multiply its length by $|s|$.

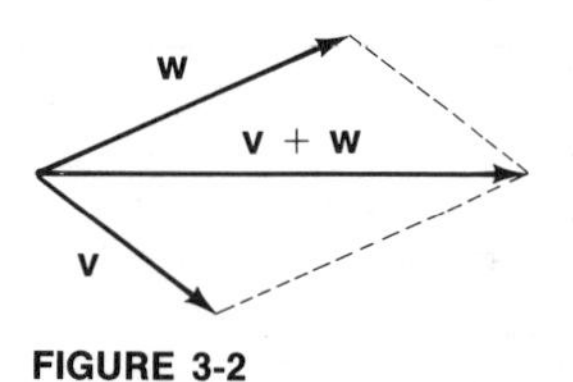

FIGURE 3-2

Although the idea of a vector as an arrow is geometrically compelling, it is neither precise enough nor sufficiently general for the needs of more advanced mathematics. By abstracting from the algebraic properties of vectors represented as arrows, mathematicians have formulated the following definition:

5.1 DEFINITION

VECTOR SPACE Let F be a field. A **vector space**, or **linear space**, **over** F is a commutative group $(V, +)$ and a mapping $(s, \mathbf{v}) \mapsto s\mathbf{v}$ from $F \times V$ into V, called **multiplication by scalars**, such that, for all $s, t \in F$ and all $\mathbf{v}, \mathbf{w} \in V$, the following conditions hold:

(i) $(st)\mathbf{v} = s(t\mathbf{v})$
(ii) $(s + t)\mathbf{v} = s\mathbf{v} + t\mathbf{v}$
(iii) $s(\mathbf{v} + \mathbf{w}) = s\mathbf{v} + s\mathbf{w}$
(iv) $1\mathbf{v} = \mathbf{v}$

Elements of V are called **vectors**, elements of F are called **scalars**, and $+: V \times V \to V$ is called **vector addition**. Condition (i) is an *associative*-type law, Conditions (ii) and (iii) are *distributive*-type laws, and Condition (iv) makes the unity element $1 \in F$ act as a neutral element on the vectors in V. The neutral element of the group $(V, +)$ is called the **zero vector**, or **null vector**, and is denoted by $\mathbf{0}$. The inverse in the group $(V, +)$ of a vector $\mathbf{v}$ is called the **negative** of $\mathbf{v}$ and is written as $-\mathbf{v}$.

In the following examples of vector spaces, F denotes a field and X denotes a nonempty set.

1. F^n, called **coordinate n-space over the field F**, is defined as the set of all ordered n-tuples

$$\mathbf{v} = (v_1, v_2, v_3, \ldots, v_n)$$

of elements $v_1, v_2, v_3, \ldots, v_n \in F$ with the following operations of vector addition and multiplication by scalars: For $\mathbf{v} = (v_1, v_2, v_3, \ldots, v_n)$, $\mathbf{w} = (w_1, w_2, w_3, \ldots, w_n) \in F^n$, and $s \in F$,

$$\mathbf{v} + \mathbf{w} = (v_1 + w_1, v_2 + w_2, v_3 + w_3, \ldots, v_n + w_n)$$

and

$$s\mathbf{v} = (sv_1, sv_2, sv_3, \ldots, sv_n).$$

If $\mathbf{v} = (v_1, v_2, v_3, \ldots, v_n)$, we call the scalars $v_1, v_2, v_3, \ldots, v_n$ the **coordinates** of the vector $\mathbf{v}$, and we refer to the operations defined above as **coordinatewise addition** and **coordinatewise multiplication by scalars**.

2. The field F *forms a vector space over itself* if multiplication by scalars is taken as the multiplication operation in F. Thus, as a vector space, F is a special case of F^n with $n = 1$.
3. F^X is the set of all functions $f:X \to F$ with the following operations of vector addition and multiplication by scalars: For $f:X \to F$, $g:X \to F$, and $s \in F$, $f + g:X \to F$ is defined by $(f + g)(x) = f(x) + g(x)$, $\forall x \in F$, and $sf:X \to F$ is defined by $(sf)(x) = sf(x)$, $\forall x \in F$. Thus, *functions from a fixed set X into a fixed field F can be regarded as vectors in the vector space F^X.*
4. $F^{[X]}$ is the subset of F^X consisting of all functions $f:X \to F$ that are **finitely nonzero** in the sense that $f(x) = 0$ for all but (possibly) a finite number of elements $x \in X$.
5. $C(D)$ is the set of all continuous functions $f:D \to \mathbb{R}$ with domain $D \subseteq \mathbb{R}$. Operations of vector addition and multiplication by scalars are defined as in example 3. (Recall from calculus that *the sum of two continuous functions is continuous*, and *a constant multiple of a continuous function is continuous.*)

By an argument that is similar to the proof of Theorem 2.3, you can prove the following theorem (Problem 7).

5.2 THEOREM **MULTIPLICATION BY ZERO IN A VECTOR SPACE** Let V be a vector space over F, let $s \in F$, and let $\mathbf{v} \in V$. Then, if either $s = 0$ or $\mathbf{v} = \mathbf{0}$ (or both), it follows that

$$s\mathbf{v} = \mathbf{0}.$$

5.3 COROLLARY

Let V be a vector space over F, let $s \in F$, and let $\mathbf{v} \in V$. Then,

$$s\mathbf{v} = \mathbf{0} \Leftrightarrow s = 0 \lor \mathbf{v} = \mathbf{0}.$$

PROOF The implication $\Leftarrow$ follows from Theorem 5.2. To prove the converse implication, suppose that $s\mathbf{v} = \mathbf{0}$. If $s = 0$, then the statement $s = 0 \lor \mathbf{v} = \mathbf{0}$ is true. On the other hand, if $s \neq 0$, then we can form the reciprocal s^{-1} of s in the field F, and we have

$$s^{-1}(s\mathbf{v}) = s^{-1}\mathbf{0} = \mathbf{0},$$

where we have used Theorem 5.2 to conclude that $s^{-1}\mathbf{0} = \mathbf{0}$. Also, by Parts (i) and (iv) of Definition 5.1,

$$s^{-1}(s\mathbf{v}) = (s^{-1}s)\mathbf{v} = 1\mathbf{v} = \mathbf{v},$$

and it follows that $\mathbf{v} = \mathbf{0}$. Thus, if $s \neq 0$, the statement $s = 0 \lor \mathbf{v} = \mathbf{0}$ is again true. Consequently, if $s\mathbf{v} = \mathbf{0}$, it follows that $s = 0 \lor \mathbf{v} = \mathbf{0}$. □

5.4 LEMMA

THE NEGATIVE OF A VECTOR Let V be a vector space over F, and let $\mathbf{v} \in V$. Then,

$$(-1)\mathbf{v} = -\mathbf{v}.$$

PROOF By Parts (iv) and (ii) of Definition 5.1 and Theorem 5.2,

$$\mathbf{v} + (-1)\mathbf{v} = 1\mathbf{v} + (-1)\mathbf{v} = (1 + (-1))\mathbf{v} = 0\mathbf{v} = \mathbf{0}.$$

Therefore, since $\mathbf{0} = \mathbf{v} + (-\mathbf{v})$, we have

$$\mathbf{v} + (-1)\mathbf{v} = \mathbf{v} + (-\mathbf{v}),$$

and it follows from the cancellation law in the group $(V, +)$ that

$$(-1)\mathbf{v} = -\mathbf{v}.$$ □

We define subtraction for vectors in much the same way that we defined subtraction for elements of a ring.

5.5 DEFINITION

SUBTRACTION IN A VECTOR SPACE Let V be a vector space over F, and let $\mathbf{v}, \mathbf{w} \in V$. We define

$$\mathbf{v} - \mathbf{w} = \mathbf{v} + (-\mathbf{w}).$$

Many of the properties of elements of a ring have analogues for vectors in a vector space V. We collect some of these properties in the following theorem, leaving the proofs as an exercise (Problem 9).

5.6 THEOREM

ALGEBRAIC PROPERTIES OF VECTORS Let V be a vector space over F, let $s \in F$, and let $\mathbf{v}, \mathbf{w} \in V$. Then:

(i) $\mathbf{v} + \mathbf{w} = \mathbf{0} \Leftrightarrow \mathbf{v} = -\mathbf{w}$
(ii) $-(-\mathbf{v}) = \mathbf{v}$
(iii) $-(\mathbf{v} + \mathbf{w}) = (-\mathbf{v}) + (-\mathbf{w})$
(iv) $s(\mathbf{v} - \mathbf{w}) = (s\mathbf{v}) - (s\mathbf{w})$

In a vector space, we adopt the standard convention that *multiplication by scalars takes precedence over addition and subtraction of vectors.* In other words, if parentheses are not present to show the order in which operations are to be carried out, first you carry out all indicated multiplications by scalars, and then add or subtract the resulting vectors. For instance, with this convention, we can write Property (iv) in Theorem 5.6 as

$$s(\mathbf{v} - \mathbf{w}) = s\mathbf{v} - s\mathbf{w}.$$

Algebraic calculations in a vector space are similar to calculations of ordinary algebra with the exception that you must be careful to distinguish between vectors and scalars. In particular, *beware of equations with a vector on one side and a scalar on the other.* Note that, whereas you can add two scalars and you can add two vectors, you should *beware of attempts to add a scalar to a vector.* (However, in the case in which a field F is regarded as a vector space over itself, there is no distinction between vectors and scalars.)

5.7 DEFINITION

LINEAR COMBINATION OF VECTORS Let V be a vector space over F, and let

$$\mathbf{v}_1, \mathbf{v}_2, \mathbf{v}_3, \ldots, \mathbf{v}_n \in V.$$

If $s_1, s_2, s_3, \ldots, s_n \in F$, then a vector of the form

$$\mathbf{v} = s_1\mathbf{v}_1 + s_2\mathbf{v}_2 + s_3\mathbf{v}_3 + \cdots + s_n\mathbf{v}_n$$

is said to be a **linear combination** of the vectors $\mathbf{v}_1, \mathbf{v}_2, \mathbf{v}_3, \ldots, \mathbf{v}_n$ with **coefficients** $s_1, s_2, s_3, \ldots, s_n$.

By convention, *the zero vector* $\mathbf{0}$ *is regarded as a linear combination of the empty set* $\varnothing$ *of vectors.*

5.8 Example Let V be a vector space over F, and let

$$\mathbf{v} = s_1\mathbf{v_1} + s_2\mathbf{v_2} + s_3\mathbf{v_3} + \cdots + s_n\mathbf{v_n}$$

be a linear combination of $\mathbf{v_1}, \mathbf{v_2}, \mathbf{v_3}, \ldots, \mathbf{v_n}$. If one of the coefficients s_i, $1 \le i \le n$, is nonzero, show that the corresponding vector $\mathbf{v_i}$ can be written as a linear combination of $\mathbf{v}$ and the remaining vectors $\mathbf{v_j}$ with $j \ne i$.

SOLUTION Multiplying both sides of the equation

$$\mathbf{v} = s_1\mathbf{v_1} + s_2\mathbf{v_2} + s_3\mathbf{v_3} + \cdots + s_n\mathbf{v_n}$$

by $1/s_i$ and solving the resulting equation for $\mathbf{v_i}$, we have

$$\mathbf{v}_i = t_0\mathbf{v} + t_1\mathbf{v_1} + t_2\mathbf{v_2} + \cdots + t_n\mathbf{v_n},$$

where $t_0 = 1/s_i$, $t_j = -s_j/s_i$ for $1 \le j \le n$ with $j \ne i$, and all vectors $\mathbf{v_1}, \mathbf{v_2}, \mathbf{v_3}, \ldots, \mathbf{v_n}$ *except* $\mathbf{v_i}$ appear on the right side of the equation. □

5.9 DEFINITION

SPANNING VECTORS Let V be a vector space over F, and let

$$\mathbf{v_1}, \mathbf{v_2}, \mathbf{v_3}, \ldots, \mathbf{v_n} \in V.$$

We say that the set of vectors $\{\mathbf{v_1}, \mathbf{v_2}, \mathbf{v_3}, \ldots, \mathbf{v_n}\}$ **spans** the vector space V if every vector $\mathbf{v} \in V$ can be written as a linear combination of the vectors $\mathbf{v_1}, \mathbf{v_2}, \mathbf{v_3}, \ldots, \mathbf{v_n}$.

Rather than saying that the set $\{\mathbf{v_1}, \mathbf{v_2}, \mathbf{v_3}, \ldots, \mathbf{v_n}\}$ spans the vector space V, we often say that *the vectors* $\mathbf{v_1}, \mathbf{v_2}, \mathbf{v_3}, \ldots, \mathbf{v_n}$ *span* V.

5.10 Example Let F be a field. In the coordinate 2-space F^2 over F, let $\mathbf{e_1} = (1, 0)$, and let $\mathbf{e_2} = (0, 1)$. Show that the vectors $\mathbf{e_1}$ and $\mathbf{e_2}$ span F^2.

SOLUTION Let $\mathbf{v} = (v_1, v_2) \in F^2$. Then $v_1\mathbf{e_1} = (v_1, 0)$ and $v_2\mathbf{e_2} = (0, v_2)$, and it follows that

$$\mathbf{v} = (v_1, v_2) = (v_1, 0) + (0, v_2) = v_1\mathbf{e_1} + v_2\mathbf{e_2}.$$ □

Example 5.10 shows that the vectors $(1, 0)$ and $(0, 1)$ span F^2. More generally, consider the vectors

$$\mathbf{e_1}, \mathbf{e_2}, \mathbf{e_3}, \ldots, \mathbf{e_n} \in F^n,$$

where, for $1 \le i \le n$, $\mathbf{e_i}$ is the ordered n-tuple

$$\mathbf{e_i} = (0, 0, 0, \ldots, 0, 1, 0, \ldots, 0)$$

with 1 in the ith position and 0 in all other positions. The vectors $\mathbf{e_1}$, $\mathbf{e_2}$, $\mathbf{e_3}, \ldots, \mathbf{e_n}$ are called the **standard basis vectors for** F^n. If

$$\mathbf{v} = (v_1, v_2, v_3, \ldots, v_n),$$

then

$$\mathbf{v} = v_1\mathbf{e_1} + v_2\mathbf{e_2} + v_3\mathbf{e_3} + \cdots + v_n\mathbf{e_n}$$

(Problem 15), and so *the vectors* $\mathbf{e_1}, \mathbf{e_2}, \mathbf{e_3}, \ldots, \mathbf{e_n}$ *span* F^n. Notice that, if

$$v_1\mathbf{e_1} + v_2\mathbf{e_2} + v_3\mathbf{e_3} + \cdots + v_n\mathbf{e_n} = \mathbf{0},$$

then $\mathbf{v} = \mathbf{0}$, and it follows that all of the coefficients $v_1, v_2, v_3, \ldots, v_n$ are equal to zero. Thus, the standard basis vectors in F^n are *linearly independent* in the sense of the following definition.

5.11 DEFINITION

LINEARLY INDEPENDENT VECTORS Let V be a vector space over F, and let

$$\mathbf{v_1}, \mathbf{v_2}, \mathbf{v_3}, \ldots, \mathbf{v_n} \in V.$$

We say that the set of vectors $\{\mathbf{v_1}, \mathbf{v_2}, \mathbf{v_3}, \ldots, \mathbf{v_n}\}$ is **linearly independent** if, when a linear combination of these vectors is equal to the zero vector,

$$s_1\mathbf{v_1} + s_2\mathbf{v_2} + s_3\mathbf{v_3} + \cdots + s_n\mathbf{v_n} = \mathbf{0},$$

then all of the coefficients are equal to the zero scalar, that is,

$$s_1 = s_2 = s_3 = \cdots = s_n = 0.$$

Rather than saying that the set $\{\mathbf{v_1}, \mathbf{v_2}, \mathbf{v_3}, \ldots, \mathbf{v_n}\}$ is linearly independent, we often say that *the vectors* $\mathbf{v_1}, \mathbf{v_2}, \mathbf{v_3}, \ldots, \mathbf{v_n}$ *are linearly independent*. By convention, *the empty set of vectors is regarded as being linearly independent.*

5.12 Example Consider the vector space $C(\mathbb{R})$ of all continuous functions $f:\mathbb{R} \to \mathbb{R}$. For each positive integer j, let $p_j \in C(\mathbb{R})$ be defined by

$$p_j(x) = x^j, \quad \text{for all } x \in \mathbb{R}.$$

Let p_0 be the constant function defined by

$$p_0(x) = 1, \quad \text{for all } x \in \mathbb{R}.$$

For any integer $n \geq 0$, show that the functions

$$p_0, p_1, p_2, \ldots, p_n,$$

regarded as vectors in $C(\mathbb{R})$, are linearly independent.

SOLUTION Suppose that we have a linear combination

$$c_0p_0 + c_1p_1 + c_2p_2 + \cdots + c_np_n = f,$$

where f is the constant function defined by

$$f(x) = 0, \quad \text{for all } x \in \mathbb{R}.$$

We must show that

$$c_0, c_1, c_2, \ldots, c_n = 0.$$

Now,

$$\begin{aligned} f(x) &= c_0p_0(x) + c_1p_1(x) + c_2p_2(x) + \cdots + c_np_n(x) \\ &= c_0 + c_1x + c_2x^2 + \cdots + c_nx^n \end{aligned}$$

is a polynomial function with coefficients $c_0, c_1, c_2, \ldots, c_n$, and $f(x) = 0$ for all real numbers x. If any one of the coefficients were nonzero, then the polynomial function f would have a degree $k \leq n$; and therefore, the equation $f(x) = 0$ would have at most k real roots. Since this equation has infinitely many roots (every real number is a root!), then all of the coefficients must be zero. □

5.13 LEMMA

LINEAR INDEPENDENCE LEMMA Let V be a vector space over F, and let $\mathbf{v_1}, \mathbf{v_2}, \mathbf{v_3}, \ldots, \mathbf{v_n} \in V$. Then $\mathbf{v_1}, \mathbf{v_2}, \mathbf{v_3}, \ldots, \mathbf{v_n}$ are linearly independent if and only if no one of the vectors $\mathbf{v_i}$ can be written as a linear combination of the remaining vectors $\mathbf{v_j}$ for $j \neq i$.

PROOF We prove the "only if" here and leave the proof of the "if" as an exercise (Problem 18). Thus, suppose that $\mathbf{v_1}, \mathbf{v_2}, \mathbf{v_3}, \ldots, \mathbf{v_n}$ are linearly independent and that one of the vectors $\mathbf{v_i}$ can be expressed as a linear combination

$$\mathbf{v_i} = t_1\mathbf{v_1} + t_2\mathbf{v_2} + \cdots + t_{i-1}\mathbf{v_{i-1}} + t_{i+1}\mathbf{v_{i+1}} + \cdots + t_n\mathbf{v_n}$$

of the remaining vectors. Then,

$$t_1\mathbf{v_1} + t_2\mathbf{v_2} + \cdots + t_{i-1}\mathbf{v_{i-1}} + (-1)\mathbf{v_i} + t_{i+1}\mathbf{v_{i+1}} + \cdots + t_n\mathbf{v_n} = \mathbf{0},$$

is a linear combination of the vectors $\mathbf{v_1}, \mathbf{v_2}, \mathbf{v_3}, \ldots, \mathbf{v_n}$ in which at least one of the coefficients [namely, (-1)] is different from 0. But this contradicts the supposition that these vectors are linearly independent. □

5.14 Example Let V be a vector space over F, and suppose that the vectors $\mathbf{v_1}, \mathbf{v_2}, \mathbf{v_3}, \ldots, \mathbf{v_n}$ are linearly independent in V. Show that all of the vectors $\mathbf{v_1}, \mathbf{v_2}, \mathbf{v_3}, \ldots, \mathbf{v_n}$ are nonzero.

SOLUTION If one of the vectors were zero, then it could be written as a linear combination of the remaining vectors (just by taking all coefficients equal to zero), contradicting Lemma 5.13. □

The following lemma gives conditions under which it is possible to *exchange* a vector $\mathbf{c}$ for one of the vectors in a spanning set and, as a result, obtain a new spanning set.

5.15 LEMMA

EXCHANGE LEMMA Let V be a vector space over F, and assume that the vectors

$$\mathbf{b_1}, \mathbf{b_2}, \mathbf{b_3}, \ldots, \mathbf{b_m}$$

span V. Suppose that $\mathbf{c} \in V$ with

$$\mathbf{c} = q_1\mathbf{b_1} + q_2\mathbf{b_2} + \cdots + q_i\mathbf{b_i} + \cdots + q_m\mathbf{b_m}$$

and that $q_i \neq 0$. Then the vectors

$$\mathbf{b_1}, \mathbf{b_2}, \mathbf{b_3}, \ldots, \mathbf{b_{i-1}}, \mathbf{c}, \mathbf{b_{i+1}}, \ldots, \mathbf{b_m}$$

span V.

PROOF By Example 5.8 and the fact that $q_i \neq 0$, we can write $\mathbf{b_i}$ as a linear combination,

$$\mathbf{b_i} = t_0\mathbf{c} + t_1\mathbf{b_1} + \cdots + t_{i-1}\mathbf{b_{i-1}} + t_{i+1}\mathbf{b_{i+1}} + \cdots + t_m\mathbf{b_m},$$

where the vector $\mathbf{b_i}$ does not appear on the right of the equation. Let $\mathbf{v}$ denote an arbitrary vector in V. Because the vectors $\mathbf{b_1}, \mathbf{b_2}, \mathbf{b_3}, \ldots, \mathbf{b_m}$ span V, we can write

$$\mathbf{v} = s_1\mathbf{b_1} + s_2\mathbf{b_2} + \cdots + s_i\mathbf{b_i} + \cdots + s_m\mathbf{b_m}.$$

Substituting the expression for $\mathbf{b_i}$ as a linear combination of $\mathbf{c}$ and the remaining vectors $\mathbf{b_j}$ for $j \neq i$ into the last equation, and collecting like terms, we obtain $\mathbf{v}$ as a linear combination of the vectors

$$\mathbf{b_1}, \mathbf{b_2}, \mathbf{b_3}, \ldots, \mathbf{b_{i-1}}, \mathbf{c}, \mathbf{b_{i+1}}, \ldots, \mathbf{b_m},$$

showing that these vectors span V. □

The method of proof used in the following theorem is called the **Steinitz exchange process** after the German mathematician Ernst Steinitz (1871–1928), who did important early work in the theory of fields.

5.16 THEOREM

STEINITZ'S THEOREM Let V be a vector space over F, suppose that the vectors

$$\mathbf{a_1}, \mathbf{a_2}, \mathbf{a_3}, \ldots, \mathbf{a_n}$$

are linearly independent in V and that the vectors

$$\mathbf{b_1}, \mathbf{b_2}, \mathbf{b_3}, \ldots, \mathbf{b_m}$$

span V. Then, $n \leq m$.

PROOF Because $\mathbf{b_1}, \mathbf{b_2}, \mathbf{b_3}, \ldots, \mathbf{b_m}$ span V, we can write $\mathbf{a_1}$ as a linear combination,

$$\mathbf{a_1} = s_1\mathbf{b_1} + s_2\mathbf{b_2} + \cdots + s_m\mathbf{b_m}.$$

We cannot have all of the coefficients $s_1, s_2, s_3, \ldots, s_m$ equal to zero because, if they were, then $\mathbf{a_1} = \mathbf{0}$, contradicting Example 5.14. Thus, at least one of these coefficients is nonzero. By relabeling the vectors $\mathbf{b_1}, \mathbf{b_2}, \mathbf{b_3}, \ldots, \mathbf{b_m}$ if necessary, we can suppose that $s_1 \neq 0$. By the exchange lemma (Lemma 5.15), it follows that

$$\mathbf{a_1}, \mathbf{b_2}, \mathbf{b_3}, \ldots, \mathbf{b_m}$$

span V. Thus, by suitably relabeling the vectors $\mathbf{b_1}, \mathbf{b_2}, \mathbf{b_3}, \ldots, \mathbf{b_m}$, if necessary, we can *exchange* $\mathbf{a_1}$ for $\mathbf{b_1}$, and the resulting vectors $\mathbf{a_1}, \mathbf{b_2}, \mathbf{b_3}, \ldots, \mathbf{b_m}$ continue to span V.

Now, since $\mathbf{a_1}, \mathbf{b_2}, \mathbf{b_3}, \ldots, \mathbf{b_m}$ span V, we can write $\mathbf{a_2}$ as a linear combination,

$$\mathbf{a_2} = u_1\mathbf{a_1} + u_2\mathbf{b_2} + \cdots + u_m\mathbf{b_m}.$$

At least one of the coefficients $u_2, u_3, \ldots, u_m$ must be nonzero, for otherwise $\mathbf{a_2} = u_1\mathbf{a_1}$, contradicting the linear independence lemma (Lemma 5.13). By relabeling the vectors $\mathbf{b_2}, \mathbf{b_3}, \ldots, \mathbf{b_m}$, if necessary, we can suppose that $u_2 \neq 0$. By the exchange lemma again, it follows that

$$\mathbf{a_1}, \mathbf{a_2}, \mathbf{b_3}, \ldots, \mathbf{b_m}$$

span V. Thus, by suitably relabeling the vectors $\mathbf{b_2}, \mathbf{b_3}, \ldots, \mathbf{b_m}$, if necessary, we can exchange $\mathbf{a_2}$ for $\mathbf{b_2}$, and the resulting vectors $\mathbf{a_1}, \mathbf{a_2}, \mathbf{b_3}, \ldots, \mathbf{b_m}$ continue to span V.

Proceeding in this way, we can exchange the vectors $\mathbf{a_1}, \mathbf{a_2}, \mathbf{a_3}, \ldots$ for corresponding vectors $\mathbf{b_1}, \mathbf{b_2}, \mathbf{b_3}, \ldots$ until either we run out of vectors $\mathbf{a_i}$ or run out of vectors $\mathbf{b_i}$. In the first case, $n \leq m$, and our proof is complete. Thus, we need only show that the second case, in which we run out of vectors $\mathbf{b_i}$ before we run out of vectors $\mathbf{a_i}$ cannot happen. However, if it did, we would arrive at the situation in which $\mathbf{a_1}, \mathbf{a_2}, \mathbf{a_3}, \ldots, \mathbf{a_m}$ have been exchanged for $\mathbf{b_1}, \mathbf{b_2}, \mathbf{b_3}, \ldots, \mathbf{b_m}$, and there is at least one vector $\mathbf{a_{m+1}}$ left over. Then $\mathbf{a_1}, \mathbf{a_2}, \mathbf{a_3}, \ldots, \mathbf{a_m}$ would span V, and we could write $\mathbf{a_{m+1}}$ as a linear combination,

$$\mathbf{a_{m+1}} = q_1\mathbf{a_1} + q_2\mathbf{a_2} + \cdots + q_m\mathbf{a_m},$$

again contradicting the linear independence lemma. □

5.17 DEFINITION

BASIS OF A VECTOR SPACE Let V be a vector space over F, and let

$$B = \{\mathbf{b_1}, \mathbf{b_2}, \mathbf{b_3}, \ldots, \mathbf{b_n}\} \subseteq V.$$

If B is both linearly independent and spans V, then B is called a **basis** for V. If there is a finite subset $B \subseteq V$ that is a basis for V, then V is said to be **finite dimensional**.

Rather than saying that the set $B = \{\mathbf{b_1}, \mathbf{b_2}, \mathbf{b_3}, \ldots, \mathbf{b_n}\}$ is a basis for V, we often say that *the vectors* $\mathbf{b_1}, \mathbf{b_2}, \mathbf{b_3}, \ldots, \mathbf{b_n}$ *form a basis for* V.

5.18 Example

(a) Find a basis for $\mathbb{R}^n$.
(b) Show that $\mathbb{R}^n$ is finite dimensional.

SOLUTION

(a) The standard basis vectors $\mathbf{e_1}, \mathbf{e_2}, \mathbf{e_3}, \ldots, \mathbf{e_n}$ (see page 226) do, in fact, form a basis for $\mathbb{R}^n$.
(b) Because $\mathbb{R}^n$ has a finite basis, it is finite dimensional. □

5.19 THEOREM

INVARIANCE OF DIMENSION Let V be a vector space over F, and let

$$A = \{\mathbf{a_1}, \mathbf{a_2}, \mathbf{a_3}, \ldots, \mathbf{a_n}\} \quad \text{and} \quad B = \{\mathbf{b_1}, \mathbf{b_2}, \mathbf{b_3}, \ldots, \mathbf{b_m}\}$$

be two bases for V. Then, $n = m$.

PROOF

The set A is linearly independent, and the set B spans V; so $n \leq m$ by Steinitz's theorem (Theorem 5.16). Also, the set B is linearly independent, and the set A spans V; so by Steinitz's theorem, $m \leq n$. Consequently, $m = n$, and the theorem is proved. □

5.20 DEFINITION

DIMENSION OF A VECTOR SPACE Let V be a finite dimensional vector space over F, and let B be a basis for V. We define the dimension of V over F, in symbols $\dim_F(V)$, to be the number of vectors in B,

$$\dim_F(V) = \#B.$$

By the theorem on invariance of dimension (Theorem 5.19), it does not matter which basis B is chosen in Definition 5.20 because they all contain the same number of vectors. If the field F is understood, we often write $\dim(V)$ rather than $\dim_F(V)$. If V is not finite dimensional, it is said to be **infinite dimensional.**

5.21 Example Find the dimension of $\mathbb{R}^n$.

SOLUTION By Example 5.18, $\dim(\mathbb{R}^n) = n$. □

PROBLEM SET 3.5

1. If F is a field and n is a positive integer, show that F^n is a vector space over F.
2. Let V be the set of all n-by-m matrices (n rows and m columns) with entries in the field F. Add two matrices by adding their corresponding entries, and multiply a matrix by a scalar by multiplying each of its entries by that scalar. Show that V is a vector space over F.
3. If F is a field and X is a nonempty set, show that F^X is a vector space over F.
4. If F is a field and X is a nonempty set, show that $F^{[X]}$ is a vector space over F.
5. Let $[a, b] \subseteq \mathbb{R}$ be the closed interval $[a, b] = \{x \mid a \leq x \leq b\}$. Show that $C([a, b])$ is a vector space over $\mathbb{R}$.
6. Let $(a, b) \subseteq \mathbb{R}$ be the open interval $(a, b) = \{x \mid a < x < b\}$, and let V be the set of all differentiable functions $f:(a, b) \to \mathbb{R}$. With the usual operations of function addition and multiplication by a constant, show that V is a vector space over $\mathbb{R}$.
7. Prove Theorem 5.2.
8. Let X be a nonempty set, and let $\mathscr{P}(X)$ denote the set of all subsets of X. For $A, B \in \mathscr{P}(X)$, define $A + B = (A \setminus B) \cup (B \setminus A)$. Consider the field $\text{GF}(2) = \{0, 1\}$, and define multiplication of elements of $\mathscr{P}(X)$ by scalars in GF(2) as follows: For $A \in \mathscr{P}(X)$, $0A = \varnothing$ and $1A = A$. Show that $\mathscr{P}(X)$ is a vector space over GF(2).
9. Prove Theorem 5.6.
10. Let F be a field, and let H be an extension field of F, so that H is a field, $F \subseteq H$, and the addition and multiplication operations in F are the restrictions to F of the corresponding operations in H. Show that H can be considered a vector space over F with the following operations: For $a \in F$ and for $x, y \in H$, $x + y$ is the sum of x and y, and ax is the product of a and x as computed in the field H. The resulting vector space is called H **over** F.

11. Let V be a vector space over F, and let $\mathbf{v_1}, \mathbf{v_2}, \mathbf{v_3}, \ldots, \mathbf{v_n} \in V$. Let S be the subset of V consisting of all linear combinations of the vectors $\mathbf{v_1}, \mathbf{v_2}, \mathbf{v_3}, \ldots, \mathbf{v_n}$.
(a) If $\mathbf{x}, \mathbf{y} \in S$, show that $\mathbf{x} + \mathbf{y} \in S$.
(b) If $t \in F$ and $\mathbf{x} \in S$, show that $t\mathbf{x} \in S$.

12. Consider the vector space $C(\mathbb{R})$ of all continuous functions $f:\mathbb{R} \to \mathbb{R}$. Let a and b be fixed real numbers, and let $f, g, h \in C(X)$ be given by $f(x) = a\cos(x - b)$, $g(x) = \cos x$, and $h(x) = \sin x$ for all $x \in \mathbb{R}$. In the vector space $C(X)$, show that f is a linear combination of g and h.

13. Let F be a field.
(a) Show that the vectors $\mathbf{e} = (1, 1)$ and $\mathbf{f} = (1, 0)$ span F^2.
(b) Do the vectors $\mathbf{e} = (1, 1)$ and $\mathbf{g} = (-1, 1)$ necessarily span F^2?

14. Let $F = \mathrm{GF}(2)$. Find a set of three vectors, *other than the standard basis vectors*, that span F^3.

15. Show that the standard basis vectors in F^n span F^n.

16. Show that the sine and cosine functions are linearly independent vectors in $C(\mathbb{R})$.

17. Show that, in a vector space V over F, *a superset of a spanning set is a spanning set*. That is, show that if $A \subseteq B \subseteq V$, and if A spans V, then B spans V. (You can assume here that B is a finite set.)

18. Complete the proof of Lemma 5.13 by showing that, if no one of the vectors $\mathbf{v_1}, \mathbf{v_2}, \mathbf{v_3}, \ldots, \mathbf{v_n}$ can be expressed as a linear combination of the remaining vectors, then these vectors are linearly independent.

19. Show that, in a vector space V over F, *a subset of a linearly independent set is a linearly independent set*. That is, show that if $A \subseteq B \subseteq V$, and if B is linearly independent, then A is linearly independent. (You can assume here that B is a finite set.)

20. Let $\mathbf{v_1}, \mathbf{v_2}, \mathbf{v_3}, \ldots, \mathbf{v_n}$ be vectors in a vector space V over F. Show that these vectors are linearly independent if and only if $\mathbf{v_1} \neq \mathbf{0}$ and, for $i = 2, 3, \ldots, n$, no one of the vectors $\mathbf{v_i}$ can be written as a linear combination of the preceding vectors $\mathbf{v_1}, \mathbf{v_2}, \mathbf{v_3}, \ldots, \mathbf{v_{i-1}}$.

21. In the proof of Lemma 5.15, it is stated that $\mathbf{v}$ is obtained as a linear combination of the vectors $\mathbf{b_1}, \mathbf{b_2}, \mathbf{b_3}, \ldots, \mathbf{b_{i-1}}, \mathbf{c}, \mathbf{b_{i+1}}, \ldots, \mathbf{b_m}$ by substituting

$$\mathbf{b_i} = t_0\mathbf{c} + t_1\mathbf{b_1} + \cdots + t_{i-1}\mathbf{b_{i-1}} + t_{i+1}\mathbf{b_{i+1}} + \cdots + t_m\mathbf{b_m}$$

into the equation

$$\mathbf{v} = s_1\mathbf{b_1} + s_2\mathbf{b_2} + \cdots + s_i\mathbf{b_i} + \cdots + s_m\mathbf{b_m}.$$

Perform this substitution, collect the terms, and find an explicit formula for the linear combination in question.

22. Show that the vector space V over F is infinite dimensional if and only if there is no finite subset B of V that spans V.

23. Suppose that $B = \{\mathbf{b_1}, \mathbf{b_2}, \mathbf{b_3}, \ldots, \mathbf{b_n}\}$ is a maximal linearly independent set of vectors in the vector space V over F; that is, suppose that B is linearly independent but that there does not exist a set $C \subseteq V$ with C linearly independent, $B \subseteq C$, and $B \neq C$. Prove that B is a basis for V.

24. Show that the vector space V over F is infinite dimensional if and only if, for every positive integer n, there exists a linearly independent subset A of V with $\# A = n$.

25. Suppose that $B = \{\mathbf{b_1}, \mathbf{b_2}, \mathbf{b_3}, \ldots, \mathbf{b_n}\}$ spans the vector space V over F, but that no proper subset C of B spans V. Show that B is a basis for V.

26. Show that $C(\mathbb{R})$ is an infinite dimensional vector space over $\mathbb{R}$. [Hint: Use Example 5.12 and Problem 24.]

27. Let V be a vector space over F, and suppose that S is a finite set of vectors that spans V. If $A \subseteq S$ and A is a linearly independent set of vectors, show that there exists a basis B for V with $A \subseteq B \subseteq S$. [Hint: Adjoin vectors from S to the set A, one at a time, until you cannot adjoin any more vectors without losing linear independence. Call the resulting set of vectors B, and use Problem 23.]

28. Regard the field of real numbers $\mathbb{R}$ as being a vector space over the field of rational numbers $\mathbb{Q}$ as in Problem 10. Show that 1 and $\sqrt{2}$ are linearly independent elements of the vector space $\mathbb{R}$ over $\mathbb{Q}$.

29. The complex numbers $\mathbb{C}$ can be regarded as a vector space over the field $\mathbb{R}$ of real numbers. In this vector space, complex numbers are added in the usual way and multiplied by real numbers in the usual way.
(a) Find a basis for $\mathbb{C}$ over $\mathbb{R}$.
(b) Find the dimension of $\mathbb{C}$ over $\mathbb{R}$.

30. In Problem 2, find $\dim_F(V)$.

31. Let V be a vector space over $F = \mathrm{GF}(p)$. If $\dim_F(V) = n$, show that $\# V = p^n$.

32. Let F be a field, let H be an extension field of F, and let K be an extension field of H. Consider H to be a vector space over F as in Problem 10. Likewise, K can be considered to be a vector space over H and also a vector space over F. Prove that $\dim_F(K) = \dim_H(K) \cdot \dim_F(H)$.

33. If V is a finite-dimensional vector space over F and $n = \dim_F(V)$, show that, for every linearly independent subset A of V, $\# A \leq n$.

*6 SUBSPACES AND LINEAR TRANSFORMATIONS

In this section we continue the study of vector spaces begun in Section 5. Here, we study *subspaces* (which are *closed* under the vector-space operations) and *linear transformations* (which *preserve* the vector-space operations).

6.1 DEFINITION

CLOSURE Let V be a vector space over the field F, and let $U \subseteq V$.

(i) U is said to be **closed under addition** if

$$\mathbf{u}, \mathbf{v} \in U \Rightarrow \mathbf{u} + \mathbf{v} \in U.$$

(ii) U is said to be **closed under subtraction** if

$$\mathbf{u}, \mathbf{v} \in U \Rightarrow \mathbf{u} - \mathbf{v} \in U.$$

(iii) U is said to be **closed under multiplication by scalars** if

$$s \in F, \mathbf{u} \in U \Rightarrow s\mathbf{u} \in U.$$

6.2 Example Let F be a field, and let $U \subseteq F^2$ be defined by $U = \{(x, x) | x \in F\}$. Show that U is closed under addition and under multiplication by scalars.

SOLUTION Let $\mathbf{u} = (x, x) \in U$, $\mathbf{v} = (y, y) \in U$, and $s \in F$. Then,

$$\mathbf{u} + \mathbf{v} = (x, x) + (y, y) = (x + y, x + y) \in U$$

and

$$s\mathbf{u} = s(y, y) = (sy, sy) \in U.$$

□

6.3 Example Let V be a vector space over F, and let $U \subseteq V$. Suppose that U is closed under addition and under multiplication by scalars. Show that U is closed under subtraction.

SOLUTION Let $\mathbf{u}, \mathbf{v}, \in U$. Because U is closed under multiplication by scalars, $-\mathbf{v} = (-1)\mathbf{v} \in U$. Therefore, because U is closed under addition,

$$\mathbf{u} - \mathbf{v} = \mathbf{u} + (-\mathbf{v}) \in U.$$

□

6.4 DEFINITION

SUBSPACE Let V be a vector space over F, and let $U \subseteq V$. We say that U is a **subspace** of V if the following three conditions hold:

(i) $\mathbf{0} \in U$.
(ii) U is closed under addition.
(iii) U is closed under multiplication by scalars.

For emphasis, a subspace U of a vector space V is often referred to as a **vector subspace**, or a **linear subspace** of V.

6.5 Example In Example 6.2, show that U is a subspace of F^2.

SOLUTION We have shown in Example 6.2 that U is closed under addition and multiplication by scalars. Since

$$\mathbf{0} = (0, 0) \in U,$$

it follows that U is a subspace of F^2. □

Suppose that U is a subspace of the vector space V over F. Because U is closed under addition and multiplication by scalars, we can consider the restriction of these operations to U. Under these operations, *the subspace U becomes a vector space in its own right*, since it "inherits" all of the properties in Definition 5.1 from the "parent" vector space V (Problem 6).

Note that the subset $\{\mathbf{0}\}$ consisting only of the zero vector is a subspace of V. Since the empty set $\varnothing$ of vectors is linearly independent, and since $\mathbf{0}$ is a linear combination of $\varnothing$, it follows that $\varnothing$ is a basis for $\{\mathbf{0}\}$; hence,

$$\dim_F(\{\mathbf{0}\}) = \#\varnothing = 0.$$

The **zero subspace** $\{\mathbf{0}\}$ is the "smallest" subspace of V in the sense that it is contained in every other subspace. Of course, V is a subspace of itself, and it is the "largest" subspace of V in the sense that it contains every other subspace.

6.6 DEFINITION

LINEAR SPAN Let V be a vector space over F, and let $M \subseteq V$. The **linear span** of M, in symbols $\text{lin}(M)$, is defined to be the set of all linear combinations of finite sets of vectors in M.

By Problem 11 in Problem Set 3.5, $\text{lin}(M)$ is closed under addition and multiplication by scalars. Obviously, $\mathbf{0} \in \text{lin}(M)$, and so $\text{lin}(M)$ *is a subspace of V*.

6.7 Example Let V be a vector space over F, and let $M \subseteq V$. Show that $\text{lin}(M)$ *is the smallest subspace of V containing M* in the sense that, if U is any other subspace of V with $M \subseteq U$, then $\text{lin}(M) \subseteq U$.

SOLUTION Suppose $M \subseteq U$ and that U is a subspace of V. Since U is closed under addition and multiplication by scalars, it follows that every linear combination of vectors in M belongs to U; hence, $\text{lin}(M) \subseteq U$. □

6.8 DEFINITION

SUM OF SUBSPACES Let V be a vector space over F, and let P and Q be subspaces of V. We define the **sum** of P and Q, in symbols $P + Q$, by

$$P + Q = \{\mathbf{p} + \mathbf{q} \mid \mathbf{p} \in P \wedge \mathbf{q} \in Q\}.$$

6.9 THEOREM

SUM AND INTERSECTION OF SUBSPACES Let V be a vector space over F, and let P and Q be subspaces of V. Then:

(i) $P + Q$ is a subspace of V.
(ii) $P \cap Q$ is a subspace of V.

PROOF We prove (i) here and leave the proof of (ii) as an exercise (Problem 9). Since P and Q are subspaces, $\mathbf{0} \in P$ and $\mathbf{0} \in Q$; hence, $\mathbf{0} = \mathbf{0} + \mathbf{0} \in P + Q$. To show that $P + Q$ is closed under addition, suppose that $\mathbf{v}, \mathbf{w} \in P + Q$. Then, there exist $\mathbf{p}_1, \mathbf{p}_2 \in P$ and $\mathbf{q}_1, \mathbf{q}_2 \in Q$ such that

$$\mathbf{v} = \mathbf{p}_1 + \mathbf{q}_1 \quad \text{and} \quad \mathbf{w} = \mathbf{p}_2 + \mathbf{q}_2.$$

Let $\mathbf{p} = \mathbf{p}_1 + \mathbf{p}_2$, and let $\mathbf{q} = \mathbf{q}_1 + \mathbf{q}_2$, noting that (since P and Q are closed under addition) $\mathbf{p} \in P$ and $\mathbf{q} \in Q$. Hence,

$$\begin{aligned}\mathbf{v} + \mathbf{w} &= (\mathbf{p}_1 + \mathbf{q}_1) + (\mathbf{p}_2 + \mathbf{q}_2) = (\mathbf{p}_1 + \mathbf{p}_2) + (\mathbf{q}_1 + \mathbf{q}_2) \\ &= \mathbf{p} + \mathbf{q} \in P + Q.\end{aligned}$$

Finally, to show that $P + Q$ is closed under multiplication, note that, for $\mathbf{p} \in P$, $\mathbf{q} \in Q$, and $s \in F$, we have $s\mathbf{p} \in P$ and $s\mathbf{q} \in Q$; hence,

$$s(\mathbf{p} + \mathbf{q}) = s\mathbf{p} + s\mathbf{q} \in P + Q. \qquad \square$$

We leave it as an exercise for you to show that, if P and Q are subspaces of the vector space V, then

$$P + Q = \operatorname{lin}(P \cup Q)$$

(Problem 11). Thus, *$P + Q$ is the smallest subspace of V containing both P and Q*; that is, if U is a subspace of V, then

$$P \subseteq U \text{ and } Q \subseteq U \Rightarrow P + Q \subseteq U$$

(Problem 12). Likewise, *$P \cap Q$ is the largest subspace of V contained in both P and Q*; that is, if U is a subspace of V, then

$$U \subseteq P \text{ and } U \subseteq Q \Rightarrow U \subseteq P \cap Q$$

(Problem 13).

6.10 DEFINITION

DIRECT SUM Let V be a vector space over F, and let P and Q be subspaces of V. If $P \cap Q = \{\mathbf{0}\}$, we refer to the sum $P + Q$ as the **direct sum** of P and Q. We write $P + Q$ as

$$P \oplus Q$$

to indicate that the sum is a direct sum.

6.11 LEMMA **UNIQUENESS IN THE DIRECT SUM** Let V be a vector space over F, and let P and Q be subspaces of V with $P \cap Q = \{\mathbf{0}\}$. Let $\mathbf{v} \in P \oplus Q$. Then, there exist unique vectors $\mathbf{p} \in P$ and $\mathbf{q} \in Q$ such that $\mathbf{v} = \mathbf{p} + \mathbf{q}$.

PROOF Suppose $\mathbf{p}_1, \mathbf{p}_2 \in P$ and $\mathbf{q}_1, \mathbf{q}_2 \in Q$ with

$$\mathbf{v} = \mathbf{p}_1 + \mathbf{q}_1 \quad \text{and} \quad \mathbf{v} = \mathbf{p}_2 + \mathbf{q}_2.$$

Then

$$\mathbf{p}_1 + \mathbf{q}_1 = \mathbf{p}_2 + \mathbf{q}_2$$

and so

$$\mathbf{p}_1 - \mathbf{p}_2 = \mathbf{q}_2 - \mathbf{q}_1.$$

Let

$$\mathbf{w} = \mathbf{p}_1 - \mathbf{p}_2 = \mathbf{q}_2 - \mathbf{q}_1.$$

Since P is closed under subtraction, we have $\mathbf{w} \in P$. Since Q is closed under subtraction, we also have $\mathbf{w} \in Q$. Therefore,

$$\mathbf{w} \in P \cap Q = \{\mathbf{0}\},$$

so that $\mathbf{w} = \mathbf{0}$. Therefore,

$$\mathbf{p}_1 = \mathbf{p}_2 \quad \text{and} \quad \mathbf{q}_1 = \mathbf{q}_2.$$

□

6.12 Example Let F be a field, and consider the vector space F^2 over F. Let

$$P = \{(x, y) \in F^2 \mid y = 0\},$$

and let

$$Q = \{(x, y) \in F^2 \mid x = 0\}.$$

Show that

$$F^2 = P \oplus Q.$$

SOLUTION Since $\mathbf{0} = (0, 0)$ belongs to both P and Q, and since P and Q are closed under addition and multiplication by scalars, they are subspaces of F^2. If $\mathbf{v} = (x, y) \in F^2$, then

$$\mathbf{v} = (x, 0) + (0, y) \in P + Q;$$

hence, $F^2 = P + Q$. Finally, since

$$P \cap Q = \{(0, 0)\} = \{\mathbf{0}\},$$

$P + Q$ is a direct sum, and we can write

$$F^2 = P \oplus Q.$$

□

6.13 DEFINITION

COMPLEMENTARY DIRECT SUMMAND Let V be a vector space over the field F, and let P be a subspace of V. If Q is a subspace of V such that

$$V = P \oplus Q,$$

then we say that Q is a **complementary direct summand** of P in V.

Note that, if P is a complementary direct summand of Q, then Q is a complementary direct summand of P. For instance, in Example 6.12, P and Q are complementary direct summands of each other in the vector space F^2.

6.14 THEOREM

EXISTENCE OF A COMPLEMENTARY DIRECT SUMMAND Let V be a vector space over the field F, and let P be a subspace of V. If V is finite dimensional, there exists a subspace Q of V such that

$$V = P \oplus Q.$$

PROOF We sketch the proof here, and leave the details as an exercise (Problem 18). Let $n = \dim_F(V)$. By Steinitz's theorem (Theorem 5.16), no linearly independent subset of V can contain more than n vectors. Let

$$A = \{\mathbf{a}_1, \mathbf{a}_2, \mathbf{a}_3, \ldots, \mathbf{a}_m\}$$

be a maximal linearly independent subset of P. Then A is a basis for P. Extend A to a maximal linearly independent subset

$$B = \{\mathbf{a}_1, \mathbf{a}_2, \mathbf{a}_3, \ldots, \mathbf{a}_m, \mathbf{a}_{m+1}, \mathbf{a}_{m+2}, \ldots, \mathbf{a}_r\}$$

of V. Then B is a basis for V. Let

$$C = B \setminus A,$$

and let

$$Q = \operatorname{lin}(C).$$

Then C is a basis for Q, $P + Q = V$, and $P \cap Q = \{\mathbf{0}\}$. □

6.15 DEFINITION

LINEAR TRANSFORMATION Let V and W be vector spaces over the same field F. A function $T: V \to W$ is called a **linear transformation** from V to W if it satisfies the following two conditions for all vectors $\mathbf{u}, \mathbf{v} \in V$ and all scalars $s \in F$:

(i) $T(\mathbf{u} + \mathbf{v}) = T(\mathbf{u}) + T(\mathbf{v})$
(ii) $T(s\mathbf{v}) = s(T(\mathbf{v}))$

A function $T: V \to W$ that satisfies Condition (i) in Definition 6.15 is said to be **additive**, or to **preserve addition**. If T satisfies Condition (ii), it is said to be **homogeneous**, or to **preserve multiplication by scalars**. Thus, *a linear transformation is a mapping from a first vector space to a second that preserves the vector-space operations.* Condition (ii) is often written in the simpler form

$$T(s\mathbf{v}) = sT(\mathbf{v}),$$

it being understood that $sT(\mathbf{v})$ means $s(T(\mathbf{v}))$. Linear transformations are also called **linear operators**, or sometimes **operators**.

6.16 Example Let $T: \mathbb{R}^3 \to \mathbb{R}^2$ be defined by

$$T((x, y, z)) = (x, y)$$

for all $(x, y, z) \in \mathbb{R}^3$. Show that T is a linear transformation.

SOLUTION We must check that T is additive and homogeneous. Let $\mathbf{u} = (u_1, u_2, u_3)$, $\mathbf{v} = (v_1, v_2, v_3)$, and let $s \in \mathbb{R}$. Then,

$$\begin{aligned} T(\mathbf{u} + \mathbf{v}) &= T((u_1 + v_1, u_2 + v_2, u_3 + v_3)) = (u_1 + v_1, u_2 + v_2) \\ &= (u_1, u_2) + (v_1, v_2) = T(\mathbf{u}) + T(\mathbf{v}), \end{aligned}$$

which shows that T is additive. Also,

$$\begin{aligned} T(s\mathbf{v}) &= T(s(v_1, v_2, v_3)) = T((sv_1, sv_2, sv_3)) \\ &= (sv_1, sv_2) = s(v_1, v_2) = sT(\mathbf{v}), \end{aligned}$$

which shows that T is homogeneous. □

6.17 LEMMA **PRESERVATION OF 0 AND SUBTRACTION** Let V and W be vector spaces over F, and let $T: V \to W$ be a linear transformation. Then:

(i) $T(\mathbf{0}) = \mathbf{0}$
(ii) $T(-\mathbf{v}) = -T(\mathbf{v})$
(iii) $T(\mathbf{u} - \mathbf{v}) = T(\mathbf{u}) - T(\mathbf{v})$ for all $\mathbf{u}, \mathbf{v} \in V$.

PROOF (i) $T(\mathbf{0}) = T(0 \cdot \mathbf{0}) = 0T(\mathbf{0}) = \mathbf{0}$.
(ii) $T(-\mathbf{v}) = T((-1)\mathbf{v}) = (-1)T(\mathbf{v}) = -T(\mathbf{v})$.
(iii) $T(\mathbf{u} - \mathbf{v}) = T(\mathbf{u} + (-\mathbf{v})) = T(\mathbf{u}) + T(-\mathbf{v})$
$= T(\mathbf{u}) + (-T(\mathbf{v})) = T(\mathbf{u}) - T(\mathbf{v})$. □

Since subspaces are closed under the vector-space operations and linear transformations preserve the vector-space operations, it should come as no surprise that there is an interesting interplay between subspaces and linear transformations. We begin to see some of this interplay in the next definition and theorem.

6.18 DEFINITION

IMAGE AND INVERSE IMAGE OF A SUBSPACE Let $T: V \to W$ be a linear transformation from the vector space V over F to the vector space W over F. Let P be a subspace of V, and let Q be a subspace of W. We define the **image of P under T** by

$$T(P) = \{T(\mathbf{v}) \mid \mathbf{v} \in P\},$$

and we define the **inverse image of Q under T** by

$$T^{-1}(Q) = \{\mathbf{v} \in V \mid T(\mathbf{v}) \in Q\}.$$

Images and inverse images of subspaces are just special cases of the images and inverse images of sets under functions introduced in Definition 7.1 in Chapter 2.

6.19 THEOREM

PRESERVATION OF SUBSPACES Let $T: V \to W$ be a linear transformation from the vector space V over F to the vector space W over F. Let P be a subspace of V, and let Q be a subspace of W. Then:

(i) $T(P)$ is a subspace of W.
(ii) $T^{-1}(Q)$ is a subspace of V.

PROOF We prove Part (ii) here and leave the proof of Part (i) as an exercise (Problem 23). By Part (i) of Lemma 6.17, $T(\mathbf{0}) = \mathbf{0}$. Because Q is a subspace of W, $\mathbf{0} \in Q$, and therefore $T(\mathbf{0}) \in Q$; hence, by Definition 6.18, $\mathbf{0} \in T^{-1}(Q)$. To prove that $T^{-1}(Q)$ is closed under addition, suppose that

$$\mathbf{v}_1, \mathbf{v}_2 \in T^{-1}(Q).$$

Then, by Definition 6.18,

$$T(\mathbf{v}_1), T(\mathbf{v}_2) \in Q,$$

and, since Q is closed under addition, it follows that

$$T(\mathbf{v}_1 + \mathbf{v}_2) = T(\mathbf{v}_1) + T(\mathbf{v}_2) \in Q.$$

Therefore, by Definition 6.18 again,

$$\mathbf{v}_1 + \mathbf{v}_2 \in T^{-1}(Q),$$

and so $T^{-1}(Q)$ is closed under addition. Finally, suppose that

$$\mathbf{v} \in T^{-1}(Q) \quad \text{and} \quad s \in F.$$

Then,

$$T(\mathbf{v}) \in Q,$$

and, since Q is closed under multiplication by scalars,

$$sT(\mathbf{v}) = T(s\mathbf{v}) \in Q.$$

Therefore,

$$s\mathbf{v} \in T^{-1}(Q),$$

which shows that $T^{-1}(Q)$ is closed under multiplication by scalars. □

6.20 DEFINITION

KERNEL OF A LINEAR TRANSFORMATION . Let $T: V \to W$ be a linear transformation from the vector space V over F to the vector space W over F. The **kernel** of T, in symbols $\ker(T)$, is defined by

$$\ker(T) = T^{-1}(\{\mathbf{0}\}).$$

The kernel of a linear transformation T is sometimes referred to as the **null space** of T. Note that

$$\ker(T) = \{\mathbf{v} \in V \mid T(\mathbf{v}) = \mathbf{0}\}.$$

Since $\{\mathbf{0}\}$ is a subspace of W, it follows from Theorem 6.19 that *the kernel of a linear transformation T is a subspace of the domain of T*. In particular, we always have $\mathbf{0} \in \ker(T)$.

6.21 THEOREM

INJECTIVITY OF A LINEAR TRANSFORMATION Let $T: V \to W$ be a linear transformation from the vector space V over F to the vector space W over F. Then, T is injective (one-to-one) if and only if $\ker(T) = \{\mathbf{0}\}$.

PROOF Suppose T is injective, and let $\mathbf{v} \in \ker(T)$. Then, $T(\mathbf{v}) = \mathbf{0}$. Because $T(\mathbf{0}) = \mathbf{0}$, we have $T(\mathbf{v}) = T(\mathbf{0})$, and it follows from the injectivity of T that $\mathbf{v} = \mathbf{0}$. Thus, $\mathbf{0}$ is the only vector in $\ker(T)$; so $\ker(T) = \{\mathbf{0}\}$. Conversely, suppose that $\ker(T) = \{\mathbf{0}\}$ and that $T(\mathbf{v}_1) = T(\mathbf{v}_2)$ with $\mathbf{v}_1, \mathbf{v}_2 \in V$. To prove that T is injective, we must show that $\mathbf{v}_1 = \mathbf{v}_2$. Let $\mathbf{v} = \mathbf{v}_1 - \mathbf{v}_2$, noting that it is sufficient to prove that $\mathbf{v} = \mathbf{0}$. Now,

$$T(\mathbf{v}) = T(\mathbf{v}_1 - \mathbf{v}_2) = T(\mathbf{v}_1) - T(\mathbf{v}_2) = \mathbf{0},$$

and so $\mathbf{v} \in \ker(T) = \{\mathbf{0}\}$, that is, $\mathbf{v} = \mathbf{0}$. □

Suppose that $T: V \to W$ is a linear transformation. Note that

$$\text{range}(T) = T(V).$$

Since V is a subspace of itself, it follows from Theorem 6.19 that *the range of a linear transformation $T: V \to W$ is a subspace of W.*

6.22 DEFINITION

RANK AND NULLITY Let $T: V \to W$ be a linear transformation from the vector space V over F to the vector space W over F. By definition, the dimension of $T(V)$ is called the **rank** of T and the dimension of $\ker(T)$ is called the **nullity** of T.

6.23 THEOREM

RANK-PLUS-NULLITY THEOREM Let $T: V \to W$ be a linear transformation from the vector space V over F to the vector space W over F. Suppose that the rank and nullity of T are finite. Then V is finite dimensional, and

$$(\text{rank of } T) + (\text{nullity of } T) = \dim_F(V).$$

PROOF Let r = rank of T and n = nullity of T. Select a basis

$$\mathbf{a_1}, \mathbf{a_2}, \mathbf{a_3}, \ldots, \mathbf{a_n}$$

for $\ker(T)$ and a basis

$$\mathbf{b_1}, \mathbf{b_2}, \mathbf{b_3}, \ldots, \mathbf{b_r}$$

for $T(V)$. By the definition of $T(V)$, each vector $\mathbf{b} \in T(V)$ has the form $\mathbf{b} = T(\mathbf{a})$ for some (not necessarily unique) vector $\mathbf{a} \in V$. Thus, for $j = 1, 2, 3, \ldots, r$, we can choose vectors $\mathbf{a_{n+j}} \in V$ such that

$$T(\mathbf{a_{n+j}}) = \mathbf{b_j}. \tag{1}$$

Suppose that $s_1, s_2, \ldots, s_n, s_{n+1}, \ldots, s_{n+r}$ are scalars in F such that

$$s_1\mathbf{a_1} + s_2\mathbf{a_2} + s_3\mathbf{a_3} + \cdots + s_n\mathbf{a_n} + s_{n+1}\mathbf{a_{n+1}} + \cdots + s_{n+r}\mathbf{a_{n+r}} = \mathbf{0}. \tag{2}$$

Applying T to both sides of the last equation, and noting that $T(\mathbf{a_i}) = \mathbf{0}$ for $i = 1, 2, 3, \ldots, n$ (since $\mathbf{a_1}, \mathbf{a_2}, \mathbf{a_3}, \ldots, \mathbf{a_n} \in \ker(T)$), we find that

$$s_{n+1}T(\mathbf{a_{n+1}}) + \cdots + s_{n+r}T(\mathbf{a_{n+r}}) = \mathbf{0},$$

so that, by Equation (1),

$$s_{n+1}\mathbf{b_1} + s_{n+2}\mathbf{b_2} + s_{n+3}\mathbf{b_3} + \cdots + s_{n+r}\mathbf{b_r} = \mathbf{0}. \tag{3}$$

Because $\mathbf{b_1}, \mathbf{b_2}, \mathbf{b_3}, \ldots, \mathbf{b_r}$ are linearly independent, it follows from Equation (3) that

$$s_{n+1} = s_{n+2} = s_{n+3} = \cdots = s_{n+r} = 0. \tag{4}$$

Therefore, we can rewrite Equation (2) as

$$s_1\mathbf{a_1} + s_2\mathbf{a_2} + s_3\mathbf{a_3} + \cdots + s_n\mathbf{a_n} = \mathbf{0}. \tag{5}$$

Because $\mathbf{a_1}, \mathbf{a_2}, \mathbf{a_3}, \ldots, \mathbf{a_n}$ are linearly independent, it follows from Equation (5) that

$$s_1 = s_2 = s_3 = \cdots = s_n = 0. \tag{6}$$

Therefore, if Equation (2) holds, then, by Equations (5) and (6), all of the coefficients of the linear combination in Equation (2) are zero; hence, the vectors

$$\mathbf{a_1}, \mathbf{a_2}, \mathbf{a_3}, \ldots, \mathbf{a_n}, \mathbf{a_{n+1}}, \mathbf{a_{n+2}}, \ldots, \mathbf{a_{n+r}}$$

are linearly independent.

Now suppose that $\mathbf{v}$ is any vector in V. Because $T(\mathbf{v}) \in T(V)$ and $\mathbf{b_1}, \mathbf{b_2}, \mathbf{b_3}, \ldots, \mathbf{b_r}$ form a basis for $T(V)$, there are scalars $s_{n+1}, s_{n+2}, s_{n+3}, \ldots, s_{n+r}$ in F such that

$$T(\mathbf{v}) = s_{n+1}\mathbf{b_1} + s_{n+2}\mathbf{b_2}\ s_{n+3}\mathbf{b_3} + \cdots + s_{n+r}\mathbf{b_r}. \tag{7}$$

Let

$$\mathbf{k} = \mathbf{v} - s_{n+1}\mathbf{a_{n+1}} - s_{n+2}\mathbf{a_{n+2}} - \cdots - s_{n+r}\mathbf{a_{n+r}}. \tag{8}$$

Applying T to both sides of Equation (8) and using Equation (1), we find that

$$T(\mathbf{k}) = T(\mathbf{v}) - s_{n+1}\mathbf{b_1} - s_{n+2}\mathbf{b_2} - \cdots - s_{n+r}\mathbf{b_r}. \tag{9}$$

Combining Equations (7) and (9), we obtain

$$T(\mathbf{k}) = T(\mathbf{v}) - T(\mathbf{v}) = \mathbf{0}, \tag{10}$$

from which we conclude that $\mathbf{k} \in \ker(T)$. Because $\mathbf{a_1}, \mathbf{a_2}, \mathbf{a_3}, \ldots, \mathbf{a_n}$ form a basis for $\ker(T)$, there are scalars $s_1, s_2, s_3, \ldots, s_n$ such that

$$\mathbf{k} = s_1\mathbf{a_1} + s_2\mathbf{a_2} + s_3\mathbf{a_3} + \cdots + s_n\mathbf{a_n}. \tag{11}$$

By Equation (8), we have

$$\mathbf{v} = \mathbf{k} + s_{n+1}\mathbf{a_{n+1}} + s_{n+2}\mathbf{a_{n+2}} + \cdots + s_{n+r}\mathbf{a_{n+r}}; \tag{12}$$

hence, by Equations (11) and (12),

$$\begin{aligned} \mathbf{v} = s_1\mathbf{a_1} &+ s_2\mathbf{a_2} + s_3\mathbf{a_3} + \cdots + s_n\mathbf{a_n} \\ &+ s_{n+1}\mathbf{a_{n+1}} + s_{n+2}\mathbf{a_{n+2}} + \cdots + s_{n+r}\mathbf{a_{n+r}}. \end{aligned} \tag{13}$$

Equation (13) shows that the vectors

$$\mathbf{a_1}, \mathbf{a_2}, \mathbf{a_3}, \ldots, \mathbf{a_n}, \mathbf{a_{n+1}}, \mathbf{a_{n+2}}, \ldots, \mathbf{a_{n+r}}$$

span V. Because these vectors are also linearly independent, they form a basis for V, and it follows that

$$\dim_F(V) = n + r = r + n = (\text{rank of } T) + (\text{nullity of } T). \qquad \square$$

6.24 DEFINITION

> **ISOMORPHISM** Let V and W be vector spaces over F. A bijective linear transformation $T: V \to W$ is said to be an **isomorphism**.

If there exists an isomorphism $T: V \to W$, then we say that the vector spaces V and W are *isomorphic*. Two vector spaces that are isomorphic have, as vector spaces, the "same algebraic structure."

6.25 Example Let V be a two-dimensional vector space over F. Show that F^2 is isomorphic to V.

SOLUTION Let $\mathbf{b}_1, \mathbf{b}_2$ form a basis for V. Define $T: F^2 \to V$ by $T((x, y)) = x\mathbf{b}_1 + y\mathbf{b}_2$. Then, T is a linear transformation (Problem 21), and it is obviously surjective. If $(x, y) \in \ker(T)$, then $x\mathbf{b}_1 + y\mathbf{b}_2 = \mathbf{0}$; hence, $x = y = 0$, and so $(x, y) = \mathbf{0}$. Consequently, by Theorem 6.21, T is injective, and therefore it is bijective. □

Generalizing the argument in Example 6.25 shows that *if V is a finite-dimensional vector space over F and if $n = \dim_F(V)$, then F^n is isomorphic to V* (Problem 35).

PROBLEM SET 3.6

1. Let V be a vector space over F, and suppose that U is a *nonempty* subset of V. If U is closed under addition and closed under multiplication by scalars, show that U is a subspace of V.
2. Regard the complex numbers $\mathbb{C}$ as a vector space over $\mathbb{C}$. Do the real numbers $\mathbb{R}$ form a subspace of $\mathbb{C}$? Explain.
3. Let F be a field, and let n be a positive integer. Let $s_1, s_2, s_3, \ldots, s_n$ be fixed scalars in F, and define $U \subseteq F^n$ to be the set of all vectors $(x_1, x_2, x_3, \ldots, x_n) \in F^n$ such that
$$s_1x_1 + s_2x_2 + s_3x_3 + \cdots + s_nx_n = 0.$$
Show that U is a subspace of F.
4. Show that the set P of all polynomial functions $p: \mathbb{R} \to \mathbb{R}$ is a subspace of $C(\mathbb{R})$.
5. Visualizing $\mathbb{R}^2$ as the xy coordinate plane, show that the subspaces of $\mathbb{R}^2$ are $\{(0, 0)\}$, $\mathbb{R}^2$ itself, and all straight lines $L \subseteq \mathbb{R}^2$ that contain the origin $(0, 0)$.
6. Prove that, if U is a subspace of the vector space V over F, then, under the restrictions to U of the vector-space operations on V, U forms a vector space in its own right.

7. Visualizing $\mathbb{R}^3$ as xyz-space, show that the subspaces of $\mathbb{R}^3$ are $\{(0, 0, 0)\}$, $\mathbb{R}^3$ itself, all straight lines $L \subseteq \mathbb{R}^3$ that contain the origin $(0, 0, 0)$, and all planes $P \subseteq \mathbb{R}^3$ that contain the origin $(0, 0, 0)$.

8. If V is a finite-dimensional vector space over F and U is a subspace of V, prove that $\dim_F(U) \leq \dim_F(V)$.

9. Prove Part (ii) of Theorem 6.9.

10. If V is a finite-dimensional vector space over F and U is a subspace of V such that $\dim_F(U) = \dim_F(V)$, show that $U = V$.

11. If P and Q are subspaces of a vector space V over F, show that $P + Q = \text{lin}(P \cup Q)$.

12. If P, Q, and U are subspaces of a vector space V over F, and if $P \subseteq U$ and $Q \subseteq U$, show that $P + Q \subseteq U$.

13. If P, Q, and U are subspaces of a vector space V over F, and if $U \subseteq P$ and $U \subseteq Q$, show that $U \subseteq P \cap Q$.

14. Let P and Q be subspaces of a vector space V over F. If $P \cup Q$ is a subspace of V, show that $P \subseteq Q$ or $Q \subseteq P$.

15. Let F be a field, let P be the subset of F^3 consisting of all (x, y, z) with $y = 0$, and let Q be the subset of F^3 consisting of all (x, y, z) with $x = z = 0$. Show that $F^3 = P \oplus Q$.

16. Let F be a field, and let P be the subset of F^3 consisting of all (x, y, z) with $z = 0$. Find a complementary direct summand Q of P in F^3.

17. If V is a finite-dimensional subspace over F and A is a linearly independent subset of V, show that A can be extended to a basis $B \supseteq A$ for V. [Hint: Use the same idea as in Problem 27 in Problem Set 3.5.]

18. Fill in the details of the proof of Theorem 6.14.

19. Let $T\colon\mathbb{R}^2 \to \mathbb{R}^3$ be defined by $T((x, y)) = (x, y, x)$ for all $(x, y) \in \mathbb{R}^2$.
(a) Show that T is a linear transformation. (b) Find $\ker(T)$.

20. Let P be the subspace of $C(\mathbb{R})$ consisting of all polynomial functions $p\colon\mathbb{R} \to \mathbb{R}$. If $p \in P$, let p' denote the derivative of p. Define $D\colon P \to C(\mathbb{R})$ by $Dp = p'$ for all $p \in P$.
(a) Show that D is a linear transformation. (b) Find $\ker(D)$.
(c) Find the image $D(P)$ of P under D.

21. Let F be a field, let n be a positive integer, and let V be a vector space over F. For each positive integer $i = 1, 2, 3, \ldots, n$, choose and fix a vector $\mathbf{b_i} \in V$. (The chosen vectors need not be different from one another.) Define $T\colon\mathbb{R}^n \to V$ by

$$T((x_1, x_2, x_3, \ldots, x_n)) = x_1\mathbf{b_1} + x_2\mathbf{b_2} + x_3\mathbf{b_3} + \cdots + x_n\mathbf{b_n}$$

for every $(x_1, x_2, x_3, \ldots, x_n) \in \mathbb{R}^n$. Show that $T\colon\mathbb{R}^n \to V$ is a linear transformation.

22. Let V be a vector space over F. A mapping $f:V \to F$ is called a **linear functional on** V if, with F regarded as a vector space over itself, f is a linear transformation. The constant function that maps every vector in V to 0 is called the **zero functional on** V. If f is a linear functional on V and f is not the zero functional, show that $f:V \to F$ is surjective.

23. Prove Part (i) of Theorem 6.19.

24. Let V be a finite-dimensional vector space over F, and let f be a nonzero linear functional on V. (See Problem 22.) Show that the nullity of f is $\dim_F(V) - 1$. [Hint: Use the rank-plus-nullity theorem.]

25. Let V and W be vector spaces over F. Show that the Cartesian product $V \times W$ becomes a vector space under the "coordinatewise" operations defined by

$$(\mathbf{v}_1, \mathbf{w}_1) + (\mathbf{v}_2, \mathbf{w}_2) = (\mathbf{v}_1 + \mathbf{v}_2, \mathbf{w}_1 + \mathbf{w}_2)$$

and

$$s(\mathbf{v}, \mathbf{w}) = (s\mathbf{v}, s\mathbf{w})$$

for $\mathbf{v}, \mathbf{v}_1, \mathbf{v}_2 \in V$, $\mathbf{w}, \mathbf{w}_1, \mathbf{w}_2 \in W$, and $s \in F$. The vector space $V \times W$ is called the **Cartesian product** of the vector spaces V and W.

26. If V and W are vector spaces over F, show that $V \times W$ is isomorphic to $W \times V$. (See Problem 25.)

27. Let V and W be vector spaces over F. Define mappings $\Pi_1: V \times W \to V$ and $\Pi_2: V \times W \to W$ by $\Pi_1((\mathbf{v}, \mathbf{w})) = \mathbf{v}$ and $\Pi_2((\mathbf{v}, \mathbf{w})) = \mathbf{w}$ for all $(\mathbf{v}, \mathbf{w}) \in V \times W$. Show that Π_1 and Π_2 are linear transformations. (See Problem 25.)

28. Let V be a vector space over F, and suppose that P and Q are complementary direct summands in V, so that $V = P \oplus Q$. Let $P \times Q$ be the Cartesian product of the vector spaces P and Q. (See Problem 25.) Show that the mapping $T: P \times Q \to P \oplus Q$ given by $T((\mathbf{p}, \mathbf{q})) = \mathbf{p} + \mathbf{q}$ for all $(\mathbf{p}, \mathbf{q}) \in P \times Q$ is an isomorphism.

29. Let V and W be vector spaces over F, let $S: V \to W$ and $T: V \to W$ be linear transformations, and let $s \in F$. Define $(S + T): V \to W$ by $(S + T)(\mathbf{v}) = S(\mathbf{v}) + T(\mathbf{v})$ for all $\mathbf{v} \in V$. Define $(sT): V \to W$ by $(sT)(\mathbf{v}) = s(T(\mathbf{v}))$ for all $\mathbf{v} \in V$.
(a) Show that $S + T$ and sT are linear transformations.
(b) Let $\mathscr{L}(V, W)$ denote the set of all linear transformations $T: V \to W$. Show that, with the operations defined above, $\mathscr{L}(V, W)$ is a vector space.

30. Let V and W be finite-dimensional vector spaces with $n = \dim_F(V)$ and $m = \dim_F(W)$. Choose and fix bases $A = \{\mathbf{a}_1, \mathbf{a}_2, \mathbf{a}_3, \ldots, \mathbf{a}_n\}$ for V and $B = \{\mathbf{b}_1, \mathbf{b}_2, \mathbf{b}_3, \ldots, \mathbf{b}_m\}$ for W. Let $T: V \to W$ be a linear transformation. For each $j = 1, 2, 3, \ldots, n$, $T(\mathbf{a}_j)$ is a vector in W, and therefore it can be written as a linear combination

$$T(\mathbf{a_j}) = \sum_{i=1}^{m} t_{ij}\mathbf{b_i}$$

for suitable scalars t_{ij}, with $1 \leq i \leq m$ and $1 \leq j \leq n$. Each vector $\mathbf{v} \in V$ and each vector $\mathbf{w} \in W$ can be written in the forms

$$\mathbf{v} = \sum_{j=1}^{n} v_j\mathbf{a_j} \quad \text{and} \quad \mathbf{w} = \sum_{i=1}^{m} w_i\mathbf{b_i}.$$

If $T(\mathbf{v}) = \mathbf{w}$, show that, for each $i = 1, 2, 3, \ldots, m$,

$$w_i = \sum_{j=1}^{n} t_{ij}v_j.$$

31. Let U, V, and W be vector spaces over F, and let $S: U \to V$ and $T: V \to W$ be linear transformations. Show that the composition $(T \circ S): U \to W$ is a linear transformation, that is, show that *the composition of linear transformations is a linear transformation.*

32. With the notation of Problem 30, the m-by-n array

$$M = \begin{bmatrix} t_{11} & t_{12} & \cdots & t_{1n} \\ t_{21} & t_{22} & \cdots & t_{2n} \\ \vdots & \vdots & & \vdots \\ t_{m1} & t_{m2} & \cdots & t_{mn} \end{bmatrix}$$

of scalars is called the **matrix of T with respect to the bases A and B**. Show that, given any m-by-n array M of scalars, there is a unique linear transformation T with M as its matrix with respect to the bases A and B.

33. Let V be a vector space over F, and let $\mathscr{L}(V)$ denote the set of all linear transformations $T: V \to V$. Show that $(\mathscr{L}(V), +, \circ)$ is a ring with unity.

34. Let U, V, and W be finite-dimensional vector spaces over F, and let $S: U \to V$ and $T: V \to W$ be linear transformations. Denote the dimensions of U, V, and W by d, n, and m, respectively. Choose and fix bases A, B, and C for U, V, and W, respectively. Let $N = [s_{pq}]$ be the matrix for S with respect to the bases A and B, and let $M = [t_{ij}]$ be the matrix for T with respect to the bases B and C. (See Problem 32.) Show that the entries in the matrix $P = [p_{rk}]$ for the composition $(T \circ S): U \to W$ with respect to the bases A and C are given by

$$p_{rk} = \sum_{j=1}^{n} t_{rj}s_{jk}$$

for $1 \leq r \leq m$ and $1 \leq k \leq d$. This is the formula for the **row-by-column** product $P = MN$ of the matrices M and N.

35. If V is a finite-dimensional vector space over F and if $n = \dim_F(V)$, show that F^n is isomorphic to V.

36. If V and W are finite-dimensional vector spaces over F, show that V is isomorphic to W if and only if $\dim_F(V) = \dim_F(W)$.

37. If V and W are vector spaces over F and if $T: V \to W$ is an isomorphism, show that $T^{-1}: W \to V$ is an isomorphism, that is, show that *the inverse of an isomorphism is an isomorphism.*

38. Let V and W be vector spaces over F, and let $T: V \to W$ be a function. Then, regarded as a set of ordered pairs,

$$T = \{(\mathbf{v}, T(\mathbf{v})) \mid \mathbf{v} \in V\} \subseteq V \times W.$$

Also, $V \times W$ can be regarded as a vector space as in Problem 25. Show that $T: V \to W$ is a linear transformation if and only if T is a subspace of the Cartesian product $V \times W$.

39. If U, V, and W are vector spaces over F, if $S: U \to V$ and $T: V \to W$ are isomorphisms, show that $(T \circ S): U \to W$ is an isomorphism, that is, show that *the composition of isomorphisms is an isomorphism.*

HISTORICAL NOTES

The theory of groups in its present abstract form came about through the fusion of developments in three distinct areas: number theory, solution of algebraic equations, and geometry. If we identify the beginning of group theory with the appearance of methods of proof that are now typical of the subject, then group theory begins with the paper of Leonhard Euler (1707–1783) on power residues (1761). Let a, b, and n be integers with n positive. Then $a \equiv b \pmod{n}$ if and only if n divides $b - a$. Euler investigated the question: If p is prime and p does not divide a, then for what λ is $a^\lambda \equiv 1 \pmod{p}$? For the case when $\lambda < p - 1$, Euler's discussion involves not only groups but subgroups and factor groups. (We will discuss this further in Chapter 5.) Though Euler never mentions groups, his methods may be taken over into modern group theory with only the most trivial modifications. Groups also lie just below the surface of Carl Friedrich Gauss's (1777–1855) investigations of the classes of quadratic forms (1801).

The second source of modern group theory appeared in efforts to find solutions to equations of fifth and higher degree in radicals. Paulo Ruffini (1765–1822), Niels Abel (1802–1829), and Evariste Galois (1811–1832), building on work of Joseph Louis Lagrange (1736–1813), showed that such solutions were usually impossible. Galois's method was to associate with each equation a certain permutation group (Section 5.5) and then to show that the equation is solvable in radicals if and only if the group has certain properties. In this context, the groups are finite groups of permutations (also referred to as substitutions), and so, for 50 years, group theory was finite-permutation group theory.

The third source of group theory was found in geometry in groups of transformations—for example, the group of rotations of a sphere in three-space. Felix Klein (1849–1925) and Marius Sophus Lie (1842–1899) did the most important work in this area.

While it was well known as early as the 1850s that all the above concepts were closely related, it took some time to realize that the efficiency and elegance of an axiomatic treatment was preferable to the greater concreteness of the older approaches. The modern axiomatic treatment was initiated by Arthur Cayley (1821–1895) and almost achieved by Walther Franz von Dyke (1856–1934) in 1882. The fully abstract definition was given by Heinrich Weber (1842–1913) in 1893, and then incorporated into his splendid textbook, *Lehrbuch der Algebra.* It is interesting that in the first decade of this century a fair number of articles appeared in American journals discussing which of the equivalent systems of axioms for a group ought to be the "official" version.

In comparison with groups, rings are relatively new structures, dating from the 1920s, but their history is closely connected to the theory of algebras. Algebras differ from rings in that the algebra elements may be multiplied by "scalars." The scalars are usually considered to belong to a field F, and we thus speak of an algebra over a field F. For example, the complex numbers form an algebra over the field $\mathbb{R}$.

The theory of algebras begins with the discovery of quaternions by William Rowan Hamilton in 1843. Hamilton's quaternions were important in the development of vector analysis and decisive in freeing algebra from the commutative law. Other important algebras investigated in the nineteenth century include the matrix algebras (1857) and group algebras (1854) of Cayley, the alternating algebra (1844, 1862) of Hermann Grassmann (1809–1877), and the Clifford algebras (1878) of William Kingdon Clifford (1845–1879). Matrix algebras are particularly important because many algebras are isomorphic to matrix algebras.

Great progress was made on the classification of algebras by G. F. Frobenius (1849–1917) and others. Starting about 1920, Emmy Noether (1882–1935) initiated the project of carrying over the classification of algebras to the theory of rings. Noether's efforts met with great success, and the subject continues to be an active area of research. Many of the concepts and methods introduced by Noether in her ring theory work have proved fruitful in other areas of mathematics.

The fields $\mathbb{Q}$ and $\mathbb{R}$ have, of course, been known since antiquity, but as abstract, axiomatically defined objects, fields date from the 1830 memoir of Galois. In this work, fields are defined by the requirement that they satisfy the "usual laws" (the field axioms) and the additional axiom that quantities multiplied by p become zero. Systems of this kind (for example, the integers modulo a prime p) had been known since Euler and were extensively investigated by Gauss, but the abstract formulation and many new examples appear here for the first time. Moreover, the habit of defining mathematical structures by axioms and investigating all examples simultaneously by deducing theorems from the axioms (the activity we call modern algebra) begins with this work.

ANNOTATED BIBLIOGRAPHY

Birkhoff, Garrett, and T. C. Bartee. *Modern Applied Algebra* (New York: McGraw-Hill), 1970.

One of the earliest and best surveys of applications of algebra to modern problems of engineering and technology.

Birkhoff, Garrett, and Saunders MacLane. *A Survey of Modern Algebra*, 3rd ed. (New York: Macmillan), 1965.

This is the book that introduced a whole generation of mathematicians to abstract algebra. First published in 1941, it had enormous impact in bringing to the United States ideas then prevalent in Europe and in introducing these ideas at the undergraduate level. Further, the text is the result of a collaboration of two American mathematicians of international reputation.

Fraleigh, John B. *A First Course in Abstract Algebra*, 3rd ed. (Reading, Mass.: Addison-Wesley), 1982.

A popular text in algebra that has many nice problems and takes up topics of some difficulty, albeit often in the direction of applications in algebraic topology.

Halmos, Paul R. *Finite-Dimensional Vector Spaces* (New York: Springer-Verlag), 1974.

A classic text in the subject; exceptionally well written.

Herstein, I. N. *Abstract Algebra* (New York: Macmillan), 1986.

An elementary text in algebra by a first-class algebraist.

Hillman, A. P., and G. L. Alexanderson. *A First Undergraduate Course in Abstract Algebra*, 4th ed. (Belmont, Ca.: Wadsworth), 1988.

A widely used text in abstract algebra treating groups, rings, and fields in some depth.

Kleiner, Israel. "The Evolution of Group Theory. A Brief Survey," *Mathematics Magazine* **59** (1986), 195–215.

A fresh look at the history of group theory. The article won the Allendoerfer Prize for expository writing in 1987.

McCoy, Neal H. *Rings and Ideals* (Carus Monograph No. 8) (Washington, D.C.: The Mathematical Association of America), 1948.

Now somewhat old, this nevertheless remains an interesting and valuable treatment of the subject by a well-known ring theorist.

4 NUMBER SYSTEMS

We have often spoken about the real number system $\mathbb{R}$, and we have given definitions, theorems, examples, and problems involving $\mathbb{R}$. However, we have never provided a formal definition of $\mathbb{R}$—instead, we have assumed an intuitive understanding and a working knowledge of the real number system.

A formal definition of the real number system that is acceptable by modern standards of mathematical rigor must be based on (axiomatic) set theory. Starting from such a basis, it is possible to build up a succession of number systems, each including its predecessor, culminating with $\mathbb{R}$. This process, which is known as *constructing the real numbers*, begins by building up the system

$$\mathbb{N} = \{1, 2, 3, \ldots\}$$

of *natural numbers*, using nothing but the machinery of set theory. Next, the ring of *integers*

$$\mathbb{Z} = \{0, \pm 1, \pm 2, \pm 3, \ldots\}$$

is constructed from the natural numbers. With $\mathbb{Z}$ at hand, it is possible to form the ordered field

$$\mathbb{Q} = \{m/n \mid m, n \in \mathbb{Z} \wedge n \neq 0\}$$

of *rational numbers*. Finally, by applying a suitable completion process to $\mathbb{Q}$, the ordered field $\mathbb{R}$ of real numbers is obtained. The succession of number systems does not stop here, but proceeds to the field of *complex numbers*

$$\mathbb{C} = \{x + yi \mid x, y \in \mathbb{R}\}$$

and beyond!

The process of building the succession of number systems

$$\mathbb{N} \subseteq \mathbb{Z} \subseteq \mathbb{Q} \subseteq \mathbb{R} \subseteq \mathbb{C} \subseteq \cdots$$

from below, starting with $\mathbb{N}$, is best left to more advanced courses, since it is somewhat involved (and even a bit tedious!). In this chapter, we shall introduce these number systems and study them with some care, but we are going to start with $\mathbb{R}$ and extract the systems $\mathbb{N}$, $\mathbb{Z}$, and $\mathbb{Q}$ from it. Thus, in Section 1, we simply postulate $\mathbb{R}$ as an ordered field satisfying an additional condition called completeness.

In Sections 2, 3, and 4, we extract the systems $\mathbb{N}$, $\mathbb{Z}$, and $\mathbb{Q}$ from $\mathbb{R}$, and in Section 5, we show how the field $\mathbb{C}$ of complex numbers can be constructed from $\mathbb{R}$. In Sections 6 and 7, we relate the natural numbers to the counting process, and in Section 8, we include a sketch of the construction of $\mathbb{R}$ and provide a brief discussion of number systems "beyond $\mathbb{C}$."

1 THE REAL NUMBERS

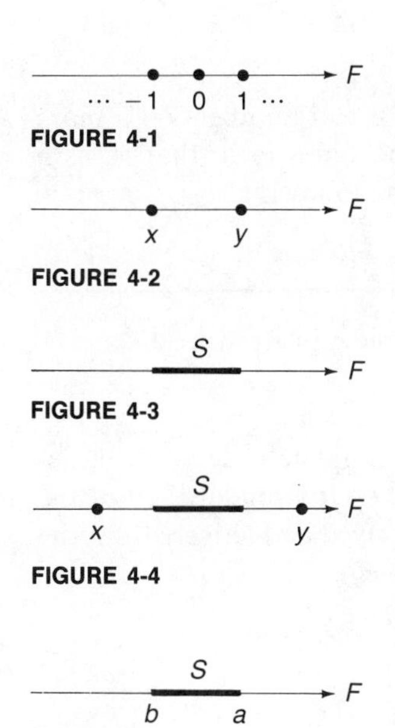

FIGURE 4-1
FIGURE 4-2
FIGURE 4-3
FIGURE 4-4
FIGURE 4-5
FIGURE 4-6

The real numbers form an ordered field $\mathbb{R}$ that satisfies a condition called *completeness.* To understand this condition, we begin by considering an arbitrary ordered field F and visualize the elements of F, in the usual way, as points on a line (Figure 4-1).

If $x, y \in F$ with $x < y$, we picture x lying to the left of y on this line (Figure 4-2).

A subset S of F can be depicted by shading the portion of the line corresponding to points in S (Figure 4-3).

Recall that an element $y \in F$ is called an *upper bound* for the set S if $s \leq y$ holds for all $s \in S$ (Definition 2.20 in Chapter 2). Similarly, an element $x \in F$ is called a *lower bound* for S if $x \leq s$ holds for all $s \in S$ (Figure 4-4).

Upper and lower bounds, when they exist, are not unique; in fact, any element of F that is larger than an upper bound for S is again an upper bound for S, and any element of F that is smaller than a lower bound for S is again a lower bound for S. When we say that a set has a lower bound or that it has an upper bound, we do not mean to imply that this lower or upper bound necessarily belongs to the set. An upper bound a for S that actually *belongs* to S is called a **maximum**, **largest**, **greatest**, or **last** element of S, and a lower bound b for S that actually *belongs* to S is called a **minimum**, **smallest**, **least**, or **first** element of S (Figure 4-5).

A set $S \subseteq F$ may or may not have a greatest element, and it may or may not have a least element. For instance, the set

$$S = \{s \in F \mid -1 < s \leq 1\}$$

has 1 as a greatest element, but it has no least element. Indeed, -1 is the only possible candidate for the least element of S, but $-1 \notin S$ (Figure 4-6).

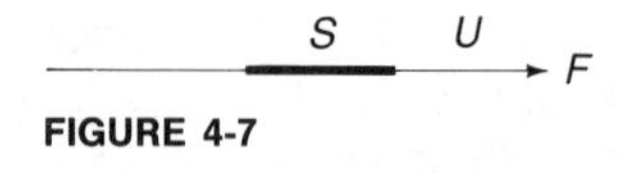

FIGURE 4-7

Now let S be any nonempty subset of F, suppose that S has an upper bound, and let U be the set of all upper bounds for S (Figure 4-7). Consider the question of whether or not U has a least element. Recall from Section 2.2 of Chapter 2 that a least element of U, if it exists, is called a **least upper bound** of S. Suppose that there is a "hole" in the field F, that all of the elements of S are to the left of the hole, and that all of the elements of U are to the right of the hole (Figure 4-8). Then U would have no least element, since the only possible least element of U would occupy the position where the hole is, and there is no element of F in this position.

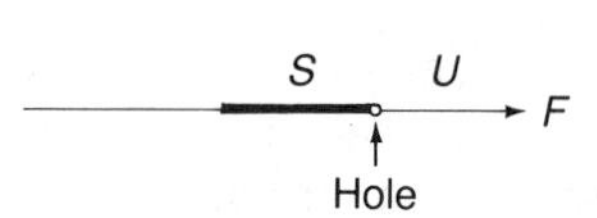

FIGURE 4-8

One way to ensure that an ordered field F has no holes is to insist that every nonempty set S that has an upper bound has a least upper bound. This brings us to the following definition:

1.1 DEFINITION

COMPLETENESS An ordered field F is said to be **complete** if every nonempty subset of F that has an upper bound, has a least upper bound.

Now we are ready to introduce the field $\mathbb{R}$ of real numbers on a more formal basis. We visualize $\mathbb{R}$ as a line with no holes in it, that is, as a complete ordered field. Therefore, we adopt the following:

1.2 AXIOM

THE FIELD OF REAL NUMBERS $\mathbb{R}$ is a complete ordered field.

Here we are postulating the existence of a complete ordered field, denoted by $\mathbb{R}$. We make no other assumptions about $\mathbb{R}$; indeed, all of the properties of the real number system can be derived deductively from the assumption that $\mathbb{R}$ forms a complete ordered field.

1.3 Example Let $x, y \in \mathbb{R}$. Prove:

(i) $x < y \Leftrightarrow -y < -x$
(ii) $x \leq y \Leftrightarrow -y \leq -x$

SOLUTION We prove (i) here and leave the proof of (ii) as an exercise (Problem 3). The idea is to use the fact that *multiplication by* -1 *reverses inequalities.* Specifically, by Part (ii) of Theorem 4.12 (Chapter 3) and the fact that $-1 < 0$,

$$x < y \Rightarrow (-1)y < (-1)x \Rightarrow -y < -x.$$

Conversely,

$$-y < -x \Rightarrow (-1)(-x) < (-1)(-y) \Rightarrow x < y. \qquad \square$$

The idea of multiplying by -1 to reverse inequalities is exploited in the proof of the next lemma.

1.4 LEMMA Let $S \subseteq \mathbb{R}$, and let

$$T = \{-s \mid s \in S\}.$$

Then:

(i) x is an upper bound for S if and only if $-x$ is a lower bound for T.
(ii) $a = \text{LUB}(S)$ if and only if $-a = \text{GLB}(T)$.

PROOF (i) If x is an upper bound for S, then $s \leq x$ holds for all $s \in S$; hence, $-x \leq -s$ holds for all $s \in S$, that is, $-x$ is a lower bound for T. Conversely, if $-x$ is a lower bound for T, then $-x \leq -s$ holds for all $s \in S$; hence, $s \leq x$ holds for all $s \in S$, that is, x is an upper bound for S.

(ii) Suppose that $a = \text{LUB}(S)$, and let U be the set of all upper bounds of S. Then a is a lower bound for U, and $a \in U$. Let

$$L = \{-x \mid x \in U\}.$$

By Part (i) above, L is the set of all lower bounds for T. Also, by Part (i) and the fact that a is a lower bound for U, we have that $-a$ is an upper bound for L. Finally, since $a \in U$, we have $-a \in L$, and we conclude that $-a = \text{GLB}(T)$. A similar argument, which we leave as an exercise (Problem 6) shows that, conversely, if $-a = \text{GLB}(T)$, then $a = \text{LUB}(S)$. $\square$

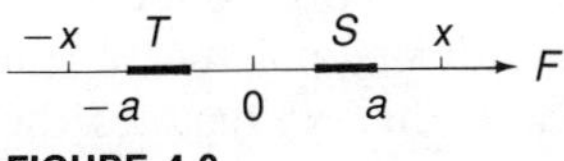

FIGURE 4-9

Lemma 1.4 is illustrated in Figure 4-9. Pictorially, what happens is just that *inequalities are reversed by a reflection through the origin.* We leave as an exercise (Problem 7) the proof of the following important corollary to Lemma 1.4:

1.5 COROLLARY Every nonempty subset of $\mathbb{R}$ that has a lower bound has a greatest lower bound.

The results in the next lemma are almost indispensable for dealing with the LUB or GLB of a set $S \subseteq \mathbb{R}$.

1.6 LEMMA Let $S \subseteq \mathbb{R}$, and let $\varepsilon \in \mathbb{R}$ with $\varepsilon > 0$. Then:

(i) If $a = \text{LUB}(S)$, there exists $s \in S$ with

$$a - \varepsilon < s \leq a.$$

(ii) If $b = \text{GLB}(S)$, there exists $s \in S$ with

$$b \leq s < b + \varepsilon.$$

PROOF We prove (i) here and leave (ii) as an exercise (Problem 9a). Since $\varepsilon > 0$, we have

$$a - \varepsilon < a.$$

Because a is the *least* upper bound of S, and $a - \varepsilon$ is *smaller* than a, it follows that $a - \varepsilon$ cannot be an upper bound for S. Therefore, there must exist at least one number $s \in S$ such that

$$a - \varepsilon < s$$

(for, otherwise, $s \leq a - \varepsilon$ would hold for all $s \in S$, and $a - \varepsilon$ would be an upper bound for S). Also, since a is an upper bound for S, $s \leq a$, and we have

$$a - \varepsilon < s \leq a. \qquad \square$$

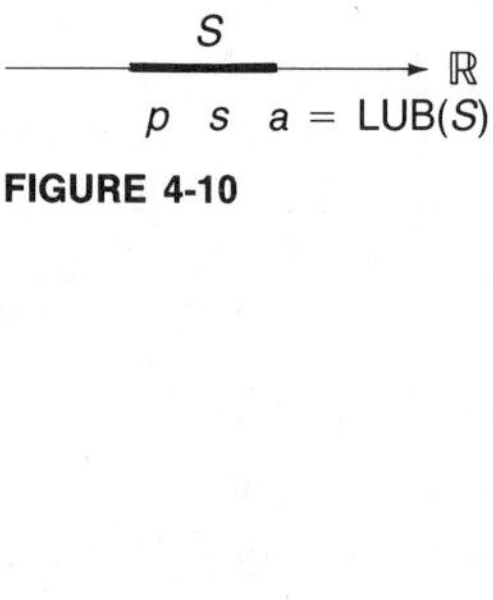

FIGURE 4-10

If you visualize $\mathbb{R}$ as a number line, then Part (i) of Lemma 1.6 says that *if you choose a point p to the left of the least upper bound a of a set S, then there is at least one point $s \in S$ that lies to the right of the chosen point but not to the right of a* (Figure 4-10). We leave it to you to formulate a similar graphical interpretation for Part (ii) of Lemma 1.6 (Problem 9b).

One useful fact about $\mathbb{R}$ is that every positive real number has a unique positive square root. We devote the remainder of this section to a careful proof of this fact. Naturally, in what follows, we understand that 2 is a symbol that represents the real number $1 + 1$ and that x^2 denotes the real number $x \cdot x$.

1.7 LEMMA Let $a \in \mathbb{R}$, and let $\varepsilon \in \mathbb{R}$ with $\varepsilon > 0$. Then there exists $\delta \in \mathbb{R}$ with $\delta > 0$ such that, for all $x \in \mathbb{R}$,

$$|x - a| < \delta \Rightarrow |x^2 - a^2| < \varepsilon.$$

PROOF Let δ be the smaller of the two real numbers 1 and $\varepsilon/(2|a| + 1)$, or their common value if they are equal. Thus,

$$\delta \leq 1 \tag{1}$$

and

$$\delta \leq \frac{\varepsilon}{2|a| + 1}. \tag{2}$$

Now, suppose that

$$|x - a| < \delta. \tag{3}$$

We have to prove that $|x^2 - a^2| < \varepsilon$. Combining Inequalities (2) and (3), we have

$$|x - a| < \frac{\varepsilon}{2|a| + 1}. \tag{4}$$

Combining Inequalities (1) and (3), we have

$$|x - a| < 1. \tag{5}$$

By the triangle inequality (Theorem 4.18 in Chapter 3),

$$|x| = |a + (x - a)| \leq |a| + |x - a|. \tag{6}$$

Adding $|a|$ to both sides of Inequality (5), we find that

$$|a| + |x - a| < |a| + 1. \tag{7}$$

From Inequalities (6) and (7), we infer that

$$|x| < |a| + 1. \tag{8}$$

By the triangle inequality again,

$$|x + a| \leq |x| + |a|. \tag{9}$$

Adding $|a|$ to both sides of Inequality (8), we obtain

$$|x| + |a| < 2|a| + 1. \tag{10}$$

As a consequence of Inequalities (9) and (10), we have

$$|x + a| < 2|a| + 1. \tag{11}$$

Also,

$$|x^2 - a^2| = |(x + a)(x - a)| = |x + a|\,|x - a|. \tag{12}$$

Now, combining Equation (12) and Inequalities (11) and (4), we conclude that

$$|x^2 - a^2| < (2|a| + 1)\frac{\varepsilon}{2|a| + 1} = \varepsilon.$$

□

From your study of calculus, you probably recognize Lemma 1.7 as stating that *the function* $f:\mathbb{R} \to \mathbb{R}$ *given by* $f(x) = x^2$ *is continuous* (see Problem 25). The proof is a typical "ε–δ proof" from calculus.

1.8 LEMMA Let $b \in \mathbb{R}$ with $b > 1$. Then there exists $a \in \mathbb{R}$ with $a > 0$ and $a^2 = b$.

PROOF Let

$$S = \{s \in \mathbb{R} \mid 0 < s \wedge s^2 < b\}.$$

We note that $1 \in S$, so $S \neq \varnothing$. Also, since $1 < b$, we have $b < b^2$ (Problem 5); hence, if $s \in S$, then $s^2 < b < b^2$, from which it follows that $s < b$ by Corollary 4.13 in Chapter 3. Thus, b is an upper bound for S. Because $\mathbb{R}$ is complete and S is a nonempty subset of $\mathbb{R}$ with an upper bound, there exists a real number a such that

$$a = \mathrm{LUB}(S).$$

Since $1 \in S$ and a is an upper bound for S, we have $1 < a$, so that $a > 0$.

We claim that $a^2 = b$. Suppose, on the contrary, that $a^2 \neq b$, and let

$$\varepsilon = |a^2 - b|,$$

noting that $\varepsilon > 0$. By Lemma 1.7, there exists $\delta > 0$ such that, for any $x \in \mathbb{R}$,

$$|x - a| < \delta \Rightarrow |x^2 - a^2| < \varepsilon.$$

By Lemma 1.6, there exists $s \in S$ with

$$a - \delta < s \leq a.$$

Thus, $a < s + \delta$, so that $a - s < \delta$. Also, $a - s \geq 0$; so

$$|s - a| = -(s - a) = a - s < \delta,$$

and it follows that

$$|s^2 - a^2| < \varepsilon.$$

But, because $0 < s \leq a$, we have $s^2 \leq a^2$, and so

$$a^2 - s^2 = |a^2 - s^2| = |s^2 - a^2| < \varepsilon.$$

Now, let

$$x = a + \frac{\delta}{2},$$

noting that $x > a$ and that $|x - a| = \delta/2 < \delta$. Therefore,

$$|x^2 - a^2| < \varepsilon.$$

Also, since $x > a > 0$, we have $x^2 > a^2$. Because $a^2 \neq b$, either $a^2 < b$ or else $b < a^2$. If, on the one hand, $a^2 < b$, then

$$x^2 - a^2 = |x^2 - a^2| < \varepsilon = |a^2 - b| = b - a^2;$$

so $x^2 < b$, and it follows that $x \in S$. However, $x \in S$ implies that $x \leq a$, since a is an upper bound for S, which contradicts the fact that $x > a$. If, on the other hand, $b < a^2$, then

$$a^2 - s^2 < \varepsilon = |a^2 - b| = a^2 - b;$$

so $-s^2 < -b$, and it follows that $b < s^2$. But, $b < s^2$ contradicts $s \in S$. Since contradictions follow from our assumption that $a^2 \neq b$, we must have $a^2 = b$. □

1.9 THEOREM

EXISTENCE OF SQUARE ROOTS Let $b \in \mathbb{R}$ with $b > 0$. Then there exists $a \in \mathbb{R}$ with $a > 0$ and $a^2 = b$.

PROOF Lemma 1.8 takes care of the case in which $b > 1$. If $b = 1$, then we simply take $a = 1$. Therefore, we only have to take care of the remaining case, in which

$$0 < b < 1.$$

Thus, assume $0 < b < 1$. By Problem 4, it follows that $1/b = b^{-1} > 1$. Therefore, we can apply Lemma 1.8 to $1/b$ and conclude that there exists $x \in \mathbb{R}$ with $x > 0$ such that $x^2 = 1/b$. Let $a = 1/x$, noting that, since $x > 0$, we have $a > 0$. Now, we have

$$a^2 = \left(\frac{1}{x}\right)^2 = \frac{1}{x^2} = \frac{1}{1/b} = b$$

as desired. □

1.10 LEMMA

UNIQUENESS OF NONNEGATIVE SQUARE ROOTS Let $b \in \mathbb{R}$ with $b \geq 0$. Then there exists a unique $a \in \mathbb{R}$ with $a \geq 0$ and $a^2 = b$.

PROOF If $b > 0$, then, by Theorem 1.9, there exists $a > 0$ such that $a^2 = b$. If $b = 0$, just take $a = 0$, and again we have $a^2 = b$. Thus, if $b \geq 0$, there exists $a \geq 0$ with $a^2 = b$. To prove that there is a *unique* such a, suppose that $c \in \mathbb{R}$ with $c \geq 0$ and $c^2 = b$. We have to prove that $c = a$. We have $c^2 = b = a^2$, so

$$c^2 - a^2 = 0,$$

that is,

$$(c - a)(c + a) = 0.$$

Because a field is an integral domain (Problem 13 in Problem Set 3.3), it follows that

$$c - a = 0 \quad \text{or} \quad c + a = 0.$$

Therefore,

$$c = a \quad \text{or} \quad c = -a.$$

If $c = a$, our proof is complete. Thus, suppose $c = -a$. Then $-a = c \geq 0$; so $a \leq 0$. But $a \geq 0$, and it follows that $a = 0$. However, if $a = 0$, then, since $c^2 = a^2$, we have $c^2 = 0$; hence $c = 0$, and again $c = a$. □

In previous chapters we have made informal use of square roots in some examples and problems. Now, as a consequence of Lemma 1.10, we can give a formal definition of square roots in the field $\mathbb{R}$ of real numbers.

1.11 DEFINITION

PRINCIPAL SQUARE ROOT Let $b \in \mathbb{R}$ with $b \geq 0$. Then the unique $a \in \mathbb{R}$ with $a \geq 0$ and $a^2 = b$ is called the **principal square root** of b and is written as $a = \sqrt{b}$.

The proof of the existence of square roots (of nonnegative numbers) in $\mathbb{R}$ illustrates the utility of the completeness condition. Typically, completeness is used to establish the existence of a number that satisfies a particular condition by forming a set S whose LUB or GLB is the desired number.

It can be proved that *up to an isomorphism, there exists at most one complete ordered field*, so that Axiom 1.2 uniquely determines the algebraic structure of $\mathbb{R}$. We shall not give the proof here.

PROBLEM SET 4.1

1. Give an example of a subset S of $\mathbb{R}$ such that S has a least element, but no greatest element.
2. The empty set $\varnothing$ is a subset of $\mathbb{R}$.
   (a) What is the set U of upper bounds of $\varnothing$?
   (b) Does $\varnothing$ have a LUB in $\mathbb{R}$? Explain.
3. Complete the solution of Example 1.3 by proving that, for $x, y \in \mathbb{R}$, $x \leq y \Leftrightarrow -y \leq -x$.

4. If $b \in \mathbb{R}$ and $0 < b < 1$, prove that $b^{-1} > 1$.

5. If $b \in \mathbb{R}$ and $1 < b$, prove that $b < b^2$.

6. (a) Complete the proof of Part (ii) of Lemma 1.4 by showing that, if $T = \{-s | s \in S\}$ and if $-a = \text{GLB}(T)$, then $a = \text{LUB}(S)$.
(b) Would the results of Lemma 1.4 hold in an arbitrary ordered field F? Explain.

7. Prove Corollary 1.5.

8. Suppose that $A \subseteq B \subseteq \mathbb{R}$, $a = \text{LUB}(A)$, $b = \text{GLB}(A)$, $c = \text{LUB}(B)$, and $d = \text{GLB}(B)$. Prove that $d \leq b \leq a \leq c$.

9. (a) Prove Part (ii) of Lemma 1.6.
(b) Formulate a graphical interpretation for Part (ii) of Lemma 1.6.

10. If F is an ordered field, then a **ray** (opening to the left) in F is defined to be a subset R of F such that $R \neq \varnothing$, $R \neq F$; and, if $r, x \in F$ with $r \in R$ and $x \leq r$, then $x \in R$.
(a) Sketch a figure showing such a ray.
(b) Suppose that $b \in F$ and define $R_b = \{x \in F | x \leq b\}$. Show that R_b is a ray.
(c) Give an example of a ray in F that does not have the form R_b as in (b) above.
(d) If $b, c \in F$, show that $b \leq c$ if and only if $R_b \subseteq R_c$.

11. Let $\varepsilon \in \mathbb{R}$ with $\varepsilon > 0$. Find δ (depending on ε) such that $\delta \in \mathbb{R}$, $\delta > 0$, and, for all $x \in \mathbb{R}$,

$$|x - 2| < \delta \Rightarrow |x^2 - 4| < \varepsilon.$$

[Hint: See the proof of Lemma 1.7.]

12. Find $\delta \in \mathbb{R}$, such that $\delta > 0$, and, for all $x \in \mathbb{R}$,

$$|x - 2| < \delta \Rightarrow |x^2 - 4| < 0.007.$$

[Hint: Use the result of Problem 11.]

13. What is wrong with the following: $\sqrt{4} = \pm 2$?

14. Let F be an ordered field, and let $\varnothing \neq S \subseteq F$. Define $R = \{r \in F | \exists s \in S, r \leq s\}$. Show that:
(a) R is a ray in F (Problem 10).
(b) R and S have the same set of upper bounds in F.

15. Let $a \in \mathbb{R}$ with $a \geq 0$. Is it correct to say that the solutions, in $\mathbb{R}$, of the equation $x^2 = a$ are given by $x = \pm\sqrt{a}$? Explain.

16. Prove that an ordered field F is complete if and only if every ray in F has a LUB. (See Problems 10 and 14.)

17. In the proof of Lemma 1.10, we reasoned that if $c \in \mathbb{R}$ with $c^2 = 0$, then $c = 0$. Justify this reasoning.

18. Let F be an ordered field. A subset C of F is called a **Dedekind cut** in F if $C \neq \varnothing$, $C \neq F$, and C contains every lower bound in F of its own set of upper bounds in F.
(a) Show that every Dedekind cut in F is a ray in F (Problem 10).
(b) With the notation of Problem 10, show that every ray of the form R_b, $b \in F$, is a Dedekind cut in F.
(c) Give an example of a ray in F that is not a Dedekind cut in F.

19. Let $x, y \in \mathbb{R}$ with $0 \leq x \leq y$. Prove that $(x-y)^2 \leq y^2 - x^2$. [Hint: Expand $(y^2 - x^2) - (x - y)^2$.]

20. Let F be an ordered field. Show that F is complete if and only if every Dedekind cut in F has a LUB. (See Problem 18.)

21. Let $x, y \in \mathbb{R}$ with $x, y \geq 0$. Prove that $|x - y|^2 \leq |x^2 - y^2|$. [Hint: Use the result of Problem 19.]

22. Let F be an ordered field. Prove that F is complete if and only if every Dedekind cut in F has the form R_b for some $b \in F$. (See Problems 10 and 18.)

23. Let $x, y \in \mathbb{R}$ with $x, y \geq 0$. Use the result of Problem 21 to prove that

$$|\sqrt{x} - \sqrt{y}| \leq \sqrt{|x - y|}.$$

24. Let F be an ordered field, and let C be a Dedekind cut in F (Problem 18). Show that $a \in F$ is the LUB of C if and only if every element $x \in F$ with $x < a$ is in C, but no element $y \in F$ with $a < y$ is in C.

25. A function $f: D \to \mathbb{R}$, with $D \subseteq \mathbb{R}$, is said to be **continuous** if, for every $x \in D$ and every $\varepsilon \in \mathbb{R}$ with $\varepsilon > 0$, there exists $\delta \in \mathbb{R}$ with $\delta > 0$ such that, for every $y \in D$,

$$|x - y| < \delta \Rightarrow |f(x) - f(y)| < \varepsilon.$$

If $D = \{x \in \mathbb{R} \mid x \geq 0\}$ and $f: D \to \mathbb{R}$ is defined by $f(x) = \sqrt{x}$ for all $x \in D$, show that f is continuous. [Hint: Use the result of Problem 23.]

26. A function $f: D \to \mathbb{R}$, with $D \subseteq \mathbb{R}$, is said to be **uniformly continuous** if, for every $\varepsilon \in \mathbb{R}$ with $\varepsilon > 0$, there exists $\delta \in \mathbb{R}$ with $\delta > 0$ such that, for every $x, y \in D$,

$$|x - y| < \delta \Rightarrow |f(x) - f(y)| < \varepsilon.$$

(a) Explain the distinction between continuity (Problem 25) and uniform continuity. [Hint: Write out the definitions more formally with the aid of quantifiers.]
(b) Use the result of Problem 23 to show that the square-root function in Problem 25 is actually uniformly continuous.
(c) Prove that the squaring function $g: \mathbb{R} \to \mathbb{R}$, $g(x) = x^2$ for all $x \in \mathbb{R}$, is not uniformly continuous.

2 THE NATURAL NUMBERS AND MATHEMATICAL INDUCTION

In this section we are going to extract the system $\mathbb{N}$ of natural numbers from the complete ordered field $\mathbb{R}$ of real numbers. One of the consequences of this process is a justification of the *principle of mathematical induction.*

Intuitively, $\mathbb{N}$ consists of the numbers 1, 2, 3, ... with which we count finite nonempty sets. Thus, $\mathbb{N}$ should be a subset of $\mathbb{R}$ that contains 1 and has the additional property that

$$n \in \mathbb{N} \Rightarrow n + 1 \in \mathbb{N}.$$

Moreover, we should expect $\mathbb{N}$ to be the "smallest" subset of $\mathbb{R}$ that satisfies these conditions. In the next two definitions, we formalize these ideas.

2.1 DEFINITION

INDUCTIVE SET A subset I of $\mathbb{R}$ is said to be **inductive** if and only if it satisfies the following two conditions:

(i) $1 \in I$
(ii) For all $n \in \mathbb{R}$, $n \in I \Rightarrow n + 1 \in I$

Note that there exists at least one inductive subset of $\mathbb{R}$; indeed, $\mathbb{R}$ is an inductive subset of itself. Of course, there are other inductive subsets of $\mathbb{R}$; for instance, the set P consisting of all positive real numbers is inductive (Problem 1).

2.2 DEFINITION

NATURAL NUMBERS A real number n is said to be a **natural number** if and only if n belongs to every inductive subset of $\mathbb{R}$. The set of all natural numbers is denoted by $\mathbb{N}$.

As our first result, we show that the natural numbers form the smallest inductive subset of $\mathbb{R}$.

2.3 LEMMA

$\mathbb{N}$ is an inductive subset of $\mathbb{R}$, and, if I is an inductive subset of $\mathbb{R}$, then $\mathbb{N} \subseteq I$.

PROOF By Definition 2.1, the real number 1 belongs to every inductive subset I of $\mathbb{R}$; hence, by Definition 2.2, we have $1 \in \mathbb{N}$. Now suppose that $n \in \mathbb{N}$. To finish the proof that $\mathbb{N}$ is inductive, we must show that $n + 1 \in \mathbb{N}$; that is, we must prove that $n + 1$ belongs to every inductive subset of $\mathbb{R}$. Thus, let I be an arbitrary inductive subset of $\mathbb{R}$. Because $n \in \mathbb{N}$, it follows that n belongs to every inductive subset of $\mathbb{R}$; hence, in particular, $n \in I$. Since I is inductive and $n \in I$, we conclude that $n + 1 \in I$. This shows that $n + 1$ belongs to every inductive subset of $\mathbb{R}$ and, hence, that $n + 1 \in \mathbb{N}$. Therefore, $\mathbb{N}$ is an inductive subset of $\mathbb{R}$.

To complete the proof, let I be an inductive subset of $\mathbb{R}$, and let $n \in \mathbb{N}$. Then, n belongs to every inductive subset of $\mathbb{R}$, and, in particular, $n \in I$. Thus, $\mathbb{N} \subseteq I$. □

Because $1 \in \mathbb{N}$ and $\mathbb{N}$ is an inductive subset of $\mathbb{R}$, it follows that $1 + 1 \in \mathbb{N}$. Naturally, we *define* $2 = 1 + 1$, so that $2 \in \mathbb{N}$. Again, since $\mathbb{N}$ is inductive and $2 \in \mathbb{N}$, it follows that $2 + 1 \in \mathbb{N}$. Naturally, we *define* $3 = 2 + 1$, so that $3 \in \mathbb{N}$. We can continue in this way as long as we please, *defining* $4 = 3 + 1, 5 = 4 + 1$, and so on, and showing that $4, 5, \ldots \in \mathbb{N}$. Note that, for all $n \in \mathbb{N}$, we have

$$n < n + 1.$$

(This follows upon adding n to both sides of the inequality $0 < 1$, which is a consequence of Corollary 4.4 in Chapter 3.) Therefore,

$$1 < 2 < 3 < 4 < 5,$$

and so on.

Lemma 2.3 is used to prove the following theorem, which is the basis for proofs by mathematical induction:

2.4 THEOREM

PRINCIPLE OF MATHEMATICAL INDUCTION For each natural number n, let P_n be a proposition. Suppose that the following two conditions hold:

(i) P_1 is true.
(ii) For all $n \in \mathbb{N}$, if P_n is true, then P_{n+1} is true.

Then P_n is true for all $n \in \mathbb{N}$.

PROOF Let I be the subset of $\mathbb{N}$ consisting of all of the natural numbers n for which P_n is true. Conditions (i) and (ii) imply that I is an inductive subset of $\mathbb{R}$. By Lemma 2.3, it follows that $\mathbb{N} \subseteq I$ and, hence, that P_n is true for all $n \in \mathbb{N}$. □

A proof that P_n is true for all $n \in \mathbb{N}$, obtained by using Theorem 2.4, is called a **proof by induction on *n***. Such a proof involves two things: First, you must prove that P_1 is true. Second, you must prove that, for all $n \in \mathbb{N}$,

$$P_n \Rightarrow P_{n+1}.$$

Naturally, in proving this implication, you are entitled to assume that P_n is true for the purpose of showing that P_{n+1} follows from this assumption. In an inductive proof, the step in which you assume that P_n is true (for an arbitrary but fixed $n \in \mathbb{N}$) is called the **inductive hypothesis**. The technique of making proofs by using the principle of mathematical induction is discussed in more detail and illustrated with further examples in Appendix I.

2.5 Example Prove that $n \geq 1$ holds for all $n \in \mathbb{N}$.

SOLUTION For each $n \in \mathbb{N}$, let P_n be the proposition asserting that $n \geq 1$. We prove that P_n is true for all $n \in \mathbb{N}$ by using the principle of mathematical induction (Theorem 2.4). First, we must prove that P_1 is true. Second, we must prove that $P_n \Rightarrow P_{n+1}$. Because $1 \geq 1$, the proposition P_1 is true. To prove that $P_n \Rightarrow P_{n+1}$, assume that P_n is true for an arbitrary but fixed $n \in \mathbb{N}$; that is, assume

$$n \geq 1. \tag{1}$$

(This is the induction hypothesis.) Adding 1 to both sides of the inequality in (1), we find that

$$n + 1 \geq 2. \tag{2}$$

Adding 1 to both sides of the inequality $1 > 0$, we also have

$$2 > 1. \tag{3}$$

Combining Inequalities (2) and (3), we obtain

$$n + 1 \geq 1, \tag{4}$$

that is, P_{n+1} is true. This shows that $P_n \Rightarrow P_{n+1}$ and completes the proof. □

As the proof of the following lemma illustrates, it is not always necessary to invoke the induction hypothesis when you make a proof by mathematical induction. The proof is also a good review of some of the features of the implication connective.

2.6 LEMMA **SUBTRACTION OF 1 FROM A NATURAL NUMBER** For every $n \in \mathbb{N}$,

$$n \neq 1 \Rightarrow n - 1 \in \mathbb{N}.$$

PROOF For each $n \in \mathbb{N}$, let P_n denote the proposition

$$n \neq 1 \Rightarrow n - 1 \in \mathbb{N}.$$

We prove, by the principle of mathematical induction, that P_n is true for all $n \in \mathbb{N}$. The proposition P_1 is true because its hypothesis, $1 \neq 1$, is false.

We must prove that

$$P_n \Rightarrow P_{n+1}$$

holds for all $n \in \mathbb{N}$. Since an implication with a true conclusion is automatically true, it suffices to prove that P_{n+1} is true for all $n \in \mathbb{N}$. Thus, we must prove that

$$n + 1 \neq 1 \Rightarrow (n + 1) - 1 \in \mathbb{N}$$

holds for all $n \in \mathbb{N}$. In other words, we must prove that

$$n + 1 \neq 1 \Rightarrow n \in \mathbb{N}$$

holds for all $n \in \mathbb{N}$. But, the last implication is true, for all $n \in \mathbb{N}$, since its conclusion is true. □

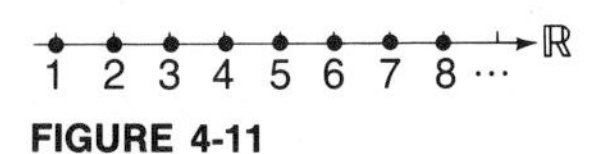

FIGURE 4-11

If we visualize the real number system $\mathbb{R}$ as a number scale, then our intuition tells us that the natural numbers 1, 2, 3, . . . are distributed along this number scale like "beads on a wire," starting at 1 and extending indefinitely to the right (Figure 4-11). The following theorem helps to confirm this.

2.7 THEOREM

DISCRETENESS OF THE NATURAL NUMBERS If $n \in \mathbb{N}$, there exists no $k \in \mathbb{N}$ with

$$n < k < n + 1.$$

PROOF The proof is by induction on n. Let P_n be the proposition asserting that there is no natural number k strictly between n and $n + 1$. We begin by proving that P_1 is true. To this end, let

$$I = \{1\} \cup \{x \in \mathbb{R} | x \geq 2\}.$$

It is not difficult to check that I is an inductive subset of $\mathbb{R}$ (Problem 2); hence, by Lemma 2.3, $\mathbb{N} \subseteq I$. Thus, if $k \in \mathbb{N}$, then either $k = 1$ or else $k \geq 2$; hence, we cannot have $1 < k < 2$, and P_1 is proved.

To prove that $P_n \Rightarrow P_{n+1}$ holds for every $n \in \mathbb{N}$, we proceed by contradiction. Thus, suppose there exists $n \in \mathbb{N}$ such that P_n is true, but P_{n+1} is false; that is, suppose there is *no* natural number strictly between n and $n + 1$, but there is a natural number $k \in \mathbb{N}$ that *is* strictly between $n + 1$ and $(n + 1) + 1$. Thus,

$$n + 1 < k < (n + 1) + 1. \tag{1}$$

Adding -1 to all members of Inequality (1), we obtain

$$n < k - 1 < n + 1. \tag{2}$$

Now, k cannot be 1; for, if it were, then the first inequality in (2) would imply that $n < 0$, contradicting Example 2.5. Since $k \in \mathbb{N}$ and $k \neq 1$, we can use Lemma 2.6 to conclude that

$$k - 1 \in \mathbb{N}. \tag{3}$$

Combining Inequality (2) and (3), we arrive at a contradiction to our supposition that there is no natural number strictly between n and $n + 1$, and our proof by contradiction is complete. □

2.8 THEOREM **DIFFERENCE OF NATURAL NUMBERS** If $m, n \in \mathbb{N}$ and $m > n$, then $m - n \in \mathbb{N}$.

PROOF Fix $m \in \mathbb{N}$. For each $n \in \mathbb{N}$, let P_n be the statement

$$m > n \Rightarrow m - n \in \mathbb{N}.$$

Then P_1 is the statement

$$m > 1 \Rightarrow m - 1 \in \mathbb{N},$$

which is true by Lemma 2.6.

Now we must prove that

$$P_n \Rightarrow P_{n+1}$$

holds for all $n \in \mathbb{N}$. We are going to prove this by contradiction, so suppose there is a $k \in \mathbb{N}$ for which $P_k \Rightarrow P_{k+1}$ is false. The only way that an implication can be false is for its hypothesis to be true and its conclusion false; hence, we have P_k true,

$$m > k \Rightarrow m - k \in \mathbb{N}, \tag{1}$$

and P_{k+1} false. To say that P_{k+1} is false is to say that its hypothesis is true,

$$m > k + 1, \tag{2}$$

and its conclusion is false,

$$m - (k + 1) \notin \mathbb{N}; \tag{3}$$

that is,

$$(m - k) - 1 \notin \mathbb{N}. \tag{4}$$

Now, $k + 1 > k$, and, therefore, by Inequality (2), we have

$$m > k. \tag{5}$$

By (1), Inequality (5), and modus ponens,

$$m - k \in \mathbb{N}. \tag{6}$$

If $m - k \neq 1$, then, by (6) and Lemma 2.6, we would have $(m - k) - 1 \in \mathbb{N}$, contradicting (4). Consequently,

$$m - k = 1. \tag{7}$$

It follows that $m = k + 1$, contradicting Inequality (2). □

After you become familiar with the technique of proving theorems by induction, you need only keep in mind the proposition corresponding to each $n \in \mathbb{N}$ rather than denoting it explicitly by P_n. This is illustrated in the proof of the next theorem.

2.9 THEOREM **SUM OF NATURAL NUMBERS** If $m, n \in \mathbb{N}$, then $m + n \in \mathbb{N}$.

PROOF Fix $m \in \mathbb{N}$. We prove that $m + n \in \mathbb{N}$ by induction on n. For $n = 1$, we obtain the statement $m + 1 \in \mathbb{N}$, which is true because $m \in \mathbb{N}$ and $\mathbb{N}$ is an inductive set. Now, we assume the inductive hypothesis

$$m + n \in \mathbb{N}$$

for some (arbitrary but fixed) $n \in \mathbb{N}$. Again, since $\mathbb{N}$ is inductive and $m + n \in \mathbb{N}$, we have $(m + n) + 1 \in \mathbb{N}$; hence,

$$m + (n + 1) \in \mathbb{N}.$$ □

We leave it as an exercise for you to prove the following theorem (Problem 11).

2.10 THEOREM **PRODUCT OF NATURAL NUMBERS** If $m, n \in \mathbb{N}$, then $mn \in \mathbb{N}$.

Note carefully how the completeness of $\mathbb{R}$ is used in the proof of the following important theorem.

2.11 THEOREM **ARCHIMEDEAN PROPERTY OF $\mathbb{R}$** Let $\varepsilon, a \in \mathbb{R}$ with $\varepsilon, a > 0$. Then there exists $n \in \mathbb{N}$ such that

$$n\varepsilon > a.$$

PROOF The proof is by contradiction. Thus, suppose that no such $n \in \mathbb{N}$ exists, and let

$$S = \{n\varepsilon \mid n \in \mathbb{N}\}.$$

Then, $\varepsilon = 1 \cdot \varepsilon \in S$; so $S \neq \varnothing$, and our supposition that $n\varepsilon > a$ is false for all $n \in \mathbb{N}$ means that a is an upper bound for S. Let $c = \text{LUB}(S)$. By Part (i) of Lemma 1.6, there exists $n \in \mathbb{N}$ such that

$$c - \frac{\varepsilon}{2} < n\varepsilon \leq c.$$

Consequently,

$$c < n\varepsilon + \frac{\varepsilon}{2} < n\varepsilon + \varepsilon = \varepsilon(n + 1).$$

But, since $n + 1 \in \mathbb{N}$, we have $\varepsilon(n + 1) \in S$; hence,

$$\varepsilon(n + 1) \leq c = \text{LUB}(S).$$

Thus,

$$c < \varepsilon(n + 1) \leq c,$$

which is a contradiction, because we cannot have $c < c$. □

In Theorem 2.11, it is useful to think of ε as a "very small" positive real number, and to think of a as being "very large." The **Archimedean property** of the field $\mathbb{R}$ of real numbers, established by Theorem 2.11, can then be paraphrased as follows:

> *By adding a small positive real number ε to itself sufficiently many times, you can obtain an arbitrarily large positive real number.*

There are ordered fields that do not have the Archimedean property (Problem 20); however, *any subfield of a field with the Archimedean property inherits the Archimedean property* (Problem 18). It can be proved (although we shall not do so) that *an ordered field is isomorphic to a subfield of $\mathbb{R}$ if and only if it has the Archimedean property.*

The following corollary of Theorem 2.11 shows that there are arbitrarily large natural numbers.

2.12 COROLLARY If $x \in \mathbb{R}$, then there exists $n \in \mathbb{N}$ such that $x < n$.

PROOF If $x \leq 0$, just take $n = 1$. Thus, we can suppose that $x > 0$. In Theorem 2.11, take $\varepsilon = 1$ and $a = x$. □

2.13 COROLLARY

Let $x \in \mathbb{R}$ with $1 \leq x$. Then there exists a unique natural number n such that $n \leq x < n + 1$.

PROOF We prove the existence of n and leave the proof of its uniqueness as an exercise (Problem 17). Our proof is by contradiction, so suppose that

$$1 \leq x, \tag{1}$$

but there does not exist $n \in \mathbb{N}$ such that $n \leq x < n + 1$. In other words, for all $n \in \mathbb{N}$,

$$n \leq x \Rightarrow n + 1 \leq x. \tag{2}$$

By Inequality (1), the implication in (2), and the principle of mathematical induction, it follows that $n \leq x$ holds for every natural number n, contradicting Corollary 2.12. □

A totally ordered set $(X, \leq)$ is said to be **well ordered** if every nonempty subset of X contains a smallest element. The next theorem tells us that the natural numbers, under the total order inherited from the real numbers, form a well-ordered set.

2.14 THEOREM

$\mathbb{N}$ IS A WELL-ORDERED SET If M is a nonempty subset of $\mathbb{N}$, then M contains a smallest element. In fact, if $m = \text{GLB}(M)$, then m is the smallest element of M.

PROOF By the result of Example 2.5, 1 is a lower bound for M; hence, by Corollary 1.5, M has a GLB. Let

$$m = \text{GLB}(M). \tag{1}$$

Because 1 is a lower bound for M, we have

$$1 \leq m. \tag{2}$$

At this point, we have no guarantee that m belongs to M, or even that it is a natural number. However, by Corollary 2.13, there exists $n \in \mathbb{N}$ such that

$$n \leq m < n + 1. \tag{3}$$

Let $\varepsilon = (n + 1) - m$, noting that $\varepsilon > 0$ by the second inequality in (3). Therefore, since $m = \text{GLB}(M)$, we can apply Part (ii) of Lemma 1.6 to conclude that there exists $k \in M$ such that

$$m \leq k < m + \varepsilon. \tag{4}$$

But, since $\varepsilon = (n + 1) - m$, we have $m + \varepsilon = n + 1$; so we can rewrite Inequality (4) as

$$m \leq k < n + 1. \tag{5}$$

Note that, by Inequalities (3) and (5),

$$n \leq m \leq k. \tag{6}$$

As a consequence of Inequalities (5) and (6), we have

$$n \leq k < n + 1. \tag{7}$$

Since $k \in M$ and $M \subseteq \mathbb{N}$, it follows that $k \in \mathbb{N}$; hence, by Theorem 2.7, we must have

$$n = k, \tag{8}$$

and so, by Inequality (6),

$$m = n = k \in M. \tag{9}$$

Since $m \in M$ and m is a lower bound for M, it follows that m is the smallest element in M. □

Theorem 2.14 has the following very useful corollary:

2.15 COROLLARY

ALTERNATIVE FORM OF MATHEMATICAL INDUCTION For each natural number n, let P_n be a proposition. Suppose that the following two conditions hold:

(i) P_1 is true.
(ii) If m is a natural number, $1 < m$, and P_k is true for all natural numbers $k < m$, then P_m is true.

Then P_n is true for all $n \in \mathbb{N}$.

PROOF The proof is by contradiction. Suppose that (i) and (ii) hold, but that there exists $q \in \mathbb{N}$ such that P_q is false. Let

$$M = \{m \in \mathbb{N} \mid P_m \text{ is false}\}.$$

Then $q \in M$, so that M is a nonempty subset of $\mathbb{N}$. By Theorem 2.14, M contains a smallest element m. Thus, m is the smallest natural number for which P_m is false. Because P_1 is true, $m \neq 1$ and, therefore, $1 < m$. Furthermore, if k is a natural number and $k < m$, then P_k must be true (since m is the smallest natural number for which P_m is false). Therefore, by (ii), P_m is true, contradicting the fact that P_m is false. □

Now we have nearly all of the elements of elementary arithmetic at our disposal in the system $\mathbb{N}$ of natural numbers. We can add and multiply natural numbers (Theorems 2.9 and 2.10), and we can subtract a smaller natural number from a larger (Theorem 2.8). In Section 3, we shall discuss the process of division, and in Section 6, we shall consider how the natural numbers are used to count finite nonempty sets.

There remain two aspects of elementary arithmetic that we do not propose to discuss in detail: *positional notation* and *computational algorithms*. As you know, in accordance with the Hindu–Arabic scheme, the natural numbers are written using the digits 0, 1, 2, 3, 4, 5, 6, 7, 8, and 9. You know, for instance, that 365 means $3 \cdot 10^2 + 6 \cdot 10^1 + 5 \cdot 10^0$. This scheme is called **positional notation to the base 10**. The only real difficulty is to prove that each natural number has a unique representation in such a form. A proof of this important fact can be found in S. Feferman, *Number Systems* (Reading, Mass.: Addison-Wesley), 1964, and will not be repeated here.

In grade-school arithmetic, you learned the basic rules, or **computational algorithms**, for working out sums, differences, products, and quotients of numbers expressed in positional notation to the base 10. Although we shall not discuss these algorithms here, you should now be able to see how and why they work.

PROBLEM SET 4.2

1. Prove that the set $P = \{x \in \mathbb{R} | x > 0\}$ is inductive.
2. Prove that the set $I = \{1\} \cup \{x \in \mathbb{R} | x \geq 2\}$ is inductive.
3. Prove that the set $J = \{x \in \mathbb{R} | x \geq 1\}$ is inductive.
4. If $\mathscr{I}$ denotes the set of all inductive subsets of $\mathbb{R}$, prove that $\mathbb{N} = \bigcap_{I \in \mathscr{I}} I$.
5. If $I \subseteq \mathbb{N}$ and I is an inductive set, prove that $I = \mathbb{N}$.
6. State and prove the converse of Lemma 2.6.
7. True or false: If I is an inductive set and $I \subseteq J \subseteq \mathbb{R}$, then J is an inductive set. If true, prove it; if false, give a counterexample.
8. Prove: If $n, m \in \mathbb{N}$, then $m > n$ if and only if there exists $k \in \mathbb{N}$ such that $m = n + k$.
9. Prove: If $n, m \in \mathbb{N}$, then $m > n \Rightarrow m \geq n + 1$. [Hint: Use Theorem 2.7.]
10. Prove the following generalized version of the principle of mathematical induction: *Let k be a fixed natural number. For each natural number $n \geq k$, let P_n be a proposition. Suppose P_k is true and suppose that, for every natural number $n \geq k$, $P_n \Rightarrow P_{n+1}$. Then P_n is true for every natural number $n \geq k$.*
11. Prove Theorem 2.10. [Hint: Fix m and prove that $mn \in \mathbb{N}$ by induction on n.]

12. Let F be a field (not necessarily ordered), and define a subset I of F to be **inductive** if $1 \in I$ and

$$x \in I \Rightarrow x + 1 \in I$$

for all $x \in F$. Define an analogue of $\mathbb{N}$ in F, call it $\mathbb{N}_F$, by the requirement that $x \in \mathbb{N}_F$ if and only if x belongs to every inductive subset of F. State and prove an analogue of Lemma 2.3 for $\mathbb{N}_F$.

13. Prove: Given any $\varepsilon \in \mathbb{R}$ with $\varepsilon > 0$, there exists $m \in \mathbb{N}$ such that, for all $n \in \mathbb{N}$, $n > m \Rightarrow 1/n < \varepsilon$.

14. As a consequence of Example 2.5, $0 \notin \mathbb{N}$. With the notation of Problem 12, show that there are fields F for which $0 \in \mathbb{N}_F$. [Hint: Try $F = \text{GF}(2)$.]

15. Prove: If $n, k \in \mathbb{N}$ and $n < k < n + 2$, then $k = n + 1$.

16. Prove that, if F is a subfield of $\mathbb{R}$, then $\mathbb{N} \subseteq F$. (See Problem 35 in Problem Set 3.3.)

17. Complete the proof of Corollary 2.13 by showing that, if $1 \leq x$, and if $n, m \in \mathbb{N}$ with $n \leq x < n + 1$ and $m \leq x < m + 1$, then $n = m$. [Hint: Begin by showing that $n < m + 1$ and $m < n + 1$, deduce that $n < m + 1 < n + 2$, and use the result of Problem 15.]

18. Prove that, if F is a subfield of $\mathbb{R}$, then F inherits the Archimedean property from $\mathbb{R}$. (See Problem 4 in Problem Set 3.4.)

19. A **Peano model**, named after the Italian mathematician Giuseppe Peano (1858–1932), is an ordered triple (N, a, s) consisting of a set N, a fixed element $a \in N$, and a mapping $s: N \to N$ satisfying the following three postulates:

(i) $s: N \to N$ is an injection.
(ii) $(\forall x)(x \in N \Rightarrow a \neq s(x))$.
(iii) If $a \in M \subseteq N$ and if $(\forall x)(x \in M \Rightarrow s(x) \in M)$, then $M = N$.

Define $s: \mathbb{N} \to \mathbb{N}$ by $s(n) = n + 1$ for all $n \in \mathbb{N}$ and show that $(\mathbb{N}, 1, s)$ is a Peano model.

20. Show that the ordered field $\mathbb{R}(X)$ of rational functions with real coefficients (page 212) does not have the Archimedean property. [Hint: Let $\varepsilon = 1$, and let a be the rational function $a = p(X)/q(X)$, where $p(X) = X$ and $q(X) = 1$. Show that, for all $n \in \mathbb{N}$, $n\varepsilon < a$.]

21. Let P denote the set of all positive real numbers. With the usual ordering $\leq$ inherited from $\mathbb{R}$, is $(P, \leq)$ a well-ordered set? Explain.

22. Discuss the usual algorithm for adding natural numbers expressed as numerals using the positional notation to the base 10. In particular, explain the procedure of *carrying*.

23. If M is a nonempty subset of $\mathbb{N}$ and if M has an upper bound in $\mathbb{R}$, prove that M contains a largest element.

24. Suppose that $m, n \in \mathbb{N}$ with $n \leq m$. Let $K = \{k \in \mathbb{N} \mid kn \leq m\}$.

(a) Show that K is a nonempty subset of $\mathbb{N}$.

(b) Show that K has an upper bound in $\mathbb{R}$. [Hint: Use Theorem 2.11.]
(c) Let q be the largest element of K (Problem 23). Show that $qn \le m < (q + 1)n$.

25. A natural number n is said to be **even** if it can be written in the form $n = 2k$ for some natural number k. If $n \in \mathbb{N}$ is not even, it is said to be **odd**.
(a) Show that, if n is even, then $n + 1$ is odd.
(b) Show that, if n is odd, then $n + 1$ is even.
(c) Show that every natural number is either even or odd.
(d) Show that the product of two odd natural numbers is odd.
(e) If n is a natural number and n^2 is even, show that n must be even.

26. Let n be a fixed natural number. Prove that every natural number m with $m < 2^n$ can be written in the form

$$m = c_0 + 2c_1 + 2^2c_2 + 2^3c_3 + \cdots + 2^{n-1}c_{n-1},$$

where $c_j \in \{0, 1\}$ for each $j = 0, 1, 2, \ldots, n - 1$.

27. Show that *there do not exist natural numbers n and m with $n^2 = 2m^2$* by carrying out the following: Suppose that such natural numbers exist.
(a) Show then that there exists a smallest natural number n such that n^2 has the form $n^2 = 2m^2$ for some $m \in \mathbb{N}$.
(b) Show that there exists $k \in \mathbb{N}$ such that $n = 2k$. [Hint: Use Part (e) of Problem 25.]
(c) Show that $m^2 = 2k^2$, and conclude that there exists $r \in \mathbb{N}$ such that $m = 2r$.
(d) Show that $k^2 = 2r^2$ and that $k < n$, thus contradicting Part (a).

3
THE INTEGERS AND DIVISIBILITY

In the previous section, we showed that the system $\mathbb{N}$ of natural numbers is **closed** under addition and multiplication; that is, the sum and product of natural numbers are again natural numbers. However, $\mathbb{N}$ is not closed under subtraction because, if $m, n \in \mathbb{N}$ with $m \le n$, then $m - n \notin \mathbb{N}$. If we enlarge $\mathbb{N}$ by appending 0 and the negative integers, we obtain the system $\mathbb{Z}$ of *integers*, which is not only closed under subtraction but is closed under addition and multiplication as well.

3.1 DEFINITION

INTEGERS We define $\mathbb{Z} = \mathbb{N} \cup \{0\} \cup \{-n \mid n \in \mathbb{N}\}$. A real number is called an **integer** if it belongs to $\mathbb{Z}$.

Note that, if $k \in \mathbb{Z}$, then $k \in \mathbb{N}$ if and only if $k > 0$; in other words, $\mathbb{N}$ *is precisely the set of positive integers* (Problem 1). Also note that

$$k \in \mathbb{Z} \Leftrightarrow -k \in \mathbb{Z}$$

(Problem 3). We leave it as an exercise for you to prove the following lemma and its corollary (Problems 2 and 4).

3.2 LEMMA **CLOSURE OF $\mathbb{Z}$ UNDER SUMS, DIFFERENCES, AND PRODUCTS** Let m and n be integers. Then $m + n$, $m - n$, and mn are integers.

3.3 COROLLARY Under the operations of addition and multiplication inherited from $\mathbb{R}$, the system $(\mathbb{Z}, +, \cdot)$ forms an integral domain (Definition 2.15 in Chapter 3).

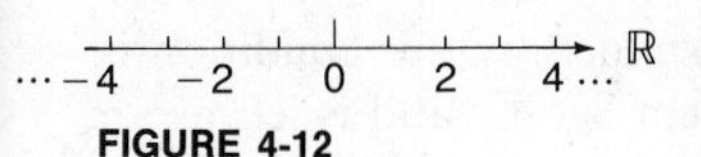

FIGURE 4-12

Like the natural numbers, the integers are distributed along the number scale like "beads on a wire," but they extend indefinitely in both directions (Figure 4-12). In particular, we have the following analogue of Theorem 2.7:

3.4 THEOREM **DISCRETENESS OF THE INTEGERS** If $n \in \mathbb{Z}$, there exists no $k \in \mathbb{Z}$ with $n < k < n + 1$.

PROOF Suppose, on the contrary, that $n, k \in \mathbb{Z}$ with

$$n < k < n + 1.$$

Adding $1 - n$ to all members of this inequality, we obtain

$$n + (1 - n) < k + (1 - n) < n + 1 + (1 - n);$$

that is,

$$1 < k + 1 - n < 2.$$

Let $q = k + 1 - n$, so that

$$1 < q < 2.$$

By Lemma 3.2, $q \in \mathbb{Z}$ and, since $1 < q$, it follows that $q \in \mathbb{N}$. But then, $1 < q < 2$ contradicts Theorem 2.7. □

We also have the following analogue for $\mathbb{Z}$ of Corollary 2.13:

3.5 THEOREM

GREATEST INTEGER LESS THAN OR EQUAL TO x Let $x \in \mathbb{R}$. Then there exists a unique $n \in \mathbb{Z}$ such that

$$n \leq x < n + 1.$$

PROOF By Corollary 2.12, there exists $q \in \mathbb{N}$ such that

$$1 - x < q.$$

Therefore,

$$1 < q + x,$$

and it follows from Corollary 2.13 that there exists $m \in \mathbb{N}$ such that

$$m \leq q + x < m + 1.$$

Let $n = m - q$ and add $-q$ to all members of the last inequality to obtain

$$n \leq x < n + 1.$$

By Lemma 3.2, $n \in \mathbb{Z}$. That n is uniquely determined by this condition is proved by arguing as in Problem 17 in Problem Set 4.2 and is left as an exercise (Problem 6). □

3.6 DEFINITION

GREATEST INTEGER FUNCTION Let $x \in \mathbb{R}$. Then the unique $n \in \mathbb{Z}$ such that

$$n \leq x < n + 1$$

is called the **greatest integer less than or equal to x** and is denoted by $[\![x]\!] = n$. The function from $\mathbb{R}$ into $\mathbb{Z}$ defined by $x \mapsto [\![x]\!]$ for all $x \in \mathbb{R}$ is called the **greatest integer function**.

In some computer languages, such as BASIC, the greatest integer function is written as INT, so that

$$\text{INT}(X) = [\![X]\!].$$

The effect of INT(X) is to round X *down* to the nearest integer. (In other computer languages, this is called the **floor** function.) A related function is the **closest integer function** CINT defined by

$$\text{CINT}(X) = \text{INT}(X + 0.5) = \left[\!\!\left[X + \frac{1}{2} \right]\!\!\right].$$

The effect of CINT(X) is to round off X to the nearest integer. If n is a natural number, then the function

$$x \mapsto [\![x \cdot 10^n + 0.5]\!] \cdot 10^{-n}$$

has the effect of *rounding x off to n decimal places* (Problem 7). In BASIC, the corresponding formula

$$\text{CINT}(X*10\wedge N)/10\wedge N$$

is used to round X off to N decimal places.

Although $\mathbb{Z}$ is closed under addition, subtraction, and multiplication, *it is not closed under division.* For instance, $1, 2 \in \mathbb{Z}$, but $1/2 \notin \mathbb{Z}$.

3.7 Example Show that $1/2 \notin \mathbb{Z}$.

SOLUTION Suppose, on the contrary, that

$$\frac{1}{2} = k \in \mathbb{Z}.$$

Then,

$$2k = 1.$$

Obviously, we cannot have $k = 0$. If $k < 0$, then, since $2 > 0$, we would have $1 < 0$, contradicting the fact that $0 < 1$. Therefore, $k > 0$; so $k \in \mathbb{N}$. By the result of Example 2.5,

$$k \geq 1,$$

and consequently,

$$1 = 2k \geq 2,$$

contradicting the fact that $1 < 2$. □

Let $m, n \in \mathbb{Z}$ with $n \neq 0$, and consider the fraction $m/n \in \mathbb{R}$. Although m/n need not be an integer, we can use long division to divide m by n and obtain an **integer quotient** q and **a remainder** r with $0 \leq r < |n|$:

$$n\overset{\displaystyle q}{\overline{\big)\,m}} \quad \text{and a remainder } r.$$

That this is always possible, and that the quotient q and remainder r are uniquely determined by m and n, are the content of the following important theorem and corollary:

3.8 THEOREM **DIVISION ALGORITHM** Let $m, n \in \mathbb{Z}$ with $n > 0$. Then there exist unique integers q and r such that

$$m = nq + r \quad \text{and} \quad 0 \leq r < n.$$

PROOF We begin by proving the existence of q and r. Let $x = m/n$, let $q = [\![x]\!]$, and let $r = m - nq$. Then q and r are integers and

$$m = nq + r. \tag{1}$$

Also, since $q = [\![x]\!]$, we have

$$q \le x < q + 1. \tag{2}$$

Multiplying each member of Inequality (2) by the positive integer n, and using the fact that $x = m/n$, we find that

$$nq \le m < nq + n. \tag{3}$$

Adding $-nq$ to each member of Inequality (3), and using the fact that $r = m - nq$, we obtain

$$0 \le r < n. \tag{4}$$

To prove the uniqueness of q and r, suppose that we also have integers q' and r' with

$$m = nq' + r' \tag{5}$$

and

$$0 \le r' < n. \tag{6}$$

We must prove that $q = q'$ and $r = r'$. From Equations (1) and (5), we have

$$(nq + r) - (nq' + r') = 0, \tag{7}$$

so that

$$n(q - q') = r' - r. \tag{8}$$

We have $0 \le q - q'$ or $q - q' \le 0$. Suppose, for instance, that $0 \le q - q'$. (The contrary case is handled by a symmetric argument; see Problem 8.) If $q - q' = 0$, then $r' - r = 0$ by Equation (8), and we have $q = q'$ and $r = r'$, as desired. Thus, we can suppose that $0 < q - q'$. By Lemma 3.2, $q - q'$ is an integer. By Theorem 3.4, we cannot have $0 < q - q' < 1$, and consequently,

$$1 \le q - q'. \tag{9}$$

From Inequality (9) and the fact that $n > 0$, we find that

$$n \le n(q - q'). \tag{10}$$

Combining Equation (8) and Inequality (10), we have

$$n \le r' - r. \tag{11}$$

From Inequality (11) and the fact that $0 \le r$, we obtain

$$n \le n + r \le r', \tag{12}$$

contradicting Inequality (6) and completing the proof. □

3.9 COROLLARY

GENERALIZED DIVISION ALGORITHM Let $m, n \in \mathbb{Z}$ with $n \neq 0$. Then there exist unique integers q and r such that

$$m = nq + r \quad \text{and} \quad 0 \le r < |n|.$$

PROOF We prove the existence of q and r and leave the proof of their uniqueness as an exercise (Problem 10). Applying Theorem 3.8 to m and $|n|$, we find that there are unique integers k and r such that

$$m = |n|k + r \quad \text{and} \quad 0 \le r < |n|.$$

Let $q = \operatorname{sgn}(n)k$. (See Definition 4.14 in Chapter 3.) Then q is an integer, and

$$m = |n|k + r = n \cdot \operatorname{sgn}(n)k + r = nq + r.$$

□

In Corollary 3.9, q is called the **integer quotient**, and r is called the **remainder** upon division of m by n, $n \neq 0$. Do not confuse the *integer quotient* q with the *quotient* m/n obtained by dividing m by n in the field $\mathbb{R}$. The relationship between these two types of quotients is expressed by the simple formula

$$\frac{m}{n} = q + \frac{r}{n}$$

obtained by dividing both sides of $m = nq + r$ by n. If $r = 0$, so that $m/n = q$, we say that the integer m is *exactly divisible* by the integer n or, for short, that n *divides* m. More generally, we make the following definition:

3.10 DEFINITION

DIVISIBILITY OF ONE INTEGER BY ANOTHER If $m, n \in \mathbb{Z}$, we say that n **divides** m and write $n|m$ if there exists $q \in \mathbb{Z}$ such that $m = nq$.

If $n \neq 0$, then $n|m$ if and only if the remainder upon division of m by n is 0. Note, however, that there is no stipulation in Definition 3.10 that $n \neq 0$. There is no necessity for such a stipulation because the equation $m = nq$ makes sense even if $n = 0$ (in which case, $m = nq$ holds only if $m = 0$). We collect some of the elementary properties of the relation of divisibility in the following theorem, the proof of which we leave as an exercise (Problems 9 and 11).

3.11 THEOREM

PROPERTIES OF DIVISIBILITY Let $m, n, k \in \mathbb{Z}$. Then:

(i) $1 \mid m$ (ii) $n \mid 0$ (iii) $n \mid n$ (iv) $-n \mid n$
(v) $n \mid m \Rightarrow -n \mid m$ (vi) $n \mid m \wedge m \mid n \Leftrightarrow n = \pm m$
(vii) $n \mid m \Rightarrow n \mid (mk)$ (viii) $n \mid m \wedge n \mid k \Rightarrow n \mid (m + k)$
(ix) $n \mid m \wedge m \mid k \Rightarrow n \mid k$

As a consequence of Parts (iii) and (ix) of Theorem 3.11, the relation $\mid$ is a preorder on the set $\mathbb{Z}$ of all integers. (See Definition 2.13 in Chapter 2.)

If $m, n \in \mathbb{Z}$ and $n \mid m$, then we say that m is an **integer multiple** of n. If $n \in \mathbb{Z}$, then the set of all integer multiples of n is given by

$$\{kn \mid k \in \mathbb{Z}\}.$$

This brings us to the following definition:

3.12 DEFINITION

IDEAL GENERATED BY AN INTEGER Let $n \in \mathbb{Z}$. We define

$$\mathbb{Z}n = \{kn \mid k \in \mathbb{Z}\},$$

and we call $\mathbb{Z}n$ the **ideal in $\mathbb{Z}$ generated by** n.

Note that, if $n, m \in \mathbb{Z}$, then

$$n \mid m \Leftrightarrow m \in \mathbb{Z}n.$$

Furthermore,

$$n \mid m \Leftrightarrow \mathbb{Z}m \subseteq \mathbb{Z}n \qquad \text{(Problem 13a).}$$

The subset $\mathbb{Z}n$ of $\mathbb{Z}$ also has the following two properties:

(i) $h, k \in \mathbb{Z}n \Rightarrow h + k \in \mathbb{Z}n$ (Problem 13b),
(ii) $h \in \mathbb{Z}n \wedge k \in \mathbb{Z} \Rightarrow hk \in \mathbb{Z}n$ (Problem 13c).

Properties (i) and (ii) suggest the following definition:

3.13 DEFINITION

IDEAL IN $\mathbb{Z}$ A nonempty subset I of $\mathbb{Z}$ is said to be an **ideal** in $\mathbb{Z}$ if it has the following two properties:

(i) $h, k \in I \Rightarrow h + k \in I$.
(ii) $h \in I,\ k \in \mathbb{Z} \Rightarrow hk \in I$.

An ideal of the form $\mathbb{Z}n$ is called a **principal ideal**. In the next theorem, we show that *every ideal in $\mathbb{Z}$ is principal.*

3.14 THEOREM

PRINCIPAL IDEAL THEOREM If I is an ideal in $\mathbb{Z}$, then there exists a unique $n \in \mathbb{Z}$ such that $n \geq 0$ and $I = \mathbb{Z}n$.

PROOF By Definition 3.13, we have

$$h, k \in I \Rightarrow h + k \in I \tag{1}$$

and

$$h \in I, k \in \mathbb{Z} \Rightarrow hk \in I. \tag{2}$$

Taking $k = -1$ in (2), we find that

$$h \in I \Rightarrow -h \in I. \tag{3}$$

Since I is nonempty, there is at least one integer $h \in I$, and, taking $k = 0$ in (2), we conclude that

$$0 \in I. \tag{4}$$

If 0 is the only integer in I, then $I = \mathbb{Z}n$ with $n = 0$.

Thus, we can assume that there exists $h \in I$ with $h \neq 0$. If $h > 0$, then $h \in \mathbb{N}$; so $h \in \mathbb{N} \cap I$. If $h < 0$, then $-h > 0$, $-h \in I$ by (3), and so $-h \in \mathbb{N} \cap I$. In either case, we have found an element in $\mathbb{N} \cap I$, and so

$$\mathbb{N} \cap I \neq \varnothing. \tag{5}$$

Because $\mathbb{N}$ is well ordered (Theorem 2.14) and $\mathbb{N} \cap I$ is a nonempty subset of $\mathbb{N}$, it follows that $\mathbb{N} \cap I$ contains a smallest natural number n. In particular,

$$n \in I \tag{6}$$

and $n \geq 1$. By (2) and (6),

$$k \in \mathbb{Z} \Rightarrow kn \in I. \tag{7}$$

Therefore,

$$\mathbb{Z}n \subseteq I. \tag{8}$$

Conversely, suppose that $m \in I$. By the division algorithm (Theorem 3.8), there exist integers q and r with

$$m = qn + r \tag{9}$$

and

$$0 \leq r < n. \tag{10}$$

By (8), $qn \in I$; hence, by (3), $-qn \in I$. Therefore, by (1) and (9),

$$r = m + (-qn) \in I. \tag{11}$$

If $0 < r$, then $r \in \mathbb{N} \cap I$, and Inequality (10) contradicts the fact that n is the smallest element in $\mathbb{N} \cap I$. Therefore, we must have $r = 0$, so that, by

Equation (9), $m = qn \in \mathbb{Z}n$. This proves that $I \subseteq \mathbb{Z}n$, which, together with (8), shows that

$$I = \mathbb{Z}n. \tag{12}$$

To show that n is unique, suppose that we also have $n' \in \mathbb{Z}$ with $n' \geq 0$ and $I = \mathbb{Z}n'$. We must prove that $n = n'$. Because $\mathbb{Z}n = \mathbb{Z}n'$, it follows that $n|n'$ and $n'|n$. By Part (vi) of Theorem 3.11, $n = \pm n'$; hence, since $n, n' \geq 0$, we have $n = n'$. □

3.15 DEFINITION

SUM OF SUBSETS Let M and N be subsets of $\mathbb{Z}$. We define

$$M + N = \{m + n | m \in M \wedge n \in N\}.$$

3.16 LEMMA

SUM OF IDEALS Let I and J be ideals in $\mathbb{Z}$. Then $I + J$ is an ideal in $\mathbb{Z}$.

The proof of Lemma 3.16 is just a matter of verifying that $I + J$ satisfies Conditions (i) and (ii) in Definition 3.13 and is left as an exercise (Problem 15). As a consequence of Lemma 3.16, if $m, n \in \mathbb{Z}$, then

$$\mathbb{Z}m + \mathbb{Z}n$$

is an ideal in $\mathbb{Z}$; hence, by Theorem 3.14, there exists a unique integer $g \geq 0$ such that

$$\mathbb{Z}m + \mathbb{Z}n = \mathbb{Z}g.$$

3.17 DEFINITION

GREATEST COMMON DIVISOR Let $m, n \in \mathbb{Z}$. The unique $g \in \mathbb{Z}$ such that $g \geq 0$ and

$$\mathbb{Z}m + \mathbb{Z}n = \mathbb{Z}g$$

is called the **greatest common divisor** of m and n, denoted by

$$\text{GCD}(m, n) = g.$$

3.18 THEOREM

PROPERTIES OF THE GCD Let $m, n, k \in \mathbb{Z}$, and let $g = \text{GCD}(m, n)$. Then:

(i) $g|m$ and $g|n$.
(ii) There exist $a, b \in \mathbb{Z}$ such that $g = am + bn$.
(iii) If $k|m$ and $k|n$, then $k|g$.

PROOF (i) By Definition 3.17, $\mathbb{Z}m + \mathbb{Z}n = \mathbb{Z}g$. Since $0 \in \mathbb{Z}n$, it follows that $\mathbb{Z}m \subseteq \mathbb{Z}m + \mathbb{Z}n = \mathbb{Z}g$; hence, $g \mid m$ (see Problem 13a). Likewise, $g \mid n$.

(ii) This follows from the fact that

$$g = 1 \cdot g \in \mathbb{Z}g = \mathbb{Z}m + \mathbb{Z}n,$$

so that $g \in \mathbb{Z}m + \mathbb{Z}n$.

(iii) Suppose that $k \mid m$ and $k \mid n$. Then, there exist $c, d \in \mathbb{Z}$ such that $m = ck$ and $n = dk$. By (ii), there exist $a, b \in \mathbb{Z}$ such that $g = am + bn$. Therefore,

$$g = ack + bdk = (ac + bd)k.$$

By Lemma 3.2, $ac + bd \in \mathbb{Z}$, and it follows that $k \mid g$. □

The **Euclidean algorithm** is a step-by-step procedure, easily implemented on a computer or programmable calculator, for computing not only $g = \text{GCD}(x, y)$ for $x, y \in \mathbb{N}$ but also the integers a, b such that $g = ax + by$. We leave it as an exercise for you to prove that the algorithm works (Problem 18).

3.19 THEOREM

EUCLIDEAN ALGORITHM Let $x, y \in \mathbb{N}$, and consider ordered triples

$$(a_n, b_n, r_n) \text{ for } n = 0, 1, 2, 3, \ldots, N$$

with

$$(a_0, b_0, r_0) = (0, 1, y), \quad (a_1, b_1, r_1) = (1, 0, x),$$

such that, for $n < N$, the triple $(a_{n+1}, b_{n+1}, r_{n+1})$ is determined from the triples $(a_{n-1}, b_{n-1}, r_{n-1})$ and (a_n, b_n, r_n) by carrying out the following procedure: Let q_{n+1} and r_{n+1} be the integer quotient and remainder obtained upon division of r_{n-1} by r_n, so that

$$r_{n-1} = r_n q_{n+1} + r_{n+1}, \quad \text{with } 0 \le r_{n+1} < r_n.$$

If $r_{n+1} = 0$, then $N = n$, and the procedure terminates. Otherwise, define

$$a_{n+1} = a_{n-1} - a_n q_{n+1} \quad \text{and} \quad b_{n+1} = b_{n-1} - b_n q_{n+1}.$$

Then,

$$r_N = \text{GCD}(x, y)$$

and

$$r_N = a_N x + b_N y.$$

3.20 Example Find $g = \text{GCD}(231, 297)$ and find integers a and b such that $g = a(231) + b(297)$.

SOLUTION We use the Euclidean algorithm (Theorem 3.19) with $x = 231$, $y = 297$. We have

$$(a_0, b_0, r_0) = (0, 1, 297), \quad (a_1, b_1, r_1) = (1, 0, 231).$$

We compute (a_2, b_2, r_2) as follows: Dividing $r_0 = 297$ by $r_1 = 231$, we obtain an integer quotient $q_2 = 1$ and a remainder $r_2 = 66$. Hence,

$$a_2 = a_0 - a_1 q_2 = 0 - 1 \cdot 1 = -1$$

and

$$b_2 = b_0 - b_1 q_2 = 1 - 0 \cdot 1 = 1.$$

Thus,

$$(a_2, b_2, r_2) = (-1, 1, 66).$$

We compute (a_3, b_3, r_3) as follows: Dividing $r_1 = 231$ by $r_2 = 66$, we obtain an integer quotient $q_3 = 3$ and a remainder $r_3 = 33$. Hence,

$$a_3 = a_1 - a_2 q_3 = 1 - (-1)(3) = 4$$

and

$$b_3 = b_1 - b_2 q_3 = 0 - 1 \cdot 3 = -3.$$

Thus,

$$(a_3, b_3, r_3) = (4, -3, 33).$$

Finally, dividing $r_2 = 66$ by $r_3 = 33$, we obtain an integer quotient 2 and a remainder 0; so the process terminates with $N = 3$. Thus,

$$g = \text{GCD}(231, 297) = r_3 = 33,$$

and with $a = a_3 = 4$, $b = b_3 = -3$, we have

$$33 = (4)(231) + (-3)(297) = ax + by.$$ □

3.21 DEFINITION

RELATIVELY PRIME INTEGERS Let $m, n \in \mathbb{Z}$. We say that m and n are **relatively prime** if and only if

$$\text{GCD}(m, n) = 1.$$

In general, just because an integer divides the product of two other integers, it need not divide either of the factors (Problem 17). However, as we prove in the following lemma, *if an integer divides a product of two other integers and is relatively prime to one of the factors, then it must divide the other factor.*

3.22 LEMMA

EUCLID'S LEMMA Let $m, h, k \in \mathbb{Z}$. Suppose that m and k are relatively prime and that $m | hk$. Then $m | h$.

PROOF Because $\mathrm{GCD}(m, k) = 1$, it follows from Part (ii) of Theorem 3.18 that there exist $a, b \in \mathbb{Z}$ such that

$$1 = am + bk. \tag{1}$$

Also, because $m | hk$, there exists $c \in \mathbb{Z}$ such that

$$hk = cm. \tag{2}$$

Multiplying both sides of Equation (1) by h, we obtain

$$h = amh + bhk. \tag{3}$$

Substitution of Equation (2) in Equation (3) yields

$$h = amh + bcm. \tag{4}$$

As a consequence of Equation (4), we have

$$h = (ah + bc)m. \tag{5}$$

By Lemma 3.2, $ah + bc \in \mathbb{Z}$, and it follows from Equation (5) that

$$m | h. \tag{6}$$

□

3.23 COROLLARY

Let $n \in \mathbb{N}$, and let $m, h, k \in \mathbb{Z}$. Suppose that m and k are relatively prime and that $m | hk^n$. Then $m | h$.

PROOF The proof is by induction on n. For $n = 1$, the result follows from Lemma 3.22. Assume that the result holds for an arbitrary but fixed value n, and suppose that $m | hk^{n+1}$. Then, $m | (hk)k^n$; hence, by the induction hypothesis, $m | hk$. Consequently, $m | h$ by Lemma 3.22. □

3.24 DEFINITION

PROPER DIVISORS AND PRIMES If $m, n \in \mathbb{Z}$ and $n | m$, we say that n is a **proper divisor** of m if $n \neq \pm 1$ and $n \neq \pm m$. A natural number p is said to be a **prime** if $p \neq 1$ and p has no proper divisors in $\mathbb{Z}$.

Note that, if p is a prime, k is any nonzero integer, and k is not a multiple of p, then p is relatively prime to k (Problem 16). Therefore, Lemma 3.22 has another important corollary.

3.25 COROLLARY

Let p be a prime, and let $h, k \in \mathbb{Z}$. Then,

$$p|hk \Rightarrow p|h \vee p|k.$$

We leave the proof of Corollary 3.25 as an exercise (Problem 19).

Because division is not always possible in the system $\mathbb{Z}$ of integers, at least not without dealing with integer quotients and remainders, it is considerably more difficult to solve equations in the system $\mathbb{Z}$ than in the field $\mathbb{R}$ of all real numbers. Equations whose solutions are required to have integer values are called **diophantine equations**, in honor of the Greek mathematician Diophantus of Alexandria, who published solutions of such equations circa A.D. 250. In modern times, diophantine equations are studied in a branch of mathematics called the **theory of numbers**, or **higher arithmetic**. The primes

$$2, 3, 5, 7, 11, 13, 17, 19, 23, 29, 31, 37, \ldots$$

play an important role in the theory of numbers.

PROBLEM SET 4.3

1. Prove that $\mathbb{N} = \{n \in \mathbb{Z} | n > 0\}$.
2. Prove Lemma 3.2.
3. Prove that, for $k \in \mathbb{R}$, $k \in \mathbb{Z} \Leftrightarrow -k \in \mathbb{Z}$.
4. Prove Corollary 3.3.
5. Suppose that $m, n \in \mathbb{Z}$ with $mn = 1$. Prove that either $m = n = 1$ or else $m = n = -1$.
6. By arguing as in Problem 17 in Problem Set 4.2, show that the integer n in Theorem 3.5 is uniquely determined by x.
7. Let $x, x^* \in \mathbb{R}$, and let $n \in \mathbb{N}$. We say that x^* approximates x **to** $\boldsymbol{n}$ **decimal places** if $|x - x^*| \leq (10^{-n})/2$. Let $x^* = [\![x \cdot 10^n + 0.5]\!] \cdot 10^{-n}$. Show that x^* approximates x to n decimal places.
8. Complete the proof of Theorem 3.8 by considering the case in which $q - q' \leq 0$.
9. Prove Parts (i), (ii), (iii), and (iv) of Theorem 3.11.

10. Complete the proof of Corollary 3.9 by showing that q and r are uniquely determined by m and n.

11. Prove Parts (v), (vi), (vii), (viii), and (ix) of Theorem 3.11.

12. Let n be a fixed positive integer. Define a $\rho:\mathbb{Z} \to \{0, 1, 2, 3, \ldots, n-1\}$ by $\rho(m) = r$, where r is the remainder upon division of m by n (Theorem 3.8). Let $h, k \in \mathbb{Z}$.
(a) Prove that $\rho(h + k) = \rho(\rho(h) + \rho(k))$.
(b) Prove that $\rho(hk) = \rho(\rho(h) \cdot \rho(k))$.

13. Let $m, n, h, k \in \mathbb{Z}$. Prove:
(a) $n|m \Leftrightarrow \mathbb{Z}m \subseteq \mathbb{Z}n$ (b) $h, k \in \mathbb{Z}n \Rightarrow h + k \in \mathbb{Z}n$
(c) $h \in \mathbb{Z}n \Rightarrow hk \in \mathbb{Z}n$

14. If I and J are ideals in $\mathbb{Z}$, prove that $I \cap J$ is an ideal in $\mathbb{Z}$.

15. Prove Lemma 3.16.

16. Suppose that p is a prime, k is a nonzero integer, and k is not a multiple of p. Prove that p is relatively prime to k.

17. Give an example of three nonzero integers m, h, and k such that m divides hk, but m divides neither h nor k.

18. In the Euclidean algorithm (Theorem 3.19), prove:
(a) For $0 < n < N$, $\text{GCD}(r_{n-1}, r_n) = \text{GCD}(r_n, r_{n+1})$.
(b) $\text{GCD}(x, y) = r_N$.
(c) For $0 \leq n \leq N$, $r_n = a_n x + b_n y$.
(d) $\text{GCD}(x, y) = a_N x + b_N y$.

19. Prove Corollary 3.25.

20. Use the Euclidean algorithm to find $g = \text{GCD}(8619, 6643)$ and integers a and b such that $g = a(8619) + b(6643)$.

21. Let $m, n \in \mathbb{Z}$ with $m, n \neq 0$. Prove that m and n are relatively prime if and only if there exist $a, b \in \mathbb{Z}$ with $am + bn = 1$.

22. Suppose that $a, b, c \in \mathbb{Z}$ with $\text{GCD}(a, b) = \text{GCD}(a, c) = 1$. Prove that $\text{GCD}(a, bc) = 1$.

23. Let $m, n \in \mathbb{Z}$ with $m, n \neq 0$, and let $g = \text{GCD}(m, n)$. Prove that the integers m/g and n/g are relatively prime. [Hint: Use the result of Problem 21.]

24. If $\text{GCD}(a, b) = 1$ and $n \in \mathbb{N}$, prove that $\text{GCD}(a, b^n) = 1$.

25. Let $m, n \in \mathbb{Z}$. If $m, n \neq 0$, we define the **least common multiple** of m and n, in symbols $\text{LCM}(m, n)$, to be the smallest natural number that is an integer multiple of both m and n. If either $m = 0$ or $n = 0$, we define $\text{LCM}(m, n) = 0$. Let $c = \text{LCM}(m, n)$, and suppose that k is an integer multiple of both m and n. Prove that $c|k$.

26. Let $m, n \in \mathbb{Z}$, and let $I = \mathbb{Z}m \cap \mathbb{Z}n$. By Problem 14, I is an ideal in $\mathbb{Z}$. By Theorem 3.14, there exists a unique $c \in \mathbb{Z}$ such that $c \geq 0$ and $I = \mathbb{Z}c$. Prove that $c = \text{LCM}(m, n)$. (See Problem 25.)

27. If $m, n \in \mathbb{N}$, prove that $mn = \text{GCD}(m, n) \cdot \text{LCM}(m, n)$.

28. Suppose that I is a nonempty subset of $\mathbb{Z}$. Show that I is an ideal in $\mathbb{Z}$ if and only if I is closed under subtraction.

29. Suppose that $a, b, c \in \mathbb{Z}$ with $\text{GCD}(a, b) = 1$. Show that $a|c \wedge b|c \Rightarrow (ab)|c$.

30. Find all solutions of the Diophantine equation $2x + 3y = 0$. (Here, x and y are required to be integers.)

31. Let $m, n \in \mathbb{Z}$ with $g = \text{GCD}(m, n)$. Show that g really is the greatest common divisor of m and n in the sense that it is the largest integer in the set C of all integers that divide both m and n.

32. (a) Using the result of Problem 27, specify a procedure for computing $c = \text{LCM}(x, y)$ for $x, y \in \mathbb{N}$.
(b) Find LCM(231, 297).

33. Let $(\mathbb{Z}, \leq)$ be the totally ordered set obtained by restricting to $\mathbb{Z}$ the order relation $\leq$ on $\mathbb{R}$. Is $(\mathbb{Z}, \leq)$ a well-ordered set? Explain.

34. Prove the **fundamental theorem of arithmetic**: If n is an integer greater than 1, then n can be factored as a product of finitely many distinct primes raised to positive integer powers; and, apart from the order in which these factors are written, this factorization is unique.

4 RATIONAL AND IRRATIONAL NUMBERS

We have seen that the system $\mathbb{Z}$ of integers is closed under addition, multiplication, negation, and subtraction, but it is not closed under division. In fact, for $m, n \in \mathbb{Z}$ with $n \neq 0$, $m/n \in \mathbb{Z}$ if and only if $n|m$. To obtain a number system that is closed under division (by nonzero numbers), we simply pass to the *rational numbers.*

4.1 DEFINITION

RATIONAL NUMBERS A real number x is called a **rational number** if it can be expressed in the form $x = m/n$, where $m, n \in \mathbb{Z}$ and $n \neq 0$. The set of all rational numbers is denoted by $\mathbb{Q}$.

Numbers of the form m/n with $m, n \in \mathbb{Z}$ and $n \neq 0$ are called *rational* because they are *ratios* of integers, not because they have anything to do with being "reasonable." The symbol $\mathbb{Q}$ is used for the set of all rational numbers to help remind us that these numbers are *quotients* (of integers). We leave it as an exercise for you to verify the following theorem (Problem 2).

4.2 THEOREM

$\mathbb{Q}$ IS AN ORDERED FIELD $\mathbb{Q}$ is closed under addition, negation, subtraction, multiplication, and division by nonzero numbers. Furthermore, $(\mathbb{Q}, +, \cdot)$ is a field, and under the order relation $\leq$ inherited from $\mathbb{R}$, it is an ordered field.

Note that $\mathbb{Z} \subseteq \mathbb{Q}$ because every integer n can be expressed in the form $n = n/1$. Thus, we have

$$\mathbb{N} \subseteq \mathbb{Z} \subseteq \mathbb{Q} \subseteq \mathbb{R}.$$

4.3 DEFINITION

IRRATIONAL NUMBER A real number that is not a rational number is called an **irrational number**.

Thus, $\mathbb{R} \setminus \mathbb{Q}$ is the set of all irrational numbers. There is no special symbol (apart from $\mathbb{R} \setminus \mathbb{Q}$) for the set of irrational numbers, and $\mathbb{R} \setminus \mathbb{Q}$ is not regarded as a "number system" because it is not closed under any of the arithmetic operations of addition, subtraction, multiplication, or division (Problem 1).

4.4 Example Show that $\sqrt{2}$ is an irrational number.

SOLUTION Suppose, on the contrary, that $\sqrt{2}$ is rational. Then, there exist integers m, n with $n \neq 0$ such that

$$\sqrt{2} = \frac{m}{n}. \tag{1}$$

By multiplying numerator and denominator of the fraction m/n by -1, if necessary, we can assume without loss of generality that both m and n are natural numbers. Multiplying both sides of Equation (1) by n, we obtain

$$n\sqrt{2} = m. \tag{2}$$

After squaring both sides of the last equation, we find that

$$m^2 = 2n^2, \tag{3}$$

which contradicts Problem 27 in Problem Set 4.2. □

4.5 LEMMA

Let $r \in \mathbb{Q}$ with $r \neq 0$. Then there exist unique integers m and n with $n > 0$, $\mathrm{GCD}(m, n) = 1$, and $r = m/n$.

PROOF Since $r \in \mathbb{Q}$, there are integers a and b such that $r = a/b$, and, because $r \neq 0$, we have $a, b \neq 0$. By multiplying both a and b by -1, if necessary, we can suppose without loss of generality that $b > 0$. Let $k = \text{GCD}(a, b)$, let $m = a/k$, and let $n = b/k$. Then $r = m/n$, $n > 0$, and, by Problem 23 in Problem Set 4.3, $\text{GCD}(m, n) = 1$. This proves the existence part of the lemma.

To prove the uniqueness part of the lemma, suppose that we also have $r = m'/n'$ with $n' > 0$ and $\text{GCD}(n', m') = 1$. Then $m/n = m'/n'$; hence,

$$mn' = m'n.$$

From the last equation, we have $n \mid mn'$. Since $\text{GCD}(n, m) = 1$, it follows from Lemma 3.22 that $n \mid n'$. A symmetric argument shows that $n' \mid n$; hence, by Part (vi) of Theorem 3.11, $n' = \pm n$. But, both n and n' are positive, and so we must have $n' = n$. Thus, the equation $mn' = m'n$ can be rewritten as $mn = m'n$, from which it follows that $m = m'$. □

4.6 DEFINITION

REDUCED FORM Let $r \in \mathbb{Q}$ with $r \neq 0$. We say that r is expressed in **reduced form** if it is written as $r = m/n$ with $n > 0$ and $\text{GCD}(m, n) = 1$. We say that 0 is expressed in reduced form if it is written as $0 = 0/1$.

4.7 THEOREM

RATIONAL-ROOTS THEOREM Let n be a positive integer, and suppose that $a_n, a_{n-1}, a_{n-2}, \ldots, a_1, a_0 \in \mathbb{Z}$ with $a_n \neq 0$. Let $x = h/k$ be a rational number in reduced form, and suppose that x is a solution of the equation

$$a_n x^n + a_{n-1} x^{n-1} + a_{n-2} x^{n-2} + \cdots + a_1 x + a_0 = 0$$

Then, $k \mid a_n$ and $h \mid a_0$.

PROOF Suppose that $x = h/k$ is a solution of the equation

$$a_n x^n + a_{n-1} x^{n-1} + a_{n-2} x^{n-2} + \cdots + a_1 x + a_0 = 0. \tag{1}$$

For any integer j with $0 \leq j \leq n$, we have

$$x^j k^n = \left(\frac{h}{k}\right)^j k^n = h^j k^{n-j}. \tag{2}$$

Multiplying both sides of Equation (1) by k^n and using Equation (2), we find that

$$a_n h^n + a_{n-1} h^{n-1} k + a_{n-2} h^{n-2} k^2 + \cdots + a_1 h k^{n-1} + a_0 k^n = 0. \tag{3}$$

Solving Equation (3) for a_nh^n, we obtain

$$a_nh^n = k(-a_{n-1}h^{n-1} - a_{n-2}h^{n-2}k - \cdots - a_1hk^{n-2} + a_0k^{n-1}), \tag{4}$$

which shows that

$$k \mid a_nh^n. \tag{5}$$

Using (5), the fact that k and h are relatively prime, and Corollary 3.23, we conclude that

$$k \mid a_n. \tag{6}$$

Likewise, solving Equation (3) for a_0k^n, we obtain

$$a_0k^n = h(-a_nh^{n-1} - a_{n-1}h^{n-2}k - \cdots - a_1k^{n-1}), \tag{7}$$

which shows that

$$h \mid a_0k^n. \tag{8}$$

Using (8), the fact that h and k are relatively prime, and Corollary 3.23, we conclude that

$$h \mid a_0. \tag{9}$$

□

We leave it to you to prove the following corollary of Theorem 4.7 (Problem 7).

4.8 COROLLARY Let n be a positive integer, and suppose that the rational number x is a solution of the polynomial equation

$$a_nx^n + a_{n-1}x^{n-1} + a_{n-2}x^{n-2} + \cdots + a_1x + a_0 = 0,$$

where the coefficients a_j are integers, $0 \le j \le n$ and

$$a_n = 1.$$

Then, x is an integer.

4.9 Example A nonnegative integer q is called a **perfect square** if there is an integer j such that $q = j^2$. If k is a nonnegative integer and k is *not* a perfect square, show that $\sqrt{k}$ is an irrational number.

SOLUTION Let $x = \sqrt{k}$, and suppose that x is a rational number. Then, $x^2 = k$; so x satisfies the polynomial equation

$$x^2 - k = 0.$$

By Corollary 4.8, x is an integer; hence, $k = x^2$ is a perfect square, in violation of our hypothesis. □

Although $\mathbb{Q}$ forms an Archimedean ordered field (Problem 12), it is *not a complete ordered field*. Indeed, if $\mathbb{Q}$ were a complete ordered field, then, by the argument in the proof of Lemma 1.8, $\sqrt{2}$ would be an element of $\mathbb{Q}$, contradicting Example 4.4.

As we have noted, the integers are spread out along the number line like beads on a wire. How are the rational numbers distributed among the real numbers? Here, the story is more complicated. To begin with, the rational numbers are **dense** in the real numbers in the sense that, *between any two real numbers, there is a rational number*. We devote the next theorem to a proof of this fact.

4.10 THEOREM

DENSITY OF $\mathbb{Q}$ IN $\mathbb{R}$ Let $a, b \in \mathbb{R}$ with $a < b$. Then, there exists a rational number $r \in \mathbb{Q}$ such that $a < r < b$.

PROOF Since $a < b$, it follows that $0 < 1/(b - a)$. By Corollary 2.12, there exists a natural number n such that

$$\frac{1}{b-a} < n, \tag{1}$$

that is,

$$0 < \frac{1}{n} < b - a. \tag{2}$$

As a consequence of Inequality (2), we have

$$a + \frac{1}{n} < b. \tag{3}$$

Now let

$$m = [\![na + 1]\!] \tag{4}$$

so that, by Definition 3.6, $m \in \mathbb{Z}$ and

$$m \le na + 1 < m + 1. \tag{5}$$

The first inequality in (5) implies that

$$\frac{m}{n} \le a + \frac{1}{n}. \tag{6}$$

Combining Inequalities (6) and (3), we obtain

$$\frac{m}{n} \le a + \frac{1}{n} < b. \tag{7}$$

Also, from the second inequality in (5), $na < m$, that is,

$$a < \frac{m}{n}. \tag{8}$$

From Inequalities (8) and (7),

$$a < \frac{m}{n} < b. \tag{9}$$

Let $r = m/n$. Since m and n are integers and $n \neq 0$, r is a rational number and Inequality (9) shows that $a < r < b$. □

The fact that there is a rational number between any two real numbers might lead you to believe that most real numbers are rational and that the irrational numbers are comparatively rare. It turns out that this is not true; in fact, the situation is quite the contrary. As a first indication of this, we shall prove that *the irrational numbers are also dense in* $\mathbb{R}$.

4.11 THEOREM

DENSITY OF $\mathbb{R} \setminus \mathbb{Q}$ IN $\mathbb{R}$ Let $a, b \in \mathbb{R}$ with $a < b$. Then, there exists an irrational number $x \in \mathbb{R} \setminus \mathbb{Q}$ such that $a < x < b$.

PROOF Let u be any fixed positive irrational number, for instance, $u = \sqrt{2}$. Dividing both sides of the inequality $a < b$ by the positive number u, we have

$$\frac{a}{u} < \frac{b}{u}. \tag{1}$$

By Theorem 4.10, there exists a rational number $r \in \mathbb{Q}$ such that

$$\frac{a}{u} < r < \frac{b}{u}. \tag{2}$$

Without loss of generality, we can assume that

$$r \neq 0. \tag{3}$$

(If $r = 0$, choose another rational number between 0 and b/u.) Multiplying all members of Inequality (2) by u, we obtain

$$a < ru < b. \tag{4}$$

Let $x = ru$. If x were rational, then $u = x/r$ would be rational by Theorem 4.2, contradicting our original choice of u as an irrational number. Hence, x is irrational and

$$a < x < b.$$

□

PROBLEM SET 4.4

1. Show that the set $\mathbb{R}\setminus\mathbb{Q}$ of irrational numbers is not closed under addition, subtraction, multiplication, or division. [Hint: If u is irrational, so are $-u$ and $1/u$.]

2. Prove Theorem 4.2.

3. Rewrite the rational number 1078/2409 in reduced form. [Hint: Begin by using the Euclidean algorithm to find GCD(1078, 2409) as in Example 3.20.]

4. A **proper fraction** is a rational number of the form m/n where $0 \le m < n$. Show that every nonnegative rational number r can be written uniquely in the form $r = k + m/n$ where k is a nonnegative integer and m/n is a proper fraction in reduced form.

5. Using Theorem 4.7, you can find all of the rational solutions of a polynomial equation of the form

$$a_n x^n + a_{n-1}x^{n-1} + a_{n-2}x^{n-2} + \cdots + a_1 x + a_0 = 0$$

with integer coefficients and with $a_n, a_0 \ne 0$ by finding all positive integer divisors h of a_0, k of a_n, and trying the rational numbers $x = \pm h/k$, one at a time, to see if they satisfy the equation. Use this procedure to find all rational solutions of the equation

$$9x^3 + 6x^2 - 5x - 2 = 0.$$

6. Suppose that m, h, and k are integers and that h and k are relatively prime. Show that there exist integers a and b such that $m/(hk) = a/h + b/k$.

7. Prove Corollary 4.8.

8. Generalize the result of Problem 6 to the case of a rational number of the form $m/(h_1 h_2 \cdots h_n)$, where $h_1, h_2, \ldots, h_n$ are integers that are relatively prime in pairs.

9. Give an alternative solution to Example 4.4 by using the result of Example 4.9.

10. Find integers a, b, c such that $1/30 = a/2 + b/3 + c/5$.

11. Prove that the cube root of 2 is an irrational number.

12. Let $\varepsilon = s/t$, $a = h/k$ be positive rational numbers expressed in reduced form. Show that there exists $n \in \mathbb{N}$ such that $n\varepsilon > a$, and find a formula involving s, t, h, k and the greatest integer function for the *smallest* such n.

13. Let $r = m/n$ be a positive rational number written in reduced form. Suppose that $\sqrt{r}$ is a rational number. Prove that both m and n are perfect squares.

14. Show that $\sqrt{2} + \sqrt{3}$ is an irrational number.

15. Let x be a positive real number with a *terminating* decimal expansion; that is, for some $n \in \mathbb{N}$, all of the digits after the nth decimal place are zeros. Show that x is a rational number.

16. A rational approximation x^* to a positive real number x can be obtained by dropping all digits after the nth digit in the decimal expansion of x. (This procedure is called *chopping*.) If x^* is obtained by chopping x in this way, show that $x^* = [\![10^n x]\!] \cdot 10^{-n}$.

17. Let x be a positive real number with a repeating decimal expansion; that is, for some $n \in \mathbb{N}$, all of the digits from the nth decimal place and beyond repeat in blocks of k digits. Let $y = 10^{n-1}x - [\![10^{n-1}x]\!]$. Show that $0 \le y < 1$ and that, in the decimal expansion of y, the digits to the right of the decimal point repeat in blocks of k digits.

18. Let x be a positive real number, and let $x^* = [\![10^n x]\!] \cdot 10^{-n}$ as in Problem 16. Prove that $0 \le x - x^* < 10^{-n}$.

19. Suppose that $0 \le y < 1$ and that, in the decimal expansion of y, the digits to the right of the decimal point repeat in blocks of k digits. Let $m = [\![10^k y]\!]$, and let $n = 10^k - 1$. Show that $y = m/n$.

20. Let $0 < x < 1$. Show that the nth digit in the decimal expansion of x is given by $[\![10^n x]\!] - 10[\![10^{n-1}x]\!]$.

21. By combining the results of Problems 17 and 19, show that a positive real number x with a repeating decimal expansion is a rational number.

22. Show that the decimal expansion of a positive rational number x either terminates as in Problem 15 or repeats as in Problem 17. [Hint: Let $x = m/n$ in reduced form, and consider the successive remainders produced by long division of m by n. If one of these remainders is zero, the decimal expansion of x terminates. Otherwise, each of the remainders is a positive integer less than n; hence, after at most n steps, one of these remainders must be repeated.]

23. Express $3.142\overline{446}$ as a rational number in reduced form. (The overbar indicates that the block of digits 446 repeats forever.) [Hint: See Problems 17 and 19.]

24. Show that the field $\mathbb{Q}$ of rational numbers has no proper subfields; that is, show that, if F is a subfield of $\mathbb{Q}$, then $F = \mathbb{Q}$.

5 THE COMPLEX NUMBERS

Although the real numbers form a complete ordered field, there are algebraic processes that cannot be carried out within this field. For instance, there are polynomial equations with real coefficients that cannot be solved within the system $\mathbb{R}$. The simplest such equation is

$$x^2 + 1 = 0,$$

which, because x^2 cannot be negative, has no solution in $\mathbb{R}$. We are going to show that it is possible to enlarge the field $\mathbb{R}$ to a field $\mathbb{C}$, called the field of *complex numbers*, in which the equation $x^2 + 1 = 0$ has a solution.

To obtain some clues as to how to construct such a field $\mathbb{C}$, we do a little experiment. *Assume* that there is a field F such that

$$\mathbb{R} \subseteq F,$$

and the operations of addition and multiplication in $\mathbb{R}$ are restrictions to $\mathbb{R}$ of the corresponding operations in F. We refer to such a field F as an **extension field** of $\mathbb{R}$. Furthermore, we *assume* that there is an element $i \in F$ such that

$$i^2 + 1 = 0,$$

that is,

$$i^2 = -1.$$

5.1 DEFINITION Let

$$G = \{x + yi \mid x, y \in \mathbb{R}\}.$$

Note that $\mathbb{R} \subseteq G$ because every element $x \in \mathbb{R}$ can be written in the form $x = x + (0)i$. Of course, every element of the form $x + yi$ with $x, y \in \mathbb{R}$ belongs to F, and, therefore, we have

$$\mathbb{R} \subseteq G \subseteq F.$$

5.2 LEMMA

(i) G is closed under addition.
(ii) $z \in G \Rightarrow -z \in G$.
(iii) G is closed under multiplication.
(iv) If $z \in G$ and $z \neq 0$, then $z^{-1} \in G$.

PROOF Let $x, y, u, v \in \mathbb{R}$, so that

$$z = x + yi \in G \quad \text{and} \quad w = u + vi \in G.$$

(i) $$z + w = x + yi + u + vi = (x + u) + (y + v)i \in G,$$

because $x + u,\ y + v \in \mathbb{R}$.

(ii) $-z = -(x + yi) = (-x) + (-y)i \in G,$

because $-x, -y \in \mathbb{R}$.

(iii)
$$\begin{aligned} zw &= (x + yi)(u + vi) = xu + xvi + yiu + yivi \\ &= xu + yvi^2 + xvi + yui \\ &= (xu - yv) + (xv + yu)i \in G, \end{aligned}$$

because $xu - yv, xv + yu \in \mathbb{R}$.

(iv) Suppose that $z = x + yi \neq 0$. Then, we cannot have both $x = 0$ and $y = 0$, and it follows that $x^2 + y^2 \neq 0$. Let

$$d = x^2 + y^2, u = \frac{x}{d}, v = \frac{-y}{d},$$

and

$$w = u + vi.$$

Then, by Part (iii) above,

$$\begin{aligned} zw &= (x + yi)(u + vi) = (xu - yv) + (xv + yu)i \\ &= \left(\frac{x^2}{d} + \frac{y^2}{d}\right) + \left(\frac{-xy}{d} + \frac{xy}{d}\right)i \\ &= \frac{x^2 + y^2}{d} \\ &= 1; \end{aligned}$$

hence, $z^{-1} = w \in G$. □

As a consequence of Lemma 5.2, G is a field in its own right (Problem 2). Also, $i = 0 + 1 \cdot i \in G$, and so G is an extension field of $\mathbb{R}$ containing a solution i of the equation $x^2 + 1 = 0$.

5.3 LEMMA Let $x, y, u, v \in \mathbb{R}$, and suppose that $x + yi = u + vi$. Then, $x = u$ and $y = v$.

PROOF The hypothesis that $x + yi = u + vi$ implies that

$$x - u = (v - y)i. \tag{1}$$

Therefore, if $y = v$, then $x - u = 0$, $x = u$, and we are done. Suppose, on the contrary, that $y \neq v$. Then, $v - y \neq 0$, and it follows from Equation (1) that

$$i = \frac{x - u}{v - y}. \tag{2}$$

However, because $x, y, u, v \in \mathbb{R}$, the fraction on the right of Equation (2) is a real number; so $i \in \mathbb{R}$, contradicting the fact that $i^2 = -1$. □

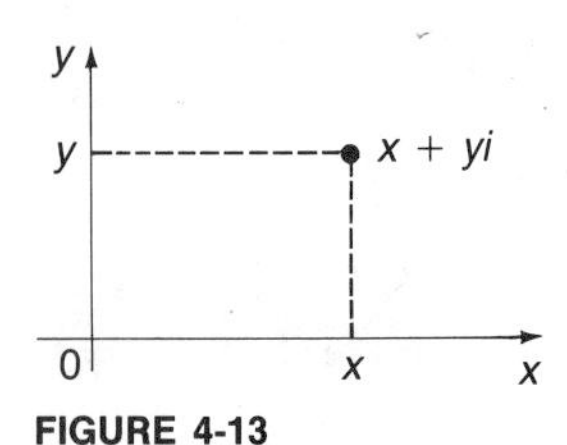

FIGURE 4-13

As a consequence of Lemma 5.3, the mapping

$$x + yi \mapsto (x, y)$$

from G to $\mathbb{R} \times \mathbb{R}$ is a well-defined injection. Evidently, it is a surjection and, hence, a bijection. This bijection enables us to regard the element $x + yi$ of the field G as being represented by a corresponding point (x, y) in the Cartesian plane $\mathbb{R} \times \mathbb{R}$ (Figure 4-13). Furthermore, it allows us to interpret the algebraic operations in G as corresponding operations on the points of the plane.

Until now, our work has been predicated on the assumption that there exists an extension field of $\mathbb{R}$ in which the equation $x^2 + 1 = 0$ has a solution. We now drop this assumption and proceed to study the idea to which it has given rise.

5.4 DEFINITION

GAUSSIAN PLANE Define binary operations of addition and multiplication on the Cartesian product $\mathbb{R} \times \mathbb{R}$ by

$$(x, y) + (u, v) = (x + u, y + v)$$

and

$$(x, y) \cdot (u, v) = (xu - yv, xv + yu)$$

for $(x, y), (u, v) \in \mathbb{R} \times \mathbb{R}$. The resulting algebraic system

$$(\mathbb{R} \times \mathbb{R}, +, \cdot)$$

is called the **Gaussian plane**.

The Gaussian plane is named in honor of the great German mathematician Carl Friedrich Gauss (1777–1855). The operations of addition and multiplication in Definition 5.4 are suggested by the proofs of Parts (i) and (iii) of Lemma 5.2. The proof of the following theorem is only a matter of verifying the postulates for a field in Theorem 3.2 of Chapter 3 and is left as an exercise (Problem 5).

5.5 THEOREM

THE GAUSSIAN PLANE IS A FIELD The algebraic system $(\mathbb{R} \times \mathbb{R}, +, \cdot)$ is a field with $(0, 0)$ as the additive neutral element and $(1, 0)$ as the multiplicative unity element.

We are thinking of the points (x, y) in the Gaussian plane as corresponding to elements $x + yi$ of an extension field of $\mathbb{R}$. Thus, points of the form $(x, 0)$, that is, points on the "x axis," ought to correspond to real numbers $x \in \mathbb{R}$:

$$(x, 0) \leftrightarrow x.$$

It is reassuring to note that this correspondence preserves the algebraic operations of addition, subtraction, multiplication, and division in the sense that, for $x, x' \in \mathbb{R}$,

$$(x, 0) + (x', 0) = (x + x', 0),$$
$$(x, 0) - (x', 0) = (x - x', 0),$$
$$(x, 0) \cdot (x', 0) = (xx', 0),$$

and, for $x' \neq 0$,

$$(x, 0)/(x', 0) = (x/x', 0)$$

(Problem 4). In view of all this, we propose to *identify* points of the form $(x, 0)$ in the Gaussian plane with the corresponding real numbers $x \in \mathbb{R}$.

Of course, $(x, 0)$ and x are mathematically distinct objects, and we cannot really set them equal. The formal procedure for making the proposed identification is carried out as follows:

5.6 DEFINITION

THE SET $\mathbb{C}$ OF COMPLEX NUMBERS Let

$$X = \{(x, 0) \mid x \in \mathbb{R}\},$$

and let

$$\mathbb{C} = ((\mathbb{R} \times \mathbb{R}) \setminus X) \cup \mathbb{R}.$$

Elements of $\mathbb{C}$ are called **complex numbers**.

Thus, $\mathbb{C}$ is the set obtained by removing the "x axis" from $\mathbb{R} \times \mathbb{R}$ and replacing it by $\mathbb{R}$.

5.7 DEFINITION

THE IDENTIFICATION MAPPING Define a mapping

$$\phi: \mathbb{R} \times \mathbb{R} \to \mathbb{C},$$

called the **identification mapping**, by

$$\phi((x, y)) = \begin{cases} (x, y), & \text{if } y \neq 0 \\ x, & \text{if } y = 0 \end{cases} \quad \text{for } (x, y) \in \mathbb{R} \times \mathbb{R}.$$

The identification mapping $\phi:\mathbb{R} \times \mathbb{R} \to \mathbb{C}$ is a bijection (Problem 6), and so it has an inverse

$$\phi^{-1}:\mathbb{C} \to \mathbb{R} \times \mathbb{R}$$

[Part (ii) of Theorem 4.18 in Chapter 2].

5.8 DEFINITION

ADDITION AND MULTIPLICATION OF COMPLEX NUMBERS Define binary operations of addition and multiplication on $\mathbb{C}$ by

$$z + w = \phi(\phi^{-1}(z) + \phi^{-1}(w))$$

and

$$z \cdot w = \phi(\phi^{-1}(z) \cdot \phi^{-1}(w))$$

for $z, w \in \mathbb{C}$.

The next theorem is a direct consequence of Theorem 5.5 and Definition 5.8 (Problem 8).

5.9 THEOREM

THE FIELD OF COMPLEX NUMBERS The algebraic system

$$(\mathbb{C}, +, \cdot)$$

is a field.

The field $\mathbb{C}$, which is called the **field of complex numbers**, is an extension field of the field $\mathbb{R}$ of real numbers (Problem 9).

5.10 DEFINITION

THE IMAGINARY UNIT We define the **imaginary unit** $i \in \mathbb{C}$ by

$$i = (0, 1).$$

Notice that there is nothing fictitious or imaginary about i at all; the word *imaginary* is merely a legacy from the early days when people had difficulty conceiving of a number whose square is -1.

5.11 LEMMA

$$i^2 = -1.$$

PROOF By Definition 5.7,

$$\phi((0, 1)) = (0, 1) = i,$$

and it follows that

$$(0, 1) = \phi^{-1}(i).$$

Also, by Definition 5.4,

$$(0, 1) \cdot (0, 1) = (0 \cdot 0 - 1 \cdot 1, 0 \cdot 1 + 1 \cdot 0) = (-1, 0).$$

Therefore, by Definition 5.8,

$$\begin{aligned} i^2 = i \cdot i &= \phi(\phi^{-1}(i) \cdot \phi^{-1}(i)) \\ &= \phi((0, 1) \cdot (0, 1)) = \phi((-1, 0)) \\ &= -1. \end{aligned}$$

□

We leave the proof of the following lemma as an exercise (Problem 11).

5.12 LEMMA

Let $(a, b) \in \mathbb{R} \times \mathbb{R}$ with $b \neq 0$, and let $x, y \in \mathbb{R}$ with $y \neq 0$. Then $(a, b) \in \mathbb{C}$, and we have:

(i) $x + (a, b) = (x + a, b)$.
(ii) $y \cdot (a, b) = (ya, yb)$.

If $x, y \in \mathbb{R}$ with $y \neq 0$, then, as a consequence of Lemma 5.12,

$$x + yi = x + y(0, 1) = x + (0, y) = (x, y).$$

Therefore, every complex number $z \in \mathbb{C}$ can be written in the form

$$z = x + yi$$

for uniquely determined real numbers $x, y \in \mathbb{R}$, and we have

$$\mathbb{C} = \{x + yi \mid x, y \in \mathbb{R}\}.$$

5.13 DEFINITION

REAL AND IMAGINARY PART If $x, y \in \mathbb{R}$, so that

$$z = x + yi \in \mathbb{C},$$

then x is called the **real part** of z, and y is called the **imaginary part** of z.

All calculations with complex numbers can be carried out by writing the numbers in the form $x + yi$ with $x, y \in \mathbb{R}$, keeping in mind that $\mathbb{C}$ is an extension field of $\mathbb{R}$, and using the fact that $i^2 = -1$.

5.14 Example Find $(3 - 2i)(4 + i)$.

SOLUTION

$$\begin{aligned} (3 - 2i)(4 + i) &= 12 + 3i - 8i - 2i^2 = 12 - 5i - 2(-1) \\ &= 14 - 5i. \end{aligned}$$

□

5.15 DEFINITION

COMPLEX CONJUGATE If $x, y \in \mathbb{R}$, so that

$$z = x + yi \in \mathbb{C},$$

we define

$$z^* = x - yi.$$

The complex number z^* is called the **complex conjugate** of z, and the mapping $z \mapsto z^*$ is called **complex conjugation**.

Some authors denote the complex conjugate of z by $\bar{z}$ rather than by z^*. Notice that

$$0^* = 0,$$
$$1^* = 1,$$

and

$$i^* = -i.$$

Furthermore, it is clear that, for any $z \in \mathbb{C}$,

$$(z^*)^* = z.$$

The following theorem shows that *complex conjugation preserves all of the algebraic operations in the field* $\mathbb{C}$.

5.16 THEOREM

COMPLEX CONJUGATION Let $z, w \in \mathbb{C}$. Then:

(i) $(z + w)^* = z^* + w^*$
(ii) $(z - w)^* = z^* - w^*$
(iii) $(zw)^* = z^*w^*$
(iv) $w \neq 0 \Rightarrow \left(\dfrac{z}{w}\right)^* = \dfrac{z^*}{w^*}$

PROOF We prove (iii) and (iv) here and leave the proofs of (i) and (ii) as an exercise (Problem 13).

(iii) Let $z = x + yi$, $w = u + vi$ with $x, y, u, v \in \mathbb{R}$. Proceeding as in Part (iii) of Lemma 5.2, we find that

$$zw = (xu - yv) + (xv + yu)i.$$

Therefore,

$$(zw)^* = (xu - yv) - (xv + yu)i.$$

Now, $z^* = x - yi$, $w^* = u - vi$, and

$$\begin{aligned} z^*w^* &= (xu - (-y)(-v)) + (x(-v) + (-y)u)i \\ &= (xu - yv) - (xv + yu)i \\ &= (zw)^*. \end{aligned}$$

(iv) Assume that $w \neq 0$. Then,

$$w\left(\frac{1}{w}\right) = 1.$$

Therefore,

$$\left(w\left(\frac{1}{w}\right)\right)^* = 1^* = 1.$$

Applying the result of Part (iii) to the left side of the last equation, we conclude that

$$w^*\left(\frac{1}{w}\right)^* = 1,$$

from which it follows that

$$\left(\frac{1}{w}\right)^* = \frac{1}{w^*}.$$

Therefore, by Part (iii) again,

$$\left(\frac{z}{w}\right)^* = \left(z\left(\frac{1}{w}\right)\right)^* = z^*\left(\frac{1}{w}\right)^* = z^*\left(\frac{1}{w^*}\right) = \frac{z^*}{w^*}. \qquad \square$$

Let $z \in \mathbb{C}$, and let x and y be the real and imaginary parts, respectively, of z. Thus, $z = x + yi$, $z^* = x - yi$, and

$$z + z^* = x + yi + x - yi = 2x;$$

hence,

$$x = \frac{1}{2}(z + z^*). \tag{1}$$

Similarly,

$$z - z^* = x + yi - (x - yi) = 2yi,$$

and it follows that

$$y = \frac{1}{2i}(z - z^*). \tag{2}$$

Formulas (1) and (2) are useful for expressing the real and imaginary parts of a complex number z in terms of z and z^*. We also have

$$zz^* = (x + yi)(x - yi) = (x^2 + y^2) + (x(-y) + yx)i,$$

so that

$$zz^* = x^2 + y^2. \tag{3}$$

Consequently,

$$zz^* \geq 0 \tag{4}$$

and

$$zz^* = 0 \Leftrightarrow z = 0. \tag{5}$$

If $z, w \in \mathbb{C}$ with $w \neq 0$, then $w^* \neq 0$ (Problem 14), and it follows that

$$\frac{z}{w} = \frac{zw^*}{ww^*}.$$

Notice that the denominator ww^* is a positive real number. Thus, to find the quotient of two complex numbers, multiply numerator and denominator by the complex conjugate of the denominator.

5.17 Example Evaluate $\dfrac{2 + 3i}{1 + i}$.

SOLUTION

$$\begin{aligned}\frac{2 + 3i}{1 + i} &= \frac{(2 + 3i)(1 - i)}{(1 + i)(1 - i)}\\ &= \frac{(2 - 3(-1)) + (-2 + 3)i}{1^2 + 1^2} = \frac{5 + i}{2}\\ &= \frac{5}{2} + \frac{1}{2}i.\end{aligned}$$

□

If $z = x + yi \in \mathbb{C}$, then, because $zz^* \geq 0$, we can form the principal square root

$$\sqrt{zz^*} = \sqrt{x^2 + y^2}$$

as in the next definition.

5.18 DEFINITION

ABSOLUTE VALUE, OR MODULUS If z is a complex number, we define the **absolute value**, or **modulus**, of z, in symbols $|z|$, to be the nonnegative real number

$$|z| = \sqrt{zz^*}.$$

Notice that, for any complex number z, we have

$$|z| \geq 0,$$
$$|z| = 0 \Leftrightarrow z = 0,$$

and

$$|z|^2 = zz^*$$

(Problem 17).

5.19 LEMMA Let $z, w \in \mathbb{C}$. Then:

(i) $|z^*| = |z|$
(ii) $|zw| = |z|\,|w|$

PROOF (i) $|z^*| = \sqrt{z^*(z^*)^*} = \sqrt{z^*z} = \sqrt{zz^*} = |z|$.
(ii) $|zw| = \sqrt{zw(zw)^*} = \sqrt{zwz^*w^*} = \sqrt{zz^*ww^*}$
$= \sqrt{zz^*}\sqrt{ww^*} = |z|\,|w|$. □

If $x \in \mathbb{R}$, then $x = x^*$ (Problem 17a), and it follows that

$$\sqrt{xx^*} = \sqrt{x^2}.$$

Therefore, the absolute value of a real number x, computed according to the formula in Definition 5.18, coincides with the absolute value of x computed according to Definition 4.15 in Chapter 3. Furthermore, the triangle inequality (Theorem 4.18 in Chapter 3) generalizes as follows:

5.20 THEOREM **TRIANGLE INEQUALITY** If $z, w \in \mathbb{C}$, then

$$|z + w| \leq |z| + |w|.$$

PROOF Let x and y be the real and imaginary parts, respectively, of zw^*, so that

$$zw^* = x + yi. \tag{1}$$

As a consequence of Equation (1), we have

$$x = \frac{1}{2}(zw^* + (zw^*)^*) = \frac{1}{2}(zw^* + z^*(w^*)^*) = \frac{1}{2}(zw^* + z^*w), \tag{2}$$

so that

$$2x = zw^* + z^*w. \tag{3}$$

Because x and y are real numbers and $0 \le y^2$, we have

$$x^2 \le x^2 + y^2, \tag{4}$$

so that

$$x \le |x| = \sqrt{x^2} \le \sqrt{x^2 + y^2} = |zw^*|. \tag{5}$$

By Lemma 5.19,

$$|zw^*| = |z|\,|w^*| = |z|\,|w|, \tag{6}$$

and it follows from (5) and (6) that

$$x \le |z|\,|w|. \tag{7}$$

As a consequence of Equation (3), we have

$$\begin{aligned} |z + w|^2 &= (z + w)(z + w)^* = (z + w)(z^* + w^*) \\ &= zz^* + zw^* + z^*w + ww^* \\ &= |z|^2 + zw^* + z^*w + |w|^2 \\ &= |z|^2 + 2x + |w|^2. \end{aligned} \tag{8}$$

Combining Equation (8) and Inequality (7), we obtain

$$|z + w|^2 \le |z|^2 + 2|z|\,|w| + |w|^2; \tag{9}$$

that is,

$$|z + w|^2 \le (|z| + |w|)^2, \tag{10}$$

from which the desired inequality follows immediately. □

In this section, we started with the field $\mathbb{R}$ of real numbers, noted that the polynomial equation $x^2 + 1 = 0$ has no solution in this field, and constructed an extension field $\mathbb{C}$ of $\mathbb{R}$ in which the equation has a solution i. In more advanced algebra courses, it is proved that an analogous construction is possible for any field F. Here is a statement of the pertinent theorem:

5.21 THEOREM

EXTENSION OF A FIELD If F is a field and if $p(x)$ is a polynomial with coefficients in F such that the equation $p(x) = 0$ has no solution in F, then there exists an extension field of F in which the equation $p(x) = 0$ has a solution.

However, to solve polynomial equations with complex coefficients, it is not necessary to invoke Theorem 5.21. Indeed, the solutions of such equations are already available in $\mathbb{C}$ as a consequence of the following important theorem:

5.22 THEOREM **THE FUNDAMENTAL THEOREM OF ALGEBRA** If $p(x)$ is any polynomial of degree $n \geq 1$ with coefficients in $\mathbb{C}$, then the equation $p(x) = 0$ has a solution in $\mathbb{C}$.

Theorem 5.22, the proof of which is beyond the scope of this textbook, shows exactly why the field $\mathbb{C}$ is so useful in algebra. It was proved in 1799 by Gauss in his doctoral dissertation. You can find a detailed proof of the Fundamental Theorem of Algebra in D. Dobbs and R. Hanks, *A Modern Course on the Theory of Equations* (Passaic, N.J.: Polygonal Publishing House), 1980.

PROBLEM SET 4.5

1. Suppose that F is an extension field of $\mathbb{R}$ and that i is a solution in F of $x^2 + 1 = 0$.
   (a) Prove that $-i$ is also a solution in F of $x^2 + 1 = 0$.
   (b) Prove that $i \neq -i$.
   (c) Prove that i and $-i$ are the only solutions in F of $x^2 + 1 = 0$.
2. Use the results of Lemma 5.2 to show that G in Definition 5.1 is an extension field of $\mathbb{R}$.
3. With the operations given in Definition 5.4, show that $\mathbb{R} \times \mathbb{R}$ is a commutative ring with unit.
4. Let $X = \{(x, 0) \in \mathbb{R} \times \mathbb{R} \mid x \in \mathbb{R}\}$. Prove that, under the operations given in Definition 5.4, X is a field that is isomorphic to the field $\mathbb{R}$ of real numbers under the mapping $(x, 0) \mapsto x \in \mathbb{R}$.
5. Prove Theorem 5.5. [Hint: To show that every nonzero element has a multiplicative inverse, see the proof of Part (iv) of Lemma 5.2.]
6. Prove that the mapping $\phi: \mathbb{R} \times \mathbb{R} \to \mathbb{C}$ of Definition 5.7 is a bijection.
7. Let $x, y \in \mathbb{R}$ with $y \neq 0$, and let $\phi: \mathbb{R} \times \mathbb{R} \to \mathbb{C}$ be the mapping of Definition 5.7.
   (a) Find $\phi^{-1}(x)$. (b) Find $\phi^{-1}((x, y))$.
8. Prove Theorem 5.9.
9. Prove that $\mathbb{C}$ is an extension field of $\mathbb{R}$.
10. Let $y \in \mathbb{R}$ with $y \neq 0$. Compute $(0, y)^2$ in $\mathbb{C}$.
11. Prove Lemma 5.12.

12. Let $z \in \mathbb{C}$, $x = (z + z^*)/2$, and $y = (z - z^*)/2i$. Prove that $x, y \in \mathbb{R}$ and $z = x + yi$.

13. Prove Parts (i) and (ii) of Theorem 5.16.

14. If $w \in \mathbb{C}$ with $w \neq 0$, prove that $w^* \neq 0$ and that $(w^*)^{-1} = (w^{-1})^*$.

15. Let $z = \frac{2}{3} + \frac{5}{4}i$ and $w = \frac{1}{7} - i$. Find:
(a) $z + w$ (b) $z - w$ (c) zw (d) z/w
(e) w/z (f) zw^* (g) $|z|$

16. Let $\rho: \mathbb{C} \times \mathbb{C} \to \mathbb{R}$ be defined by $\rho(z, w) = |z - w|$. The nonnegative real number $\rho(z, w)$ is called the **distance between z and w**. Let $z, z', w \in \mathbb{C}$. Prove:
(a) $\rho(z, z') \geq 0$ with equality if and only if $z = z'$
(b) $\rho(z, z') = \rho(z', z)$
(c) $\rho(z, z') \leq \rho(z, w) + \rho(w, z')$

17. Let $z \in \mathbb{C}$. Prove:
(a) $z \in \mathbb{R}$ if and only if $z = z^*$
(b) $|z| \geq 0$ with equality if and only if $z = 0$
(c) $|z|^2 = zz^*$

18. If

$$e = \lim_{n \to \infty} \left(1 + \frac{1}{n}\right)^n$$

is the base of the natural logarithms and $z = x + yi$ with $x, y \in \mathbb{R}$, the complex number e^z is defined by $e^z = e^{x+yi} = e^x(\cos y + i \sin y)$. If $z, w \in \mathbb{C}$, prove that $e^z e^w = e^{z+w}$.

19. By identifying $\mathbb{R}$ with the x axis, we can visualize complex numbers as points in the Gaussian plane $\mathbb{R} \times \mathbb{R}$. With this in mind, give geometric interpretations of each of the following for $z, w \in \mathbb{C}$:
(a) $z + w$ (b) $z - w$ (c) $-z$ (d) $|z|$

20. If $z \in \mathbb{C}$, show that z can be written in the form $z = re^{i\theta}$ for $r \in \mathbb{R}$ with $r \geq 0$ and $\theta \in \mathbb{R}$. (See Problem 18.) The representation $z = re^{i\theta}$ is called the **polar form** for the complex number z. The real number θ, which is determined up to an additive integer multiple of 2π, is called the **argument** of z.

21. If $z, w \in \mathbb{C}$ with $w \neq 0$, prove that

$$\left|\frac{z}{w}\right| = \frac{|z|}{|w|}.$$

22. If $z = re^{i\theta}$ as in Problem 20, give a geometric interpretation of r and θ as in Problem 19. [Hint: Think of polar coordinates.]

23. Let $w \in \mathbb{C}$ with $w = a + bi$, $a, b \in \mathbb{R}$, and let $r = \sqrt{a^2 + b^2}$.
(a) Show that $r + a \geq 0$.
(b) Show that $r - a \geq 0$.

Let $x = \sqrt{(r + a)/2}$, and let $y = \text{sgn}(b)\sqrt{(r - a)/2}$.
(c) Show that $x + yi$ and $x - yi$ are complex square roots of w.

24. Using the polar form (Problems 20 and 22), give a geometric interpretation of the product of two complex numbers in terms of the absolute values and arguments of these numbers.

25. If $p(x)$ is a polynomial with real coefficients, and $z \in \mathbb{C}$ is a solution of the equation $p(x) = 0$, show that z^* is also a solution of $p(x) = 0$.

26. (a) Show that there is no complex number z such that $e^z = 0$.
(b) Find all complex numbers z such that $e^z = 1$. (See Problem 20.)

27. Using the fundamental theorem of algebra (Theorem 5.22), show that every polynomial $p(x)$ of degree $n \geq 1$ with complex coefficients can be factored into a product of first-degree polynomials with complex coefficients.

28. True or false: $(e^z)^* = e^{z^*}$. Explain.

29. By combining Problems 25 and 27, show that every polynomial of degree $n \geq 1$ with real coefficients can be factored into a product of first-degree and second-degree polynomials with real coefficients.

30. Generalize the "identification procedure" employed in the construction of $\mathbb{C}$ from the Gaussian plane as follows: Let H be a field, and let G be an extension field of H. Let H' be a field, and suppose that $\psi: H \to H'$ is an isomorphism; that is, ψ is a bijection that preserves the field operations. Show that there is an extension field G' of G and an isomorphism $\Psi: G \to G'$ such that Ψ restricted to H is ψ.

31. Let $S^1 = \{z \in \mathbb{C} \mid |z| = 1\}$.
(a) Show that S^1 is closed under multiplication.
(b) Show that $(S^1, \cdot)$ is a commutative group.
(c) Explain why $(S^1, \cdot)$ is called the **circle group**.

32. The remarkable equation $e^{i\pi} + 1 = 0$ connects the five most important constants in mathematics, namely, 0, 1, π, e, and i. Prove that this equation is true.

*6 COUNTING AND THE NATURAL NUMBERS

In previous chapters, we have spoken of the number of elements in a finite set M, denoted as $\#M$. But we have never said exactly what we meant by $\#M$. Instead, we have assumed an intuitive idea of $\#M$ based on the idea of counting, a process with which we are all familiar. In this section, we propose to reconsider the idea of counting from a more formal point of view.

How do we actually count a finite set M? We point to the elements of M, one at a time, in some definite order. As we point to each element, we name a natural number, starting from 1 and proceeding with the successive natural numbers until we have exhausted the set M. At the end of this counting process, a certain natural number n is assigned to the last element of M; and we conclude that $\#M = n$. The effect of the counting process is to set up a bijection

$$f:\{1, 2, 3, \ldots, n\} \to M$$

from an *initial segment* $\{1, 2, 3, \ldots, n\}$ of the set $\mathbb{N}$ of all natural numbers onto M.

Thus, we begin our study of the counting process by formalizing the notion of an initial segment of $\mathbb{N}$.

6.1 DEFINITION

INITIAL SEGMENT If $n \in \mathbb{N}$, we define

$$I_n = \{m \in \mathbb{N} \mid m \leq n\}.$$

A subset of $\mathbb{N}$ having the form I_n for some $n \in \mathbb{N}$ is called an **initial segment** of $\mathbb{N}$.

6.2 DEFINITION

COUNTING A FINITE SET A set M is said to be **finite** if M is empty or if there exists $n \in \mathbb{N}$ and a bijection $f:I_n \to M$. To **count** the set M means to find such a bijection.

As the German mathematician and physicist Hermann Weyl (1885–1955) remarked in his influential book, *Theory of Groups and Quantum Mechanics* (New York: Dover), 1950, p. 131, the fact that if a finite set M is counted in two different ways, one always obtains the same number is "perhaps the most fundamental theorem of mathematics." This "most fundamental theorem" (Theorem 6.8 below) states that *if*

$$f:I_n \to M \quad \text{and} \quad g:I_m \to M$$

are bijections (that is, if M is counted in two different ways), *then* $n = m$. To prove this, we begin by developing a series of lemmas.

6.3 LEMMA

If $n \in \mathbb{N}$, then $I_{n+1} = I_n \cup \{n + 1\}$.

PROOF Evidently,

$$I_n \cup \{n + 1\} \subseteq I_{n+1}.$$

To prove the opposite inclusion, suppose that $k \in I_{n+1}$, so that $k \leq n + 1$. On the one hand, if $k = n + 1$, then $k \in \{n + 1\}$. On the other hand, if

$k < n + 1$, then, by Theorem 2.7, we cannot have $n < k < n + 1$, so it follows that $k \leq n$, so $k \in I_n$. In either case, $k \in I_n \cup \{n + 1\}$ and, therefore,

$$I_{n+1} \subseteq I_n \cup \{n + 1\}. \qquad \square$$

6.4 LEMMA Let B be a set, and let $b, c \in B$. Then, there exists a bijection $j:B \to B$ such that $j(b) = c$ and $j(c) = b$.

PROOF We sketch the proof and leave the details as an exercise (Problem 2). For $x \in B$, define

$$j(x) = \left\{\begin{array}{ll} x, & \text{if } x \neq b \text{ and } x \neq c \\ c, & \text{if } x = b \\ b, & \text{if } x = c \end{array}\right\}.$$

If $b = c$, then $j(x) = x$ for all $x \in N$. If $b \neq c$, then j interchanges b with c and leaves all other elements of B fixed. In either case, $j:B \to B$ is a bijection. $\square$

6.5 LEMMA Let A and B be sets, and let $g:A \to B$ be an injection (respectively, a bijection). Let $a \in A$ and $b \in B$. Then, there exists an injection (respectively, a bijection) $h:A \to B$ such that $h(a) = b$.

PROOF Let $c = g(a)$. By Lemma 6.4, there exists a bijection $j:B \to B$ such that $j(c) = b$. Let $h:A \to B$ be defined by $h = j \circ g$. Then, we have

$$h(a) = j(g(a)) = j(c) = b.$$

By Corollary 5.11 in Chapter 2, if g is an injection (respectively, a bijection), then h is an injection (respectively, a bijection). $\square$

6.6 LEMMA Let M and N be sets, and suppose that

$$A = M \cup \{a\} \quad \text{and} \quad B = N \cup \{b\},$$

where $a \notin M$ and $b \notin N$. Let

$$g:A \to B$$

be an injection (respectively, a bijection). Then:

(i) If $N = \varnothing$, then $M = \varnothing$.
(ii) If $M \neq \varnothing$, and $N \neq \varnothing$, then there exists an injection (respectively, a bijection) $f:M \to N$.

PROOF We sketch the proof and leave the details as an exercise (Problem 4). By Lemma 6.5, there exists an injection (respectively, a bijection) $h: A \to B$ such that $h(a) = b$. If $N = \varnothing$, then $g: A \to \{b\}$, and since g is an injection, it follows that $A = \{a\}$; so $M = \varnothing$. Suppose that $M \neq \varnothing$, $N \neq \varnothing$, and let f be the restriction of h to M, so that

$$f(x) = h(x), \quad \text{for all } x \in M.$$

Then $f: M \to N$, and f is an injection (respectively, a bijection). □

6.7 LEMMA If $m, n \in \mathbb{N}$ and $f: I_n \to I_m$ is an injection, then $n \leq m$.

PROOF The proof is by induction on n. If $n = 1$, then $n \leq m$ holds for all $m \in \mathbb{N}$ and there is nothing further to prove. Thus, we assume as our induction hypothesis that $n \in \mathbb{N}$ has the following property:

For every $m \in \mathbb{N}$, if there is an injection $f: I_n \to I_m$, then $n \leq m$.

We must prove that $n + 1$ has the same property. Thus, suppose $k \in \mathbb{N}$ and that

$$g: I_{n+1} \to I_k$$

is an injection. We have to prove that $n + 1 \leq k$.

If $k = 1$, then $I_k = \{1\}$, and thus $g(n) = g(n + 1) = 1$, contradicting the fact that g is an injection. Therefore, $k \neq 1$, and it follows from Lemma 2.6 that $k - 1 \in \mathbb{N}$. Let $m = k - 1$, so that $k = m + 1$ and

$$g: I_{n+1} \to I_{m+1}$$

is an injection. By Lemma 6.3, we have

$$g: I_n \cup \{n + 1\} \to I_m \cup \{m + 1\}.$$

By Part (ii) of Lemma 6.6, there exists an injection

$$f: I_n \to I_m,$$

and it follows from our induction hypothesis that $n \leq m$. Therefore, $n + 1 \leq m + 1 = k$. □

We are now in a position to state and prove the **Fundamental Theorem of Counting**.

6.8 THEOREM

FUNDAMENTAL THEOREM OF COUNTING Let M be a set, and let $m, n \in \mathbb{N}$. If there exist bijections

$$f: I_n \to M \quad \text{and} \quad g: I_m \to M,$$

then $n = m$.

PROOF By Part (ii) of Theorem 4.18 in Chapter 2,

$$g^{-1}: M \to I_m$$

and

$$f^{-1}: M \to I_n$$

are bijections. Hence, by Part (iii) of Corollary 5.11 in Chapter 2,

$$(g^{-1}) \circ f: I_n \to I_m$$

and

$$(f^{-1}) \circ g: I_m \to I_n$$

are bijections. In particular, $(g^{-1}) \circ f$ is an injection, and it follows from Lemma 6.7 that $n \leq m$. Likewise, $(f^{-1}) \circ g$ is an injection, and so $m \leq n$. Thus, $n = m$. □

If M is a finite nonempty set, then, by Definition 6.2, there exists a natural number n and a bijection $f: I_n \to M$. By Theorem 6.8, n is uniquely determined by M. Thus, we can make the following definition:

6.9 DEFINITION

THE CARDINAL NUMBER OF A FINITE SET Let M be a finite, nonempty set. The unique natural number n such that there is a bijection $f: I_n \to M$ is called the **cardinal number** of M, written

$$n = \#M.$$

We define 0 to be the cardinal number of the empty set,

$$0 = \#\varnothing.$$

The cardinal number of a set is also referred to as the **cardinality** of the set.

6.10 LEMMA

Let M be a finite set and let $a \in M$. Then $M \setminus \{a\}$ is a finite set and

$$\#(M \setminus \{a\}) = \#M - 1.$$

PROOF Let $\#M = m$. Since $a \in M$, it follows that $M \neq \varnothing$, $m \in \mathbb{N}$, and there exists a bijection

$$g: I_m \to M.$$

If $m = 1$, then $I_m = \{1\}$, and since g is a bijection, $M = \{a\}$ (Problem 1). Hence, if $m = 1$, we have

$$\#(M \setminus \{a\}) = \#\varnothing = 0 = m - 1 = \#M - 1.$$

Thus, we can assume that $m > 1$. By Lemma 2.6, $m - 1 \in \mathbb{N}$. Let $k = m - 1$, so that $m = k + 1$. By Lemma 6.3,

$$I_m = I_{k+1} = I_k \cup \{k + 1\} = I_k \cup \{m\},$$

and we have a bijection

$$g: I_k \cup \{m\} \to (M \setminus \{a\}) \cup \{a\}.$$

Since $I_k \neq \varnothing$, it follows from Part (i) of Lemma 6.6, that $M \setminus \{a\} \neq \varnothing$; hence, by Part (ii) of Lemma 6.6, there exists a bijection $f: I_k \to M \setminus \{a\}$. Consequently,

$$\#(M \setminus \{a\}) = k = m - 1 = \#M - 1.$$

□

6.11 COROLLARY

Any subset of a finite set is a finite set.

PROOF Suppose, on the contrary, that N is not a finite set, but there exists a finite set M such that $N \subseteq M$. Since N is not a finite set, it follows that $N \neq \varnothing$; hence, $M \neq \varnothing$, so $\#M \in \mathbb{N}$. Among all such sets M, choose one for which $m = \#M$ is the smallest (Theorem 2.14). Since M is finite, and N is not, N is a proper subset of M; hence there exists an element $a \in M$ with $a \notin N$. Thus,

$$N \subseteq M \setminus \{a\}.$$

By Lemma 6.10,

$$\#(M \setminus \{a\}) = m - 1 < m.$$

Thus, there exists a finite set $M \setminus \{a\}$, containing N, and having $m - 1$ elements, This contradicts our choice of m as being the smallest such natural number.

□

6.12 LEMMA

If M is a finite set and $a \notin M$, then $M \cup \{a\}$ is a finite set and

$$\#(M \cup \{a\}) = \#M + 1.$$

PROOF The statement to be proved is obvious if $M = \varnothing$; so we can assume that $M \neq \varnothing$. Suppose that $\# M = m$; so there exists a bijection

$$f: I_m \to M.$$

By Lemma 6.3, $I_{m+1} = I_m \cup \{m+1\}$. Define

$$g: I_{m+1} \to M \cup \{a\}$$

by

$$g(n) = \left.\begin{cases} f(n), & \text{if } n \in I_m \\ a, & \text{if } n = m+1 \end{cases}\right\} \text{ for all } n \in I_{m+1}.$$

We leave it as an exercise for you to prove that g is a bijection (Problem 11). □

6.13 THEOREM

ADDITIVITY THEOREM If M and N are finite sets with $M \cap N = \varnothing$, then $M \cup N$ is a finite set and

$$\#(M \cup N) = \# M + \# N.$$

PROOF Let M be a fixed but arbitrary finite set. If $N = \varnothing$, the statement to be proved is obvious. Thus, we can assume that $N \neq \varnothing$. The remainder of the proof is by induction on $n = \# N$. If $n = 1$, then N has the form $N = \{a\}$ for some element $a \notin M$, $M \cup N$ is a finite set, and the equation

$$\#(M \cup N) = \# M + \# N$$

holds by Lemma 6.12. Now, assume as the induction hypothesis that n is an arbitrary but fixed natural number with the following property:

For every set N such that $\# N = n$ and $M \cap N = \varnothing$, the set $M \cup N$ is finite and $\#(M \cup N) = \# M + \# N = \# M + n$.

We must prove that $n + 1$ has the same property.

Thus, suppose K is a finite set with $\# K = n + 1$ and $M \cap K = \varnothing$. We have to prove that $M \cup K$ is a finite set and

$$\#(M \cup K) = \# M + (n+1).$$

Since $\# K > 0$, it follows that $K \neq \varnothing$. Let $a \in K$, and let $N = K \setminus \{a\}$. Because $M \cap K = \varnothing$ and $N \subseteq K$, we have $M \cap N = \varnothing$. Furthermore, by Lemma 6.10, N is a finite set and

$$\# N = \#(K \setminus \{a\}) = \# K - 1 = (n+1) - 1 = n.$$

Therefore, by the induction hypothesis, $M \cup N$ is finite and

$$\#(M \cup N) = \#M + n.$$

Now, $K = N \cup \{a\}$ with $a \notin N$. Also, since $M \cap K = \varnothing$ and $a \in K$, we have $a \notin M$. Therefore, $a \notin M \cup N$; so, by Lemma 6.12,

$$M \cup K = M \cup (N \cup \{a\}) = (M \cup N) \cup \{a\}$$

is a finite set and

$$\begin{aligned} \#(M \cup K) &= \#((M \cup N) \cup \{a\}) = \#(M \cup N) + 1 \\ &= (\#M + n) + 1 = \#M + (n + 1). \end{aligned}$$

□

6.14 THEOREM

GENERALIZED ADDITIVITY THEOREM Let M and N be finite sets. Then, $M \cup N$ and $M \cap N$ are finite sets and

$$\#(M \cup N) + \#(M \cap N) = \#M + \#N.$$

PROOF Let $A = M \cap N$, and let $B = M \setminus N$. Since A and B are subsets of the finite set M, it follows from Corollary 6.11 that A and B are finite sets. By set algebra, we have

$$M \cup N = B \cup N \quad \text{and} \quad B \cap N = \varnothing. \tag{1}$$

Also,

$$M = A \cup B \quad \text{and} \quad A \cap B = \varnothing \tag{2}$$

(Problem 12). By Equations (1) and Theorem 6.13, we have

$$\#(M \cup N) = \#B + \#N. \tag{3}$$

Also, by Equations (2) and Theorem 6.13, we have

$$\#M = \#A + \#B. \tag{4}$$

Combining Equations (3) and (4), we obtain

$$\begin{aligned} \#(M \cup N) + \#(M \cap N) &= \#(M \cup N) + \#A \\ &= (\#B + \#N) + \#A \\ &= (\#A + \#B) + \#N \\ &= \#M + \#N. \end{aligned} \tag{5}$$

□

Note that Theorem 6.14 is just a restatement of Theorem 5.9 in Chapter 1. (In Chapter 1, we could offer only an informal proof of this theorem.)

The proofs of Theorem 5.12 and Corollary 5.13 in Chapter 1 can now be made on a formal basis; and so we have the following results:

1. If M and N are sets and M is finite, then $M \cap N$ and $M \setminus N$ are finite sets and

$$\#(M \setminus N) = \#M - \#(M \cap N).$$

2. If M is a finite set and $N \subseteq M$, then N and $M \setminus N$ are finite sets and

$$\#(M \setminus N) = \#M - \#N.$$

6.15 DEFINITION

INFINITE SET, COUNTABLY INFINITE SET A set M that is not finite is said to be **infinite**. If there exists a bijection

$$f: \mathbb{N} \to M,$$

then M is said to be a **countably infinite** set. A set M is said to be **countable** if it is either finite or countably infinite. If M is not countable, it is said to be **uncountable**.

Thus, a set is countably infinite if its elements can be placed in one-to-one correspondence with the natural numbers 1, 2, 3, 4, ... In particular, $\mathbb{N}$ itself is a countably infinite set. Some authors prefer to use the word **denumerable**, rather than the word *countable*, in Definition 6.15. Then, a set that is not denumerable is called **nondenumerable**.

6.16 Example Show that the set E of all even positive integers is countably infinite.

SOLUTION The mapping

$$f: \mathbb{N} \to E$$

given by $f(n) = 2n$ is a bijection. □

6.17 Example Show that the set $\mathscr{P}(\mathbb{N})$ of all subsets of $\mathbb{N}$ is uncountable.

SOLUTION Clearly, $\mathscr{P}(\mathbb{N})$ is not a finite set (Problem 15). We must show that $\mathscr{P}(\mathbb{N})$ is not countably infinite. Suppose, on the contrary, that there exists a bijection

$$f: \mathbb{N} \to \mathscr{P}(\mathbb{N}). \tag{1}$$

Thus, for each $k \in \mathbb{N}$, $f(k)$ is a subset of $\mathbb{N}$. Let

$$K = \{k \in \mathbb{N} \mid k \notin f(k)\}. \tag{2}$$

Because $K \in \mathscr{P}(\mathbb{N})$ and $f: \mathbb{N} \to \mathscr{P}(\mathbb{N})$ is surjective, there exists $n \in \mathbb{N}$ such that

$$f(n) = K. \tag{3}$$

If $n \in K$, then, by Equation (2), $n \notin f(n) = K$, a contradiction. Therefore, we must have $n \notin K$. But then, $n \notin f(n)$, and it follows from Equation (2) that $n \in K$, contradicting $n \notin K$. □

6.18 THEOREM

COUNTABILITY OF $\mathbb{N} \times \mathbb{N}$ $\mathbb{N} \times \mathbb{N}$ is a countably infinite set.

INFORMAL PROOF Arrange the elements of $\mathbb{N} \times \mathbb{N}$ in an infinite array as follows:

$$\begin{matrix} (1,1) & (1,2) & (1,3) & (1,4) & (1,5) & \cdots \\ (2,1) & (2,2) & (2,3) & (2,4) & (2,5) & \cdots \\ (3,1) & (3,2) & (3,3) & (3,4) & (3,5) & \cdots \\ (4,1) & (4,2) & (4,3) & (4,4) & (4,5) & \cdots \\ \vdots & \vdots & \vdots & \vdots & \vdots & \end{matrix}$$

Now we "count" the elements in this array by following the successive diagonals as indicated, so that the ordered pairs in $\mathbb{N} \times \mathbb{N}$ are rearranged in a sequence: (1, 1), (2, 1), (1, 2), (3, 1), (2, 2), (1, 3), (4, 1), (3, 2), (2, 3), (1, 4), (5, 1), and so on. In other words, we have a bijection

$$f: \mathbb{N} \to \mathbb{N} \times \mathbb{N}$$

such that

$$f(1) = (1, 1),\ f(2) = (2, 1),\ f(3) = (1, 2),\ f(4) = (3, 1), \text{ and so forth.}$$ □

This scheme for setting up the bijection $f: \mathbb{N} \times \mathbb{N} \to \mathbb{N}$ is called the **Cantor diagonal process** in honor of its discoverer, Georg Cantor (1845–1918), the founder of set theory. With a bit of work, a formal proof of Theorem 6.18 can be constructed, based on the idea of the Cantor diagonal process (Problems 18 and 20).

6.19 THEOREM

UNCOUNTABILITY OF THE UNIT INTERVAL
Let $[0, 1] = \{x \in \mathbb{R} \mid 0 \le x \le 1\}$. Then, $[0, 1]$ is an uncountable set.

PROOF Each number $x \in [0, 1]$ can be written in decimal form

$$x = 0.d_1 d_2 d_3 \ldots.$$

Here, for each $n \in \mathbb{N}$, d_n is the nth digit in the decimal expansion of x, so that d_n is an integer and $0 \le d_n \le 9$. Note that it is possible for x to have two different decimal expansions; indeed, if $d_k \neq 9$ and $d_n = 9$ for all $n > k$, then we have

$$x = 0.d_1 d_2 d_3 \ldots d_{k-1} d_k 999 \ldots = 0.d_1 d_2 d_3 \ldots d_{k-1} q000 \ldots,$$

where $q = d_k + 1$. When we have the option of writing x in one of these two forms, let us agree to use the form with the infinite string of 0's.

Now, suppose that $[0, 1]$ is a countable set. Evidently, this set is infinite (Problem 17); hence, we can suppose that it is countably infinite. Therefore, there exists a bijection

$$f: \mathbb{N} \to [0, 1].$$

For each $n \in \mathbb{N}$, let d_n denote the nth digit in the decimal expansion of $f(n)$. For each $n \in \mathbb{N}$, let

$$e_n = \begin{Bmatrix} 0, & \text{if } d_n = 1 \\ 1, & \text{if } d_n \neq 1 \end{Bmatrix}.$$

Then, for every $n \in \mathbb{N}$,

$$d_n \neq e_n.$$

Now, let

$$x = 0.e_1 e_2 e_3 \ldots$$

be the real number with e_n as the nth digit in its decimal expansion for every $n \in \mathbb{N}$. Because $x \in [0, 1]$ and $f: \mathbb{N} \to [0, 1]$ is a surjection, there exists $n \in \mathbb{N}$ such that $f(n) = x$. Thus, $d_n = e_n$, contradicting the fact that $d_n \neq e_n$. □

PROBLEM SET 4.6

1. If M is a set, prove that M is a finite set with $\#M = 1$ if and only if M has the form $M = \{a\}$ for some element a.
2. Complete the proof of Lemma 6.4 by showing that $j: B \to B$ is a bijection.
3. If M is a set, $A = M \cup \{a\}$, $B = \{b\}$, and $g: A \to B$ is an injection, prove that $M = \varnothing$.
4. Fill in the details of the proof of Lemma 6.6. [Note that Part (i) is a consequence of the result in Problem 3.]
5. If M and N are sets, N is a finite set, and $g: M \to N$ is an injection, prove that M is a finite set and that $\#M \leq \#N$. [Hint: Use Corollary 6.11 to prove that M is a finite set. Let $m = \#M$, $n = \#N$, so that there exist bijections $j: I_m \to M$ and $k: I_n \to N$. Show that $(k^{-1} \circ g \circ j): I_m \to I_n$ is an injection, and use Lemma 6.7.]
6. If M and N are sets, N is a finite set, and $f: N \to M$ is a surjection, prove that M is a finite set and that $\#M \leq \#N$. [Hint: For each element $b \in M$,

choose an element $a_b \in N$ such that $f(a_b) = b$. Define $g:M \to N$ by $g(b) = a_b$ for all $b \in M$, and use the result of Problem 5.]

7. Let M and N be finite sets. Prove that:
(a) $\#M \leq \#N$ if and only if there exists an injection $f:M \to N$.
(b) $\#M = \#N$ if and only if there exists a bijection $f:M \to N$.

8. Give an example of a set X, a proper subset $M \subseteq X$, and a bijection $f:M \to X$. Explain why such a set X cannot be finite.

9. If M and N are sets, $M \subseteq N$, and M is an infinite set, prove that N is an infinite set. [Hint: Use Corollary 6.11.]

10. Suppose that I is a proper subset of $\mathbb{N}$, $I \neq \varnothing$, and that I has the property that, if $j, k \in \mathbb{N}$, $j \leq k$, and $k \in I$, then $j \in I$. Prove that there exists a unique $n \in \mathbb{N}$ such that $I = I_n$.

11. Complete the proof of Lemma 6.12 by showing that the mapping $g:I_{m+1} \to M \cup \{a\}$ is a bijection.

12. Let M and N be sets. As in the proof of Theorem 6.14, let $A = M \cap N$, $B = M \setminus N$.
(a) Show that $M \cup N = B \cup N$ with $B \cap N = \varnothing$.
(b) Show that $M = A \cup B$ with $A \cap B = \varnothing$.

13. If M and N are sets, $M \subseteq N$, N is a finite set, and $\#M = \#N$, prove that $M = N$.

14. Let M be a finite set, and let $f:M \to M$. Prove that f is an injection if and only if it is a bijection.

15. If X is an infinite set, show that $\mathscr{P}(X)$, the set of all subsets of X, is an infinite set.

16. If X is a set, show that there cannot exist a bijection $f:X \to \mathscr{P}(X)$. [Hint: Use the idea in the solution of Example 6.17.]

17. Let $[0, 1]$ be the subset of $\mathbb{R}$ consisting of all real numbers x with $0 \leq x \leq 1$. The set $[0, 1]$ is called the **closed unit interval** in $\mathbb{R}$. Show that $[0, 1]$ is an infinite set.

18. Let $g:\mathbb{N} \times \mathbb{N} \to \mathbb{N}$ be defined for $(x, y) \in \mathbb{N} \times \mathbb{N}$ by

$$g(x, y) = \frac{(x + y - 1)(x + y - 2)}{2} + y.$$

Show that g is a bijection and that $g^{-1}:\mathbb{N} \to \mathbb{N} \times \mathbb{N}$ is the bijection $f:\mathbb{N} \to \mathbb{N} \times \mathbb{N}$ in the proof of Theorem 6.18.

19. If M and N are finite sets, prove that $M \times N$ is a finite set and that $\#(M \times N) = \#M \cdot \#N$.

20. In Problem 18, show that $g^{-1}:\mathbb{N} \to \mathbb{N} \times \mathbb{N}$ is given by the following rule: For $n \in \mathbb{N}$, let $b_n = [\![(\sqrt{8n} - 1)/2]\!]$, $x_n = b_n + b_n(b_n + 1)/2 + 2 - n$, and $y_n = n - b_n(b_n + 1)/2$. Then, for $n \in \mathbb{N}$, $g^{-1}(n) = (x_n, y_n)$.

21. If M and N are countably infinite sets, prove that $M \times N$ is a countably infinite set.

22. If M is a finite set and N is a countably infinite set, prove that $M \times N$ is a countably infinite set.

23. If M and N are countably infinite sets and $M \cap N = \varnothing$, prove that $M \cup N$ is a countably infinite set. [Hint: By hypothesis, there exist bijections $h:\mathbb{N} \to M$ and $k:\mathbb{N} \to N$. Define $f:\mathbb{N} \to M \cup N$ by

$$f(n) = \begin{Bmatrix} h(n/2), & \text{if } n \text{ is even} \\ k((n+1)/2), & \text{if } n \text{ is odd} \end{Bmatrix}.$$

Prove that f is a bijection.]

24. Let $\mathscr{F}$ be a nonempty set of finite nonempty sets. Suppose that, for $M, N \in \mathscr{F}$, $M \subseteq N$, or $N \subseteq M$. Prove that $\bigcap_{M \in \mathscr{F}} M \neq \varnothing$.

25. Show that $\mathbb{Z}$ is a countably infinite set by exhibiting a bijection $f:\mathbb{N} \to \mathbb{Z}$. [See Problem 23.]

26. If $n \in \mathbb{N}$ and $M_1, M_2, M_3, \ldots, M_n$ are finite sets such that $M_i \cap M_j = \varnothing$ for $i \neq j$, prove that

$$\#(M_1 \cup M_2 \cup M_3 \cup \cdots \cup M_n) = \#M_1 + \#M_2 + \#M_3 + \cdots + \#M_n.$$

*7 MORE ABOUT COUNTABLE AND UNCOUNTABLE SETS

In this section we continue the study of countable and uncountable sets initiated in Section 6. By Problem 7b in Problem Set 4.6, we can state that *if M and N are finite sets, then $\#M = \#N$ if and only if there is a bijection $f:M \to N$.* In the following definition, we extend this idea to sets that are not necessarily finite.

7.1 DEFINITION **EQUINUMEROUS SETS** If M and N are sets, we say that M and N are **equinumerous**, or that they have the same **cardinality**, and write

$$\#M = \#N$$

if there exists a bijection $f:M \to N$.

Note that, if M and N are infinite sets, Definition 7.1 does not specify what $\#M$ and $\#N$ *are*, it merely establishes the notation $\#M = \#N$ as shorthand for the statement that there is a bijection $f:M \to N$. In more advanced textbooks, a suitable meaning is assigned to the symbol $\#M$ for an infinite set M, and $\#M$ is referred to as a **transfinite cardinal number**.

Although we cannot go into the arithmetic of transfinite cardinal numbers in this textbook, it will be convenient for us to introduce the symbol $\aleph_0$, called **aleph-subscript-zero**, which is traditionally used to denote the *smallest* transfinite cardinal number. (Aleph, written $\aleph$, is the first letter of the Hebrew alphabet.)

7.2 DEFINITION

ALEPH-SUBSCRIPT-ZERO The symbol $\aleph_0$ denotes the cardinal number of the set $\mathbb{N}$ of all positive integers,

$$\#\mathbb{N} = \aleph_0.$$

Thus, the condition that a set M is countably infinite can be written as $\#M = \aleph_0$. For instance, using this notation, the result of Problem 21 in Problem Set 4.6 can be expressed as

$$\#M = \aleph_0 \quad \text{and} \quad \#N = \aleph_0 \Rightarrow \#(M \times N) = \aleph_0.$$

Likewise, the result of Problem 25 in Problem Set 4.6 can be restated as

$$\#\mathbb{Z} = \aleph_0.$$

The next lemma, which is of some interest in its own right, will be used in the proof of Theorem 7.5.

7.3 LEMMA

Suppose that M and N are sets, where $M = A \cup B$, $A \cap B = \varnothing$, $N = C \cup D$, $C \cap D = \varnothing$, $\#A = \#C$, and $\#B = \#D$. Then $\#M = \#N$.

PROOF We sketch the proof and leave the details as an exercise (Problem 1). By hypothesis, there exist bijections $f: A \to C$ and $g: B \to D$. Define $h: M \to N$ by

$$h(x) = \begin{cases} f(x), & \text{if } x \in A \\ g(x), & \text{if } x \in B \end{cases}$$

for all $x \in M$. Then h is a bijection; so $\#M = \#N$. □

The following definition generalizes the result of Problem 7a in Problem Set 4.6 to sets that are not necessarily finite.

7.4 DEFINITION

THE ORDER RELATION FOR CARDINAL NUMBERS If M and N are sets, we write $\#M \leq \#N$ to mean that there is an injection $f:M \to N$.

Again, if either M or N is an infinite set, Definition 7.4 merely establishes the notation $\#M \leq \#N$ as shorthand for the statement that there exists an injection $f:M \to N$. However, the expression $\aleph_0 \leq \#N$ can be interpreted as meaning that there exists an injection $f:\mathbb{N} \to N$; likewise, the expression $\#M \leq \aleph_0$ can be interpreted as meaning that there exists an injection $f:M \to \mathbb{N}$. We leave it as an exercise for you to show that

$$\#M \leq \#M$$

and that

$$\#M \leq \#N \quad \text{and} \quad \#N \leq \#P \Rightarrow \#M \leq \#P;$$

that is, $\leq$ is "reflexive" and "transitive" (Problems 5 and 4). The next theorem shows that $\leq$ is also "antisymmetric."

7.5 THEOREM

CANTOR–SCHROEDER–BERNSTEIN THEOREM If M and N are sets such that $\#M \leq \#N$ and $\#N \leq \#M$, then

$$\#M = \#N.$$

PROOF Let $\mathscr{P}(M)$ and $\mathscr{P}(N)$ denote the set of all subsets of M and N, respectively. If $S \subseteq M$, we write the complementary set $M \setminus S$ as S'; likewise, if $T \subseteq N$, we write the complementary set $N \setminus T$ as T'. By hypothesis, there exist injections $f:M \to N$ and $g:N \to M$. We use the notation of Definition 7.1 in Chapter 2 for the image of a set under a function. Define a mapping $\phi:\mathscr{P}(M) \to \mathscr{P}(M)$ by

$$\phi(S) = g(f(S)')' \tag{1}$$

for all $S \in \mathscr{P}(M)$. We claim that, for $U, V \in \mathscr{P}(M)$,

$$U \subseteq V \Rightarrow \phi(U) \subseteq \phi(V). \tag{2}$$

Indeed, if $U \subseteq V$, then $f(U) \subseteq f(V)$ by Problem 7 in Problem Set 2.7. Consequently, $f(V)' \subseteq f(U)'$ (see Problem 6d in Problem Set 1.5), and it follows that $g(f(V)') \subseteq g(f(U)')$. Therefore, $g(f(U)')' \subseteq g(f(V)')'$, so that (2) holds.

Next, we are going to show that there exists a set $A \subseteq M$ such that $\phi(A) = A$, that is, $g(f(A)')' = A$. To this end, let $\mathscr{F}$ be the subset of $\mathscr{P}(M)$

defined by

$$\mathscr{F} = \{S \in \mathscr{P}(M) \mid \phi(S) \subseteq S\}, \tag{3}$$

and let

$$A = \bigcap_{S \in \mathscr{F}} S. \tag{4}$$

If $S \in \mathscr{F}$, then we have $A \subseteq S$; and therefore, by (2), $\phi(A) \subseteq \phi(S) \subseteq S$. This shows that $\phi(A) \subseteq S$ holds for all $S \in \mathscr{F}$, and it follows that

$$\phi(A) \subseteq \bigcap_{S \in \mathscr{F}} S = A. \tag{5}$$

By (5), $\phi(A) \subseteq A$, and it follows from (2) that

$$\phi(\phi(A)) \subseteq \phi(A). \tag{6}$$

By (6) and (3), we have

$$\phi(A) \in \mathscr{F}. \tag{7}$$

As a consequence of (7) and (4),

$$A \subseteq \phi(A). \tag{8}$$

Combining (5) and (8), we conclude that

$$\phi(A) = A, \tag{9}$$

that is,

$$g(f(A)')' = A. \tag{10}$$

Now let

$$C = f(A), \quad D = f(A)', \quad \text{and} \quad B = A'. \tag{11}$$

By Equations (10) and (11), we have $g(D)' = A$; so $g(D) = A'$, that is,

$$g(D) = B. \tag{12}$$

Note that

$$M = A \cup B, \quad A \cap B = \varnothing, \quad N = C \cup D, \quad \text{and} \quad C \cap D = \varnothing. \tag{13}$$

Let $f|_A$ denote the restriction of f to A, and let $g|_D$ denote the restriction of g to D (see Definition 8.5 in Chapter 2). Then, by Equations (11) and (12),

$$f|_A : A \to C \quad \text{and} \quad g|_D : D \to B \tag{14}$$

are subjections. Furthermore, since $f: M \to N$ and $g: N \to M$ are injections, so are $f|_A$ and $g|_D$, and it follows that the mappings $f|_A$ and $g|_D$ in (14) are bijections. Consequently,

$$\# A = \# C \quad \text{and} \quad \# D = \# B. \tag{15}$$

Therefore, by Equations (13) and (15) and Lemma 7.3, we have

$$\#M = \#N. \tag{16}$$

□

7.6 Example Show that $\#\mathbb{Q} = \aleph_0$.

SOLUTION By Problem 25 in Problem Set 4.6, $\#\mathbb{Z} = \aleph_0$; so by Problem 21 in Problem Set 4.6, $\#(\mathbb{Z} \times \mathbb{Z}) = \aleph_0$. The mapping

$$f:\mathbb{Q} \to \mathbb{Z} \times \mathbb{Z}$$

defined for each $r \in \mathbb{Q}$ by $f(r) = (m, n)$, where $r = m/n$ in reduced form, is an injection. (See Lemma 4.5 and Definition 4.6.) Therefore,

$$\#\mathbb{Q} \le \aleph_0.$$

Also, the mapping $g:\mathbb{N} \to \mathbb{Q}$ defined by $g(n) = n$ for all $n \in \mathbb{N}$ is an injection; so

$$\aleph_0 \le \#\mathbb{Q}.$$

Since we have both $\#\mathbb{Q} \le \aleph_0$ and $\aleph_0 \le \#\mathbb{Q}$, it follows from the Cantor–Schroeder–Bernstein theorem (Theorem 7.5) that

$$\#\mathbb{Q} = \aleph_0.$$

□

In Example 7.6, we showed that $\aleph_0 \le \#\mathbb{Q}$ by exhibiting an injection $g:\mathbb{N} \to \mathbb{Q}$. More generally, if X is a set, then (as mentioned earlier) the condition $\aleph_0 \le \#X$ is equivalent to the existence of an injection $s:\mathbb{N} \to X$. A mapping (injective or not)

$$s:\mathbb{N} \to X$$

is called a **sequence** in X; and, for $n \in \mathbb{N}$, $s(n)$ is called the ***n*th term** of the sequence. Often, the nth term $s(n)$ of the sequence $s:\mathbb{N} \to X$ is written in the alternative form $s_n = s(n)$; and it is imagined that these terms are arranged in an endless progression,

$$s_1, s_2, s_3, \ldots.$$

Although this helps us to think about sequences, it is essential to keep in mind that a sequence is really a mapping $s:\mathbb{N} \to X$.

We can define a particular sequence $s:\mathbb{N} \to X$ by giving a formula, rule, or prescription that determines the value of $s_n = s(n)$ for every $n \in \mathbb{N}$. If this is done by means of a formula, we say that we have a description of the sequence in **closed form**. For instance,

$$s_n = \frac{1}{\pi}\int_{-\pi}^{\pi} x^2 \sin nx \, dx$$

would provide a closed-form description of a sequence

$$s:\mathbb{N} \to \mathbb{R}.$$

Another important technique for constructing a sequence

$$s_1, s_2, s_3, \ldots$$

is by **recursion**, that is, by specifying the first k terms $s_1, s_2, s_3, \ldots, s_k$ for some $k \in \mathbb{N}$ and giving a rule whereby, for each $n \in \mathbb{N}$ with $n \geq k$, s_{n+1} is determined by the preceding terms $s_1, s_2, s_3, \ldots, s_k, \ldots, s_n$. The case in which $k = 1$ and a rule is given that determines s_{n+1} from s_n for each $n \geq 1$ is called **simple recursion**.

In constructing a sequence by simple recursion, we often find that the rule for determining s_{n+1} from s_n is of the form

$$s_{n+1} = g(s_n),$$

where $g:X \to X$ is a specified mapping. (See Example 7.10 below.) In what follows, we propose to show that this method of simple recursion really works and produces a unique sequence $s:\mathbb{N} \to X$ depending only on the first term s_1 and the mapping $g:X \to X$ used to specify $s_{n+1} = g(s_n)$ in terms of s_n (Theorem 7.9 below). We begin with two technical lemmas. Recall that, for $n \in \mathbb{N}$, I_n denotes the corresponding initial segment (Definition 6.1).

7.7 LEMMA

Let X be a set, let $a \in X$, and let $g:X \to X$. Then, for each $n \in \mathbb{N}$, there exists a unique mapping $f_n:I_n \to X$ such that:

(i) $f_n(1) = a$.
(ii) For all $j \in \mathbb{N}$, $j + 1 \leq n \Rightarrow f_n(j + 1) = g(f_n(j))$.

PROOF We prove the existence of $f_n:I_n \to X$ satisfying (i) and (ii) for each $n \in \mathbb{N}$, and leave the uniqueness of f_n as an exercise (Problem 6). The existence proof is by induction on n.

For $n = 1$, the mapping $f_1:\{1\} \to X$ given by $f_1(1) = a$ satisfies (i), and it satisfies (ii) because the hypothesis $j + 1 \leq 1$ of (ii) is false for all $j \in \mathbb{N}$.

Thus, we assume (as our induction hypothesis) that, for some arbitrary but fixed $n \in \mathbb{N}$, there exists a mapping

$$f_n:I_n \to X$$

that satisfies (i) and (ii). Hence,

$$f_n(1) = a, \tag{1}$$

and, for all $j \in \mathbb{N}$,

$$j + 1 \leq n \Rightarrow f_n(j + 1) = g(f_n(j)). \tag{2}$$

We must find a mapping

$$f_{n+1}: I_{n+1} \to X$$

such that the following two conditions hold:

$$f_{n+1}(1) = a \tag{1*}$$

and, for all $j \in \mathbb{N}$,

$$j + 1 \le n + 1 \Rightarrow f_{n+1}(j + 1) = g(f_{n+1}(j)). \tag{2*}$$

By Lemma 6.3, we have

$$I_{n+1} = I_n \cup \{n + 1\}. \tag{3}$$

Therefore, we can define $f_{n+1}: I_{n+1} \to X$ by

$$f_{n+1}(j) = \left.\begin{cases} f_n(j), & \text{if } j \in I_n \\ g(f_n(n)), & \text{if } j = n + 1 \end{cases}\right\} \text{ for all } j \in I_{n+1}. \tag{4}$$

Since $1 \in I_n$, (4) and (1) imply that

$$f_{n+1}(1) = f_n(1) = a, \tag{5}$$

and so f_{n+1} satisfies Equation (1*).

To complete our induction argument, we must prove that f_{n+1} satisfies (2*). Thus, supposing that $j \in \mathbb{N}$ with

$$j + 1 \le n + 1, \tag{6}$$

we must show that the conclusion

$$f_{n+1}(j + 1) = g(f_{n+1}(j)) \tag{*}$$

of the implication in (2*) holds. From Inequality (6), we have

$$j \le n. \tag{7}$$

Thus, $j \in I_n$, and it follows from (4) that

$$f_{n+1}(j) = f_n(j). \tag{8}$$

Likewise, because $n \in I_n$, (4) implies that

$$f_{n+1}(n) = f_n(n). \tag{9}$$

By (4) and (9), we have

$$f_{n+1}(n + 1) = g(f_n(n)) = g(f_{n+1}(n)). \tag{10}$$

By Inequality (7), either $j = n$ or else $j < n$. If $j = n$, then (10) implies that

$$f_{n+1}(j + 1) = f_{n+1}(n + 1) = g(f_{n+1}(n)) = g(f_{n+1}(j)),$$

and so Equation (*) holds in this case.

Thus, we have only to settle the remaining case in which $j < n$. In this case, Theorem 2.7 implies that $j + 1 \leq n$, and it follows that

$$j + 1 \in I_n. \tag{11}$$

By (11) and (4), we have

$$f_{n+1}(j + 1) = f_n(j + 1). \tag{12}$$

Because $j + 1 \leq n$, (2) implies that

$$f_n(j + 1) = g(f_n(j)). \tag{13}$$

Combining Equations (12) and (13), we find that

$$f_{n+1}(j + 1) = g(f_n(j)). \tag{14}$$

Combining Equations (14) and (8), we conclude that

$$f_{n+1}(j + 1) = g(f_{n+1}(j)), \tag{15}$$

and hence, that Equation (*) also holds when $j < n$. □

7.8 LEMMA Let X be a set, let $a \in X$, and let $g:X \to X$. For each $n \in \mathbb{N}$, let $f_n:I_n \to X$ be the unique mapping satisfying Conditions (i) and (ii) of Lemma 7.7. Then, for every $n \in \mathbb{N}$,

$$f_{n+1}(n) = f_n(n).$$

PROOF Let $n \in \mathbb{N}$, and consider $f_{n+1}:I_{n+1} \to X$. Note that $I_n \subseteq I_{n+1}$, and observe that the restriction of f_{n+1} to I_n satisfies Conditions (i) and (ii) of Lemma 7.7. By the uniqueness of f_n, it follows that f_n is the restriction to I_n of f_{n+1}. Therefore, for every $j \leq n$,

$$f_n(j) = f_{n+1}(j).$$

Taking $j = n$ in the last equation, we find that

$$f_n(n) = f_{n+1}(n).$$

□

7.9 THEOREM **SIMPLE RECURSION** Let X be a set, let $a \in X$, and let $g:X \to X$. Then there exists a unique mapping $s:\mathbb{N} \to X$ such that

(i) $s(1) = a$, and
(ii) for every $n \in \mathbb{N}$, $s(n + 1) = g(s(n))$.

PROOF Again, we prove the existence of a mapping

$$s:\mathbb{N} \to X$$

satisfying (i) and (ii) and leave the proof of the uniqueness of s as an exercise (Problem 8). For each $n \in \mathbb{N}$, let

$$f_n : I_n \to X$$

be the unique mapping satisfying Conditions (i) and (ii) of Lemma 7.7. Define $s : \mathbb{N} \to X$ by

$$s(n) = f_n(n)$$

for every $n \in \mathbb{N}$. Then,

$$s(1) = f_1(1) = a,$$

and so Condition (i) of the present theorem is satisfied. Now, let $n \in \mathbb{N}$. To prove Condition (ii) of the present theorem, we begin by noting that, according to Lemma 7.8,

$$s(n) = f_n(n) = f_{n+1}(n).$$

Also, by Part (ii) of Lemma 7.7,

$$f_{n+1}(n + 1) = g(f_{n+1}(n)).$$

Thus,

$$s(n + 1) = f_{n+1}(n + 1) = g(f_{n+1}(n)) = g(s(n)).$$

□

7.10 Example Let $g : \mathbb{R} \to \mathbb{R}$ be defined by $g(x) = 2x$. Define the sequence $s : \mathbb{N} \to \mathbb{R}$ by simple recursion as follows:

$$s_1 = 2 \quad \text{and} \quad s_{n+1} = g(s_n), \quad \text{for all } n \in \mathbb{N}.$$

Find a description in closed form for this sequence.

SOLUTION Let $s : \mathbb{N} \to \mathbb{R}$ be defined by $s(n) = 2^n$ for all $n \in \mathbb{N}$. Then, $s(1) = 2^1 = 2$ holds. Also, for all $n \in \mathbb{N}$,

$$s(n + 1) = 2^{n+1} = 2 \cdot 2^n = g(2^n) = g(s(n)),$$

and it follows from the uniqueness part of Theorem 7.9 that

$$s_n = s(n) = 2^n, \quad \text{for all } n \in \mathbb{N}.$$

□

7.11 Example Let P denote the set of all prime natural numbers. Show that $\#P = \aleph_0$.

SOLUTION Given any prime number q, it can be shown that there exists a unique prime number q' such that q' is the smallest prime number greater than q (Problem 9). We refer to q' as the *next prime after* q. Let $g : P \to P$ be defined by

$$g(q) = \text{the next prime after } q,$$

for each $q \in P$. Define a sequence $p : \mathbb{N} \to P$ by simple recursion as follows:

$$p_1 = 2 \quad \text{and} \quad p_{n+1} = g(p_n), \quad \text{for all } n \in \mathbb{N}.$$

By Theorem 7.9, there is a unique sequence $p:\mathbb{N} \to P$ satisfying these conditions. Furthermore, since $p_n < p_{n+1}$ holds for all n, it follows that $p:\mathbb{N} \to P$ is an injection (Problem 10). Therefore, $\aleph_0 \le \#P$. Because the mapping $j:P \to \mathbb{N}$ defined by $j(q) = q$ for all $q \in P$ is an injection, we also have $\#P \le \aleph_0$; so, by Theorem 7.5, $\#P = \aleph_0$. □

The sequence $p:\mathbb{N} \to P$ of Example 7.11 is not only an injective mapping, it is also surjective (Problem 13); that is, all of the prime numbers appear in the corresponding endless progression

$$2, 3, 5, 7, 11, 13, 17, 19, 23, 29, 31, 37, \ldots.$$

It would be of the highest interest to find a description of this sequence in closed form, that is, a tractable formula giving the nth prime number p_n in terms of n. Unfortunately, no one has yet found such a formula.

The rest of our work in this section will make use of the *axiom of choice*. Roughly speaking, this axiom stipulates that, if $\mathscr{F}$ is a nonempty family of nonempty sets, you can choose an element from each set in the family. More accurately, it says that there is a function γ that makes this choice for you in that it assigns to each set $M \in \mathscr{F}$ an element $\gamma(M)$ in M. The axiom of choice is one of the commonly accepted axioms of set theory. To prove Lemma 7.13 and Theorem 7.14 below, we need to assume this axiom.

7.12 AXIOM

THE AXIOM OF CHOICE Let $\mathscr{F}$ be a nonempty set of nonempty sets, and let

$$X = \bigcup_{M \in \mathscr{F}} M.$$

Then there exists a function $\gamma:\mathscr{F} \to X$ such that, for every $M \in \mathscr{F}$,

$$\gamma(M) \in M.$$

A function $\gamma:\mathscr{F} \to X$ such that $\gamma(M) \in M$ for all $M \in \mathscr{F}$ is called a **choice function** for the set of sets $\mathscr{F}$.

7.13 LEMMA

Let Y be an infinite set, and let $\mathscr{P}(Y)$ denote the set of all subsets of Y. Then, there exists a mapping $S:\mathbb{N} \to \mathscr{P}(Y)$ such that the following conditions hold:

(i) $S(1) = \varnothing$.
(ii) For all $n \in \mathbb{N}$, $S(n) \subseteq S(n+1)$.
(iii) For all $n \in \mathbb{N}$, $S(n)$ is a finite subset of Y.
(iv) For all $n \in \mathbb{N}$, $\#S(n+1) = \#S(n) + 1$.

PROOF Let $\mathscr{F}$ be the set of all nonempty subsets of Y. By Axiom 7.12, there exists a choice function $\gamma:\mathscr{F} \to Y$ for $\mathscr{F}$. Thus, for every nonempty subset M of Y,

$$\gamma(M) \in M.$$

Define $G:\mathscr{P}(Y) \to \mathscr{P}(Y)$ by

$$G(M) = \begin{cases} M \cup \{\gamma(Y \setminus M)\}, & \text{if } M \neq Y \\ M, & \text{if } M = Y \end{cases}.$$

Note that

$$M \subseteq G(M)$$

holds for all $M \in \mathscr{P}(X)$. Define $S:\mathbb{N} \to \mathscr{P}(Y)$ by simple recursion (Theorem 7.9) in such a way that

$$S(1) = \varnothing$$

and, for every $n \in \mathbb{N}$,

$$S(n + 1) = G(S(n)).$$

Thus, we have (i) and (ii). We leave it as an exercise for you to prove by induction on n that (iii) holds.

To prove (iv), let $n \in \mathbb{N}$. By (iii), $S(n)$ is a finite subset of Y. Therefore, since Y is an infinite set, $S(n)$ is a proper subset of Y. Let $a = \gamma(Y \setminus S(n))$, so that $a \in Y$ and $a \notin S(n)$. Thus,

$$S(n + 1) = G(S(n)) = S(n) \cup \{y\},$$

and it follows from Lemma 6.12 that

$$\#S(n + 1) = \#S(n) + 1.$$ □

7.14 THEOREM **$\aleph_0$ IS THE SMALLEST TRANSFINITE CARDINAL** If Y is an infinite set, then $\aleph_0 \leq \#Y$.

PROOF Let $S: \mathbb{N} \to \mathscr{P}(Y)$ be a mapping having properties (i)–(iv) of Lemma 7.13. Then, for each $n \in \mathbb{N}$,

$$\#(S(n + 1) \setminus S(n)) = 1; \tag{1}$$

and it follows that there exists $s_n \in Y$ such that

$$S(n + 1) \setminus S(n) = \{s_n\}. \tag{2}$$

Define $s:\mathbb{N} \to Y$ by

$$s(n) = s_n \tag{3}$$

for all $n \in \mathbb{N}$. By Equations (2) and (3), we have

$$s(n) \in S(n+1) \tag{4}$$

and

$$s(n) \notin S(n) \tag{5}$$

for all $n \in \mathbb{N}$.

We complete the proof of the theorem by showing that $s:\mathbb{N} \to Y$ is an injection. An induction argument, which we leave as an exercise (Problem 15), shows that, for $n, k \in \mathbb{N}$,

$$k \le n \Rightarrow S(k) \subseteq S(n). \tag{6}$$

Suppose that $n, m \in \mathbb{N}$ with $m \ne n$. We must prove that $s(m) \ne s(n)$. Without loss of generality, we can assume that $m < n$. By Theorem 2.7, $m + 1 \le n$; therefore, as a consequence of (6), we have

$$S(m+1) \subseteq S(n). \tag{7}$$

By (4), $s(m) \in S(m+1)$; so by (7),

$$s(m) \in S(n). \tag{8}$$

But, by (5), $s(n) \notin S(n)$; and therefore, $s(m) \ne s(n)$. □

7.15 THEOREM

SUBSETS OF A COUNTABLE SET If X and Y are sets, $X \subseteq Y$, and Y is countable, then X is countable.

PROOF Since Y is countable, it is either finite or countably infinite. If Y is finite, then X is finite, hence countable, by Corollary 6.11. Thus, we can suppose that Y is countably infinite. Because $X \subseteq Y$, it follows that

$$\#X \le \#Y = \aleph_0 \tag{1}$$

(Problem 5). If X is finite, then it is countable, and we are done. Therefore, we can assume that X is infinite. Hence, by Theorem 7.14,

$$\aleph_0 \le \#X. \tag{2}$$

By (1), (2), and Theorem 7.5, we have $\#X = \aleph_0$; that is, X is countably infinite. □

7.16 COROLLARY

The set $\mathbb{R}$ of all real numbers is uncountable.

PROOF By Theorem 6.19, $\mathbb{R}$ contains an uncountable set, namely, $[0, 1] = \{x \in \mathbb{R} \mid 0 \le x \le 1\}$. Therefore, by Theorem 7.15, $\mathbb{R}$ cannot be a countable set. □

If M and N are two sets with $\#M \le \#N$ and $\#M \ne \#N$, then we write

$$\#M < \#N.$$

For instance, the result of Corollary 7.16 can be written as

$$\aleph_0 < \#\mathbb{R}.$$

Using the axiom of choice (Axiom 7.12), it can be shown that, if M and N are any two sets, then either $\#M \le \#N$ or $\#N < \#M$.

PROBLEM SET 4.7

1. Complete the proof of Lemma 7.3 by showing that $h:M \to N$ is a bijection.
2. If X is any set, $\mathscr{P}(X)$ is the set of all subsets of X, and $\phi:\mathscr{P}(X) \to \mathscr{P}(X)$ is a mapping such that, for $U, V \in \mathscr{P}(X)$, $U \subseteq V \Rightarrow \phi(U) \subseteq \phi(V)$, prove that there exists $A \in \mathscr{P}(X)$ such that $\phi(A) = A$. [Hint: Argue as in the proof of Theorem 7.5.]
3. Let M, N, and P be sets. Using Definitions 7.1 and 7.4, prove the following:
   (a) $\#M = \#M$ (b) $\#M = \#N \Rightarrow \#N = \#M$
   (c) $\#M = \#N$ and $\#N = \#P \Rightarrow \#M = \#P$
   (d) $\#M = \#N \Rightarrow \#M \le \#N$
4. If M, N, and P are sets with $\#M \le \#N$ and $\#N \le \#P$, prove that $\#M \le \#P$.
5. If M and N are sets and $M \subseteq N$, prove that $\#M \le \#N$. In particular, conclude that $\#M \le \#M$.
6. Prove that the mapping $f_n:I_n \to X$ in Lemma 7.7 is uniquely determined by Conditions (i) and (ii). [Hint: Suppose there were a second function $h:I_n \to X$ satisfying (i) and (ii). Fix $n \in \mathbb{N}$, and use induction on $m \in \mathbb{N}$ to prove that $m \in I_n \Rightarrow f_n(m) = h(m)$.]
7. Define a sequence $s:\mathbb{N} \to \mathbb{N}$ by simple recursion as follows: $s_1 = 1$ and $s_{n+1} = s_n + n$ for all $n \in \mathbb{N}$.
   (a) Write the first ten terms of this sequence.
   (b) Find a formula for this sequence in closed form.
8. Prove that the mapping $s:\mathbb{N} \to X$ in Theorem 7.9 is uniquely determined by Conditions (i) and (ii).

9. Let P be the set of all prime natural numbers.
(a) Prove that P is an infinite subset of $\mathbb{N}$. [Hint: If P were finite, with $\# P = n$, then we could write $P = \{p_1, p_2, p_3, \ldots, p_n\}$. Show that no p in P divides $k = p_1 p_2 p_3 \cdots p_n + 1$ and thus derive a contradiction.]
(b) Conclude that, for each $q \in P$, there exists a unique $q' \in P$ such that q' is the smallest prime number greater than q. [Hint: Use Theorem 2.14.]

10. Suppose that $s: \mathbb{N} \to \mathbb{R}$ is a sequence such that, for all $n \in \mathbb{N}$, $s_n < s_{n+1}$.
(a) Prove that, if $m, n \in \mathbb{N}$ with $m < n$, then $s_m < s_n$.
(b) Conclude that $s: \mathbb{N} \to \mathbb{R}$ is an injection.

11. Let X be any infinite subset of $\mathbb{N}$. Show that there exists a bijection $s: \mathbb{N} \to X$ such that, for all $n \in \mathbb{N}$, $s(n) < s(n+1)$. [Hint: Show that, for every $x \in X$, there exists a unique $x' \in X$ such that x' is the smallest element of X that is greater than x, and proceed as in Example 7.11.]

12. Let (N, a, s) be a Peano model (see Problem 19 in Problem Set 4.2). Show that there exists a unique bijection $f: \mathbb{N} \to N$ such that $f(1) = a$ and, for all $n \in \mathbb{N}$, $s(f(n)) = f(n+1)$.

13. Prove that the mapping $p: \mathbb{N} \to P$ constructed in Example 7.11 is a surjection.

14. If (N, a, s) is a Peano model, show that $\# N = \aleph_0$.

15. In the proof of Theorem 7.14, prove that, for $n, k \in \mathbb{N}$, $k \le n \Rightarrow S(k) \subseteq S(n)$.

16. Let X be a set, let $a, b \in X$, and let $g: X \times X \to X$. Prove that there exists a unique sequence $s: \mathbb{N} \to X$ such that $s_1 = a$, $s_2 = b$, and, for all $n \in \mathbb{N}$ with $n > 1$, $s_{n+1} = g(s_n, s_{n-1})$.

17. Using the axiom of choice, show that, if $f: X \to Y$ is a surjection, then there exists an injection $g: Y \to X$ such that, for all $y \in Y$, $f(g(y)) = y$. [Hint: For each $y \in Y$, let $M_y = \{x \in X \mid f(x) = y\}$ and let $\mathscr{F} = \{M_y \mid y \in Y\}$.]

18. Let $[0, 1) = \{x \in \mathbb{R} \mid 0 \le x < 1\}$ and $[0, \infty) = \{x \in \mathbb{R} \mid 0 \le x\}$. Show that $\#[0, 1) = \#[0, \infty)$. [Hint: Consider the mapping $f: [0, 1) \to [0, \infty)$ given by $f(x) = x/(1 - x)$.]

19. Assuming the axiom of choice, use the result of Problem 17 to show that $\# Y \le \# X$ if and only if there exists a surjection $f: X \to Y$.

20. Let $(-1, 0) = \{x \in \mathbb{R} \mid -1 < x < 0\}$ and $(-\infty, 0) = \{x \in \mathbb{R} \mid x < 0\}$. Show that $\#(-1, 0) = \#(-\infty, 0)$. [Hint: See Problem 18.]

21. Let $M \cap N = \varnothing$ with $\# M = \aleph_0$.
(a) If $\# N = 1$, prove that $\#(M \cup N) = \aleph_0$.
(b) If N is a finite set, prove that $\#(M \cup N) = \aleph_0$.

22. Let $(-1, 1) = \{x \in \mathbb{R} \mid -1 < x < 1\}$. By combining the results of Problems 18 and 20, show that $\#(-1, 1) = \#\mathbb{R}$.

23. Let $A = \{n \in \mathbb{N} \mid n \text{ is even}\}$ and let $B = \{n \in \mathbb{N} \mid n \text{ is odd}\}$. Prove that $\#A = \#B = \aleph_0$.

24. Let $a, b \in \mathbb{R}$ with $a < b$, and let $(a, b) = \{x \in \mathbb{R} \mid a < x < b\}$. Find a bijection $f:(-1, 1) \to (a, b)$ and thus prove that $\#(-1, 1) = \#(a, b)$. Then, use the result of Problem 22 to conclude that $\#(a, b) = \#\mathbb{R}$.

25. If $M \cap N = \varnothing$, $\#M = \#N = \aleph_0$, prove that $\#(M \cup N) = \aleph_0$. [Hint: Use the result of Problem 23.]

26. It is a fact that every real number x in the closed unit interval $[0, 1]$ can be written in the form

$$x = \sum_{k=1}^{\infty} \frac{\delta_k}{2^k},$$

where $\delta_k = 0$ or $\delta_k = 1$ for each $k \in \mathbb{N}$. This is called the **dyadic** expansion of x, and δ_k is called the ***k*th digit** in the dyadic expansion of x. Conversely, every sequence $\delta: \mathbb{N} \to \{0, 1\}$ determines a unique real number $x \in [0, 1]$ such that $\delta_k = \delta(k)$ is the kth digit in the dyadic expansion of x for every $k \in \mathbb{N}$. Use this fact to set up a surjection $f: \mathscr{P}(\mathbb{N}) \to [0, 1]$, where $\mathscr{P}(\mathbb{N})$ is the set of all subsets of $\mathbb{N}$. [Hint: A subset M of $\mathbb{N}$ determines a sequence $\delta: \mathbb{N} \to \{0, 1\}$ such that $\delta(k)$ is 1 or 0 according to whether k belongs or does not belong to M, respectively, for all $k \in \mathbb{N}$.]

27. If $M \cap N = \varnothing$ and M and N are countable sets, prove that $M \cup N$ is a countable set.

28. Combine the results of Problems 19 and 26 to show that, assuming the axiom of choice, $\#[0, 1] \leq \#\mathscr{P}(\mathbb{N})$.

29. If $A \cap B = \varnothing$, A is an infinite set, and B is a countable set, prove that $\#(A \cup B) = \#A$. [Hint: By Theorem 7.14, there exists a countable set C with $C \subseteq A$. Let $D = A \setminus C$ and note that $A \cup B = D \cup (C \cup B)$ with $D \cap (C \cup B) = \varnothing$ and $A = C \cup D$ with $C \cap D = \varnothing$. By Problem 27, $C \cup B$ is a countable set; hence, $\#(C \cup B) = \#C$. Now use Lemma 7.3.]

30. Obtain an injection $g: \mathscr{P}(\mathbb{N}) \to [0, 1]$ by using an idea similar to that in Problem 26, but using decimal expansions rather than dyadic expansions. Conclude that $\#\mathscr{P}(\mathbb{N}) \leq \#[0, 1]$.

31. Show that $\#([0, 1] \times [0, 1]) = \#[0, 1]$. [Hint: Consider "intertwined" decimal expansions; that is, for $x, y \in [0, 1]$, let $z \in [0, 1]$ be the number whose decimal expansion is obtained by alternating the digits in the decimal expansions of x and y.]

32. If M is a set and $\#M = \#\mathbb{R}$, then M is said to have the **power of the continuum.** ($\#M$ is sometimes called the **power** of M rather than the cardinal number of M.) Show that $\#\mathscr{P}(\mathbb{N})$ has the power of the continuum.

[Hint: Combine Problems 28 and 30 to obtain $\#\mathscr{P}(\mathbb{N}) = \#[0, 1]$. Combine Problems 24 and 29 to obtain $\#\mathbb{R} = \#(0, 1) = \#[0, 1]$.]

33. Show that $\#\mathbb{C} = \#\mathbb{R}$.

34. Prove the following generalization of Theorem 7.9: Let X be a set, let $a \in X$, and let $g:\mathbb{N} \times X \to X$. Then there exists a unique mapping $s:\mathbb{N} \to X$ such that (i) $s(1) = a$ and (ii) $s(n + 1) = g(n, s(n))$.

35. Show that $\#(\mathbb{N} \times \mathbb{R}) = \#\mathbb{R}$.

36. Let $X = \{x \in \mathbb{Q} \mid x > 0\}$, and let $s:\mathbb{N} \to X$ be defined recursively by $s(1) = 1$, $s(n) = 1 + s(n/2)$ if n is even, and by $s(n) = 1/s(n - 1)$ if n is odd. Show that $s:\mathbb{N} \to X$ is a bijection.

*8 NUMBER SYSTEMS: CLASSICAL AND CONTEMPORARY

We might regard the number systems

$$\mathbb{N} \subseteq \mathbb{Z} \subseteq \mathbb{Q} \subseteq \mathbb{R} \subseteq \mathbb{C}$$

as the *classical number systems* of mathematics. The systems $\mathbb{Q}$, $\mathbb{R}$, and $\mathbb{C}$ are fields, $\mathbb{Z}$ is an integral domain, and $(\mathbb{N}, 1, s)$ forms a Peano model (Problem 19 in Problem Set 4.2) in that $1 \in \mathbb{N}$ and $s:\mathbb{N} \to \mathbb{N}$ given by $s(n) = n + 1$ for all $n \in \mathbb{N}$ satisfies the following conditions:

(i) $s:\mathbb{N} \to \mathbb{N}$ is an injection.
(ii) $(\forall x)(x \in \mathbb{N} \Rightarrow 1 \neq s(x))$.
(iii) $1 \in M \subseteq \mathbb{N} \wedge (\forall x)(x \in M \Rightarrow s(x) \in M) \Rightarrow M = \mathbb{N}$.

The system $\mathbb{Q}$ forms an Archimedean ordered field (Problem 12 in Problem Set 4.4), but it is not a complete ordered field. The system $\mathbb{R}$ is a complete and Archimedean ordered field, but it is not **algebraically complete**; that is, there are polynomials $p(x)$ of degree $n > 1$ with coefficients in $\mathbb{R}$ that have no zeros in $\mathbb{R}$. The field $\mathbb{C}$ is algebraically complete (Theorem 5.22), but it cannot be made into an ordered field (Example 4.6 in Chapter 3); so it makes no sense to ask whether $\mathbb{C}$ is complete in the sense of Definition 1.1.

The absolute-value function $z \mapsto |z|$, an extension of the absolute-value function in the ordered field $\mathbb{R}$, is defined on the field $\mathbb{C}$ (Definition 5.18). The function $z \mapsto |z|$ is used to introduce a distance measure, or **metric**, on either $\mathbb{R}$ or $\mathbb{C}$ by means of the formula

$$\rho(s, t) = |s - t|,$$

which gives the *distance* between the numbers s and t (Problem 16 in Problem Set 4.5). Using this metric, we can define the concept of a **limit**:

$\lim_{x \to a} f(x) = L$ *means that, for every positive real number* ε, *there exists a positive real number* δ *such that*

$$0 < \rho(x, a) < \delta \Rightarrow \rho(f(x), f(a)) < \varepsilon.$$

The limit concept enables us to do **analysis** in the system $\mathbb{C}$, as well as in its subsystem $\mathbb{R}$; that is, we can introduce and study the important ideas of *continuity*, the *derivative*, the *integral*, and *infinite series.*

All of our work in the present chapter has been based on the assumption that a complete ordered field $\mathbb{R}$ exists (Axiom 1.2). This axiom can be dispensed with if one is willing to indulge in a somewhat protracted construction process in which the complete ordered field $\mathbb{R}$ is actually built up, step by step, from a purely set-theoretic foundation. We devote the next few pages to a sketch, *without proofs*, of this construction process.

For convenience, we arrange the pertinent information in a definition–theorem format, even though we do not give the proofs. By scanning this material, you can gain some appreciation of the techniques by which $\mathbb{R}$ is constructed set-theoretically. (A worthwhile project for the more ambitious reader would be to supply the missing proofs.)

To begin with, it is necessary to have an adequate and rigorously developed theory of sets. This set theory must be sufficiently rich to imply the existence of a Peano model $(N, 1, s)$. For an exposition of a set-theoretic proof of the existence of a Peano model, see Paul R. Halmos, *Naive Set Theory* (New York: Springer-Verlag), 1974. Assume, then, that we have a Peano model $(N, 1, s)$.

8.1 THEOREM

ADDITION IN A PEANO MODEL There exists a unique binary operation $+$ on N such that, for all $m, n \in N$,

$$n + 1 = s(n)$$

and

$$n + s(m) = s(n + m).$$

8.2 THEOREM

MULTIPLICATION IN A PEANO MODEL There exists a unique binary operation $\cdot$ on N such that, for all $m, n \in N$,

$$n \cdot 1 = n$$

and

$$n \cdot s(m) = (n \cdot m) + n.$$

8.3 THEOREM

PROPERTIES OF + AND · IN A PEANO MODEL

(i) $+$ and $\cdot$ are associative and commutative.
(ii) The distributive law $m \cdot (n + k) = (m \cdot n) + (m \cdot k)$ holds for all $m, n, k \in N$.
(iii) The cancellation law $m + k = n + k \Rightarrow m = n$ holds for all $m, n, k \in N$.
(iv) The cancellation law $m \cdot k = n \cdot k \Rightarrow m = n$ holds for all $m, n, k \in N$.

8.4 DEFINITION

THE ORDER RELATION ON *N* For $m, n \in N$, define $m < n$ if and only if there exists $k \in N$ such that $m + k = n$. Define $m \leq n$ if and only if either $m < n$ or $m = n$.

8.5 THEOREM

PROPERTIES OF THE ORDER RELATION ON *N*

(i) N is well ordered by $\leq$.
(ii) For $m, n, k \in N$, $m \leq n \Leftrightarrow m + k \leq n + k$.
(iii) For $m, n, k \in N$, $m \leq n \Leftrightarrow m \cdot k \leq n \cdot k$.
(iv) If $n \in \mathbb{N}$, there exists no $k \in N$ with $n < k < n + 1$.

As a consequence of these results, the system N acquires all of the expected features of the system of natural numbers. Our next task is to use N to construct a system Z with all of the properties of the system of integers. The plan is to construct objects that behave algebraically like *differences* $m - n$ of elements $m, n \in N$, and the key idea is that

$$m - n = h - k \Leftrightarrow m + k = h + n$$

should hold for all $m, n, h, k \in N$. Thus, in what follows, think of an ordered pair (m, n) as representing a difference $m - n$ of the two natural numbers m and n.

8.6 DEFINITION

DIFFERENCE-EQUIVALENT PAIRS Let $X = N \times N$. If $(m, n), (h, k) \in X$, we say that (m, n) and (h, k) are **difference equivalent**, and we write $(m, n)\ E\ (h, k)$, if

$$m + k = h + n.$$

8.7 LEMMA

E is an equivalence relation on $X = N \times N$.

8.8 DEFINITION

THE SYSTEM Z For $(m, n) \in X = \mathbb{N} \times \mathbb{N}$, let $[m, n]$ denote the E-equivalence class determined by (m, n), so that

$$\begin{aligned}[m, n] &= \{(h, k) \in X \mid (m, n)\, E\, (h, k)\} \\ &= \{(h, k) \in X \mid m + k = h + n\}.\end{aligned}$$

Define $Z = X/E$, so that

$$\begin{aligned}Z &= \{[m, n] \mid (m, n) \in X\} \\ &= \{[m, n] \mid m, n \in N\}.\end{aligned}$$

8.9 THEOREM

ADDITION IN Z There exists a unique binary operation $+$ on Z such that, for all $[m, n], [h, k] \in Z$,

$$[m, n] + [h, k] = [m + h, n + k].$$

8.10 THEOREM

MULTIPLICATION IN Z There exists a unique binary operation $\cdot$ on Z such that, for all $[m, n], [h, k] \in Z$,

$$[m, n] \cdot [h, k] = [mh + nk, mk + nh].$$

8.11 THEOREM

Z IS AN INTEGRAL DOMAIN With the addition and multiplication operations of Theorems 8.9 and 8.10, the system $(Z, +, \cdot)$ forms an integral domain.

Note that $[1, 1]$ (intuitively, $1 - 1$) is effective as the zero element, and $[1 + 1, 1]$ (intuitively, $2 - 1$) is effective as the unity element of the integral domain Z.

8.12 THEOREM

THE ORDER RELATION ON Z There exists a unique binary relation $\leq$ on Z such that, for $[m, n], [h, k] \in Z$,

$$[m, n] \leq [h, k] \Leftrightarrow m + h \leq n + k.$$

In what follows, we use letters a, b, c, and so on to denote arbitrary elements of the integral domain Z. The zero element of Z is written as 0, and the unit element of Z is written as 1. Also, when convenient, we write ab for the product $a \cdot b$ of elements $a, b \in Z$.

8.13 THEOREM

PROPERTIES OF THE ORDER RELATION ON Z

(i) Z is totally ordered by $\leq$.
(ii) For $a, b, c \in Z$, $a \leq b \Leftrightarrow a + c \leq b + c$.
(iii) For $a, b, c \in Z$, $a \leq b$ and $0 \leq c \Rightarrow ac \leq bc$.

Next, we propose to use the ordered integral domain Z to construct a system Q with all of the properties of the system of rational numbers. The plan is to construct objects that behave algebraically like *quotients* a/b of elements $a, b \in Z$ (with $b \neq 0$, of course), and the key idea is that

$$\frac{a}{b} = \frac{c}{d} \Leftrightarrow ad = bc$$

should hold for all $a, b, c, d \in Z$ with $b, d \neq 0$. Thus, in what follows, think of an ordered pair (a, b) as representing a quotient a/b of the two integers a and b with $b \neq 0$.

8.14 DEFINITION

QUOTIENT-EQUIVALENT PAIRS Let $Y = Z \times (Z \setminus \{0\})$. Suppose that $(a, b), (c, d) \in Y$. We say that (a, b) and (c, d) are **quotient equivalent**, and we write $(a, b)\ F\ (c, d)$, if

$$ad = bc.$$

8.15 LEMMA

F is an equivalence relation on $Y = Z \times (Z \setminus \{0\})$.

8.16 DEFINITION

THE SYSTEM Q For $(a, b) \in Y = Z \times (Z \setminus \{0\})$, denote by $[a, b]$ the F-equivalence class determined by (a, b), so that

$$\begin{aligned} [a, b] &= \{(c, d) \in Y \mid (a, b)\ F\ (c, d)\} \\ &= \{(c, d) \in Y \mid ad = bc\}. \end{aligned}$$

Define $Q = Y/F$, so that

$$\begin{aligned} Q &= \{[a, b] \mid (a, b) \in Y\} \\ &= \{[a, b] \mid a, b \in Z \wedge b \neq 0\}. \end{aligned}$$

8.17 THEOREM

ADDITION IN Q There exists a unique binary operation $+$ on Q such that, for all $[a, b]$, $[c, d] \in Q$,

$$[a, b] + [c, d] = [ad + bc, bd].$$

8.18 THEOREM

MULTIPLICATION IN Q There exists a unique binary operation $\cdot$ on Q such that, for all $[a, b]$, $[c, d] \in Q$,

$$[a, b] \cdot [c, d] = [ac, bd].$$

8.19 THEOREM

Q IS A FIELD With the addition and multiplication operations of Theorems 8.17 and 8.18, the system $(Q, +, \cdot)$ forms a field.

Note that $[0, 1]$ (intuitively, $0/1$) is effective as the zero element, and $[1, 1]$ (intuitively, $1/1$) is effective as the unity element of the field Q.

8.20 DEFINITION

THE POSITIVE ELEMENTS IN Q Define P_Q, the set of **positive elements** in Q, by

$$P_Q = \{[a, b] \in Q \mid ab > 0 \text{ in } Z\}.$$

8.21 THEOREM

Q IS AN ORDERED FIELD With P_Q as its set of positive elements, Q forms an Archimedean ordered field.

In what follows, we use letters r, s, t, and so on to denote arbitrary elements of the ordered field Q. The zero element of Q is written as 0, and the unit element of Q is written as 1. Also, when convenient, we write rs for the product $r \cdot s$ of elements $r, s \in Q$. The Archimedean ordered field Q is not complete, and our next task is to construct from it an ordered field that is complete. The resulting complete ordered field is the field of real numbers. There are a number of standard ways to construct a *completion* of Q; we propose to use the **method of Dedekind cuts**, named in honor of the German mathematician Richard Dedekind (1831–1916). (See Problem 18 in Problem Set 4.1.)

8.22 DEFINITION

DEDEKIND CUT A **Dedekind cut** in the ordered field Q is a set $C \subseteq Q$ such that $C \neq \varnothing$, $C \neq Q$, and C contains all of the lower bounds in Q of its own set of upper bounds in Q.

In what follows, we refer to a Dedekind cut in Q simply as a *cut*.

8.23 DEFINITION

$\mathbb{R}$ We define $\mathbb{R}$, called the set of **real numbers**, to be the set of all cuts in Q.

8.24 DEFINITION

ADDITION AND NEGATION IN $\mathbb{R}$ If $C, D \in \mathbb{R}$, we define

$$C + D = \{r + s \mid r \in C \wedge s \in D\}.$$

We also define

$$-C = \{r \in Q \mid (\forall s)(s \in C \Rightarrow r \leq -s)\}.$$

8.25 THEOREM

($\mathbb{R}$, +) IS A COMMUTATIVE GROUP If $C \in \mathbb{R}$, then $-C \in \mathbb{R}$ and $(\mathbb{R}, +)$ is a commutative group with $-C$ as the additive inverse of C for every $C \in \mathbb{R}$. The additive neutral element of this group is given by

$$0 = \{r \in Q \mid r \leq 0\}.$$

Note that we propose to use the same symbol, 0, for the zero element in the field Q and for the zero cut $0 = \{r \in Q \mid r \leq 0\}$. This should not cause any confusion, since you can always tell from the context what is intended.

8.26 DEFINITION

POSITIVE AND NEGATIVE CUTS A cut $C \in \mathbb{R}$ is called **positive** if there exists $r \in C$ with $r > 0$ in Q. If $-C$ is positive, we say that C is **negative**.

8.27 THEOREM

LAW OF TRICHOTOMY If $C \in \mathbb{R}$, then one and only one of the following three conditions holds:

(i) C is positive.
(ii) $C = 0$.
(iii) C is negative.

8.28 DEFINITION

PRODUCT OF NONNEGATIVE CUTS Let $C, D \in \mathbb{R}$ be nonnegative cuts. Define

$$C \cdot D = \{rs \in Q \mid 0 \leq r \in C \wedge 0 \leq s \in D\} \cup \{t \in Q \mid t \leq 0\}.$$

8.29 LEMMA

If $C, D \in \mathbb{R}$ are nonnegative cuts, then $C \cdot D \in \mathbb{R}$ and $C \cdot D$ is a nonnegative cut. If C is a negative cut, then $-C$ is a positive cut. If 0 is the zero cut, then $-0 = 0$.

8.30 DEFINITION

PRODUCT OF CUTS Let C, D be cuts, at least one of which is negative. If C and D are both negative, we define

$$C \cdot D = (-C) \cdot (-D).$$

If C is negative and D is nonnegative, we define

$$C \cdot D = -((-C) \cdot D),$$

and, if C is nonnegative and D is negative, we define

$$C \cdot D = -(C \cdot (-D)).$$

8.31 THEOREM

$\mathbb{R}$ IS A COMPLETE ORDERED FIELD With the set P of all positive cuts as its set of positive elements, and with the operations of addition and multiplication in Definitions 8.24 and 8.30, $\mathbb{R}$ is a complete ordered field.

The classical number systems $\mathbb{N}$, $\mathbb{Z}$, $\mathbb{Q}$, $\mathbb{R}$, and $\mathbb{C}$ served well enough for the needs of mathematicians and natural scientists until about the middle of the nineteenth century. In 1843, the Irish mathematician and physicist William Rowan Hamilton (1805–1865) succeeded in constructing the system $(\mathbb{H}, +, \cdot)$ of *quaternions*, a new number system!

Hamilton was led to the construction of $\mathbb{H}$ as follows: We can represent a complex number $z = x + yi$ by the vector in the Cartesian plane that starts at the origin and terminates at the point (x, y). If $w = e^{i\theta}$ (Problem 20 in Problem Set 4.5), then the product wz is represented by the vector obtained by *rotating* the vector that represents z about the origin through the angle $|\theta|$, counterclockwise if $\theta > 0$ and clockwise if $\theta < 0$ (Problem 22 in Problem Set 4.5). In this way, the field $\mathbb{C}$ of complex numbers can be used as a tool to study rotations in the plane. Hamilton set for himself the task of finding a field that could be used analogously to study rotations in three-dimensional space.

Now, rotations in three-dimensional space do not commute under composition, whereas in a field, the multiplication is commutative. After fifteen years of frustrated effort, Hamilton suddenly realized that there is nothing sacred about the commutative law of multiplication and that, simply by dropping it, a self-consistent number system $\mathbb{H}$ can be constructed and three-dimensional rotations can thereby be represented. (There is evidence that Gauss may have anticipated the construction of such a noncommutative number system as early as 1817.)

Hamilton's system $\mathbb{H}$ can be constructed as follows: We consider vectors $(s, x, y, z) \in \mathbb{R}^4$, the coordinate four-dimensional vector space over the field $\mathbb{R}$. We introduce the special notation $\mathbf{l}$, $\mathbf{i}$, $\mathbf{j}$, and $\mathbf{k}$ for the four basis vectors:

$$\mathbf{l} = (1, 0, 0, 0),$$
$$\mathbf{i} = (0, 1, 0, 0),$$
$$\mathbf{j} = (0, 0, 1, 0),$$
$$\mathbf{k} = (0, 0, 0, 1).$$

Then, every vector in $\mathbb{R}^4$ can be written uniquely in the form

$$s\mathbf{l} + x\mathbf{i} + y\mathbf{j} + z\mathbf{k}.$$

These vectors, now called **quaternions**, are added and multiplied by real scalars in the usual way. Hamilton's idea was to introduce a multiplication operation on the quaternions by stipulating how the basis vectors $\mathbf{l}$, $\mathbf{i}$, $\mathbf{j}$, and $\mathbf{k}$ multiply among themselves. Of course, it is agreed that $\mathbf{l}$ multiplies in the usual way so that it serves as a multiplicative unit, that is, $\mathbf{l} \cdot \mathbf{l} = \mathbf{l}$, $\mathbf{l} \cdot \mathbf{i} = \mathbf{i}$, and so forth. The multiplication table for the remaining three **quaternionic units** $\mathbf{i}$, $\mathbf{j}$, and $\mathbf{k}$ is as follows:

$\cdot$	$\mathbf{i}$	$\mathbf{j}$	$\mathbf{k}$
$\mathbf{i}$	$-\mathbf{l}$	$\mathbf{k}$	$-\mathbf{j}$
$\mathbf{j}$	$-\mathbf{k}$	$-\mathbf{l}$	$\mathbf{i}$
$\mathbf{k}$	$\mathbf{j}$	$-\mathbf{i}$	$-\mathbf{l}$

Notice that this multiplication is *not commutative*; for instance, $\mathbf{ij} \neq \mathbf{ji}$, $\mathbf{ik} \neq \mathbf{ki}$, and $\mathbf{jk} \neq \mathbf{kj}$.

Since $\mathbf{l} = (1, 0, 0, 0)$ multiplies in the expected way, it is traditional to drop it from the notation and write the quaternion $s\mathbf{l} + x\mathbf{i} + y\mathbf{j} + z\mathbf{k}$ in the simpler form $s + x\mathbf{i} + y\mathbf{j} + z\mathbf{k}$. The set of all quaternions is denoted by $\mathbb{H}$, so that

$$\mathbb{H} = \{s + x\mathbf{i} + y\mathbf{j} + z\mathbf{k} \,|\, s, x, y, z \in \mathbb{R}\}.$$

Two quaternions are multiplied by using the distributive laws and the table above for the multiplication of the quaternionic units. The resulting algebraic system $(\mathbb{H}, +, \cdot)$ is a division ring (Definition 3.1 in Chapter 3); but the multiplication is not commutative, so it is not a field.

By using dot products $\mathbf{v} \cdot \mathbf{u}$ and cross products $\mathbf{v} \times \mathbf{u}$ of vectors $\mathbf{v}$ and $\mathbf{u}$ in three-dimensional (oriented) Euclidean space E^3, it is possible to simplify quaternionic computations. We begin by selecting a (right-hand oriented) basis $\mathbf{i}, \mathbf{j}, \mathbf{k}$ of pairwise orthogonal unit vectors in E^3; so every vector $\mathbf{v}$ in E^3 can be written uniquely in the form

$$\mathbf{v} = x\mathbf{i} + y\mathbf{j} + z\mathbf{k},$$

with $x, y, z \in \mathbb{R}$. In ordinary vector algebra, we are not permitted to add scalars and vectors; however, if s is a (real) scalar, let us *define* $s + \mathbf{v}$ to be the quaternion

$$s + \mathbf{v} = s + x\mathbf{i} + y\mathbf{j} + z\mathbf{k} \in \mathbb{H}.$$

We refer to s as the **real part** and to $\mathbf{v}$ as the **quaternionic part** of the quaternion $s + \mathbf{v}$. If $s = 0$, we call $s + \mathbf{v}$ a **pure quaternion**. Thus, pure quaternions can be identified with vectors in E^3.

With the notation introduced above, the quaternionic operations of addition and multiplication take the following forms:

$$(s + \mathbf{v}) + (t + \mathbf{u}) = (s + t) + (\mathbf{v} + \mathbf{u})$$

and

$$(s + \mathbf{v})(t + \mathbf{u}) = (st - \mathbf{v} \cdot \mathbf{u}) + (s\mathbf{u} + t\mathbf{v} + \mathbf{u} \times \mathbf{v})$$

for $s + \mathbf{v}, t + \mathbf{u} \in \mathbb{H}$.

The following theorem, whose proof we omit, shows just how Hamilton succeeded in representing three-dimensional rotations by using quaternions.

8.32 THEOREM

ROTATIONS IN E^3 Let $\mathbf{a}, \mathbf{w} \in E^3$ with $\|\mathbf{a}\| = 1$, and let $\theta \in \mathbb{R}$. Let $\mathbf{v} \in E^3$ be the vector that results when $\mathbf{w}$ is rotated through the angle θ about the vector $\mathbf{a}$ as an axis. Let $q \in \mathbb{H}$ be defined by

$$q = \cos(\theta/2) + \sin(\theta/2)\mathbf{a}.$$

Then, $\mathbf{v} = q\mathbf{w}q^{-1}$.

If we identify the complex number $a + bi$ with the quaternion $a + b\mathbf{i}$, it turns out that all of the algebraic operations on complex numbers are preserved; hence, we can regard $\mathbb{C}$ as a *subsystem* of $\mathbb{H}$. Hence, for the number systems discussed so far, we have

$$\mathbb{N} \subseteq \mathbb{Z} \subseteq \mathbb{Q} \subseteq \mathbb{R} \subseteq \mathbb{C} \subseteq \mathbb{H}.$$

How much further can we go? Are there division rings extending the quaternions, and then extensions of the extensions, and so forth? To answer this question, we begin by noting that the division rings $\mathbb{R}$, $\mathbb{C}$, and $\mathbb{H}$ are also *vector spaces* over the field $\mathbb{R}$ of dimensions 1, 2, and 4, respectively. This leads us to the following definition:

8.33 DEFINITION

LINEAR ASSOCIATIVE ALGEBRA Let F be a field. A **linear associative algebra over** F is a vector space A over F that is equipped with a binary operation of multiplication $(\mathbf{a}, \mathbf{b}) \mapsto \mathbf{a} \cdot \mathbf{b} = \mathbf{ab}$ in such a way that $(A, +, \cdot)$ is a ring; and, for all scalars $s, t \in F$ and all vectors $\mathbf{a}, \mathbf{b} \in A$,

$$s(t\mathbf{a}) = (st)\mathbf{a}$$

and

$$(s\mathbf{a})\mathbf{b} = s(\mathbf{ab}) = \mathbf{a}(s\mathbf{b}).$$

If A is a linear associative algebra over F, we understand that the *dimension* of A over F is the dimension of A, regarded as a vector space over the field F. A linear associative algebra over F that, as a ring, is a division ring, is called a **division algebra** over F. Thus, $\mathbb{R}$, $\mathbb{C}$, and $\mathbb{H}$ are division algebras of dimensions 1, 2, and 4, respectively, over $\mathbb{R}$. The answer to the question posed above, which was obtained independently by the American philosopher C. S. Peirce (1839–1914) and the German mathematician G. F. Frobenius (1849–1917) is stated in the following theorem:

8.34 THEOREM

PEIRCE–FROBENIUS THEOREM If A is a finite-dimensional division algebra over $\mathbb{R}$, then A is isomorphic to one of $\mathbb{R}$, $\mathbb{C}$, or $\mathbb{H}$.

Thus, as far as finite-dimensional division algebras over $\mathbb{R}$ go, $\mathbb{H}$ is the end of the line. (However, there are plenty of infinite-dimensional division algebras over $\mathbb{R}$.)

Hamilton's brilliant creation of a useful and self-consistent number system in which one of the field axioms (commutativity of multiplication) was violated opened the floodgates, and mathematicians began to manufacture a deluge of "number systems" that violated one or more of the field axioms. In 1858, the English mathematician Arthur Cayley (1821–1895) invented **matrix algebras**, which include Hamilton's quaternions as a special case. Even today, new kinds of algebraic systems are being created so rapidly that algebraists have difficulty sorting them out and deciding which ones are useful and which should be consigned to the mathematical scrap heap.

Among the nonclassical algebraic systems that have proved useful and interesting, certain **nonassociative linear algebras** are of special importance for both pure and applied mathematics. A **linear algebra** A over a field F is defined to have all the properties of a linear associative algebra in Definition 8.33, but the associative law

$$\mathbf{a}(\mathbf{b}\mathbf{c}) = (\mathbf{a}\mathbf{b})\mathbf{c}$$

for multiplication of vectors $\mathbf{a}$, $\mathbf{b}$, and $\mathbf{c}$ in A is dropped. (However, the conditions $s(t\mathbf{a}) = (st)\mathbf{a}$ and $(s\mathbf{a})\mathbf{b} = s(\mathbf{ab}) = \mathbf{a}(s\mathbf{b})$ in Definition 8.33 are retained.)

A linear algebra is called a **Lie algebra**, in honor of the Norwegian mathematician Sophus Lie (1842–1899) (pronounced *lee*) if it satisfies

$$\mathbf{aa} = \mathbf{0} \quad \text{and} \quad \mathbf{a}(\mathbf{bc}) + \mathbf{b}(\mathbf{ca}) + \mathbf{c}(\mathbf{ab}) = \mathbf{0}$$

for all $\mathbf{a}, \mathbf{b}, \mathbf{c} \in A$. (The second condition is known as the **Jacobi identity**.) A linear algebra is called a **Jordan algebra**, in honor of the German physicist Pascual Jordan (1902–1980) if it satisfies

$$\mathbf{ab} = \mathbf{ba} \quad \text{and} \quad (\mathbf{aa})(\mathbf{ba}) = ((\mathbf{aa})\mathbf{b})\mathbf{a}$$

for all $\mathbf{a}, \mathbf{b} \in A$.

PROBLEM SET 4.8

1. Prove Lemma 8.7.
2. Prove Theorem 8.9.
3. Prove Lemma 8.14.
4. Prove Theorem 8.17.
5. Let $(D, +, \cdot)$ be an integral domain, and let $Y = D \times (D \setminus \{0\})$. Define a relation F of *quotient equivalence* on Y by

   $$(a, b)\ F\ (c, d) \quad \text{if and only if} \quad ad = bc$$

   for $(a, b), (c, d) \in Y$. Prove that F is an equivalence relation on Y.
6. Prove Theorem 8.25.

7. Continuing with the notation of Problem 5, define for each $(a, b) \in Y$,

$$[a, b] = \{(c, d) \in Y \mid (a, b)\ F\ (c, d)\},$$

and let $\mathscr{F} = \{[a, b] \mid (a, b) \in Y\}$. Prove that there are unique binary operations $+$ and $\cdot$ on $\mathscr{F}$ such that

$$[a, b] + [c, d] = [ad + bc, bd]$$

and

$$[a, b] \cdot [c, d] = [ac, bd]$$

for all $[a, b], [c, d] \in \mathscr{F}$.

8. Prove Theorem 8.27.

9. Continuing with the notation of Problems 5 and 7, show that $(\mathscr{F}, +, \cdot)$ is a field. This field is called the **field of quotients** of the integral domain $(D, +, \cdot)$.

10. Prove Lemma 8.29.

11. Continuing with the notation of Problems 5, 7, and 9, let $\mathscr{D}$ be the subset of $\mathscr{F}$ defined by $\mathscr{D} = \{[a, b] \in \mathscr{F} \mid b = 1\}$.
(a) Show that $\mathscr{D}$ contains the zero element and the unity element of the field $\mathscr{F}$ and that $\mathscr{D}$ is closed under addition, subtraction, and multiplication.
(b) Show that, under the operations of addition and multiplication inherited from $\mathscr{F}$, $(\mathscr{D}, +, \cdot)$ forms an integral domain isomorphic to $(D, +, \cdot)$.
(c) Show that every element x in the field $\mathscr{F}$ can be written in the form $x = y/z$ where $y, z \in \mathscr{D}$ and $z \neq 0$.

12. Prove that the product of two cuts is always a cut.

13. If $q = s + \mathbf{v} \in \mathbb{H}$, define the **quaternionic conjugate** of q, in symbols q^*, by $q^* = s - \mathbf{v}$. For $q, p \in \mathbb{H}$, prove the following:
(a) qq^* is a nonnegative real number.
(b) $(q^*)^* = q$.
(c) $(p + q)^* = p^* + q^*$.
(d) $(pq)^* = q^*p^*$. (Note the order exchange!)
(e) q is a real number if and only if $q^* = q$.
(f) q is a pure quaternion if and only if $q^* = -q$.

14. Prove that $(\mathbb{H}, +, \cdot)$ is a ring with unity.

15. Let $q = s + \mathbf{v}$ be a nonzero quaternion, and let $t = qq^*$ (see Problem 13).
(a) Prove that t is a positive real number.
(b) Let $p = t^{-1}s - t^{-1}\mathbf{v}$. Prove that $pq = 1$.

16. Show that $(\mathbb{H}, +, \cdot)$ is a division algebra over $\mathbb{R}$. [Hint: Make use of the results of Problems 14 and 15.]

17. The **absolute value** of a quaternion q is defined by $|q| = \sqrt{qq^*}$ (see Problem 13). If q and p are quaternions, prove that $|qp| = |q|\,|p|$.

18. Prove the **triangle inequality** for quaternions q and p:

$$|q + p| \leq |q| + |p|.$$

(See Problem 17.) [Hint: Pattern your proof after the proof of Theorem 5.20.]

19. Let $q = s + \mathbf{v}$ be a quaternion, and let $t = qq^*$ (see Problem 13). Show that q is a solution of

$$x^2 - 2sx + t = 0.$$

20. Show that there are infinitely many quaternions q such that $q^2 = -1$.

21. Show that $p \in \mathbb{H}$ is a pure quaternion if and only if p^2 is a real number with $p^2 \leq 0$.

22. The **center** of a ring is defined to be the set of all elements of the ring that commute multiplicatively with all of the other elements of the ring. Show that the center of the ring $\mathbb{H}$ is $\mathbb{R}$.

23. If $\mathbf{w} \in E^3$ is a pure quaternion, and q is a nonzero quaternion, prove that $q\mathbf{w}q^{-1}$ is a pure quaternion. [Hint: Use the result of Problem 21.]

24. The **3-sphere** S^3 can be defined by $S^3 = \{q \in \mathbb{H} \mid |q| = 1\}$ (see Problem 17).
(a) Show that, under quaternionic multiplication, S^3 forms a noncommutative group.
(b) If $\theta \in \mathbb{R}$, $\mathbf{a} \in E^3$ with $\|\mathbf{a}\| = 1$, and $q = \cos(\theta/2) + \sin(\theta/2)\mathbf{a}$, show that $q \in S^3$.
(c) Show that every $q \in S^3$ can be written in the form
$q = \cos(\theta/2) + \sin(\theta/2)\mathbf{a}$ with $\theta \in \mathbb{R}$, $\mathbf{a} \in E^3$, and $\|\mathbf{a}\| = 1$.

25. Let $q, p \in \mathbb{H}$ with $q \neq 0$. Prove that:
(a) $|q^{-1}| = |q|$ (b) $|qpq^{-1}| = |p|$
(See Problem 17.)

26. The set $SO(3)$ of all rigid rotations of a three-dimensional Euclidean space E^3 about the origin forms a noncommutative group, called the **special orthogonal group**, under the operation of composition. Identify E^3 with the set of all pure quaternions. As a consequence of Theorem 8.32 and Problem 24, the mapping $\Phi: S^3 \to SO(3)$ defined by $(\Phi(q))(\mathbf{w}) = q\mathbf{w}q^{-1}$ for all $q \in S^3$ and all $\mathbf{w} \in E^3$ is a surjection. Prove that, for $q, p \in S^3$, $\Phi(qp) = \Phi(q)\Phi(p)$.

27. Consider the ring $\mathcal{M}_2(\mathbb{R})$ of Example 2.13 in Chapter 3. Let C denote the subset of $\mathcal{M}_2(\mathbb{R})$ consisting of all matrices of the form

$$\alpha = \begin{bmatrix} a & b \\ -b & a \end{bmatrix}, \quad \text{with } a, b \in \mathbb{R}.$$

(a) Prove that C is closed under addition, subtraction, and multiplication.
(b) Prove that, under the restrictions to C of the addition and multiplication operations in $\mathscr{M}_2(\mathbb{R})$, $(C, +, \cdot)$ is a field.
(c) Prove that $(C, +, \cdot)$ is isomorphic to the field $(\mathbb{C}, +, \cdot)$ of complex numbers. Thus, we have a **matrix representation** of the field $\mathbb{C}$ of complex numbers.

28. The mapping $\Phi: S^3 \to SO(3)$ of Problem 26 is said to be a **double covering** of $SO(3)$ by S^3. Explain why this terminology is used. [Hint: For $R \in SO(3)$, how many elements $q \in S^3$ are there such that $\Phi(q) = R$?]

29. Two-by-two matrices with complex entries can be added and multiplied in the same way as two-by-two matrices with real entries, and they thus form a ring $\mathscr{M}_2(\mathbb{C})$. Let H denote the subset of $\mathscr{M}_2(\mathbb{C})$ consisting of all matrices of the form

$$q = \begin{bmatrix} \alpha & \beta \\ -\beta^* & \alpha^* \end{bmatrix}, \quad \text{with } \alpha, \beta \in \mathbb{C}.$$

(a) Prove that H is closed under addition, subtraction, and multiplication.
(b) Prove that, under the restrictions to H of the addition and multiplication operations in $\mathscr{M}_2(\mathbb{C})$, $(H, +, \cdot)$ is a division ring.
(c) Prove that $(H, +, \cdot)$ is isomorphic to the division ring $(\mathbb{H}, +, \cdot)$ of quaternions. Thus, we have a matrix representation of the division ring $\mathbb{H}$ of quaternions.

30. Let $\mathbf{v}, \mathbf{w} \in E^3$ be regarded as (pure) quaternions, and let $\mathbf{v} \times \mathbf{w}$ and $\mathbf{v} \cdot \mathbf{w}$ denote the cross and dot products of $\mathbf{v}$ and $\mathbf{w}$.
(a) Show that $\mathbf{v} \times \mathbf{w} = (\mathbf{vw} - \mathbf{wv})/2$.
(b) Show that $\mathbf{v} \cdot \mathbf{w} = -(\mathbf{vw} + \mathbf{wv})/2$.

31. Consider the linear algebra $(E^3, +, \times)$, where $(\mathbf{a}, \mathbf{b}) \to \mathbf{a} \times \mathbf{b}$ is the vector cross product on E^3. Show that $(E^3, +, \times)$ is a Lie algebra.

32. Let $(A, +, \cdot)$ be any linear associative algebra over a field F. Define a binary operation $\star$ on A by $a \star b = a \cdot b - b \cdot a$ for $a, b \in A$. Show that $(A, +, \star)$ is a Lie algebra. The element $a \cdot b - b \cdot a$ is called the **commutator** of a and b in the algebra $(A, +, \cdot)$.

33. Let $(A, +, \cdot)$ be any linear associative algebra over a field F in which $2 = 1 + 1 \neq 0$. Define a binary operation $*$ on A by $a * b = (a \cdot b + b \cdot a)/2$ for $a, b \in A$. Show that $(A, +, *)$ is a Jordan algebra.

HISTORICAL NOTES

Although a clear understanding of the real numbers only dates from about 1885, the numbers themselves are very ancient. Babylonian astronomers were using a system quite similar to modern decimals to predict the positions of planets for astrological purposes before 1500 B.C. This Babylonian system and its modern descendent, the decimal system, effectively disguise the difficulties inherent in the real number system. Thus, the Babylonians never were troubled by doubts, and modern Europeans evaded the problems for 200 years.

It was otherwise with the ancient Greeks, whose clumsy system of numeration exposed these difficulties immediately; the Greeks could not write down anything that plausibly represented an irrational number. This had grave consequences for mathematics. The Greeks—and thus mathematical thought—turned away from algebra and the concept of number and took refuge in geometric line segments, where the difficulties again were effectively concealed. For 1000 years, all of mathematics was dressed up in geometric clothing.

When Arab work in algebra, including Arabic numerals, had found its way into Europe, there was a reawakening of interest in the number concept. Simon Stevin (1548–1620) reformulated the Babylonian system (known from its continued use in astronomy) into the modern decimal system, and once again irrational numbers could be written down in a compact form as infinite decimals.

In the course of clarifying the foundations of calculus during the 1800s, two theorems appeared which actually characterize the real numbers: the *least upper bound principle* and the *Cauchy completeness criterion.*

In the context of an Archimedean ordered field, each of these concepts implies the other. It was this insight that led Karl Weierstrass (1815–1897), Julius Richard Dedekind (1831–1916), Georg Cantor (1845–1918), and Leopold Kronecker (1823–1891) to attempt to construct the real numbers from the rational numbers by some well-understood process.

Dedekind (1872) chose the least upper bound principle to be the defining property of the real numbers, and he defined a real number to be a cut, as we do in this book. Cantor's approach to the real number system used the other fundamental principle, the Cauchy criterion.

The advantages of Cantor's approach are that it uses everyday objects of analysis (Cauchy sequences) and that it meshes well with modern algebra. The disadvantages are that it is a rather abstract construction, which makes it difficult for the beginner, and it uses some of the ε-δ methodology (Cauchy sequences), which again makes it a little difficult to get started. Also, it has no clear connection with our feeling for the geometric line.

Kronecker saw the problem in a completely different light than did Dedekind or Cantor, for he rejected the real number system as too imprecise. In 1882, Carl Lindemann (1852–1939) showed that π is not a root of any polynomial with rational coefficients. When told of this result, Kronecker remarked: "Very interesting. Then π does not exist." This comment typifies Kronecker's attitude. "God made the positive integers," he said, "all else is the work of man." Kronecker would allow humankind the use of only algebraic processes in the manufacture of new numbers.

Thus, π, e, and the great majority of other real numbers, which can be defiined only by means of limiting processes, lie beyond the pale. (But $\sqrt{2}$, as a root of $x^2 - 2 = 0$, remains respectable.)

Kronecker's objection is easily seen in the case of the least upper bound principle. A bounded set of reals must have a least upper bound, but there may be no way to find it. We illustrate this as follows: Goldbach's conjecture states that every positive even integer greater than 2 is a sum of two prime numbers. No one knows whether this is true or false. If it is true, the set $A = \{0\}$. If it is false, $A = \{0, 1\}$. Clearly, 1 is an upper bound. Is it the least upper bound? No one can say.

Kronecker claimed it was nonsense to talk about numbers no one could compute. There was not much of a response until around 1900, when Luitzen E. J. Brouwer (1881–1966) and other Dutch mathematicians began to develop Kronecker's ideas, partly in response to the paradoxes of set theory, which made standard mathematics look a lot less solid. The followers of Brouwer are called "intuitionists" or "constructivists," because they insist that the existence of a mathematical object is assured if and only if there is a way to construct it explicitly in a finite number of steps. (This cannot be done for the least upper bound of the set A mentioned earlier.) Brouwer's followers remained a small minority among mathematicians until the invention of the digital computer. Then it was discovered that the mathematics performed by a digital computer is exactly the mathematics acceptable to Kronecker.

ANNOTATED BIBLIOGRAPHY

Gardiner, A. *Infinite Processes: Background to Analysis* (New York: Springer-Verlag), 1982.

An interesting book with a variety of pre-analysis topics, and an exceptionally good treatment of number systems.

Hardy, G. H., and E. M. Wright. *An Introduction to the Theory of Numbers*, 5th ed. (Oxford: Clarendon Press), 1979.

The classic book on the theory of numbers, concerned primarily with the integers, but with some sections on irrational numbers as well.

Knuth, Donald E. *The Art of Computer Programming:* Volume 1, *Fundamental Algorithms* (Reading, Mass.: Addison-Wesley), 1973.

One of many volumes in a series on computer science by this author (one of the leading international experts on computer science), this book has a nice selection of topics out of elementary mathematics along with some fairly sophisticated applications to computer science.

Niven, Ivan. *Irrational Numbers* (Carus Monograph No. 11) (Washington, D.C.: The Mathematical Association of America), 1956.

A beautiful treatment of some of the classic problems of deciding whether certain numbers are irrational and, if so, whether they are algebraic or transcendental. Somewhat mathematically demanding, but rewarding nonetheless.

*5

SPECIAL TOPICS

In this final chapter, we present a few selected special topics that not only illustrate the use of the concepts introduced in the previous chapters, but are of considerable interest in their own right. Some of this material is standard (subgroups, cosets, homomorphisms), some of it is well known but not usually considered at this level (affine geometry, transformation groups), and some of it appears here for the first time in a textbook (relations as group morphisms).

*1

AFFINE GEOMETRY

Affine geometry is the geometry of points, lines, planes, and higher-dimensional analogues thereof. Although it is possible to study affine geometry synthetically—that is, by developing the consequences of a suitable set of postulates governing points, lines, planes, and so forth—our approach will be more algebraic.

To set the stage, consider the three-dimensional coordinate vector space $\mathbb{R}^3$. A vector $(x, y, z) \in \mathbb{R}^3$ can be visualized as a *point* with coordinates x, y, z in the usual way (Figure 5-1a). A one-dimensional linear subspace

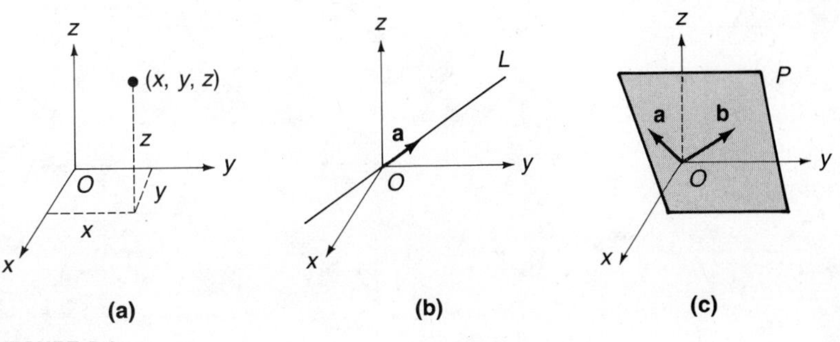

FIGURE 5-1

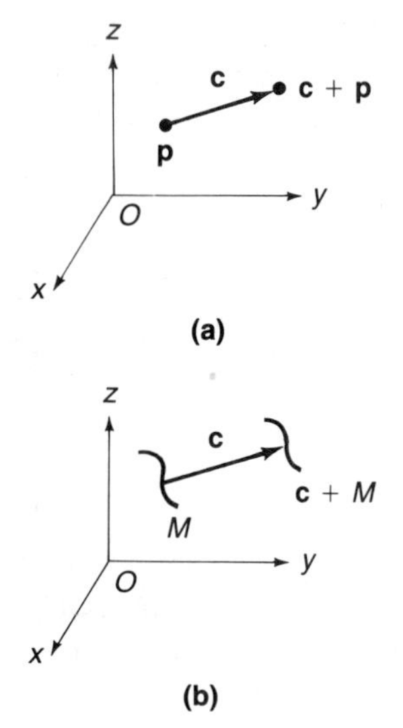

FIGURE 5-2

L of $\mathbb{R}^3$ consists of all multiples of a fixed nonzero basis vector $\mathbf{a}$, and it can be visualized as all points on a *straight line* passing through the origin in the direction of the vector $\mathbf{a}$ (Figure 5-1b). A two-dimensional linear subspace P of $\mathbb{R}^3$ consists of all linear combinations of two fixed basis vectors $\mathbf{a}$ and $\mathbf{b}$, and it can be pictured as all points on a *plane* passing through the origin and containing the vectors $\mathbf{a}$ and $\mathbf{b}$ (Figure 5-1c).

In our approach to the geometry of points, lines, and planes in $\mathbb{R}^3$, the vectors $(x, y, z) \in \mathbb{R}^3$ have a dual role to play: Not only are they used to denote points, but also they help us to describe translations in the space $\mathbb{R}^3$. Figure 5-2a shows a point $\mathbf{p} \in \mathbb{R}^3$, a *translation vector* $\mathbf{c}$, and the point $\mathbf{c} + \mathbf{p}$ obtained by *translating* $\mathbf{p}$ *by the vector* $\mathbf{c}$. If M is a set of points in $\mathbb{R}^3$, then a *translation of* M *by the vector* $\mathbf{c}$ is accomplished by translating each point of M by the vector $\mathbf{c}$ (Figure 5-2b). We write the translated set as

$$\mathbf{c} + M = \{\mathbf{c} + \mathbf{p} \mid \mathbf{p} \in M\}.$$

Thus, in visualizing affine geometry, we think of vectors representing points as actual geometric points in the space, but we picture translation vectors as arrows.

We have seen that linear subspaces of dimensions one and two can be regarded as lines and planes passing through the *origin.* By translating these lines and planes, we can obtain lines and planes passing through arbitrary points of the space (Figure 5-3).

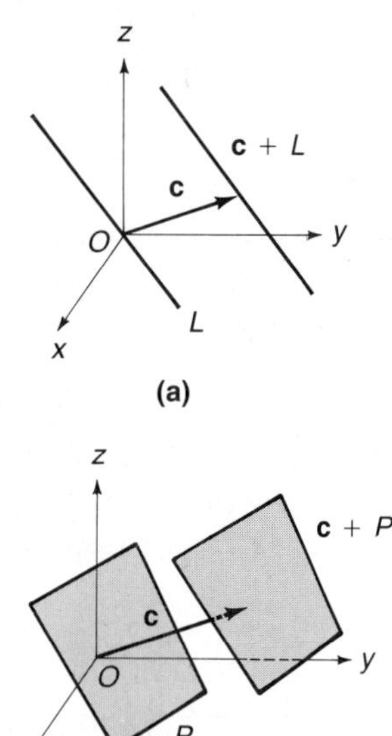

FIGURE 5-3

If L is a one-dimensional linear subspace of $\mathbb{R}^3$ and $\mathbf{c} \in \mathbb{R}^3$, then $\mathbf{c} + L$ is a straight line in $\mathbb{R}^3$ that is *parallel to* L and that *contains the point* $\mathbf{c}$. (Note the dual role of the vector $\mathbf{c}$ here, both as a geometric point on the line $\mathbf{c} + L$ and as a translation vector that produces this line by translating the linear subspace L.) The collection of all straight lines in $\mathbb{R}^3$ that are parallel to L forms a bundle of parallel lines, one through each point $\mathbf{c}$ in the space $\mathbb{R}^3$ (Figure 5-4a). This collection of parallel lines is denoted by $\mathbb{R}^3/L$, read "$\mathbb{R}^3$ *mod* L." Thus,

$$\mathbb{R}^3/L = \{\mathbf{c} + L \mid \mathbf{c} \in \mathbb{R}^3\}.$$

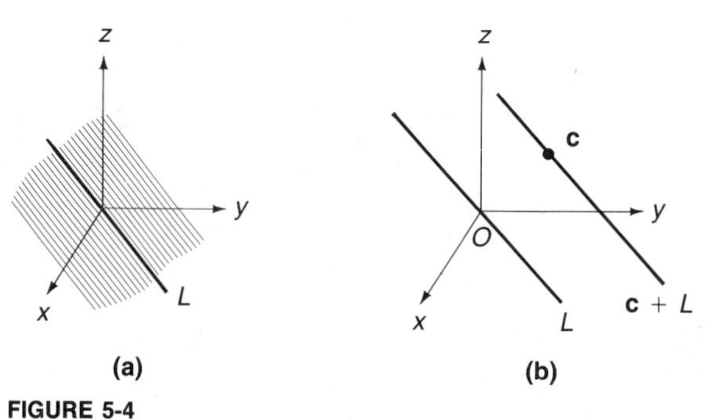

FIGURE 5-4

Choice of the point $\mathbf{c} \in \mathbb{R}^3$ determines the unique line $\mathbf{c} + L$ in the bundle $\mathbb{R}^3/L$ of parallel lines that contains the point $\mathbf{c}$ (Figure 5-4b).

If P is a two-dimensional linear subspace of $\mathbb{R}^3$ and $\mathbf{c} \in \mathbb{R}^3$, then $\mathbf{c} + P$ is a plane in $\mathbb{R}^3$ that is *parallel to* P and *contains the point* $\mathbf{c}$. The collection of all planes in $\mathbb{R}^3$ that are parallel to P forms a stack of parallel planes, one through each point $\mathbf{c}$ in the space $\mathbb{R}^3$ (Figure 5-5a). This collection of parallel planes is denoted by $\mathbb{R}^3/P$, read "$\mathbb{R}^3$ *mod* P." Thus,

$$\mathbb{R}^3/P = \{\mathbf{c} + P \mid \mathbf{c} \in \mathbb{R}^3\}.$$

Choice of the point $\mathbf{c} \in \mathbb{R}^3$ determines the unique plane $\mathbf{c} + P$ in the stack $\mathbb{R}^3/P$ of parallel planes that contains the point $\mathbf{c}$ (Figure 5-5b).

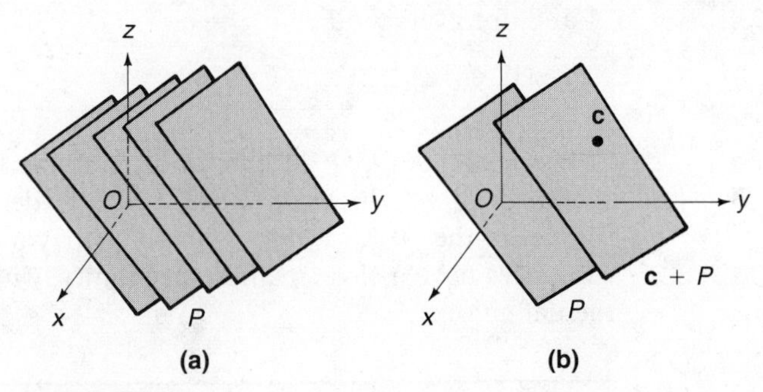

FIGURE 5-5

In what follows, we generalize the simple geometric ideas presented above. Thus, *for the remainder of this section, we assume that* V *is a vector space over a field* F. (As an aid to intuition, think of V as $\mathbb{R}^3$ and F as $\mathbb{R}$.)

1.1 DEFINITION

OPERATIONS ON SUBSETS OF *V* Let $s \in F$, $\mathbf{c} \in V$, and let M and N be subsets of V. We define:

(i) $M + N = \{\mathbf{v} + \mathbf{w} \mid \mathbf{v} \in M \wedge \mathbf{w} \in N\}$.
(ii) $M - N = \{\mathbf{v} - \mathbf{w} \mid \mathbf{v} \in M \wedge \mathbf{w} \in N\}$.
(iii) $sM = \{s\mathbf{v} \mid \mathbf{v} \in M\}$.
(iv) $\mathbf{c} + M = \{\mathbf{c} + \mathbf{v} \mid \mathbf{v} \in M\}$.

Note that

$$\mathbf{c} + M = \{\mathbf{c}\} + M.$$

The set $\mathbf{c} + M$ is called the **translation** of M by the vector $\mathbf{c}$. We leave the proof of the following two lemmas as exercises (Problems 1 and 2).

1.2 LEMMA

Let $M, N, K \subseteq V$, $s \in F$, and $\mathbf{c} \in V$. Then:

(i) $(M + N) + K = M + (N + K)$
(ii) $(M + N) - K = M + (N - K)$
(iii) $M - (N + K) = (M - N) - K$
(iv) $s(M + N) = sM + sN$
(v) $M + N = N + M$
(vi) $\mathbf{c} + M = \mathbf{c} + N \Rightarrow M = N$

1.3 LEMMA

Let $W \subseteq V$. Then W is a linear subspace of V if and only if $W \neq \varnothing$ and, for every $s \in F$,

$$s(W + W) \subseteq W.$$

The analogues in V of lines and planes in $\mathbb{R}^3$ are referred to as *affine subspaces* of V. The word *affine*, which is derived from the same Latin source as the word *affinity*, should help you keep in mind that affine subspaces have a close relation, or affinity, with linear subspaces. Here is the definition:

1.4 DEFINITION

AFFINE SUBSPACE By an **affine subspace** of V, we mean either the empty set $\varnothing$ or a subset of V having the form

$$\mathbf{c} + W$$

for some vector $\mathbf{c} \in V$ and some linear subspace W of V.

If W is a linear subspace of V, then $\mathbf{0} \in W$, and it follows that, for any $\mathbf{c} \in V$,

$$\mathbf{c} = \mathbf{c} + \mathbf{0} \in \mathbf{c} + W;$$

that is, *the point* $\mathbf{c}$ *belongs to the affine subspace* $\mathbf{c} + W$.

1.5 LEMMA

Let U and W be linear subspaces of V, and let $\mathbf{b}, \mathbf{c} \in V$. Then

$$\mathbf{b} + U \subseteq \mathbf{c} + W \Leftrightarrow U \subseteq W \quad \text{and} \quad \mathbf{b} - \mathbf{c} \in W.$$

PROOF Suppose that $\mathbf{b} + U \subseteq \mathbf{c} + W$. Then, since $\mathbf{b} \in \mathbf{b} + U$, we have $\mathbf{b} \in \mathbf{c} + W$; that is, $\mathbf{b} = \mathbf{c} + \mathbf{w}_0$ for some $\mathbf{w}_0 \in W$. Consequently,

$$\mathbf{b} - \mathbf{c} = \mathbf{w}_0 \in W.$$

Now let $\mathbf{u} \in U$. Then

$$\mathbf{b} + \mathbf{u} \in \mathbf{b} + U \subseteq \mathbf{c} + W,$$

and it follows that

$$\mathbf{b} + \mathbf{u} = \mathbf{c} + \mathbf{w}$$

for some $\mathbf{w} \in W$. Therefore, since $\mathbf{w}_0, \mathbf{w} \in W$,

$$\mathbf{u} = \mathbf{c} - \mathbf{b} + \mathbf{w} = -(\mathbf{b} - \mathbf{c}) + \mathbf{w} = -\mathbf{w}_0 + \mathbf{w} \in W.$$

Thus, we have $U \subseteq W$ and $\mathbf{b} - \mathbf{c} \in W$.

Conversely, suppose that $U \subseteq W$ and $\mathbf{b} - \mathbf{c} \in W$. Let $\mathbf{b} - \mathbf{c} = \mathbf{w}_0 \in W$. We must prove that $\mathbf{b} + U \subseteq \mathbf{c} + W$. Thus, let $\mathbf{x} \in \mathbf{b} + U$, so that $\mathbf{x} = \mathbf{b} + \mathbf{u}$ for some $\mathbf{u} \in U$. By hypothesis, $U \subseteq W$, and so $\mathbf{u} \in W$. Since $\mathbf{b} - \mathbf{c} = \mathbf{w}_0$, we have

$$\mathbf{b} = \mathbf{c} + \mathbf{w}_0.$$

Therefore, since $\mathbf{w}_0, \mathbf{u} \in W$,

$$\mathbf{x} = \mathbf{b} + \mathbf{u} = \mathbf{c} + \mathbf{w}_0 + \mathbf{u} \in \mathbf{c} + W.$$

Consequently, $\mathbf{b} + U \subseteq \mathbf{c} + W$. □

1.6 COROLLARY If A is an affine subspace of V and $A \neq \varnothing$, there exists a unique linear subspace W of V and there exists $\mathbf{c} \in V$ such that

$$A = \mathbf{c} + W.$$

PROOF Let A be a nonempty affine subspace of V. Then, by Definition 1.4, there exists $\mathbf{c} \in V$ and there exists a linear subspace W of V such that

$$A = \mathbf{c} + W.$$

We must show that W is uniquely determined by A. Suppose that we also have

$$A = \mathbf{b} + U$$

for some $\mathbf{b} \in V$ and some linear subspace U of V. Then,

$$\mathbf{b} + U = A = \mathbf{c} + W,$$

and it follows from Lemma 1.5 that $U = W$. □

In general, the vector $\mathbf{c}$ in Corollary 1.6 is not uniquely determined by the affine subspace A.

1.7 DEFINITION

If A is a nonempty affine subspace of V, then the unique linear subspace W of V such that

$$A = \mathbf{c} + W$$

for some (not necessarily unique) $\mathbf{c} \in V$ is called the **directing subspace** for A.

We note that the empty affine subspace $\varnothing$ is the only affine subspace that has no directing subspace. Furthermore, if A is a nonempty affine subspace of V and if W is the directing subspace for A, then, for $\mathbf{c} \in V$,

$$A = \mathbf{c} + W \Leftrightarrow \mathbf{c} \in A$$

(Problem 7).

1.8 DEFINITION

DIMENSION OF AN AFFINE SUBSPACE Let A be a nonempty affine subspace of V, and let W be the directing subspace of A. We define the **dimension** of A, in symbols, $\dim(A)$, by

$$\dim(A) = \dim(W).$$

By a special convention, we also define

$$\dim(\varnothing) = -1.$$

An affine subspace is called a **point** if its dimension is zero, a **line** if its dimension is one, and a **plane** if its dimension is two.

1.9 DEFINITION

AFFINE COMBINATION Let $\mathbf{v_1}, \mathbf{v_2}, \mathbf{v_3}, \ldots, \mathbf{v_n} \in V$, and let $s_1, s_2, s_3, \ldots, s_n \in F$. If

$$\mathbf{v} = s_1\mathbf{v_1} + s_2\mathbf{v_2} + s_3\mathbf{v_3} + \cdots + s_n\mathbf{v_n}$$

and

$$s_1 + s_2 + s_3 + \cdots + s_n = 1,$$

then we say that $\mathbf{v}$ is an **affine combination** of the vectors $\mathbf{v_1}, \mathbf{v_2}, \mathbf{v_3}, \ldots,$ and $\mathbf{v_n}$.

In words, *an affine combination is a linear combination in which the sum of the coefficients is 1.* Although the zero vector $\mathbf{0}$ is regarded as a linear combination of the empty set $\varnothing$, it is *not* regarded as an affine combination of $\varnothing$. A subset M of V is said to be **closed** under affine combinations if and only if every affine combination of vectors in M is again a vector in M.

We leave the proof of the following two lemmas as exercises (Problems 9 and 12).

1.10 LEMMA Let $M \subseteq V$, and let $\mathbf{c} \in V$.

(i) M is closed under affine combinations if and only if $\mathbf{c} + M$ is closed under affine combinations.
(ii) M is a linear subspace of V if and only if $\mathbf{0} \in M$ and M is closed under affine combinations.
(iii) M is an affine subspace of V if and only if M is closed under affine combinations.

1.11 LEMMA Let $M \subseteq V$, and let A be the set of all affine combinations of vectors in M. Then A is an affine subspace of V.

The affine subspace of V consisting of all affine combinations of vectors in M is called the **affine span** of M.

1.12 LEMMA Let A and B be affine subspaces, and suppose that $\mathbf{c} \in A \cap B$. Let U and W be the directing subspaces for A and B, respectively. Then

$$A \cap B = \mathbf{c} + (U \cap W).$$

PROOF Because $\mathbf{c} \in A$ and $\mathbf{c} \in B$, we have $A = \mathbf{c} + U$ and $B = \mathbf{c} + W$. Suppose that $\mathbf{x} \in A \cap B$. Then there exist $\mathbf{u} \in U$ and $\mathbf{w} \in W$ such that

$$\mathbf{x} = \mathbf{c} + \mathbf{u} = \mathbf{c} + \mathbf{w}.$$

Therefore, $\mathbf{u} = \mathbf{w}$, and it follows that

$$\mathbf{x} = \mathbf{c} + \mathbf{u} \quad \text{with } \mathbf{u} \in U \cap W.$$

Hence, we have

$$A \cap B \subseteq \mathbf{c} + (U \cap W).$$

Conversely, suppose that $x \in \mathbf{c} + (U \cap W)$. Then there exists $\mathbf{d} \in U \cap W$ such that

$$\mathbf{x} = \mathbf{c} + \mathbf{d}.$$

Because $\mathbf{d} \in U$, we have $\mathbf{x} \in \mathbf{c} + U = A$, and because $\mathbf{d} \in W$, we also have $\mathbf{x} \in \mathbf{c} + W = B$. Therefore, $\mathbf{x} \in A \cap B$, and so

$$\mathbf{c} + (U \cap W) \subseteq A \cap B. \qquad \square$$

Let A and B be two affine subspaces of V. On the one hand, if $A \cap B = \varnothing$, then $A \cap B$ is an affine subspace of V. On the other hand, if $A \cap B \neq \varnothing$, then $A \cap B$ is an affine subspace of V by Lemma 1.12. Furthermore, if $A \cap B \neq \varnothing$, then the dimension of $A \cap B$ is the same as the dimension of the intersection of the directing subspaces for A and B.

1.13 DEFINITION

PARALLEL AND SKEW AFFINE SUBSPACES Let A and B be nonempty affine subspaces of V, and let U and W be the directing subspaces for A and B, respectively. We say that A and B are **parallel** if $U \subseteq W$ or $W \subseteq U$. If A and B are not parallel and $A \cap B = \varnothing$, then we say that A and B are **skew**.

For example, two straight lines in $\mathbb{R}^3$ are skew if they are not parallel and do not intersect each other. Note that any nonempty affine subspace is regarded as being parallel to itself. If A is an affine subspace of V and $\mathbf{c} \in V$, then the affine subspace $\mathbf{c} + A$ is parallel to A (Problem 17). We leave the proof of the following lemma as an exercise (Problem 19).

1.14 LEMMA

Let W be a linear subspace of V, and let $\mathbf{a}, \mathbf{b} \in V$.

(i) $\mathbf{a} + W$ and $\mathbf{b} + W$ are parallel affine subspaces of V.
(ii) $(\mathbf{a} + W) \cap (\mathbf{b} + W) \neq \varnothing \Rightarrow \mathbf{a} + W = \mathbf{b} + W$.
(iii) $\mathbf{a} + W = \mathbf{b} + W \Leftrightarrow \mathbf{a} - \mathbf{b} \in W$.

According to Lemma 1.14, if W is a linear subspace of V, then the affine subspaces of V that have the form $\mathbf{c} + W$ are all parallel to W and to each other. Moreover, two of these affine subspaces are either disjoint, or they are identical.

1.15 DEFINITION

If W is a linear subspace of V, we define

$$V/W = \{\mathbf{c} + W \mid \mathbf{c} \in V\}.$$

Thus, if W is a linear subspace of V, then V/W is the set of all affine subspaces that can be obtained by translating W by vectors in V. We read V/W as "*V modulo W*," or "*V mod W*" for short. We can picture V/W as a stack of affine subspaces, all parallel to W and to each other. Given any point $\mathbf{c} \in V$, there is one and only one affine subspace in the stack that contains $\mathbf{c}$, namely, $\mathbf{c} + W$.

Suppose that W and U are linear subspaces of V and that U is a complementary direct summand of W in V (Definition 6.13 in Chapter 3). Then, as we prove in the next lemma, each affine subspace A in the stack V/W intersects U in a uniquely determined point $\mathbf{u}$.

1.16 LEMMA Let W and U be linear subspaces of V, and suppose that $V = W \oplus U$. Then, for each $A \in V/W$, there exists $\mathbf{u} \in U$ such that

$$A \cap U = \{\mathbf{u}\}.$$

PROOF Because $A \in V/W$, there exists $\mathbf{c} \in V$ such that

$$A = \mathbf{c} + W.$$

Because $V = W \oplus U$, there exist *unique* vectors $\mathbf{w} \in W$ and $\mathbf{u} \in U$ such that

$$\mathbf{c} = \mathbf{w} + \mathbf{u}$$

(Lemma 6.11 in Chapter 3). Thus,

$$\mathbf{u} = \mathbf{c} + (-\mathbf{w}) \in \mathbf{c} + W = A,$$

and therefore,

$$\mathbf{u} \in A \cap U.$$

To complete the proof, we must show that $\mathbf{u}$ is the *only* vector in $A \cap U$. Suppose that $\mathbf{u}'$ is another such vector, so that

$$\mathbf{u}' \in A \cap U.$$

Then, because $\mathbf{u}' \in A = \mathbf{c} + W$, there exists $\mathbf{w}' \in W$ such that

$$\mathbf{u}' = \mathbf{c} + \mathbf{w}'.$$

It follows that

$$\mathbf{c} = (-\mathbf{w}') + \mathbf{u}',$$

and therefore, by the uniqueness of $\mathbf{w}$ and $\mathbf{u}$, we have $\mathbf{w} = -\mathbf{w}'$ and $\mathbf{u} = \mathbf{u}'$. □

We leave the proof of the following corollary as an exercise (Problem 21).

1.17 COROLLARY Let W and U be linear subspaces of V, and suppose that $V = W \oplus U$. Define $\phi: U \to V/W$ by

$$\phi(\mathbf{u}) = \mathbf{u} + W$$

for all $\mathbf{u} \in U$. Then $\phi: U \to V/W$ is a bijection.

To conclude this section, we show that, if W is a linear subspace of V, then V/W can be made into a vector space over F in such a way that the mapping $\phi: U \to V/W$ of Corollary 1.17 is an isomorphism.

1.18 LEMMA Let $\mathbf{a}, \mathbf{b} \in V$, $s \in F$, and let W be a linear subspace of V. Then:

(i) $(\mathbf{a} + W) + (\mathbf{b} + W) = (\mathbf{a} + \mathbf{b}) + W$
(ii) $s(\mathbf{a} + W) = s\mathbf{a} + W$

PROOF We prove (i) and leave (ii) as an exercise (Problem 23). Because W is a linear subspace of V, we have

$$W + W = W$$

(Problem 3). Therefore,

$$\begin{aligned}(\mathbf{a} + W) + (\mathbf{b} + W) &= \{\mathbf{a}\} + W + \{\mathbf{b}\} + W \\ &= \{\mathbf{a}\} + \{\mathbf{b}\} + W + W \\ &= \{\mathbf{a}\} + \{\mathbf{b}\} + W \\ &= \{\mathbf{a} + \mathbf{b}\} + W \\ &= (\mathbf{a} + \mathbf{b}) + W.\end{aligned}$$

□

Using the results of Corollary 1.17 and Lemma 1.18, it is an easy matter to prove the following theorem (Problem 24).

1.19 THEOREM ***V/W* IS A VECTOR SPACE** Let W be a linear subspace of V. Then V/W is a vector space under the operations of addition and multiplication by scalars given by

$$(A, B) \mapsto A + B \quad \text{and} \quad (s, A) \mapsto sA$$

for $A, B \in V/W$ and $s \in F$. Furthermore, if U is a complementary direct summand of W in V, then the mapping

$$\phi: U \to V/W$$

defined for all $\mathbf{u} \in U$ by

$$\phi(\mathbf{u}) = \mathbf{u} + W$$

is an isomorphism.

PROBLEM SET 5.1

1. Prove Lemma 1.2.
2. Prove Lemma 1.3.
3. If W is a linear subspace of V, prove that $W = W + W$.
4. Show that a subset A of V is an affine subspace if and only if, for every $s \in F$, $s(A - A) + A \subseteq A$.
5. If A is an affine subspace of V and $\mathbf{a} \in A$, prove that $A - \mathbf{a}$ is the directing subspace for A.
6. Let $A, B \subseteq V$, and suppose there exists $\mathbf{c} \in V$ with $B = \mathbf{c} + A$. Assume that, for all $s \in F$, $sA + (1 - s)A \subseteq A$. Prove that, for all $s \in F$, we have $sB + (1 - s)B \subseteq B$.
7. If A is a nonempty affine subspace of V, W is the directing subspace for A, and $\mathbf{c} \in V$, prove that $A = \mathbf{c} + W \Leftrightarrow \mathbf{c} \in A$.
8. Assume that, in the field F, $1 + 1 \neq 0$. If $\mathbf{0} \in W \subseteq V$ and W has the property that, for every $s \in F$, $sW + (1 - s)W \subseteq W$, prove that W is a linear subspace of V. [Hint: Let $2 = 1 + 1$, $1/2 = 2^{-1}$, and note that, for $\mathbf{x}, \mathbf{y} \in V$, $\mathbf{x} + \mathbf{y} = 2((1/2)\mathbf{x} + (1/2)\mathbf{y})$.]
9. Prove Lemma 1.10.
10. If $A \subseteq V$ and if, for every $s \in F$, $sA + (1 - s)A \subseteq A$, prove that A is an affine subspace of V. [Hint: See Problems 6 and 8.]
11. If A is an affine subspace of V and $L: V \to V$ is a linear transformation, then the **image** of A under L, denoted by $L(A)$, is defined by $L(A) = \{L(\mathbf{a}) \mid \mathbf{a} \in A\}$. Prove that $L(A)$ is an affine subspace of V.
12. Prove Lemma 1.11.
13. For $\mathbf{a} \in V$, the mapping $T_{\mathbf{a}}: V \to V$ defined by $T_{\mathbf{a}}(\mathbf{v}) = \mathbf{a} + \mathbf{v}$ for all $\mathbf{v} \in V$ is called the **translation** determined by $\mathbf{a}$. A mapping $S: V \to V$ is called an **affine transformation** if it can be written as a composition $S = T_{\mathbf{a}} \circ L$, where $T_{\mathbf{a}}$ is the translation determined by some $\mathbf{a} \in V$ and L is a linear transformation. If L is a linear transformation and $\mathbf{b} \in V$, show that $L \circ T_{\mathbf{b}}$ is an affine transformation.
14. If $M \subseteq V$, write the affine span of M as $\text{aff}(M)$. Let $\mathscr{A}(V)$ be the set of all affine subspaces of V, partially ordered by set-theoretic inclusion. Let $A, B \in \mathscr{A}(V)$.
    (a) Show that $A \cap B$ is the greatest lower bound of A and B in $\mathscr{A}(V)$.
    (b) Show that $\text{aff}(A \cup B)$ is the least upper bound of A and B in $\mathscr{A}(V)$.
15. If A is an affine subspace of V and $S: V \to V$ is an affine transformation (Problem 13), then the **image** under S, denoted by $S(A)$, is defined by $S(A) = \{S(\mathbf{a}) \mid \mathbf{a} \in A\}$. Prove that $S(A)$ is an affine subspace of V.

16. True or false: If A and B are affine subspaces of V, and if C is the affine span of $A \cup B$, then C consists of all vectors of the form $s\mathbf{a} + (1 - s)\mathbf{b}$ for $s \in F$, $\mathbf{a} \in A$, and $\mathbf{b} \in B$.

17. If A is an affine subspace of V and $\mathbf{c} \in V$, prove that the affine subspace $\mathbf{c} + A$ is parallel to A.

18. If A and B are affine subspaces of V, if U and W are the directing subspaces for A and B, respectively, and if $\mathbf{c} \in A \cap B$, prove that the affine span of $A \cup B$ is the same as $\mathbf{c} + U + W$.

19. Prove Lemma 1.14.

20. Suppose that A and B are nonempty affine subspaces of V and that A is parallel to B. If $A \cap B \neq \varnothing$, prove that $A \subseteq B$ or $B \subseteq A$.

21. Prove Corollary 1.17.

22. If A is a zero-dimensional affine subspace of V (a point) and B is any nonempty affine subspace of V, show that A is parallel to B.

23. Prove Part (ii) of Lemma 1.18.

24. Prove Theorem 1.19.

25. The vectors $\mathbf{v_1}, \mathbf{v_2}, \mathbf{v_3}, \ldots, \mathbf{v_n} \in V$ are said to be **affine independent** if, whenever $s_1, s_2, s_3, \ldots, s_n \in F$ with $s_1\mathbf{v_1} + s_2\mathbf{v_2} + s_3\mathbf{v_3} + \cdots + s_n\mathbf{v_n} = \mathbf{0}$ and $s_1 + s_2 + s_3 + \cdots + s_n = 0$, it follows that $s_1 = s_2 = s_3 = \cdots = s_n = 0$. Show that $\mathbf{v_1}, \mathbf{v_2}, \mathbf{v_3}, \ldots, \mathbf{v_n}$ are affine independent if and only if $\mathbf{v_2} - \mathbf{v_1}, \mathbf{v_3} - \mathbf{v_1}, \ldots, \mathbf{v_n} - \mathbf{v_1}$ are linearly independent.

26. Prove that the vectors $\mathbf{v_1}, \mathbf{v_2}, \mathbf{v_3}, \ldots, \mathbf{v_n} \in V$ are affine independent (Problem 25) if and only if the affine span of $\{\mathbf{v_1}, \mathbf{v_2}, \mathbf{v_3}, \ldots, \mathbf{v_n}\}$ has dimension $n - 1$.

27. Three vectors $\mathbf{a}, \mathbf{b}, \mathbf{c} \in V$ are said to be **collinear** if there exists an affine line A (that is, an affine subspace A with $\dim(A) = 1$) such that $\mathbf{a}, \mathbf{b}, \mathbf{c} \in A$. Show that $\mathbf{a}, \mathbf{b}, \mathbf{c}$ are collinear if and only if they are *not* affine independent (see Problem 25).

28. Let A be a finite dimensional affine subspace of V, and let $k = \dim(A)$. An **affine basis** for A is, by definition, a finite set of affine independent vectors (Problem 25) whose affine span is A. Prove:
(a) A has an affine basis.
(b) Any affine basis for A consists of $k + 1$ vectors.
(c) If $\{\mathbf{v_1}, \mathbf{v_2}, \mathbf{v_3}, \ldots, \mathbf{v_{k+1}}\}$ is an affine basis for A, then every vector $\mathbf{a} \in A$ can be written uniquely as an affine combination of the vectors $\mathbf{v_1}, \mathbf{v_2}, \mathbf{v_3}, \ldots, \mathbf{v_n}$.

29. Four vectors $\mathbf{a}, \mathbf{b}, \mathbf{c}, \mathbf{d}$ are said to be **coplanar** if there exists an affine plane A (that is, an affine subspace A with $\dim(A) = 2$) such that $\mathbf{a}, \mathbf{b}, \mathbf{c}, \mathbf{d} \in A$. Show that $\mathbf{a}, \mathbf{b}, \mathbf{c}, \mathbf{d}$ are coplanar if and only if they are *not* affine independent (see Problem 25.)

30. (a) Show that $V/\{\mathbf{0}\}$ is isomorphic to V.
(b) What is the dimension of the vector space V/V?

31. If W is a linear subspace of V, show that V/W is a partition of V.

32. If W is a linear subspace of V and E is the equivalence relation determined by the partition V/W, show that, for $\mathbf{u}, \mathbf{v} \in V$, $\mathbf{u}\, E\, \mathbf{v} \Leftrightarrow \mathbf{u} - \mathbf{v} \in W$.

*2 SUBGROUPS AND COSETS

If V is a vector space over F, then $(V, +)$ is a group with neutral element $\mathbf{0}$ in which the inverse of a vector $\mathbf{v}$ is the vector $-\mathbf{v}$. This leads us to the following definition.

2.1 DEFINITION

ADDITIVELY WRITTEN GROUP An **additively written** group is a group $(G, +)$ in which the binary group operation $+$ is called **addition**. If $x, y \in G$, we refer to $x + y$ as the **sum** of x and y. The neutral element of an additively written group is denoted by 0, and the inverse of an element $x \in G$ is written as $-x$.

For simplicity, we often say that G *is an additive group*, when what we really mean is that $(G, +)$ *is an additively written group*. Note that, if G is an additive group and $x \in G$, we have

$$0 + x = x + 0 = x$$

and

$$(-x) + x = x + (-x) = 0.$$

Although there is no automatic presumption that an additive group G is commutative, it is traditional to write only commutative groups additively. Thus, if G is commutative, we also have

$$x + y = y + x$$

for all $x, y \in G$. A commutative, additively written group $(G, +)$ generalizes the additive group $(V, +)$ of a vector space V.

2.2 DEFINITION

MULTIPLICATIVELY WRITTEN GROUP A **multiplicatively written** group is a group $(G, \cdot)$ in which the binary group operation $\cdot$ is called **multiplication**. If $x, y \in G$, we refer to $x \cdot y$ as the **product** of x and y. The neutral element of a multiplicatively written group is denoted by 1, and the inverse of an element $x \in G$ is written as x^{-1}.

For simplicity, we often say that G *is a multiplicative group*, when what we really mean is that $(G, \cdot)$ *is a multiplicatively written group*. Also, for $x, y \in G$, we often denote the product $x \cdot y$ by simple juxtaposition, so that

$$xy = x \cdot y.$$

Note that, if G is a multiplicative group and $x \in G$, then

$$1x = x1 = x$$

and

$$x^{-1}x = xx^{-1} = 1.$$

Multiplicative notation is used for both commutative and noncommutative groups, and general definitions of group-theoretic notions are usually given in terms of multiplicatively written groups. Thus, in what follows, we use multiplicative notation unless there are good reasons for doing otherwise.

Many important group-theoretic notions are generalizations of vector-space concepts such as subspaces, affine subspaces, linear transformations, and so forth. These concepts are generalized to groups by first writing them (as far as possible) in additive notation, and then translating them into multiplicative notation using the following "dictionary":

Additive notation	*Multiplicative notation*
0	1
$x + y$	xy
$-x$	x^{-1}

If V is a vector space and W is a linear subspace of V, then $\mathbf{0} \in W$, W is closed under addition, and W is closed under negation. The analogous conditions for a subset H of a multiplicative group G define what is called a *subgroup* of G.

2.3 DEFINITION

SUBGROUP A subset H of a group G is said to be a **subgroup** of G if:

(i) $1 \in H$
(ii) $x, y \in H \Rightarrow xy \in H$
(iii) $x \in H \Rightarrow x^{-1} \in H$

Thus, *subgroups are to groups as linear subspaces are to vector spaces.* (The analogy is not perfect, but it is very far reaching.) If Condition (ii) in Definition 2.3 holds, we say that H is **closed under multiplication**. If Condition (iii) holds, we say that H is **closed under the formation of inverses**. Just as a linear subspace of a vector space is a vector space in its own right, a subgroup H of a group G is a group in its own right under the restrictions to H of the operations of multiplication and the formation of inverses in the "parent" group G (Problem 3). Note that the singleton set $\{1\}$ and the entire group G are subgroups of G. We refer to $\{1\}$ and G as the **trivial subgroups** of G. A subgroup H of G is called a **proper subgroup** if $H \neq G$.

2.4 Example Rewrite the group AS14 in Section 6 of Chapter 2, page 145, as a multiplicative group G, and find a subgroup H of G with $\#H = 2$.

SOLUTION In AS14, the group S consists of elements

$$S = \{e, a, b, c, d, f\},$$

and the group operation $*$ is given by a table. The element $e \in S$ is effective as the neutral element. We rewrite this group using multiplicative notation, denoting the neutral element by 1 and renaming the group

$$G = \{1, a, b, c, d, f\}.$$

The multiplication table for G is

$\cdot$	1	a	b	c	d	f
1	1	a	b	c	d	f
a	a	b	1	d	f	c
b	b	1	a	f	c	d
c	c	f	d	1	b	a
d	d	c	f	a	1	b
f	f	d	c	b	a	1

Here we have

$$1^{-1} = 1 \qquad a^{-1} = b \qquad b^{-1} = a$$
$$c^{-1} = c \qquad d^{-1} = d \qquad f^{-1} = f$$

Because $cc = 1$ and $c^{-1} = c$, the set

$$H = \{1, c\}$$

is a subgroup of G with $\#H = 2$. □

In the example above, $\{1, d\}$ and $\{1, f\}$ are also two-element subgroups of G; however $\{1, a\}$ is not a subgroup of G because $aa = b \notin \{1, a\}$, so it is not closed under multiplication. Likewise, $\{1, b\}$ is not a subgroup of G. Notice that $\{1, a, b\}$ is a three-element subgroup of G (Problem 7).

If a nonempty subset S of a group G is closed under multiplication, then S forms a semigroup under the restriction to S of the multiplication operation on G (see Definition 6.10 in Chapter 2). Furthermore, S automatically inherits the cancellation laws from G (Theorem 1.13 in Chapter

3). Therefore, as a consequence of Corollary 1.16 in Chapter 3, we have the following result (Problem 11):

A finite nonempty subset S of a group G is a subgroup if and only if it is closed under multiplication.

2.5 DEFINITION

CENTER If G is a group, then the set

$$Z = \{z \in G \mid zg = gz \quad \text{for all } g \in G\}$$

is called the **center** of the group G.

In words, *the center of a group is the set of all elements of the group that commute with every other element of the group.* We leave the proof of the next theorem as an exercise (Problem 17).

2.6 THEOREM

THE CENTER IS A SUBGROUP Let G be a group, and let Z be the center of G. Then Z is a subgroup of G.

The following is the analogue for (multiplicatively written) groups of Definition 1.1 for vector spaces.

2.7 DEFINITION

OPERATIONS ON SUBSETS OF G Let G be a group, let $g \in G$, and let $A, B \subseteq G$. We define:

(i) $AB = \{ab \mid a \in A \wedge b \in B\}$
(ii) $A^{-1} = \{a^{-1} \mid a \in A\}$
(iii) $gA = \{ga \mid a \in A\}$
(iv) $Ag = \{ag \mid a \in A\}$

Note that

$$gA = \{g\}A \quad \text{and} \quad Ag = A\{g\}.$$

Because G need not be commutative, it is quite possible that gA and Ag represent different subsets of G. We refer to gA as the **left translate** and to Ag as the **right translate** of the set A by the group element g. The proofs of the following two lemmas, which are similar to Lemmas 1.2 and 1.3, are left as exercises (Problems 18a and 18b).

2.8 LEMMA

Let G be a group, let $g \in G$, and let $A, B, C \subseteq G$. Then:

(i) $(AB)C = A(BC)$
(ii) $(A^{-1})^{-1} = A$
(iii) $(AB)^{-1} = B^{-1}A^{-1}$
(iv) $gA = gB \Rightarrow A = B$
(v) $Ag = Bg \Rightarrow A = B$

2.9 LEMMA

Let G be a group, and let $H \subseteq G$. Then H is a subgroup of G if and only if $H \neq \varnothing$ and $HH^{-1} \subseteq H$.

If H is a subgroup of the group G, then we have

$$HH = H$$

(Problem 19a) and

$$H^{-1} = H$$

(Problem 19b).

The analogues for groups of affine subspaces of vector spaces are called *cosets*.

2.10 DEFINITION

COSETS Let H be a subgroup of the group G. If $g \in G$, we call

$$gH = \{gh \mid h \in H\}$$

a **left coset of G modulo H** and we call

$$Hg = \{hg \mid h \in H\}$$

a **right coset of G modulo H**.

In this definition, the left coset gH and, likewise, the right coset Hg, are said to be **determined**, or **generated**, by g. Because $1 \in H$, we always have

$$g \in gH \quad \text{and} \quad g \in Hg.$$

Furthermore, if $h \in H$, then

$$hH = H = Hh$$

(Problem 19c). Note that cosets are to groups as affine subspaces are to vector spaces.

2.11 Example Let $G = \{1, a, b, c, d, f\}$ be the group of Example 2.4, page 367, and let H be the subgroup $H = \{1, c\}$. Find all left cosets and all right cosets of G modulo H.

SOLUTION

Left cosets	*Right cosets*
$1H = \{1, c\}$	$H1 = \{1, c\}$
$aH = \{a, d\}$	$Ha = \{a, f\}$
$bH = \{b, f\}$	$Hb = \{b, d\}$
$cH = \{c, 1\}$	$Hc = \{c, 1\}$
$dH = \{d, a\}$	$Hd = \{d, b\}$
$fH = \{f, b\}$	$Hf = \{f, a\}$

□

In Example 2.11, notice that the left and right cosets generated by an element may or may not be the same. Also, two different elements of G can generate the same left coset (or the same right coset). For instance, $aH = dH$ and $Hb = Hd$.

Although the left cosets and the right cosets of a group modulo a subgroup need not be the same, there is a natural one-to-one correspondence between them.

2.12 LEMMA Let G be a group, let H be a subgroup of G, let $\mathscr{L}$ denote the set of all left cosets of G modulo H, and let $\mathscr{R}$ denote the set of all right cosets of G modulo H. Then, for each $g \in G$,

$$(gH)^{-1} = Hg^{-1} \in \mathscr{R},$$

and the mapping

$$gH \mapsto Hg^{-1}$$

is a bijection from $\mathscr{L}$ onto $\mathscr{R}$.

PROOF Using Part (iii) of Lemma 2.8 and the fact that, for a subgroup H of G, $H^{-1} = H$, we have

$$(gH)^{-1} = H^{-1}g^{-1} = Hg^{-1}.$$

That $gH \mapsto Hg^{-1}$ maps $\mathscr{L}$ surjectively onto $\mathscr{R}$ follows from the fact that

$$(g^{-1}H)^{-1} = H(g^{-1})^{-1} = Hg.$$

That $gH \mapsto Hg^{-1}$ is injective is established as follows: Suppose that $a, b \in G$ and that

$$Ha^{-1} = Hb^{-1}.$$

Then,

$$aH = (a^{-1})^{-1}H^{-1} = (Ha^{-1})^{-1} = (Hb^{-1})^{-1} = (b^{-1})^{-1}H^{-1} = bH. \qquad \square$$

As a consequence of Lemma 2.12, $\#\mathscr{L} = \#\mathscr{R}$, that is, the number of distinct left cosets of G modulo H is the same as the number of distinct right cosets of G modulo H. Lemma 2.12 establishes a "symmetry" between the theory of left and right cosets, so that every general statement about left cosets gives rise to a corresponding general statement about right cosets, and vice versa. Because of this, we shall focus our attention on *left* cosets.

If H is a subgroup of a group G, then the set of all left cosets of G modulo H is the analogue for groups of the set V/W of all affine subspaces of a vector space V that are translates of a linear subspace W (Definition 1.15). This analogy is emphasized by using a similar notation.

2.13 DEFINITION

QUOTIENT SPACE If G is a group and H is a subgroup of G, then the set of all left cosets of G modulo H is called the (left) **quotient space of G modulo H** and is denoted by G/H. Thus,

$$G/H = \{gH \mid g \in G\}.$$

For a vector space V and a linear subspace W, V/W is a partition of V (Problem 31 in Problem Set 5.1). We show that, likewise, for a group G and a subgroup H, the set G/H of all left cosets of G modulo H is a partition of G. We begin with the following lemma:

2.14 LEMMA

Let G be a group, let H be a subgroup of G, and let $a, b \in G$. Then:

(i) $aH \neq \varnothing$
(ii) $a \in bH \Leftrightarrow b^{-1}a \in H$
(iii) $aH \cap bH \neq \varnothing \Leftrightarrow a \in bH$
(iv) $aH \cap bH \neq \varnothing \Leftrightarrow aH = bH$

PROOF

(i) We have already noted that $a \in aH$; hence, $aH \neq \varnothing$.
(ii) Suppose $a \in bH$. Then $a = bh$ for some $h \in H$, and it follows that $b^{-1}a = b^{-1}bh = 1h = h \in H$. Conversely, suppose $b^{-1}a = h \in H$. Then $a = 1a = bb^{-1}a = bh$, and so $a \in bH$.

(iii) Suppose $aH \cap bH \neq \varnothing$. Let $c \in aH \cap bH$. Then there exist elements $h, h' \in H$ such that $c = ah$ and $c = bh'$. Thus, $ah = bh'$, so $a = a1 = ahh^{-1} = bh'h^{-1}$. Since H is a subgroup of G, we have $h'h^{-1} \in H$, and it follows that $a \in bH$. Conversely, if $a \in bH$, then $aH \cap bH \neq \varnothing$, owing to the fact that $a \in aH$.

(iv) Suppose $aH \cap bH \neq \varnothing$. Then, by Part (iii), $a \in bH$. Thus, there exists $h \in H$ such that $a = bh$. Consequently, $aH = bhH$. But, since H is a subgroup of G and $h \in H$, we have $hH = H$, and it follows that $aH = bH$. Conversely, if $aH = bH$, then $aH \cap bH = aH \cap aH = aH \neq \varnothing$. □

2.15 THEOREM

G/H IS A PARTITION OF G If G is a group and H is a subgroup of G, then G/H is a partition of G.

PROOF By Part (i) of Lemma 2.14, G/H is a set of nonempty subsets of G. By Part (iv) of the same lemma, the sets in G/H are mutually exclusive. For each element $g \in G$, we have $g \in gH$, and so the sets in G/H are exhaustive. □

2.16 LEMMA

Let G be a group, let H be a subgroup of G, and let $a \in G$. Then the mapping from H into aH given by

$$h \mapsto ah$$

is a bijection.

PROOF By the very definition of aH, the mapping $h \mapsto ah$ is a surjection from H onto aH. To show that this mapping is an injection, suppose that $h, h' \in H$ with $ah = ah'$. Then, by the left-cancellation law (Theorem 1.13 in Chapter 3), $h = h'$. □

2.17 COROLLARY

Let G be a group, let H be a subgroup of G, and let $a, b \in G$. Then,

$$\#(aH) = \#H = \#(bH).$$

PROOF By Lemma 2.16, $\#H = \#(aH)$. Likewise, $\#H = \#(bH)$. □

2.18 DEFINITION

INDEX OF A SUBGROUP If H is a subgroup of the group G, then $\#(G/H)$ is called the **index** of H in G.

Thus, the index of H in G is the number of distinct left cosets of G modulo H. By Lemma 2.12, this is the same as the number of distinct right cosets of G modulo H.

2.19 THEOREM

THE ORDER-INDEX THEOREM Let G be a finite group, and let H be a subgroup of G. Then,

$$\#G = \#(G/H) \cdot \#H.$$

PROOF Let $k = \#(G/H)$ and $q = \#H$. By Corollary 2.17, all of the left cosets in G/H have the same number of elements, namely, q. There are k of these left cosets, and they are disjoint from one another, so they contain a total of kq elements. But the union of all of the left cosets in G/H is G; hence, $\#G = kq = \#(G/H) \cdot \#H$. □

Recall that $\#G$ is called the *order* of the group G. Therefore, Theorem 2.19 can be stated in words as follows: *If G is a finite group and H is a subgroup of G, then the order of G is the product of the index of H in G and the order of H.* This can also be rewritten in the equivalent and easily remembered form

$$\#(G/H) = \#G/\#H.$$

As an immediate consequence, we obtain the following basic theorem, named in honor of its discoverer, the French mathematician Joseph-Louis Lagrange (1736–1813).

2.20 THEOREM

LAGRANGE'S THEOREM If G is a finite group, then the order of any subgroup of G divides the order of G.

For instance, a group G of order 6 can have only subgroups of orders 1, 2, 3, and 6. Of course, the subgroups of orders 1 and 6 are the trivial subgroups $\{1\}$ and G. Thus, in Example 2.4, the group $G = \{1, a, b, c, d, f\}$ of order 6 has three nontrivial subgroups $\{1, c\}$, $\{1, d\}$, and $\{1, f\}$ of order 2 and one nontrivial subgroup $\{1, a, b\}$ of order 3; there cannot be any subgroups of orders 4 or 5 by Lagrange's theorem. The "converse" of Lagrange's theorem is *false*; just because a natural number k divides the order of a group G, there is no guarantee that G contains a subgroup of order k. For instance, there is a group of order 12—called the *alternating group* A_4 (see Definition 5.21)—that contains no subgroup of order 6.

If H is a subgroup of a *commutative* group G, then, for each element $g \in G$, the left coset gH is the same as the right coset Hg. There are subgroups of *noncommutative* groups that have the same property. This leads us to our next definition.

2.21 DEFINITION

NORMAL SUBGROUP Let G be a group. A subgroup N of G is called a **normal** subgroup of G if, for every $g \in G$,

$$gN = Ng.$$

Some authors refer to normal subgroups as **invariant**, or **self-conjugate**, subgroups. Note that the two trivial subgroups $\{1\}$ and G are always normal subgroups of G.

It is important to keep the following fact in mind:

If N is a normal subgroup of a group G and $g \in G$, the condition $gN = Ng$ does not necessarily imply that $gx = xg$ for all elements $x \in N$.

2.22 Example Give an example of a group G, a normal subgroup N, an element $g \in G$, and an element $x \in N$ such that $gx \neq xg$.

SOLUTION Let $G = \{1, a, b, c, d, f\}$ be the group in Example 2.4, and let $N = \{1, a, b\}$. By Problem 23, N is a normal subgroup of G. In particular, then, $cN = Nc$; in fact,

$$cN = \{c, d, f\} = Nc.$$

Also, $a \in N$, but

$$ca = f \neq d = ac. \qquad \square$$

In Example 2.22, notice that $\#G = 6$ and $\#N = 3$, so N has index 2 in G. Therefore, the fact that N is a normal subgroup of G is a consequence of the following general result:

2.23 THEOREM

SUBGROUPS OF INDEX 2 If G is a group, N is a subgroup of G, and $\#(G/N) = 2$, then N is a normal subgroup of G.

PROOF By hypothesis, there are just two left cosets of G modulo N. One of them is N itself, and since G/N is a partition of G, the other one must be $G \setminus N$. Since the number of left cosets is the same as the number of right cosets, there are just two right cosets of G modulo N. Again, one of the right cosets is N itself, so the other one must be $G \setminus N$. Now, let $g \in G$. If $g \in N$, then

$$gN = N = Ng.$$

If $g \notin N$, then $gN \neq N$, and so $gN = G \setminus N$. Likewise, if $g \notin N$, then $Ng = G \setminus N$. Hence, if $g \notin N$,

$$gN = G \setminus N = Ng.$$

Therefore, in any case, $gN = Ng$, so N is a normal subgroup of G. □

We leave the proof of the following lemma as an exercise (Problem 25).

2.24 LEMMA If G is a group, Z is the center of G, and N is a subgroup of Z, then N is a normal subgroup of G.

For a particular group G, there may or may not be nontrivial normal subgroups of G, that is, normal subgroups other than $\{1\}$ and G itself. A group G is said to be **simple** if it contains no nontrivial normal subgroups. Simple groups are the "building blocks" with which all other finite groups can be constructed; they are the analogues in group theory of the primes in number theory. One of the triumphs of modern algebra has been the complete classification of all finite simple groups, a monumental undertaking that was completed in the 1970s.

PROBLEM SET 5.2

1. Write the definition of a subgroup of an additively written group $(G, +)$.
2. Consider the two-dimensional coordinate vector space $\mathbb{R}^2$ over the field $\mathbb{R}$. Find a subgroup of the additive group $(\mathbb{R}^2, +)$ that is *not* a linear subspace of the vector space $\mathbb{R}^2$. [Hint: Make use of the additive group $(\mathbb{Z}, +)$.]
3. If H is a subgroup of a multiplicative group G, prove that, under the restrictions to H of the operations of multiplication and the formation of inverses, H is a group in its own right.
4. Give an example of a (necessarily infinite) multiplicative group G and a nonempty subset M of G that is closed under multiplication but is not a subgroup of G.
5. Rewrite the group AS13 in Section 6 of Chapter 2, page 145, as a multiplicative group G, and find all (trivial and nontrivial) subgroups of G.
6. Consider the multiplicative group S^1 of all complex numbers z with $|z| = 1$. If n is a positive integer, show that the set $H = \{z \in \mathbb{C} \mid z^n = 1\}$ is a subgroup of S^1, and show that H has order n. [Hint: Use the polar form of complex numbers; see Problem 20 in Problem Set 4.5.]

7. In Example 2.4, show that $\{1, a, b\}$ is a subgroup of G.

8. Consider the multiplicative group G of all nonzero quaternions. Find the center Z of G.

9. Let G be a multiplicative group, let $\varnothing \neq S \subseteq G$, suppose that S is closed under multiplication, and suppose that, under the restriction to S of the multiplication operation on G, S forms a monoid. Prove that the neutral element of the monoid S must be the neutral element $1 \in G$.

10. Consider the multiplicative group S^3 of all quaternions q with $|q| = 1$. [See Problem 24 in Problem Set 4.8.] Find the center Z of S^3.

11. Prove that a finite nonempty subset S of a multiplicative group G is a subgroup if and only if it is closed under multiplication.

12. Let G be a finite group, let M be a nonempty subset of G, and let S be the subset of G consisting of all finite products of elements of M. Prove that:
    (a) S is a subgroup of G.
    (b) If H is a subgroup of G such that $M \subseteq H$, then S is a subgroup of H.

13. If G is a group, H is a subgroup of G, and, with H regarded as a group in its own right, K is a subgroup of H, prove that K is a subgroup of G.

14. Show that $\{1, c\}$, $\{1, d\}$, $\{1, f\}$, and $\{1, a, b\}$ are *all* of the nontrivial subgroups of the group G of Example 2.4.

15. If G is a group and H, K are subgroups of G, prove that $H \cap K$ is a subgroup of G.

16. Prove that the intersection of any set of subgroups of a group G is again a subgroup of G.

17. Prove Theorem 2.6.

18. (a) Prove Lemma 2.8. (b) Prove Lemma 2.9.

19. If H is a subgroup of the multiplicative group G and $g \in G$, prove:
    (a) $HH = H$ (b) $H^{-1} = H$ (c) $gH = H \Leftrightarrow g \in H$

20. Let $\mathcal{Q}$ be the set consisting of 1, the three quaternionic units **i**, **j**, and **k**, and the negatives of these four elements.
    (a) Show that $\mathcal{Q}$ is a subgroup of the group S^3. [See Problem 10.] ($\mathcal{Q}$ is called the **quaternion group**.)
    (b) Show that $H = \{1, -1\}$ is a subgroup of $\mathcal{Q}$.
    (c) Find all left cosets and all right cosets of $\mathcal{Q}$ modulo H.
    (d) Is H a normal subgroup of $\mathcal{Q}$? Explain.

21. Let $G = \{1, a, b, c, d, f\}$ be the group of Example 2.4, and let K be the subgroup $K = \{1, d\}$.
    (a) Find all left cosets and all right cosets of G modulo K.
    (b) Is K a normal subgroup of G? Explain.

22. Find a subgroup N of the quaternion group $\mathcal{Q}$ (Problem 20) such that N has index 2 in $\mathcal{Q}$, and prove by direct computation (without invoking Theorem 2.23) that N is a normal subgroup of $\mathcal{Q}$.

23. Show by a direct calculation (without invoking Theorem 2.23) that $N = \{1, a, b\}$ is a normal subgroup of the group G of Example 2.4.

24. If H is a subgroup of a group G, the **normalizer** of H in G is defined to be the set $S = \{g \in G \mid gHg^{-1} \subseteq H\}$.
(a) Show that S is a subgroup of G.
(b) Show that H is a normal subgroup of S.

25. Prove Lemma 2.24.

26. Let G be the group of Example 2.4, and let H be the subgroup $H = \{1, c\}$. Find the normalizer S of H in G (Problem 24).

27. If p is a prime number and G is a group of order p, show that G has no nontrivial subgroups.

28. Prove that every subgroup of the quaternion group $\mathcal{Q}$ (Problem 20) is a normal subgroup.

29. Let H be a subgroup of the group G. Prove that H is a normal subgroup of G if and only if $gHg^{-1} = H$ holds for every $g \in G$.

30. Reformulate the following for the case of an additively written commutative group $(G, +)$:
(a) Definition 2.7 (b) Lemma 2.8 (c) Lemma 2.9
(d) Definition 2.10 (e) Definition 2.13 (f) Lemma 2.14
(g) Lemma 2.16 (h) Corollary 2.17

31. Explain why a simple commutative group can have no nontrivial subgroups.

*3 QUOTIENT GROUPS AND HOMOMORPHISMS

In Theorem 1.19 we showed that, if W is a linear subspace of a vector space V, then, in a natural way, the set V/W forms a vector space in its own right. We now show that, analogously, if N is a *normal* subgroup of a group G, then, in a natural way, G/N forms a group. In what follows, we use multiplicative notation; however, all of our results hold for additively written groups as well, provided that we switch from multiplicative to additive notation. We begin with the following lemma:

3.1 LEMMA Let N be a normal subgroup of the group G. Then, for $a, b \in G$, $(aH)(bH) = (ab)H$.

PROOF Using Part (i) of Lemma 2.8, the fact that N is a subgroup of G so that $NN = N$, and the supposition that N is normal so that $Nb = bN$, we have

$$\begin{aligned}(aN)(bN) &= a(N(bN)) = a((Nb)N) = a((bN)N) \\ &= a(b(NN)) = a(bN) \\ &= (ab)N.\end{aligned}$$

□

3.2 THEOREM

QUOTIENT GROUP If N is a normal subgroup of the group G, then, under the binary operation

$$(aN, bN) \mapsto (aN)(bN) = (ab)N,$$

the set G/N of all left cosets of G modulo N forms a group.

PROOF By Lemma 3.1, the product $(aN)(bN)$ of two left cosets of G modulo N is again a left coset of G modulo N. That this multiplication is associative follows from Part (i) of Lemma 2.8. Because, for every $g \in G$,

$$(1N)(gN) = (1g)N = gN \quad \text{and} \quad (gN)(1N) = (g1)N = gN,$$

it follows that the left coset $1N = N$ serves as the neutral element for the semigroup $(G/N, \cdot)$. For every $g \in G$,

$$(gN)(g^{-1}N) = (gg^{-1})N = 1N \quad \text{and} \quad (g^{-1}N)(gN) = (g^{-1}g)N = 1N,$$

and it follows that every element gN in the monoid $(G/N, \cdot)$ is invertible, with

$$(gN)^{-1} = g^{-1}N.$$

□

The group G/N of Theorem 3.2 is called the **quotient group**, or **factor group**, of G modulo N. In working with this group, you must keep in mind that its elements are *subsets* (that is, cosets modulo N) of the group G. In particular, the neutral element of the group G/N is the normal subgroup $N = 1N$ itself. Also, as the proof of Theorem 3.2 shows, for $g \in G$, the inverse of the element gN of G/N is given by

$$(gN)^{-1} = g^{-1}N.$$

If G is a finite group and N is a normal subgroup of G, then, by Theorem 2.19,

$$\#(G/N) = \#G/\#N.$$

The following lemma shows that, if H is a subgroup of a group G and the set G/H of left cosets of G modulo H is closed under multiplication, then H must be a normal subgroup of G (and so G/H is, in fact, a group).

3.3 LEMMA Let H be a subgroup of the group G, and suppose that, for every $a, b \in G$, there exists $c \in G$ such that

$$(aH)(bH) = cH.$$

Then H is a normal subgroup of G.

PROOF We must prove that $gH = Hg$ holds for all $g \in G$. Let $g \in G$. We begin by proving that $Hg \subseteq gH$. Thus, let $h \in H$. By the hypothesis of the lemma with $a = 1$ and $b = g$, there exists $c \in G$ such that

$$HgH = (1H)(gH) = cH.$$

Now, since $1 \in H$, we have

$$hg, g \in HgH = cH.$$

Therefore, because $g \in gH$, we have

$$cH = gH, \quad \text{so} \quad hg \in gH.$$

This proves that, for every $g \in G$,

$$Hg \subseteq gH.$$

If $g \in G$, then $g^{-1} \in G$, and so, by the inclusion above,

$$Hg^{-1} \subseteq g^{-1}H.$$

Therefore, the set of all inverses of elements in Hg^{-1} is contained in the set of all inverses of elements in $g^{-1}H$, that is,

$$(Hg^{-1})^{-1} \subseteq (g^{-1}H)^{-1}.$$

Consequently,

$$(g^{-1})^{-1}H^{-1} \subseteq H^{-1}(g^{-1})^{-1},$$

and, in view of the fact that $H^{-1} = H$, we have

$$gH \subseteq Hg.$$

Combining the last inclusion with the previously obtained inclusion $Hg \subseteq gH$, we find that $gH = Hg$. □

The analogue for groups of the concept of a linear transformation for vector spaces (Definition 6.15 in Chapter 3) is called a *homomorphism* (from the Greek *homos*, meaning *alike*, and *morphe*, meaning *form* or *structure*).

3.4 DEFINITION

HOMOMORPHISM If G and G' are multiplicative groups, then a mapping $\phi: G \to G'$ is called a **group homomorphism** if and only if, for every $a, b \in G$,

$$\phi(ab) = \phi(a)\phi(b).$$

For short, a group homomorphism is usually referred to simply as a **homomorphism**. Notice that *an isomorphism* $\phi: G \to G'$ (Definition 6.12 in Chapter 2) *is the same thing as a bijective homomorphism.* In other words, the concept of a group homomorphism is a generalization of the concept of a group isomorphism.

3.5 LEMMA

Let G and G' be groups, and let $\phi: G \to G'$ be a homomorphism. Let 1 and $1'$ denote the neutral elements in G and G', respectively, and let $g \in G$. Then:

(i) $\phi(1) = 1'$
(ii) $\phi(g^{-1}) = (\phi(g))^{-1}$

PROOF (i) Let $e = \phi(1)$. Then,

$$ee = \phi(1)\phi(1) = \phi(1 \cdot 1) = \phi(1) = e = e1'.$$

Applying the left-cancellation law to the equation

$$ee = e1',$$

we conclude that

$$\phi(1) = e = 1'.$$

(ii) By Part (i), we have

$$1' = \phi(1) = \phi(gg^{-1}) = \phi(g)\phi(g^{-1}),$$

from which it follows that

$$\phi(g^{-1}) = (\phi(g))^{-1}.$$

□

In Lemma 3.5, we have used different notation (1 and $1'$) for the neutral elements in the groups G and G'. Ordinarily, this is not necessary, since it is usually possible to tell from the context which neutral element is intended. Thus, we can rewrite Part (i) of Lemma 3.5 as

$$\phi(1) = 1,$$

it being understood that the 1 on the left is the neutral element of G and the 1 on the right is the neutral element of G'.

We leave the proof of the following theorem as an exercise (Problem 9).

3.6 THEOREM

THE COMPOSITION OF HOMOMORPHISMS Let G, G', and G'' be groups, and let

$$\phi: G \to G' \quad \text{and} \quad \theta: G' \to G''$$

be homomorphisms. Then,

$$\theta \circ \phi: G \to G''$$

is a homomorphism.

In words, Theorem 3.6 says that *the composition of homomorphisms is again a homomorphism.*

3.7 DEFINITION

IMAGE AND INVERSE IMAGE Let G and G' be groups, and let $\phi: G \to G'$ be a homomorphism. Let $A \subseteq G$, and let $B \subseteq G'$.

(i) The **direct image**, or simply the **image**, of A under ϕ, in symbols $\phi(A)$, is defined by

$$\phi(A) = \{\phi(a) \mid a \in A\}.$$

(ii) The **inverse image** of B under ϕ, in symbols $\phi^{-1}(B)$, is defined by

$$\phi^{-1}(B) = \{g \in G \mid \phi(g) \in B\}.$$

Images and inverse images of sets under homomorphisms are just special cases of the images and inverse images of sets under functions introduced in Section 2.7. The next theorem is the analogue for groups of Theorem 6.19 in Chapter 3 for vector spaces.

3.8 THEOREM

PRESERVATION OF SUBGROUPS Let G, G' be groups, and let $\phi: G \to G'$ be a homomorphism.

(i) If H is a subgroup of G, then $\phi(H)$ is a subgroup of G'.
(ii) If K is a subgroup of G', then $\phi^{-1}(K)$ is a subgroup of G.

PROOF We prove Part (ii) here and leave Part (i) as an exercise (Problem 11). Let K be a subgroup of G'. We must show that $1 \in \phi^{-1}(K)$ and that $\phi^{-1}(K)$ is closed under multiplication and the formation of inverses. By Part (i) of Lemma 3.5, $\phi(1) = 1$. Since K is a subgroup of G', it follows that $1 \in K$, and therefore $\phi(1) \in K$, which shows that $1 \in \phi^{-1}(K)$. Let $a, b \in \phi^{-1}(K)$. Then $\phi(a), \phi(b) \in K$; hence, $\phi(ab) = \phi(a)\phi(b) \in K$, and so $ab \in \phi^{-1}(K)$. Finally, suppose that $a \in \phi^{-1}(K)$. Then $\phi(a) \in K$; so, by Part (ii) of Lemma 3.5, $\phi(a^{-1}) = (\phi(a))^{-1} \in K$, and it follows that $a^{-1} \in \phi^{-1}(K)$. □

The next definition and theorem are analogues for groups of Definition 6.20 and Theorem 6.21 in Chapter 3.

3.9 DEFINITION

KERNEL OF A HOMOMORPHISM Let $\phi: G \to G'$ be a homomorphism from the group G into the group G'. The **kernel** of ϕ, in symbols, $\ker(\phi)$, is defined by

$$\ker(\phi) = \phi^{-1}(\{1\}).$$

Thus, the kernel of $\phi: G \to G'$ is the inverse image under ϕ of the trivial subgroup $\{1\}$ of G'. Therefore, by Part (ii) of Theorem 3.8, *the kernel of a group homomorphism* $\phi: G \to G'$ *is a subgroup of* G. For simplicity, $\phi^{-1}(\{1\})$ is often written as $\phi^{-1}(1)$.

3.10 THEOREM

INJECTIVITY OF A HOMOMORPHISM Let $\phi: G \to G'$ be a homomorphism from the group G into the group G'. Then ϕ is injective (one-to-one) if and only if $\ker(\phi) = \{1\}$.

PROOF Suppose ϕ is injective, and let $a \in \ker(\phi) = \phi^{-1}(1)$. Then $\phi(a) = 1 = \phi(1)$, and it follows from the injectivity of ϕ that $a = 1$. Therefore, $\ker(\phi) \subseteq \{1\}$. That $\{1\} \subseteq \ker(\phi)$ follows from the fact that $\ker(\phi)$ is a subgroup of G. Hence, if ϕ is injective, then $\ker(\phi) = \{1\}$.

Conversely, suppose that $\ker(\phi) = \{1\}$, and let $a, b \in G$ with $\phi(a) = \phi(b)$. To show that ϕ is injective, we must prove that $a = b$. Using Part (ii) of Lemma 3.5, we have

$$\phi(ab^{-1}) = \phi(a)\phi(b^{-1}) = \phi(a)(\phi(b))^{-1} = \phi(a)(\phi(a))^{-1} = 1,$$

and it follows that

$$ab^{-1} \in \phi^{-1}(1) = \ker(\phi) = \{1\}.$$

Therefore, $ab^{-1} = 1$. Multiplying both sides of the last equation on the right by b, we conclude that $a = b$. □

There are close connections between quotient groups and group homomorphisms. For instance, we have the following result, the proof of which we leave as an exercise (Problem 13).

3.11 LEMMA Let N be a normal subgroup of the group G. Define

$$\eta: G \to G/N$$

by $\eta(g) = gN$ for all $g \in G$. Then $\eta: G \to G/N$ is a group homomorphism.

Note that the homomorphism $\eta: G \to G/N$ in Lemma 3.11 is actually a *surjection*, that is, it is a homomorphism of the group G *onto* the quotient group G/N (Problem 15).

3.12 DEFINITION **NATURAL HOMOMORPHISM** If N is a normal subgroup of the group G and if

$$\eta: G \to G/N$$

is defined by $\eta(g) = gN$ for all $g \in G$, then $\eta: G \to G/N$ is called the **natural homomorphism**, or the **canonical homomorphism**, of G onto G/N.

If N is a normal subgroup of the group G, then the kernel of the natural homomorphism $\eta: G \to G/N$ is given by

$$\ker(\eta) = N$$

(Problem 17). More generally, as the next lemma shows, the kernel of any group homomorphism is a normal subgroup of its domain.

3.13 LEMMA Let $\phi: G \to G'$ be a homomorphism from the group G into the group G'. Then $\ker(\phi)$ is a normal subgroup of G.

PROOF Let $K = \ker(\phi)$, and let $g \in G$. We must show that $gK = Kg$. We prove that $gK \subseteq Kg$; the proof that $Kg \subseteq gK$ is analogous. (Alternatively, $Kg \subseteq gK$ follows from the result in Problem 18.) Thus, let

$$x \in gK,$$

so that $x = gk$ for some element $k \in K = \ker(\phi)$. We must prove that $x \in Kg$. Since $\phi(k) = 1$, we have

$$\begin{aligned}\phi(xg^{-1}) = \phi(gkg^{-1}) &= \phi(g)\phi(k)\phi(g^{-1}) \\ &= \phi(g) \cdot 1 \cdot \phi(g^{-1}) = \phi(g)(\phi(g))^{-1} \\ &= 1,\end{aligned}$$

and it follows that $xg^{-1} \in K$. Let $xg^{-1} = k' \in K$. Then,

$$x = k'g \in Kg. \qquad \square$$

3.14 DEFINITION

HOMOMORPHIC IMAGE If G and G' are groups, then G' is said to be a **homomorphic image** of G if there exists a surjective homomorphism

$$\phi: G \to G'$$

from G onto G'.

A surjective homomorphism $\phi: G \to G'$ is called an **epimorphism**. (The Greek prefix *epi* means *on* or *upon*.) If N is a normal subgroup of a group G, then the natural homomorphism

$$\eta: G \to G/N$$

is an epimorphism (Problem 15), and so G/N is a homomorphic image of G. Thus, any quotient group of a group modulo a normal subgroup is a homomorphic image of the group. The next theorem shows that, conversely, every homomorphic image of a group is isomorphic to a quotient group of the group modulo a normal subgroup.

3.15 THEOREM

FUNDAMENTAL HOMOMORPHISM THEOREM Let G and G' be groups, and let $\phi: G \to G'$ be a surjective homomorphism of G onto G' (that is, an epimorphism). Let $N = \ker(\phi)$, and let $\eta: G \to G/N$ be the natural homomorphism. Then there exists a unique isomorphism $\Phi: G/N \to G'$ such that $\Phi \circ \eta = \phi$.

PROOF Let $a, b \in G$, and suppose that $aN = bN$. Since $a \in aN = bN$, it follows that $a = bk$ with $k \in N = \ker(\phi)$. Thus, $\phi(k) = 1$, and we have

$$\phi(a) = \phi(bk) = \phi(b)\phi(k) = \phi(b) \cdot 1 = \phi(b).$$

This proves that, for all $a, b \in G$,

$$aN = bN \Rightarrow \phi(a) = \phi(b).$$

Consequently, the mapping

$$\Phi: G/N \to G'$$

defined for all $a \in G$ by

$$\Phi(aN) = \phi(a)$$

is well defined. Also, if $x, y \in G$, then

$$\Phi((xN)(yN)) = \Phi(xyN) = \phi(xy) = \phi(x)\phi(y) = \Phi(xN)\Phi(yN),$$

so Φ is a homomorphism.

Suppose that $xN \in \ker(\Phi)$. Then,

$$\Phi(xN) = \phi(x) = 1,$$

and therefore, $x \in N$, so that $xN = N$. Thus,

$$\ker(\Phi) = \{N\}.$$

By Theorem 3.10, the last equation shows that Φ is an injection. (Recall that N is the neutral element of G/N.)

Suppose that $g' \in G'$. Since $\phi: G \to G'$ is surjective, there exists $g \in G$ such that $\phi(g) = g'$. Therefore,

$$\Phi(gN) = \phi(g) = g',$$

which shows that $\Phi: G/N \to G'$ is surjective. Consequently, $\Phi: G/N \to G'$ is an isomorphism.

For every $g \in G$, we have

$$(\Phi \circ \eta)(g) = \Phi(\eta(g)) = \Phi(gN) = \phi(g),$$

and therefore,

$$\Phi \circ \eta = \phi.$$

To show that Φ is unique, suppose $\Phi': G/N \to G'$ is an isomorphism such that

$$\Phi' \circ \eta = \phi.$$

Then, for every $g \in G$, we have

$$\Phi'(gN) = \Phi'(\eta(g)) = (\Phi' \circ \eta)(g) = \phi(g) = \Phi(gN),$$

and therefore,

$$\Phi' = \Phi. \qquad \square$$

Theorem 3.15 has the following important consequence:

> *We get a survey of all possible homomorphic images, up to isomorphism, of a group G by finding all normal subgroups N of G and forming the corresponding quotient groups G/N.*

PROBLEM SET 5.3

1. If G is a group, H is a subgroup of G, and $a, b \in G$, show that $aH = bH \Leftrightarrow b^{-1}a \in H$.
2. Suppose that G is a group and that $\mathscr{D}$ is a partition of the set G such that, under the operation of set multiplication given in Part (i) of Definition

2.7, $\mathscr{D}$ forms a semigroup. Show that there exists a normal subgroup N of G such that $\mathscr{D} = G/N$.

3. If G is a group and H is a subgroup of G, show that H is a normal subgroup if and only if $gHg^{-1} \subseteq H$ holds for every $g \in G$.

4. Let $G = \{1, a, b, c, d, f\}$ be the group of Example 2.4, and let $G' = \{1, -1\} \subseteq \mathbb{R}$ with the usual multiplication. Show that the mapping $\phi: G \to G'$ given by $g(1) = 1$, $g(a) = 1$, $g(b) = 1$, $g(c) = -1$, $g(d) = -1$, and $g(f) = -1$ is a surjective homomorphism of G onto G' and find $\ker(\phi)$.

5. Formulate the concept of a group homomorphism $\phi: G \to G'$ for additively written groups $(G, +)$ and $(G', +)$.

6. Let $G = \mathbb{R} \setminus \{0\}$, so that G is a group under ordinary multiplication of real numbers.
   (a) Show that the mapping $\phi: G \to G$ given by $\phi(x) = |x|$ for all $x \in G$ is a homomorphism.
   (b) Find $\ker(\phi)$.
   (c) Find $\phi(G)$.

7. Formulate the concept of a group homomorphism $\phi: G \to G'$ if $(G, +)$ is an additively written group and $(G, \cdot)$ is a multiplicatively written group.

8. Let F be a field, let G be the multiplicative group of all nonzero elements of F, and let $GL(2, F)$ be the set of all two-by-two matrices

$$M = \begin{bmatrix} a & b \\ c & d \end{bmatrix}$$

   with $a, b, c, d \in F$ and $ad - bc \neq 0$.
   (a) Show that $GL(2, F)$ forms a group under matrix multiplication (as in Example 2.13 in Chapter 3).
   (b) Show that the mapping $\delta: GL(2, F) \to G$ given by $\delta(M) = ad - bc$ is a homomorphism.
   (c) Show that $\delta: GL(2, F) \to G$ is surjective.

9. Prove Theorem 3.6.

10. Let $(\mathbb{R}, +)$ be the additive group of real numbers, and let $GL(2, \mathbb{R})$ be the group defined in Problem 8.
   (a) Show that the mapping $\rho: \mathbb{R} \to GL(2, \mathbb{R})$ of the additive group $\mathbb{R}$ into the multiplicative group $GL(2, \mathbb{R})$ defined for all $\theta \in \mathbb{R}$ by

$$\rho(\theta) = \begin{bmatrix} \cos\theta & \sin\theta \\ -\sin\theta & \cos\theta \end{bmatrix}$$

   is a homomorphism.
   (b) Find $\ker(\theta)$.

11. Prove Part (i) of Theorem 3.8.

12. In Problem 8, let $F = \mathbb{R}$, and let $SL(2, \mathbb{R}) = \ker(\delta)$. If ρ is the homomorphism of Problem 10, show that $\rho(\mathbb{R})$ is a subgroup of $SL(2, \mathbb{R})$.

13. Prove Lemma 3.11.

14. Let S^1 denote the multiplicative group of all complex numbers of the form $z = e^{i\theta}$ (see Problem 20 in Problem Set 4.5). Let n be a fixed positive integer, and define $\psi_n:\mathbb{Z} \to S^1$ by $\psi_n(m) = e^{2\pi im/n}$ for all $m \in \mathbb{Z}$.
(a) Prove that ψ_n is a homomorphism of the additive group $(\mathbb{Z}, +)$ into the multiplicative group $(S^1, \cdot)$.
(b) Find $\ker(\psi_n)$.

15. If N is a normal subgroup of the group G and $\eta:G \to G/N$ is the natural homomorphism, prove that η is surjective; that is, show that η is an epimorphism.

16. In Problem 14, show that $\psi_n(\mathbb{Z}) = \{z \in \mathbb{C} \mid z^n = 1\}$ (see Problem 6 in Problem Set 5.2).

17. In Problem 15, show that $N = \ker(\eta)$. [Hint: Keep in mind that the elements of G/N are *subsets* of G and that the neutral element of G/N is N.]

18. Let K be a subgroup of a group G, and suppose that $gK \subseteq Kg$ holds for every $g \in G$. Prove that K is a normal subgroup of G.

19. If G is a group, then an isomorphism $\alpha:G \to G$ of G onto itself is called an **automorphism** of G. For the group $G = \{1, a, b, c, d, f\}$ of Example 2.4, show that the mapping $\alpha:G \to G$ given by $\alpha(1) = 1$, $\alpha(a) = b$, $\alpha(b) = a$, $\alpha(c) = c$, $\alpha(d) = f$, and $\alpha(f) = d$ is an automorphism of G.

20. Find all homomorphic images (up to isomorphism) of the group $G = \{1, a, b, c, d, f\}$ of Example 2.4.

21. If G is a group and c is a fixed element of G, show that the mapping $\alpha_c:G \to G$ defined for each element $g \in G$ by $\alpha_c(g) = cgc^{-1}$ is an automorphism of G (see Problem 19). Such an automorphism is called an **inner automorphism** of G.

22. Let G, G', and G'' be groups, and let $\phi:G \to G'$ and $\theta:G' \to G''$ be homomorphisms. Prove that $\ker(\theta \circ \phi) = \phi^{-1}(\ker(\theta))$.

23. If G is a finite group and G' is a homomorphic image of G, show that the order of G' divides the order of G.

24. In Theorem 8.1 of Chapter 2, let X and Y be groups, and let $f:X \to Y$ be a homomorphism.
(a) Show that $X/E_f = X/\ker(f)$.
(b) Show that $p:X \to X/E_f$ is the natural homomorphism.
(c) Show that range(f) is a subgroup of Y.
(d) Show that $g:X/E_f \to$ range(f) is an isomorphism.
(e) Show that i:range$(f) \to Y$ is an injective homomorphism.

25. Let G, G', and G'' be groups, let $\alpha:G \to G''$ be an epimorphism, and let $\phi:G \to G'$ be a homomorphism. Show that the necessary and sufficient condition that there exists a homomorphism $\Phi: G'' \to G'$ such that $\Phi \circ \alpha = \phi$ is that $\ker(\alpha) \subseteq \ker(\phi)$.

26. In Problem 25, if $\ker(\alpha) \subseteq \ker(\phi)$, show that there is a *unique* homomorphism $\Phi: G'' \to G'$ such that $\Phi \circ \alpha = \phi$.

27. In Problem 25, if $\phi: G \to G'$ is an epimorphism and $\ker(\alpha) \subseteq \ker(\phi)$, show that there exists an epimorphism $\Phi: G'' \to G'$ such that $\Phi \circ \alpha = \phi$.

28. Let G be a group, let H and K be normal subgroups of G, and suppose that $\alpha: G \to G/H$ and $\beta: G \to G/K$ are the natural homomorphisms. Show that there exists an epimorphism $\gamma: G/H \to G/K$ such that $\gamma \circ \alpha = \beta$ if and only if $H \subseteq K$. [Hint: Use the results of Problems 25 and 27.]

29. Use the results of Problems 25 and 27 to give an alternative proof of the fundamental homomorphism theorem.

*4 CYCLIC GROUPS

Integer exponents or powers are used in a multiplicative group G in much the same way they are used in ordinary arithmetic. Thus, if $a \in G$ and n is a positive integer, we understand that

$$a^n = aaa \cdots a,$$

where there are n factors on the right. Thus, $a^1 = a$, $a^2 = aa$, $a^3 = aaa$, and so forth. A rigorous definition of a^n can be given by invoking Theorem 7.9 in Chapter 4 with $X = G$ and $g: G \to G$ given by $g(x) = xa$ for all $x \in G$. If $s: \mathbb{N} \to G$ is the resulting mapping, then, by definition,

$$a^n = s(n), \quad \text{for all } n \in \mathbb{N}.$$

In other words, for an element $a \in G$, positive integer exponents are characterized by simple recursion as follows:

(i) $a^1 = a$
(ii) $a^{n+1} = a^n a$, for every $n \in \mathbb{N}$.

If $a \in G$, we follow the convention of ordinary arithmetic and define

$$a^0 = 1, \quad \text{the neutral element of } G.$$

Also, if n is a positive integer, we define

$$a^{-n} = (a^{-1})^n.$$

We leave it for you to prove the rules of exponents given in the following lemma (Problem 1).

4.1 LEMMA **RULES OF EXPONENTS IN A GROUP** Let G be a multiplicative group, let $a, b \in G$, and let $m, n \in \mathbb{Z}$. Then:

(i) $a^{m+n} = a^m a^n$
(ii) $(a^m)^n = a^{mn}$
(iii) $(a^{-1})^n = (a^n)^{-1} = a^{-n}$
(iv) If $ab = ba$, then $(ab)^n = a^n b^n$
(v) $1^n = 1$

The hypothesis that a commutes with b in Part (iv) of Lemma 4.1 is *essential.* Indeed, if $a, b \in G$ with $ab \neq ba$, then, in general, $(ab)^n$ is not equal to $a^n b^n$ (Problem 2).

For an additive group G, the integers are used as *coefficients* rather than as exponents, so that, if $a \in G$ and n is a positive integer, we understand that

$$na = a + a + \cdots + a,$$

where there are n summands on the right. More rigorously, positive integer coefficients in an additive group are characterized by simple recursion as follows:

(i) $1a = a$
(ii) $(n + 1)a = (na) + a, \quad$ for every $n \in \mathbb{N}$.

Naturally, we define

$$0a = 0, \quad \text{the neutral element in } G,$$

and, if n is a positive integer, we define

$$(-n)a = -(na).$$

Thus, the correspondence

Additive notation	*Multiplicative notation*
na	a^n

should be appended to the "dictionary" in Section 2 (page 366). We leave it for you to write out the rules for integer coefficients in an additive group G corresponding to the exponent rules in Lemma 4.1 (Problem 3).

In what follows, we denote by $\mathbb{Z}$ the group of integers under the operation $+$ of ordinary arithmetic addition. Note that $\mathbb{Z}$ is an infinite, additively written, commutative (that is, abelian) group. Also, we denote by G an arbitrary group, and we suppose, for definiteness, that G is multiplicatively written. In those cases in which we do not rephrase our work for an additively written group G, we invite you to do so for yourself.

4.2 LEMMA

Let $a \in G$, and let

$$H = \{a^m \mid m \in \mathbb{Z}\}.$$

Then H is a subgroup of G.

PROOF We verify the three conditions in Definition 2.3. That $1 \in H$ follows from the fact that $1 = a^0$. If $x = a^m \in H$ and $y = a^n \in H$, then $xy = a^m a^n = a^{m+n} \in H$. Finally, if $x = a^m \in H$, then $x^{-1} = (a^m)^{-1} = a^{-m} \in H$. □

4.3 DEFINITION

CYCLIC SUBGROUP If $a \in G$, we define

$$\langle a \rangle = \{a^m \mid m \in \mathbb{Z}\},$$

and we refer to $\langle a \rangle$ as the **cyclic subgroup of G generated by a.**

If G is an additive group and $a \in G$, then, of course,

$$\langle a \rangle = \{ma \mid m \in \mathbb{Z}\}.$$

4.4 DEFINITION

CYCLIC GROUP AND GENERATOR If there exists an element $a \in G$ such that

$$G = \langle a \rangle,$$

then we say that G is a **cyclic group** and that a is a **generator** of G.

Since

$$\mathbb{Z} = \{m \cdot 1 \mid m \in \mathbb{Z}\},$$

it follows that the additive group $\mathbb{Z}$ is cyclic with 1 as a generator. Note that -1 is also a generator of $\mathbb{Z}$ (Problem 5); therefore:

Generators of cyclic groups are not necessarily unique.

4.5 LEMMA

If G is a cyclic group, then G is a commutative (that is, abelian) group.

PROOF Suppose that $G = \langle a \rangle$, and let $x, y \in G$. Then there exist $m, n \in \mathbb{Z}$ such that

$$x = a^m \quad \text{and} \quad y = a^n,$$

and it follows that

$$xy = a^m a^n = a^{m+n} = a^{n+m} = a^n a^m = yx.$$ □

4.6 LEMMA

Let $a \in G$. If there is a nonzero integer m such that

$$a^m = 1,$$

then there is a smallest positive integer n such that

$$a^n = 1.$$

Furthermore, $n \mid m$.

PROOF Let $m \in \mathbb{Z}$ with $m \neq 0$ and $a^m = 1$. Note that

$$a^{-m} = (a^m)^{-1} = 1^{-1} = 1;$$

so there is a positive integer k such that

$$a^k = 1.$$

(Just take $k = m$ or $k = -m$, whichever is positive.) By Theorem 2.14 in Chapter 4, it follows that there is a smallest positive integer n such that

$$a^n = 1.$$

By the division algorithm (Theorem 3.8 in Chapter 4), there exist integers q and r with $0 \leq r < n$ such that

$$m = qn + r.$$

Therefore,

$$1 = a^m = a^{qn+r} = a^{qn}a^r = (a^n)^q a^r = 1^q a^r = 1a^r = a^r.$$

Since $a^r = 1$, $0 \leq r$, and n is the smallest positive integer such that $a^n = 1$, it follows that $r = 0$; hence, $n \mid m$. □

4.7 DEFINITION

ORDER OF AN ELEMENT OF A GROUP Let $a \in G$. If there exists a nonzero integer m such that

$$a^m = 1,$$

then a is said to have **finite order**; otherwise, a is said to have **infinite order**. If a has finite order, then the smallest positive integer n such that

$$a^n = 1$$

is called the **order** of a.

4.8 Example Let $G = \{1, a, b, c, d, f\}$ be the multiplicative group of Example 2.4 (page 367). Find the order of $a \in G$.

SOLUTION From the group table in Example 2.4, we find that

$$a^2 = b \quad \text{and} \quad a^3 = a^2 a = ba = 1;$$

so the smallest positive integer n such that $a^n = 1$ is $n = 3$, and, therefore, a has order 3. □

4.9 THEOREM

REPETITION OF POWERS Let $a \in G$. If the integer powers of a,

$$\ldots, a^{-3}, a^{-2}, a^{-1}, a^0, a^1, a^2, a^3, \ldots,$$

are all distinct, then a has infinite order. If

$$a^h = a^k$$

for $h, k \in \mathbb{Z}$ with $h \neq k$, then a has finite order n and

$$n \mid (k - h).$$

PROOF If a has finite order n, then

$$a^0 = 1 = a^n, \quad \text{with } 0 < n;$$

so the integer powers of a are not all distinct. Hence, if the integer powers of a are all distinct, then a must have infinite order. Now suppose that $h, k \in \mathbb{Z}$ with $h \neq k$ and

$$a^h = a^k.$$

Multiplying both sides of the last equation by a^{-h} and using the rules of exponents, we find that

$$1 = a^{k-h}$$

and that $k - h$ is a nonzero integer. It follows from Lemma 4.6 that a has finite order n and that $n \mid (k - h)$. □

4.10 THEOREM

ELEMENTS OF FINITE ORDER Let $a \in G$, and suppose that a has finite order n. Then,

$$\langle a \rangle = \{1, a, a^2, a^3, \ldots, a^{n-1}\}$$

and

$$\# \langle a \rangle = n.$$

PROOF Evidently,

$$\{1, a, a^2, a^3, \ldots, a^{n-1}\} \subseteq \{a^m \mid m \in \mathbb{Z}\} = \langle a \rangle.$$

To prove the opposite inclusion, suppose that $x \in \langle a \rangle$. Then there exists $m \in \mathbb{Z}$ such that

$$x = a^m.$$

By the division algorithm, there exist integers q and r with $0 \le r \le n - 1$ such that

$$m = nq + r.$$

Consequently,

$$\begin{aligned} x &= a^m = a^{nq+r} = a^{nq}a^r = (a^n)^q a^r \\ &= 1^q a^r = 1a^r = a^r \in \{1, a, a^2, a^3, \ldots, a^{n-1}\}. \end{aligned}$$

Therefore,

$$\{1, a, a^2, a^3, \ldots, a^{n-1}\} = \langle a \rangle.$$

To prove that $\#\langle a \rangle = n$, it will suffice to prove that the elements

$$1, a, a^2, a^3, \ldots, a^{n-1}$$

are all distinct from one another. Suppose, on the contrary, that

$$a^h = a^k$$

with $0 \le h < k \le n - 1$. Then, by Theorem 4.9,

$$n \mid (k - h).$$

Because both n and $k - h$ are positive integers, it follows that

$$n \le k - h.$$

But, $0 \le h$; so

$$k - h \le k \le n - 1,$$

from which we obtain the contradiction $n \le n - 1$. □

Combining Theorems 4.9 and 4.10, we see that:

> *The order of an element in a group is the same as the order of the cyclic subgroup generated by that element.*

4.11 Example Let $G = \{1, a, b, c, d, f\}$ be the multiplicative group of Example 2.4 (page 367). Find the cyclic subgroup $\langle a \rangle$ generated by the element $a \in G$.

SOLUTION By Example 4.8, we know that $a \in G$ has order 3 and that

$$a^2 = b.$$

Hence, by Theorem 4.10,

$$\langle a \rangle = \{1, a, a^2\} = \{1, a, b\}. \qquad \square$$

Combining Theorem 4.10 with Lagrange's theorem (Theorem 2.20), we obtain the following corollary:

4.12 COROLLARY

If G is a finite group and $a \in G$, then the order of a is a divisor of the order of G.

We leave as an exercise (Problem 11) the proof of the following additional corollary of Theorem 4.10:

4.13 COROLLARY

If G is a finite group of order n, then G is a cyclic group if and only if G contains an element of order n.

As a consequence of the next theorem, the proof of which we leave as an exercise (Problem 15), a homomorphic image of a cyclic group is again a cyclic group.

4.14 THEOREM

HOMOMORPHIC IMAGES OF CYCLIC GROUPS Let G be a cyclic group, let a be a generator of G, let G' be a group, and suppose that

$$\theta: G \to G'$$

is a surjective homomorphism (that is, an epimorphism). Then G' is a cyclic group with $\theta(a)$ as a generator.

If $n \in \mathbb{Z}$, then the cyclic subgroup of $\mathbb{Z}$ generated by n consists of all integer multiples of n,

$$\langle n \rangle = \{kn \mid k \in \mathbb{Z}\} = \mathbb{Z}n.$$

(See Definition 3.12 in Chapter 4.) Note that

$$\langle n \rangle = \langle -n \rangle$$

for all $n \in \mathbb{Z}$ (Problem 17); hence, every cyclic subgroup of $\mathbb{Z}$ can be written in the form

$$\langle n \rangle = \mathbb{Z}n,$$

with $n \geq 0$. The following theorem shows that every subgroup of $\mathbb{Z}$ has this form.

4.15 THEOREM

SUBGROUPS OF $\mathbb{Z}$ If N is a subgroup of $\mathbb{Z}$, then there is a unique integer $n \geq 0$ such that $N = \mathbb{Z}n$.

PROOF Let N be a subgroup of $\mathbb{Z}$. By the principal ideal theorem (Theorem 3.14 in Chapter 4), it will be sufficient to prove that N is an ideal in $\mathbb{Z}$ (Definition 3.13 in Chapter 4). Thus, we must prove:

(i) $h, k \in N \Rightarrow h + k \in N$,
(ii) $h \in N,\ k \in \mathbb{Z} \Rightarrow hk \in N$.

Condition (i) holds since N, being a subgroup of $\mathbb{Z}$, is closed under addition. That (ii) holds when k is a positive integer is proved by mathematical induction on k (Problem 24). That it holds when k is negative then follows from the observation that

$$hk = -(h(-k))$$

and that N, being a subgroup of $\mathbb{Z}$, is closed under negation. Finally, because $0 \in N$, (ii) also holds when $k = 0$. □

4.16 COROLLARY

Every subgroup of the cyclic group $\mathbb{Z}$ is a cyclic group.

4.17 DEFINITION

$\mathbb{Z}_n$ If n is an integer and $n \geq 0$, we define

$$\mathbb{Z}_n = \mathbb{Z}/\mathbb{Z}n.$$

4.18 THEOREM

THE CYCLIC GROUP $\mathbb{Z}_n$ Let $n \in \mathbb{Z}$ with $n \geq 0$. If $n = 0$, then $\mathbb{Z}_n$ is isomorphic to the infinite cyclic group $\mathbb{Z}$. If $n > 0$, then $\mathbb{Z}_n$ is a finite cyclic group of order n.

PROOF $\mathbb{Z}_n$ is a homomorphic image of the cyclic group $\mathbb{Z}$ under the natural homomorphism

$$\eta:\mathbb{Z} \to \mathbb{Z}/\mathbb{Z}n = \mathbb{Z}_n$$

(Definition 3.12); hence, by Theorem 4.14, $\mathbb{Z}_n$ is a cyclic group. Note that

$$\ker(\eta) = \mathbb{Z}_n.$$

If $n = 0$, then

$$\ker(\eta) = \mathbb{Z}0 = \{0\},$$

and it follows from Theorem 3.10 that η is an isomorphism of $\mathbb{Z}$ onto $\mathbb{Z}_n = \mathbb{Z}_0$. (Of course, Theorem 3.10 is written multiplicatively, whereas we are here dealing with additively written groups.)

Now, suppose that $n > 0$. Because 1 is a generator of the cyclic group $\mathbb{Z}$, Theorem 4.14 implies that $\eta(1)$ is a generator of the cyclic group $\mathbb{Z}_n$. Because

$$n \in \mathbb{Z}n = \ker(\eta),$$

it follows that

$$n\eta(1) = \eta(n \cdot 1) = \eta(n) = 0;$$

so $\eta(1)$ has finite order. Let k be the order of $\eta(1)$. Then k is a positive integer and $k|n$ by Lemma 4.6. Now,

$$0 = k\eta(1) = \eta(k \cdot 1) = \eta(k),$$

and so

$$k \in \ker(\eta) = \mathbb{Z}n,$$

that is, $n|k$. Because n and k are positive integers and each divides the other, it follows that $n = k$; hence, n is the order of the generator $\eta(1)$ of the cyclic group $\mathbb{Z}_n$. Therefore, by Theorem 4.10, $\mathbb{Z}_n$ is a cyclic group of order n. □

4.19 THEOREM

STRUCTURE OF CYCLIC GROUPS Let G be a cyclic group. If G is infinite, then G is isomorphic to $\mathbb{Z}$. If G is finite of order n, then G is isomorphic to $\mathbb{Z}_n$.

PROOF For definiteness, we assume that G is written multiplicatively. Let $a \in G$ be a generator for G. Define

$$\theta:\mathbb{Z} \to G = \langle a \rangle = \{a^m | m \in \mathbb{Z}\}$$

by

$$\theta(m) = a^m, \quad \text{for all } m \in \mathbb{Z}.$$

Then θ is a surjective homomorphism of the additive group $\mathbb{Z}$ onto the multiplicative group G (Problem 25). By Theorem 4.15, there is a unique integer $n \geq 0$ such that

$$\ker(\theta) = \mathbb{Z}n.$$

By the fundamental homomorphism theorem (Theorem 3.15), G is isomorphic to

$$\mathbb{Z}/\ker(\theta) = \mathbb{Z}/\mathbb{Z}n = \mathbb{Z}_n.$$

If G is infinite, then $\mathbb{Z}_n$ is infinite; so $n = 0$ and G is isomorphic to $\mathbb{Z}$ by Theorem 4.18. If G is finite, then $\mathbb{Z}_n$ is finite; so $n > 0$ and

$$\#G = \#\mathbb{Z}_n = n$$

by Theorem 4.18. □

4.20 THEOREM

SUBGROUPS OF CYCLIC GROUPS If G is a cyclic group and K is a subgroup of G, then K is a cyclic group.

PROOF As we showed in the proof of Theorem 4.19, there is a surjective homomorphism

$$\theta : \mathbb{Z} \to G.$$

By Part (ii) of Theorem 3.8, $\theta^{-1}(K)$ is a subgroup of $\mathbb{Z}$. By Theorem 4.15, there is a unique integer $m \geq 0$ such that

$$\theta^{-1}(K) = \mathbb{Z}m.$$

Because $\theta : \mathbb{Z} \to G$ is surjective, it follows from Problem 10 in Problem Set 2.7 that

$$\theta(\mathbb{Z}m) = \theta(\theta^{-1}(K)) = K.$$

Let

$$\psi : \mathbb{Z}m \to K$$

be defined by

$$\psi(h) = \theta(h), \quad \text{for all } h \in \mathbb{Z}m$$

(that is, ψ is the restriction of the mapping θ to $\mathbb{Z}m$). Since θ is a homomorphism, so is ψ. Furthermore, the fact that $\theta(\mathbb{Z}m) = K$ implies that $\psi : \mathbb{Z}m \to K$ is surjective. Therefore, K is a homomorphic image (under ψ) of the cyclic group $\mathbb{Z}m$; hence, by Theorem 4.14, K is a cyclic group. □

We leave the proof of the following lemma as an exercise (Problem 33).

4.21 LEMMA

CARTESIAN PRODUCT OF GROUPS Let G and H be groups. For $(g_1, h_1), (g_2, h_2) \in G \times H$, define

$$(g_1, h_1)(g_2, h_2) = (g_1g_2, h_1h_2).$$

Then, under this binary operation, $G \times H$ forms a group. The neutral element in $G \times H$ is $(1, 1)$, and, for $(g, h) \in G \times H$, we have

$$(g, h)^{-1} = (g^{-1}, h^{-1}).$$

The Cartesian product $G \times H$ of the groups G and H, with the "coordinatewise" operations defined in Lemma 4.21, is often called the **direct product** of the groups G and H. Note that, if $(g, h) \in G \times H$ and $n \in \mathbb{Z}$, then

$$(g, h)^n = (g^n, h^n).$$

Consequently, we have the following lemma and theorem (Problems 35 and 37).

4.22 LEMMA

ORDER OF AN ELEMENT IN A CARTESIAN PRODUCT Let G and H be groups, and let $(g, h) \in G \times H$. Suppose that g has finite order n in G and that h has finite order m in H. Then the order of (g, h) in $G \times H$ is the least common multiple of m and n.

4.23 THEOREM

CARTESIAN PRODUCT OF CYCLIC GROUPS Let G and H be finite cyclic groups with $n = \#G$ and $m = \#H$. Then $G \times H$ is a cyclic group if and only if n and m are relatively prime.

If G and H are finite groups, then

$$\#(G \times H) = \#G \cdot \#H.$$

For instance, the group

$$V = \mathbb{Z}_2 \times \mathbb{Z}_2$$

is a group of order 4, but, by Theorem 4.23, it is *not* cyclic. The group V is called **Klein's four-group**, in honor of the German mathematician Felix Klein (1849–1925) who was one of the first people to appreciate the profound connections between group theory and geometry. By Theorem 4.23, the group $\mathbb{Z}_2 \times \mathbb{Z}_3$ *is* cyclic; in fact, it is isomorphic to the group $\mathbb{Z}_6$.

The following theorem is proved in more advanced courses:

> *Every finite commutative (that is, abelian) group is isomorphic to a direct product of a finite number of cyclic groups.*

This important result is called the **fundamental theorem of finite abelian groups.**

PROBLEM SET 5.4

1. Prove Lemma 4.1. [Hint: First establish the rules of exponents for positive integers; then extend them to all integers.]
2. Let G be a multiplicative group, and suppose that $a, b \in G$. If $(ab)^2 = a^2b^2$, show that $ab = ba$.
3. Rewrite the rules in Lemma 4.1 for an additive group G.
4. Let G be a multiplicative group in which every element, other than 1, has order 2. Prove that G is commutative.
5. Show that -1 is a generator of $\mathbb{Z}$.
6. Show that 1 and -1 are the *only* generators of $\mathbb{Z}$.
7. For the group G of Example 2.4 on page 367, find the orders of the elements b, c, d, and f.
8. (a) If G is a group of order 2, show that G is isomorphic to $\mathbb{Z}_2$.
   (b) If G is a group of order 3, show that G is isomorphic to $\mathbb{Z}_3$.
9. For the group G of Problem 7, find $\langle b \rangle$, $\langle c \rangle$, $\langle d \rangle$, and $\langle f \rangle$.
10. If G and H are multiplicative groups, $\theta: G \to H$ is a homomorphism, $a \in G$, and $n \in \mathbb{Z}$, prove that $\theta(a^n) = (\theta(a))^n$.
11. Prove Corollary 4.13.
12. (a) Rewrite the result in Problem 10 for the case in which G is an additive group and H is a multiplicative group.
    (b) Interpret the result in Part (a) for the case in which G is the additive group $\mathbb{R}$ of real numbers, H is the multiplicative group of positive real numbers, and $\theta(x) = e^x$ for all $x \in \mathbb{R}$.
13. Prove that the group G of Problem 7 is *not* a cyclic group.
14. Suppose that $G = \{1, a, b, c\}$ is a group of order 4 and that G is *not* a cyclic group.
    (a) Show that a, b and c must have order 2.
    (b) Write the multiplication table for G (only one is possible!).
    (c) Conclude that G is isomorphic to Klein's four-group V.
    (d) Conclude that, up to isomorphism, there are only two different groups of order 4: $\mathbb{Z}_4$ and V.

15. Prove Theorem 4.14.

16. Let $\mathscr{Q}$ be the quaternion group (Problem 20 in Problem Set 5.2).
(a) Show that $\mathscr{Q}$ is *not* a cyclic group.
(b) Is every proper subgroup of $\mathscr{Q}$ a cyclic group? Explain.

17. Let n be a positive integer.
(a) Show that n and $-n$ are generators of the cyclic group $\mathbb{Z}n$.
(b) Show that n and $-n$ are the *only* generators of $\mathbb{Z}n$.

18. Let n be a positive integer. Show that the set C_n of all complex numbers of the form $e^{2\pi im/n}$, $m = 0, 1, 2, \ldots, n-1$, forms a cyclic subgroup of the **circle group** $S^1 = \{e^{i\theta} | \theta \in \mathbb{R}\}$ (see Problem 14 in Problem Set 5.3).

19. Let G be a finite group, and suppose that $\# G = p$ is a prime number. Prove that G is cyclic.

20. By Problem 18, the circle group S^1 contains cyclic subgroups of every possible finite order. Does S^1 contain an *infinite* cyclic subgroup? Explain.

21. If G is a group of order less than 6, show that G must be commutative.

22. Show that every finite subgroup of the circle group S^1 (Problem 18) is a cyclic group.

23. Prove that two cyclic groups are isomorphic if and only if they have the same order.

24. Complete the proof of Theorem 4.15 by using mathematical induction on k to show that, if N is a subgroup of $\mathbb{Z}$, $h \in N$, and $k \in \mathbb{N}$, then $hk \in N$.

25. Let G be a multiplicative cyclic group with $G = \langle a \rangle$, and define $\theta : \mathbb{Z} \to G$ by $\theta(m) = a^m$ for all $m \in \mathbb{Z}$. Show that θ is a homomorphism of the additive group $\mathbb{Z}$ onto the cyclic group G.

26. In Problem 27 of Problem Set 3.2, we define the **ring of integers mod** n and used the notation $(\mathbb{Z}_n, +, \cdot)$ for this ring. In Definition 4.17, we used the notation $\mathbb{Z}_n$ for the cyclic group $\mathbb{Z}/\mathbb{Z}n$. Reconcile these two definitions by showing that the additive group $(\mathbb{Z}_n, +)$ of the ring of integers mod n is isomorphic to the quotient group $\mathbb{Z}/\mathbb{Z}n$.

27. Let G be a cyclic group of order n, let a be a generator of G, and let k be a positive integer with $k \leq n$. Prove that the order of a^k is $n/\text{GCD}(n, k)$.

28. The **Euler phi function** $\phi : \mathbb{N} \to \mathbb{N}$ is defined by

$$\phi(n) = \textit{the number of positive integers that are less than } n \textit{ and relatively prime to } n,$$

with the special understanding that $\phi(1) = 1$. If G is a cyclic group of order n, show that $\phi(n)$ is the number of generators of G. [Hint: See Problem 27.]

29. If G is a cyclic group of order n and if d is a positive integer with $d | n$, show that there is a cyclic subgroup D of G with $\# D = d$. [Hint: Let $k = n/d$ in Problem 27.]

30. In Problem 29, show that there is *exactly one* subgroup D of G with $\#D = d$.

31. Suppose that G is a group that contains no nontrivial subgroup. Prove that G must be a cyclic group of prime order.

32. By combining Problems 28 and 30, prove that the identity

$$n = \sum_{d|n} \phi(d)$$

holds for every $n \in \mathbb{N}$, where the summation extends over all integer divisors d of n with $1 \leq d \leq n$.

33. Prove Lemma 4.21.

34. Prove that the direct product $G \times H$ of commutative groups G and H is again a commutative group.

35. Prove Lemma 4.22.

36. If G is a cyclic group of order n and if $d \in \mathbb{N}$ with $d|n$, prove that the equation $x^d = 1$ has exactly d solutions in G.

37. Prove Theorem 4.23.

38. Let G be a finite group of order n, and suppose that, for each $d \in \mathbb{N}$ with $d|n$, the equation $x^d = 1$ has no more than d solutions in G. Prove that G is cyclic.

39. Give a direct proof of Theorem 4.20 without using a homomorphism.

40. Prove that any *finite* subgroup G of the multiplicative group $F \setminus \{0\}$ of nonzero elements of a field F is necessarily a cyclic group. [Hint: Use Problem 38.]

41. True or false: A group is cyclic if and only if it is a homomorphic image of $\mathbb{Z}$. Explain.

*5 PERMUTATION GROUPS

In Section 6 of Chapter 2, we introduced the set $\mathscr{B}(X)$ of all bijections from a nonempty set X onto itself as an example of an algebraic system under the operation of function composition (Example AS9 on page 145). Our purpose here is to prove that this algebraic system is a group and to initiate a study of this group.

5.1 DEFINITION

PERMUTATION Let X be a nonempty set. A bijection $\sigma: X \to X$ is called a **permutation** of X. In what follows we denote by $\mathscr{B}(X)$ the set of all permutations of X.

By Part (iii) of Corollary 5.11 in Chapter 2, the composition $\sigma \circ \tau$ of two permutations σ and τ of the set X is again a permutation of X; hence, by Theorem 5.7 in Chapter 2, $(\mathscr{B}(X), \circ)$ forms a semigroup. The semigroup $(\mathscr{B}(X), \circ)$ is a monoid with the **identity permutation** $\varepsilon: X \to X$, given by

$$\varepsilon(x) = x, \quad \text{for all } x \in X,$$

as its neutral element. If $\sigma \in \mathscr{B}(X)$, then $\sigma^{-1} \in \mathscr{B}(X)$ by Part (ii) of Theorem 4.18 in Chapter 2, and since $\sigma \circ \sigma^{-1} = \varepsilon = \sigma^{-1} \circ \sigma$, it follows that $(\mathscr{B}(X), \circ)$ is a group. The group $(\mathscr{B}(X), \circ)$ is usually written multiplicatively, so that, for $\sigma, \tau \in \mathscr{B}(X)$,

$$\sigma\tau = \sigma \circ \tau$$

and

$$1 = \varepsilon.$$

For the remainder of this section, we assume that X is a nonempty set and regard $\mathscr{B}(X)$ as a multiplicative group under the operation of function composition.

5.2 DEFINITION

ACTIVE SET Let $\sigma \in \mathscr{B}(X)$. Define the set $A_\sigma \subseteq X$, called the **active set** of σ, by

$$A_\sigma = \{a \in X \mid \sigma(a) \neq a\}.$$

If $\sigma \in \mathscr{B}(X)$ and A_σ is the active set of σ, then each element $a \in A_\sigma$ is **moved** by σ in the sense that $\sigma(a) \neq a$. If $a \notin A_\sigma$, so that $\sigma(a) = a$, we say that a is **fixed** by σ. We note that the active set of σ is the same as the active set of σ^{-1} (Problem 1).

5.3 LEMMA

Let $\sigma \in \mathscr{B}(X)$, and let A_σ be the active set of σ. Then,

$$a \in A_\sigma \Rightarrow \sigma(a) \in A_\sigma.$$

PROOF Let $a \in A_\sigma$, and let $\sigma(a) = b$, so that $a \neq b$. We must prove that $\sigma(b) \neq b$. Suppose, on the contrary, that $\sigma(b) = b$. Then, $\sigma(a) = \sigma(b)$, and since σ is injective, it follows that $a = b$, contradicting $a \neq b$. □

5.4 DEFINITION

DISJOINT PERMUTATIONS Two permutations in $\mathscr{B}(X)$ are said to be **disjoint** if their active sets are disjoint.

5.5 THEOREM **DISJOINT PERMUTATIONS COMMUTE** Let $\sigma, \tau \in \mathscr{B}(X)$, and suppose that σ and τ are disjoint. Then

$$\sigma\tau = \tau\sigma.$$

PROOF Let A_σ and A_τ be the active sets of σ and τ, respectively. By hypothesis,

$$A_\sigma \cap A_\tau = \varnothing. \tag{1}$$

Let $x \in X$. We must prove that

$$(\sigma\tau)(x) = (\tau\sigma)(x). \tag{*}$$

Because $\sigma\tau = \sigma \circ \tau$ and $\tau\sigma = \tau \circ \sigma$, Equation (*) can be rewritten as

$$\sigma(\tau(x)) = \tau(\sigma(x)). \tag{**}$$

By Equation (1), x must be fixed either by σ or by τ or by both. We suppose that x is fixed by σ, so that

$$\sigma(x) = x. \tag{2}$$

(If x is fixed by τ, we make a similar argument with σ and τ interchanged.) If x is also fixed by τ, so that $\tau(x) = x$, then $\sigma(\tau(x)) = \sigma(x) = x$ and $\tau(\sigma(x)) = \tau(x) = x$, so Equation (**) holds. Thus, we can suppose that

$$x \in A_\tau, \tag{3}$$

and it follows from Lemma 5.3 that

$$\tau(x) \in A_\tau. \tag{4}$$

By Equation (1) and (4), $\tau(x) \notin A_\sigma$, and therefore,

$$\sigma(\tau(x)) = \tau(x). \tag{5}$$

Combining Equations (2) and (5), we again obtain Equation (**). □

We leave the proof of the following lemma as an exercise (Problem 3).

5.6 LEMMA Let $\alpha, \sigma, \tau \in \mathscr{B}(X)$ with $\alpha = \sigma\tau$, and let A_α, A_σ, A_τ be the active sets of α, σ, and τ, respectively. Then:

(i) $A_\alpha \subseteq A_\sigma \cup A_\tau$,
(ii) $A_\sigma \cap A_\tau = \varnothing \Rightarrow A_\alpha = A_\sigma \cup A_\tau$.

Let $\alpha \in \mathscr{B}(X)$, and suppose that $x, y \in X$. If

$$\sigma(x) = y,$$

then we say that σ **sends** x **into** y. Often it is convenient to use the mapping notation

$$x \mapsto y$$

to indicate that the permutation under discussion sends x into y. Likewise, the notation

$$x \mapsto y \mapsto z$$

would show that the permutation sends x into y and that, in turn, it sends y into z. More generally,

$$a_1 \mapsto a_2 \mapsto a_3 \mapsto \cdots \mapsto a_q$$

would indicate that a_j is sent into a_{j+1} for $1 \leq j \leq q-1$.

5.7 DEFINITION

CYCLE NOTATION Let X be a set, and let $a_1, a_2, a_3, \ldots, a_k$ be k distinct elements X. The notation

$$(a_1 \quad a_2 \quad a_3 \quad \cdots \quad a_k)$$

denotes the permutation in $\mathscr{B}(X)$ such that

$$a_1 \mapsto a_2 \mapsto a_3 \mapsto \cdots \mapsto a_{k-1} \mapsto a_k \mapsto a_1,$$

it being understood that all elements of the set X other than $a_1, a_2, a_3, \ldots,$ and a_k remain fixed.

A permutation of the form $(a_1 \; a_2 \; a_3 \; \cdots \; a_k)$ is called a **cyclic permutation**, a **cycle**, or a ***k*-cycle**. A 1-cycle (a_1) is understood to be alternative notation for the identity permutation.

5.8 Example Let $X = \{1, 2, 3, 4\}$, and let $\sigma \in \mathscr{B}(X)$ be the 3-cycle

$$\sigma = (2 \quad 4 \quad 1).$$

Find $\sigma(j)$ for $j = 1, 2, 3$, and 4.

SOLUTION For $\sigma = (2\;\,4\;\,1)$, we have

$$2 \mapsto 4 \mapsto 1 \mapsto 2,$$

and 3 remains fixed,

$$3 \mapsto 3.$$

Thus,

$$\sigma(1) = 2, \sigma(2) = 4, \sigma(3) = 3, \text{ and } \sigma(4) = 1.$$

□

In Example 5.8, we note that the 3-cycle $\sigma = (2\ \ 4\ \ 1)$ can also be written as $\sigma = (1\ \ 2\ \ 4)$ or as $\sigma = (4\ \ 1\ \ 2)$. More generally, a k-cycle

$$\sigma = (a_1\ \ a_2\ \ a_3\ \ \cdots\ \ a_k)$$

can be rewritten in the equivalent form

$$\sigma = (a_k\ \ a_1\ \ a_2\ \ a_3\ \ \cdots\ \ a_{k-1})$$

by moving the last entry a_k into the first position and shifting the remaining entries to the right. This process, called *cyclic permutation* of the entries in the k-cycle, can be repeated. Hence, by cyclic permutations of its entries, a k-cycle can be written in k different, but equivalent, forms. For instance,

$$(a\ \ b\ \ c\ \ d) = (d\ \ a\ \ b\ \ c) = (c\ \ d\ \ a\ \ b) = (b\ \ c\ \ d\ \ a).$$

Note that, for $k > 1$, the active set of a k-cycle

$$(a_1\ \ a_2\ \ a_3\ \ \cdots\ \ a_k)$$

is the set

$$\{a_1, a_2, a_3, \ldots, a_k\}$$

consisting of the entries in the cycle. Thus, two such cycles are disjoint if and only if they involve no common entry. As a consequence of Theorem 5.5, *disjoint cycles commute.*

5.9 LEMMA Let X be a finite set, let $\sigma \in \mathscr{B}(X)$, and suppose that $a \in X$ with $\sigma(a) \neq a$. Then, σ can be factored as

$$\sigma = \gamma\delta,$$

where γ and δ are disjoint permutations in $\mathscr{B}(X)$, γ is a cycle, and $\gamma(a) \neq a$.

PROOF Suppose that $\#X = n$, and consider the elements

$$a, \sigma(a), \sigma^2(a), \sigma^3(a), \ldots, \sigma^{n+1}(a)$$

in X. Although $a \neq \sigma(a)$, these elements cannot *all* be different from one another, otherwise there would be at least $n + 1$ elements in X. Let k be the smallest positive integer such that $\sigma^k(a)$ is the same as one of the previous elements

$$a, \sigma(a), \sigma^2(a), \ldots, \sigma^{k-1}(a).$$

Then $1 < k$ and there is an integer h with $0 \leq h \leq k - 1$ such that

$$\sigma^h(a) = \sigma^k(a).$$

Applying the permutation σ^{-h} to both sides of the last equation, we find that

$$a = \sigma^{k-h}(a),$$

and so $\sigma^{k-h}(a)$ is the same as one of the previous elements

$$a, \sigma(a), \sigma^2(a), \ldots, \sigma^{k-h-1}(a).$$

Because of our choice of k, we must have

$$k \leq k - h,$$

and it follows that $h = 0$, so that

$$a = \sigma^k(a).$$

Again, because of our choice of k, the elements

$$a, \sigma(a), \sigma^2(a), \sigma^3(a), \ldots, \sigma^{k-1}(a)$$

must all be different from one another. Furthermore,

$$\sigma(\sigma^{k-1}(a)) = \sigma^k(a) = a.$$

Let γ be the k-cycle,

$$\gamma = (a \quad \sigma(a) \quad \sigma^2(a) \quad \cdots \quad \sigma^{k-1}(a)).$$

Note that, if j is an integer with $0 \leq j \leq k - 1$, then

$$\sigma^{j+1}(a) = \gamma(\sigma^j(a)).$$

Applying the permutation γ^{-1} to both sides of the last equation, we find that

$$\gamma^{-1}(\sigma^{j+1}(a)) = \sigma^j(a).$$

Now let

$$\delta = \gamma^{-1}\sigma.$$

Then $\sigma = \gamma\delta$, and the proof will be complete as soon as we show that γ and δ are disjoint. We do this by showing that every element x in the active set A_γ of γ is fixed by δ. An element $x \in A_\gamma$ can be written as

$$x = \sigma^j(a)$$

for some integer j with $0 \leq j \leq k - 1$. Now,

$$\delta(x) = (\gamma^{-1}\sigma)(\sigma^j(a)) = \gamma^{-1}(\sigma^{j+1}(a)) = \sigma^j(a) = x,$$

and so x is indeed fixed by δ. □

5.10 THEOREM

FACTORIZATION INTO DISJOINT CYCLES Let X be a finite set, and let $\sigma \in \mathscr{B}(X)$. Then σ is either a cycle or it is a product of a finite number of disjoint cycles.

PROOF Our proof is by contradiction. Suppose there is a permutation $\sigma \in \mathscr{B}(X)$ that is not a cycle and cannot be factored as a product of a finite number of disjoint cycles. Among all such permutations σ, choose one whose active set A_σ contains the smallest possible number of elements. Note that $A_\sigma \neq \varnothing$; otherwise, σ would be the identity permutation, which can be written as a 1-cycle.

Choose $a \in A_\sigma$. By Lemma 5.9, σ can be factored as

$$\sigma = \gamma\delta,$$

where γ and δ are disjoint permutations in $\mathscr{B}(X)$, γ is a cycle, and $\gamma(a) \neq a$. Since σ cannot be factored into disjoint cycles, δ is not a cycle, nor can it be factored into disjoint cycles. If A_δ is the active set of δ, then, by Part (ii) of Lemma 5.6,

$$A_\delta \subseteq A_\sigma.$$

Since a is moved by γ, it is fixed by δ; so

$$a \notin A_\delta.$$

Hence, A_δ is a proper subset of A_σ, and consequently, A_δ contains fewer elements than A_σ, contradicting our choice of σ. □

5.11 Example Let $X = \{1, 2, 3, 4, 5\}$, $\alpha = (1\ \ 2\ \ 5)$, and $\beta = (1\ \ 3\ \ 5\ \ 4)$. Write $\alpha\beta$ as a cycle or as a product of disjoint cycles.

SOLUTION Notice that $\alpha\beta$ is a product of cycles, but the cycles α and β are not disjoint, since 1 and 5 belong to both of their active sets. In computing $(\alpha\beta)(x)$, we must keep in mind that

$$(\alpha\beta)(x) = \alpha(\beta(x)),$$

so that $\beta(x)$ must be computed *first*, and then $\alpha(\beta(x))$ can be computed. In other words, as far as the effect of the permutation $\alpha\beta$ on an element $x \in X$ is concerned, we must read $\alpha\beta$ *from right to left*. (Be careful. Many authors use a different convention and read products of permutations from left to right.)

Now β sends 1 into 3, and α fixes 3; so $\alpha\beta$ sends 1 into 3, and we write

$$\alpha\beta = (1\ \ 3\ \ \cdots,$$

where the incomplete cycle simply indicates that $1 \mapsto 3$. Next, we notice that β sends 3 into 5 and that α sends 5 into 1; so $\alpha\beta$ sends 3 into 1, closing the cycle $(1\ \ 3)$, and we have

$$\alpha\beta = (1\ \ 3)\cdots.$$

The effect of $\alpha\beta$ on the remaining elements 2, 4, and 5 of X has yet to be calculated. Notice that β fixes 2 and that α sends 2 into 5; so $\alpha\beta$ sends 2 into 5, and we have

$$\alpha\beta = (1\ \ 3)(2\ \ 5\ \ \cdots.$$

Again, the incomplete cycle shows the results of our work so far. The permutation β sends 5 into 4, and α fixes 4; so $\alpha\beta$ sends 5 into 4, and we have

$$\alpha\beta = (1\ \ 3)(2\ \ 5\ \ 4\ \ \cdots.$$

Now β sends 4 into 1, and α sends 1 into 2; so $\alpha\beta$ sends 4 into 2, closing the cycle $(2\ \ 5\ \ 4)$, and we have

$$\alpha\beta = (1\ \ 3)(2\ \ 5\ \ 4)\cdots.$$

Since all of the elements in $X = \{1, 2, 3, 4, 5\}$ are now accounted for, our computation is finished, and we conclude that

$$\alpha\beta = (1\ \ 3)(2\ \ 5\ \ 4).$$

□

If, in a computation such as that in Example 5.11, a 1-cycle appears as a factor in the final answer, then it can be dropped (since it represents the identity permutation). For instance,

$$(1\ \ 2\ \ 3)(1\ \ 3\ \ 4) = (1)(2\ \ 3\ \ 4) = (2\ \ 3\ \ 4).$$

5.12 DEFINITION

TRANSPOSITION A 2-cycle $(a\ \ b)$ is also called a **transposition**.

A transposition $(a\ \ b) \in \mathscr{B}(X)$ simply interchanges a and b, leaving all other elements of X fixed. Evidently, a transposition is its own inverse,

$$(a\ \ b)^{-1} = (a\ \ b)$$

(Problem 7a). Note that a 3-cycle $(a\ \ b\ \ c)$ can be written as a product of two transpositions:

$$(a\ \ b\ \ c) = (b\ \ c)(a\ \ c)$$

(Problem 8). Also, a 4-cycle $(a\ \ b\ \ c\ \ d)$ can be written as a product of three transpositions:

$$(a\ \ b\ \ c\ \ d) = (c\ \ d)(b\ \ d)(a\ \ d)$$

(Problem 10). (Notice that these transpositions are not disjoint.) More generally, we have the following result, the proof of which we leave as an exercise (Problem 17).

5.13 LEMMA A k-cycle $(a_1\ a_2\ a_3\ \cdots\ a_k)$ can be factored as a product of $k-1$ transpositions:

$$(a_1\ a_2\ a_3\ \cdots\ a_k) = (a_{k-1}\ a_k)(a_{k-2}\ a_k)\cdots(a_3\ a_k)(a_2\ a_k)(a_1\ a_k).$$

If $a, b \in X$ with $a \neq b$, then the identity permutation 1 can be factored as a product of transpositions as follows:

$$1 = (a\ b)(a\ b).$$

Combining this observation, Theorem 5.10, and Lemma 5.13, we obtain the following important result:

5.14 THEOREM **FACTORIZATION INTO TRANSPOSITIONS** Let X be a finite set with $\#X \geq 2$. Then every permutation in $\mathscr{B}(X)$ can be factored into a product of a finite number of transpositions.

5.15 DEFINITION **EVEN AND ODD PERMUTATIONS** A permutation that can be factored into an even number of transpositions is said to be **even**. A permutation that can be factored into an odd number of transpositions is said to be **odd**.

Since a transposition is its own inverse, and since the inverse of a product is the product of the inverses in the opposite order, it is clear that *the inverse of an even permutation is even* and *the inverse of an odd permutation is odd*. The evenness or oddness of a permutation is referred to as its **parity**.

5.16 THEOREM **UNIQUENESS OF PARITY** No permutation can be both even and odd.

PROOF The following proof by contradiction is due to Hans Liebeck [*American Mathematical Monthly*, **76** (6) (1969), p. 688]. Let $\sigma \in \mathscr{B}(X)$, and assume that σ is both even and odd. Then, the same is true of σ^{-1}. Factoring σ into an even number h of transpositions and σ^{-1} into an odd number k of transpositions, we find that the identity permutation 1 can be factored into

$n = h + k$ permutations:

$$1 = \sigma\sigma^{-1} = (a_1 \quad b_1)(a_2 \quad b_2) \cdots (a_n \quad b_n), \tag{1}$$

where n, being a sum of an even and an odd integer, is an odd integer.

Choose and fix an element $a \in X$. If the element a does not occur in the active set of one of the transpositions $(a_j \quad b_j)$ in Equation (1), $1 \leq j \leq n$, then $(a_j \quad b_j)$ can be replaced by

$$(a_j \quad b_j) = (a \quad a_j)(a \quad b_j)(a \quad a_j), \tag{2}$$

and this will increase the number of transpositions in the factorization in Equation (1) by 2; hence, there will still be an odd number of transpositions in the factorization. Repeating this process as many times as may be necessary, we obtain a factorization of the form

$$1 = (a \quad c_1)(a \quad c_2) \cdots (a \quad c_m), \tag{3}$$

where m is an odd integer. But, each c_j must occur an even number of times in Equation (3) (Problem 18), contradicting the fact that m is an odd integer. □

5.17 COROLLARY

If X is a finite set and $\# X \geq 2$, then every permutation in $\mathscr{B}(X)$ is either even or odd, but not both.

5.18 DEFINITION

$\mathscr{A}$(X) If $\# X = 1$, we define $\mathscr{A}(X) = \mathscr{B}(X)$. If $\# X > 1$, we define $\mathscr{A}(X)$ to be the set of all even permutations in $\mathscr{B}(X)$.

It is easy to check that $\mathscr{A}(X)$ is a subgroup of $\mathscr{B}(X)$ (Problem 21). We refer to $\mathscr{A}(X)$ as the **alternating** subgroup of $\mathscr{B}(X)$. In the following theorem, we use the notation $\mathbb{Z}_2 = \{0, 1\}$ for the additive cyclic group of order 2, so that 0 is the additive neutral element of $\mathbb{Z}_2$ and 1 is the generator. Note that $1 + 1 = 0$ in $\mathbb{Z}_2$.

5.19 THEOREM

THE PARITY HOMOMORPHISM Let X be a finite set with $\# X \geq 2$, and define $\rho: \mathscr{B}(X) \to \mathbb{Z}_2$ by

$$\rho(\sigma) = \begin{cases} 0, & \text{if } \sigma \text{ is even} \\ 1, & \text{if } \sigma \text{ is odd} \end{cases} \quad \text{for all } \sigma \in \mathscr{B}(X).$$

Then ρ is a surjective homomorphism of $\mathscr{B}(X)$ onto $\mathbb{Z}_2$ and

$$\ker(\rho) = \mathscr{A}(X).$$

The proof of Theorem 5.19 is not difficult, and we leave it as an exercise (Problem 23).

5.20 COROLLARY Let X be a finite set with $\#X \geq 2$. Then $\mathscr{A}(X)$ is a normal subgroup of $\mathscr{B}(X)$ and $\#\mathscr{A}(X) = \#\mathscr{B}(X)/2$.

PROOF By Theorem 5.19 and Lemma 3.13, $\mathscr{A}(X)$ is a normal subgroup of $\mathscr{B}(X)$. By Theorem 5.19 and the fundamental homomorphism theorem, $\mathscr{B}(X)/\mathscr{A}(X)$ is isomorphic to $\mathbb{Z}_2$, and it follows that $\#(\mathscr{B}(X)/\mathscr{A}(X)) = 2$; that is, $\mathscr{A}(X)$ has index 2 in $\mathscr{B}(X)$. (See Definition 2.18.) Hence, by the order-index theorem (Theorem 2.19), $\#\mathscr{A}(X) = \#\mathscr{B}(X)/2$. □

In elementary algebra, it is proved that there are $n!$ permutations of a set X of n elements, that is,

$$\#X = n \Rightarrow \#\mathscr{B}(X) = n!.$$

(Recall that, if $n \in \mathbb{N}$, then $n!$, called ***n* factorial**, is defined to be the product

$$n! = 1 \cdot 2 \cdot 3 \cdot \cdots \cdot n$$

of all the positive integers less than or equal to n.) Thus, if X is a finite set with $\#X \geq 2$, then, by Corollary 5.20,

$$\#\mathscr{A}(X) = n!/2.$$

5.21 DEFINITION **SYMMETRIC GROUP AND ALTERNATING GROUP** If $n \in \mathbb{N}$ and $X = \{1, 2, 3, \ldots, n\}$, then the group $\mathscr{B}(X)$ is also denoted by S_n and is called the **symmetric group of degree *n***. The normal subgroup $\mathscr{A}(X)$ is also denoted by A_n and is called the **alternating group of degree *n***.

Note that $\#S_n = n!$ and that, for $n \geq 2$, $\#A_n = n!/2$. Thus, $\#S_2 = 2! = 2$, $\#S_3 = 3! = 6$, $\#S_4 = 4! = 24$, $\#A_2 = 2/2 = 1$, $\#A_3 = 6/2 = 3$, $\#A_4 = 24/2 = 12$, and so forth.

PROBLEM SET 5.5

1. If $\sigma \in \mathscr{B}(X)$, prove that the active set A_σ of σ is the same as the active set $A_{\sigma^{-1}}$ of σ^{-1}.
2. If $\sigma \in \mathscr{B}(X)$, is it true that the active set A_σ of σ is the same as the active set A_{σ^2} of σ^2? Explain.

3. Prove Lemma 5.6.

4. Let $\sigma \in \mathscr{B}(X)$, and suppose that σ is a k-cycle, $k \geq 2$. Let A_σ be the active set of σ, and let $a, b \in A_\sigma$. Prove that there is an integer q with $1 \leq q \leq k$ such that $\sigma^q(a) = b$.

5. If $\sigma \in \mathscr{B}(X)$ and σ is a k-cycle, prove that k is the order of σ.

6. Let $\sigma \in \mathscr{B}(X)$, let A_σ be the active set of σ, suppose that A_σ is a finite set, and assume that, for every choice of $a, b \in A_\sigma$ there exists a positive integer q such that $\sigma^q(a) = b$. Prove that σ is a cycle.

7. (a) Show that $(a\ \ b)^{-1} = (b\ \ a)$.
(b) Show that $(a\ \ b\ \ c)^{-1} = (c\ \ b\ \ a)$.
(c) Generalize Parts (a) and (b) for an arbitrary k-cycle $(a_1\ \ a_2\ \ a_3\ \ \cdots\ \ a_k)$.

8. If a, b, and c are three distinct elements of X, show that $(a\ \ b\ \ c) = (b\ \ c)(a\ \ c)$.

9. Factor $(1\ \ 2\ \ 3\ \ 4)(2\ \ 4\ \ 5)$ as a cycle or a product of disjoint cycles.

10. If a, b, c, and d are four distinct elements of X, show that $(a\ \ b\ \ c\ \ d) = (c\ \ d)(b\ \ d)(a\ \ d)$.

11. Factor $(1\ \ 2\ \ 3)^2$ as a cycle or a product of disjoint cycles.

12. If a, b, c, and d are four distinct elements of X, show that $(a\ \ b)(c\ \ d) = (c\ \ b\ \ a)(c\ \ d\ \ a)$.

13. Factor $(1\ \ 2\ \ 3\ \ 4)^2$ as a cycle or a product of disjoint cycles.

14. Prove: If $\#X \geq 3$, then each element of $\mathscr{A}(X)$ is a 3-cycle or a product of (not necessarily disjoint) 3-cycles. [Hint: Combine Problems 8 and 12.]

15. Prove that Lemma 5.9 holds even if X is an infinite set, provided that the active set A_σ of σ is finite.

16. Prove that Theorem 5.10 holds even if X is an infinite set, provided that the active set A_σ of σ is finite.

17. Prove Lemma 5.13.

18. Let $a \in X$, and let $c_1, c_2, c_3, \ldots, c_m \in X \setminus \{a\}$. (The elements $c_1, c_2, c_3, \ldots, c_m$ are not assumed to be distinct from one another.) Suppose that $1 = (a\ \ c_1)(a\ \ c_2)(a\ \ c_3) \cdots (a\ \ c_m)$. Let $c \in X \setminus \{a\}$. Prove that $\#\{j \mid c_j = c\}$ is an even integer.

19. Let σ be a k-cycle. Prove: If k is odd, then σ is even; and, if k is even, then σ is odd.

20. Let $\sigma = \gamma_1\gamma_2\gamma_3 \cdots \gamma_n$, where γ_j is a k_j-cycle for $1 \leq j \leq n$, and the cycles $\gamma_1, \gamma_2, \gamma_3, \ldots, \gamma_n$ are disjoint. Prove that the order of σ is the least common multiple of $k_1, k_2, k_3, \ldots$, and k_n.

21. Prove that $\mathscr{A}(X)$ is a subgroup of $\mathscr{B}(X)$.

22. Let $\mathscr{B}_0(X)$ denote the set of all permutations σ in $\mathscr{B}(X)$ such that the active set A_σ is finite. Prove that $\mathscr{B}_0(X)$ is a subgroup of $\mathscr{B}(X)$.

23. Prove Theorem 5.19.

24. If $\sigma, \gamma \in \mathscr{B}(X)$, prove that $A_{\sigma\gamma\sigma^{-1}} = \{\sigma(x) | x \in A_\gamma\}$.

25. Show that $S_3 = \{(1), (1\ 2), (1\ 3), (2\ 3), (1\ 2\ 3), (1\ 3\ 2)\}$, and write the multiplication table for S_3.

26. Let $\sigma, \gamma \in \mathscr{B}(X)$, and suppose that γ is a k-cycle,

$$\gamma = (a_1\quad a_2\quad a_3\quad \cdots\quad a_k).$$

Prove that $\sigma\gamma\sigma^{-1}$ is the k-cycle

$$\sigma\gamma\sigma^{-1} = (\sigma(a_1)\quad \sigma(a_2)\quad \sigma(a_3)\quad \cdots\quad \sigma(a_k)).$$

[Hint: See Problem 24.]

27. Show that the group G of Example 2.4 on page 367 is isomorphic to S_3. [Hint: See Problem 25.]

28. Prove that, even if X is an infinite set, $\mathscr{A}(X)$ is a normal subgroup of $\mathscr{B}(X)$. [Hint: Use the result of Problem 26.]

29. Show that there are exactly eight different 3-cycles in A_4.

30. Let a, b, c, d, e, and f be six distinct elements of X. Find $\sigma \in \mathscr{B}(X)$ with $A_\sigma = \{a, b, c, d, e, f\}$ such that $\sigma(a\ b\ c)\sigma^{-1} = (d\ e\ f)$.

31. Show that $H = \{(1), (1\ 2)(3\ 4), (1\ 3)(2\ 4), (1\ 4)(2\ 3)\}$ is a subgroup of A_4 and that A_4 consists of these four permutations together with the eight 3-cycles found in Problem 29.

32. In Problem 31, show that H is a normal subgroup of A_4 and that H is isomorphic to Klein's four-group V.

33. Write the multiplication table for the alternating group A_4. [See Problem 31.]

34. Let $\#X = \#Y$, so that there exists a bijection $f: X \to Y$. Define a mapping $\Phi: \mathscr{B}(X) \to \mathscr{B}(Y)$ by $\Phi(\sigma) = f \circ \sigma \circ f^{-1}$ for all $\sigma \in \mathscr{B}(X)$.
(a) Prove that Φ is well-defined, that is, if $\sigma \in \mathscr{B}(X)$, prove that $f \circ \sigma \circ f^{-1} \in \mathscr{B}(Y)$.
(b) Prove that $\Phi: \mathscr{B}(X) \to \mathscr{B}(Y)$ is a group isomorphism.

*6

TRANSFORMATION GROUPS

The concept of an abstract group was developed in the late nineteenth century after a long evolution. Prior to that time, mathematicians had studied only specific groups of permutations, substitutions, or transformations. After the definition of an abstract group became generally accepted, it was found that the resulting theory could be used to enhance the more concrete theories from which the abstract definition had emerged. The result of this discovery is the modern theory of *transformation groups.* In this section, we introduce the notion of a transformation group and indicate just how it generalizes the idea of a specific group of permutations of a set.

In Section 5.5, we studied the group $\mathscr{B}(X)$ of all permutations on a nonempty set X. By **a permutation group on X**, we mean any subgroup of $\mathscr{B}(X)$. For instance, the alternating group $\mathscr{A}(X)$ (Definition 5.18) is a permutation group on X. If G is a permutation group on X, then an element $\sigma \in G$ "acts on" an element $x \in X$ to produce a "new" element $\sigma(x) \in X$. Because of this action, G is more than just an abstract group. This leads us to the following definition.

6.1 DEFINITION

ACTION OF A GROUP ON A SET Let G be a multiplicative group, and let X be a nonempty set. By an **action** of G on X we mean a mapping $*: G \times X \to X$ such that, for all $g, h \in G$ and all $x \in X$:

(i) $(gh) * x = g * (h * x)$.
(ii) $1 * x = x$.

In Definition 6.1, we have used the *infix notation* $g * x$ as an abbreviation for the more cumbersome expression $*((g, x))$.

6.2 Example Let G be a permutation group on X, and define $*: G \times X \to X$ by

$$\sigma * x = \sigma(x),$$

for all $\sigma \in G$ and all $x \in X$. Show that $*$ is an action of G on X.

SOLUTION We check Conditions (i) and (ii) of Definition 6.1. Let $\sigma, \tau \in G$, and let $x \in X$. Then:

(i) $(\sigma\tau) * x = (\sigma\tau)(x) = \sigma(\tau(x)) = \sigma * (\tau * x)$.
(ii) $1 * x = \varepsilon(x) = x$. □

If G is a permutation group on X, then the action $*$ defined in Example 6.2 is called the **natural action** of G on X.

6.3 DEFINITION

TRANSFORMATION GROUP A **transformation group** (abbreviated TG) is a triple $(G, X, *)$ consisting of a group G, a nonempty set X, and an action $*$ of G on X.

For instance, if G is a permutation group on X and $*$ is the natural action of G on X, then $(G, X, *)$ is a TG. We borrow the following terminology from the theory of permutation groups: If $(G, X, *)$ is a TG, $g \in G$, and $x \in X$, we say that g **moves** x if $g * x \neq x$ and that g **fixes** x if $g * x = x$. The set

$$A_g = \{x \in X \mid g * x \neq x\}$$

consisting of all elements of X that are moved by g is called the **active set** of g.

6.4 DEFINITION

EFFECTIVE AND FREE TRANSFORMATION GROUPS Let $(G, X, *)$ be a TG.

(i) If $1 \in G$ is the only element of G that fixes every element $x \in X$, then $(G, X, *)$ is said to be **effective**.
(ii) If every element $g \in G$ with $g \neq 1$ moves every element $x \in X$, then $(G, X, *)$ is said to be **free**.

If G is a permutation group on X and $*$ is the corresponding natural action of G on X, then $(G, X, *)$ is an effective TG (Problem 9).

6.5 Example Let $X = \{1, 2, 3\}$, and let $\sigma \in \mathscr{B}(X)$ be the 3-cycle $\sigma = (1\ 2\ 3)$. Let $G = \langle\sigma\rangle = \{1, \sigma, \sigma^2\}$ be the cyclic subgroup of G generated by σ, and let $*$ be the natural action of G on X. Show that the TG $(G, X, *)$ is free.

SOLUTION Clearly, σ moves every element in $X = \{1, 2, 3\}$. Likewise, $\sigma^2 = (1\ 3\ 2)$ moves every element in X. □

6.6 DEFINITION

TRANSITIVE TRANSFORMATION GROUPS A TG $(G, X, *)$ is said to be **transitive** if, for every pair of elements $x, y \in X$, there exists at least one $g \in G$ such that

$$g * x = y.$$

For instance, in Example 6.5, $(G, X, *)$ is transitive. Also, in this example, G is commutative; so the fact that $(G, X, *)$ is free could have been deduced from the following lemma:

6.7 LEMMA

If $(G, X, *)$ is effective and transitive, and if G is a commutative group, then $(G, X, *)$ is free.

PROOF Let $g \in G$, and suppose that g fixes some element $a \in X$, so that $g * a = a$. We must prove that $g = 1$. Let x be an arbitrary element in X. Because $*$ is transitive, there exists $h \in G$ such that $h * a = x$. Because G is commutative, we have $gh = hg$. Therefore,

$$g * x = g * (h * a) = (gh) * a = (hg) * a = h * (g * a) = h * a = x;$$

that is, g fixes every element $x \in X$. Because $*$ is effective, it follows that $g = 1$. □

6.8 Example Let G be any group, and let H be a (not necessarily normal) subgroup of G. Let G/H be the set of left cosets of G modulo H (see Definition 2.13). We define the action of G on the quotient space G/H by **left translation** as follows: Let $g \in G$. Then, if $u \in G$, the action of g on the left coset $uH \in G/H$ is defined by

$$g * (uH) = (gu)H.$$

Show that $(G, G/H, *)$ is a transitive TG.

SOLUTION To show that $*$ is an action of G on G/H, we check Conditions (i) and (ii) of Definition 6.1. Let $g_1, g_2, u \in G$. Then,

$$(g_1g_2) * (uH) = (g_1g_2u)H = g_1 * ((g_2u)H) = g_1 * (g_2 * (uH));$$

so Condition (i) holds. Also,

$$1 * (uH) = (1 \cdot u)H = uH;$$

so Condition (ii) also holds. To show that $(G, G/H, *)$ is transitive, let $u, v \in G$, so that $uH, vH \in G/H$. Then, with $g = vu^{-1}$, we have

$$g * (uH) = (gu)H = (vu^{-1}u)H = vH.$$

□

6.9 DEFINITION

STABILITY, OR ISOTROPY, GROUP If $(G, X, *)$ is a TG and $x \in X$, we call

$$G_x = \{g \in G \mid g * x = x\}$$

the **stability group**, or the **isotropy group**, of x.

The following lemma shows that the stability group G_x is, in fact, a subgroup of G.

6.10 LEMMA

Let $(G, X, *)$ be a TG, let $x \in G$, and let G_x be the stability group of x. Then G_x is a subgroup of G.

PROOF Clearly, $1 \in G_x$. If $g, h \in G_x$, then

$$(gh) * x = g * (h * x) = g * x = x;$$

so $gh \in G_x$. Finally, suppose that $g \in G_x$. Then,

$$g^{-1} * (x) = g^{-1} * (g * x) = (g^{-1}g) * x = 1 * x = x;$$

so $g^{-1} \in G_x$. □

Note that a TG $(G, X, *)$ is free if and only if $G_x = \{1\}$ for every $x \in X$ (Problem 11b). Also note that $(G, X, *)$ is effective if and only if

$$\bigcap_{x \in X} G_x = \{1\}$$

(Problem 7b).

For transformation groups, the notion of two algebraic systems being isomorphic has an analogue called *equivalence.*

6.11 DEFINITION

EQUIVALENT TRANSFORMATION GROUPS An **equivalence** of the TG $(G, X, *)$ and the TG $(G, Y, \star)$ is a bijection

$$f: X \to Y$$

such that, for all $g \in G$ and all $x \in X$,

$$f(g * x) = g \star f(x).$$

Two transformation groups $(G, X, *)$ and $(G, Y, \star)$ are said to be **equivalent** if there exists an equivalence $f: X \to Y$. The following theorem shows that any transitive TG is equivalent to a TG in which the action is left translation on a quotient space, as in Example 6.8.

6.12 THEOREM

STRUCTURE OF A TRANSITIVE TG Let $(G, X, *)$ be a transitive TG. Choose and fix $a \in X$, and let $H = G_a$. Let $(G, G/H, *)$ be the transitive TG in which G acts on G/H by left translation. Then there exists an equivalence

$$f: X \to G/H$$

of $(G, X, *)$ and $(G, G/H, *)$.

PROOF Let $x \in X$. Since $(G, X, *)$ is transitive, there is at least one element $u \in G$ such that $u * a = x$. Suppose that $u' \in G$ is another such element, so that

$u' * a = x$. Then,

$$(u^{-1}u') * a = u^{-1} * (u' * a) = u^{-1} * x = u^{-1} * (u * a)$$
$$= (u^{-1}u) * a = 1 * a = a,$$

and therefore,

$$u^{-1}u' \in G_a = H.$$

Consequently, by Problem 1 in Problem Set 5.3,

$$uH = u'H.$$

Now, define

$$f : X \to G/H$$

for each $x \in X$ by

$$f(x) = uH,$$

where $u \in G$ is chosen so that

$$u * a = x.$$

The argument above shows that f is well-defined.

Suppose that $x, y \in X$ with

$$f(x) = f(y).$$

Select $u, v \in G$ with

$$u * a = x \quad \text{and} \quad v * a = y.$$

Then, since $f(x) = f(y)$, we have

$$uH = vH,$$

and it follows that

$$u^{-1}v \in H = G_a.$$

Consequently,

$$(u^{-1}v) * a = a.$$

Therefore,

$$x = u * a = u * ((u^{-1}v) * a) = (uu^{-1}v) * a = v * a = y,$$

which shows that $f : X \to G/H$ is an injection.

To show that $f : X \to G/H$ is a surjection, note that a left coset in G/H can be written in the form uH for some $u \in G$. Hence, with $x = u * a$, we have

$$f(x) = uH.$$

Thus, $f : X \to G/H$ is a surjection; hence, a bijection.

Finally, to show that f is an equivalence of the TG $(G, X, *)$ and the TG $(G, G/H, *)$, let $g \in G$ and let $x \in X$. Select $u \in G$ with $u * a = x$, so that

$$f(x) = uH.$$

Then,

$$(gu) * a = g * (u * a) = g * x,$$

and it follows from the definition of f that

$$f(g * x) = (gu)H.$$

But, by the definition of the action of G on G/H by left translation,

$$(gu)H = g * (uH) = g * f(x);$$

hence,

$$f(g * x) = g * (f(x)),$$

which shows that f is an equivalence and completes the proof. □

Theorem 6.12 is important because it gives us a survey, up to equivalence, of all possible transitive actions of the group G on sets X. We can paraphrase this fundamental result as follows:

Up to equivalence, the only possible transitive actions of a group G are obtained by letting G act by left translation on quotient spaces G/H, as in Example 6.8.

In particular, if G is finite, we can combine Theorem 6.12 and the order-index theorem to obtain the following corollary:

6.13 COROLLARY If $(G, X, *)$ is a transitive TG and G is a finite group, then X is a finite set, and, for any $x \in X$, $\#X$ is the index of G_x in G. In particular, $\#X$ divides $\#G$.

PROOF Let $x \in X$. By Theorem 6.12, there is a bijection

$$f: X \to G/G_x,$$

and therefore,

$$\#X = \#(G/G_x) = \text{index of } G_x \text{ in } G.$$

Hence, by Theorem 2.19, $\#X$ divides $\#G$. □

6.14 DEFINITION

ORBIT Let $(G, X, *)$ be a TG, and let $x \in X$. Define

$$G * x = \{g * x | g \in G\}.$$

We call $G * x$ the **orbit** of x.

Note that $x \in G * x$; that is, every element x of X belongs to its own orbit (Problem 17a). A subset M of X is called an orbit of $(G, X, *)$ if it is the orbit of some $x \in X$, that is, if $M = G * x$ for some $x \in X$. The following lemma shows that either two orbits are disjoint or they coincide.

6.15 LEMMA

Let $(G, X, *)$ be a TG. If $x, y \in X$ and $G * X \cap G * y \neq \varnothing$, then $G * x = G * y$.

PROOF Suppose that $z \in G * x \cap G * y$. We begin by proving that $G * x \subseteq G * y$. Because $z \in G * x$, there exists $u \in G$ such that

$$u * x = z.$$

Likewise, because $z \in G * y$, there exists $v \in G$ such that

$$v * y = z.$$

Therefore, letting

$$w = u^{-1}v,$$

we have

$$\begin{aligned} w * y &= (u^{-1}v) * y = u^{-1} * (v * y) = u^{-1} * z \\ &= u^{-1} * (u * x) = (u^{-1}u) * x = 1 * x = x. \end{aligned}$$

Consequently, for any $g \in G$, we have

$$g * x = g * (w * y) = (gw) * y \in G * y,$$

which shows that $G * x \subseteq G * y$. A similar argument with x and y interchanged shows that $G * y \subseteq G * x$ and completes the proof. □

6.16 DEFINITION

ORBIT SPACE If $(G, X, *)$ is a TG, we define

$$X/G = \{G * x | x \in X\},$$

and we call X/G the **orbit space** of $(G, X, *)$.

Thus, the orbit space of $(G, X, *)$ is the set of all of its orbits. As a consequence of Lemma 6.15 and the fact that each $x \in X$ belongs to its own orbit $G * x$, we see that X/G is a partition of X into equivalence classes (Definition 3.1 in Chapter 2). Note that

$$(G, X, *) \text{ is transitive} \Leftrightarrow \#(X/G) = 1$$

(Problem 17e).

6.17 DEFINITION

INVARIANT SUBSET Let $(G, X, *)$ be a TG, and let $M \subseteq X$. We say that M is **invariant** under G if, for every $g \in G$,

$$x \in M \Rightarrow g * x \in M.$$

We note that each orbit of $(G, X, *)$ is invariant under G (Problem 17b) and that a subset of X is invariant under G if and only if it is a union of orbits (Problem 17d).

6.18 DEFINITION

RESTRICTED ACTION Let $(G, X, *)$ be a TG, let $\varnothing \neq M \subseteq X$, and suppose that M is invariant under G. We define the **restricted action**

$$\star : G \times M \to M$$

of G on M by

$$g \star m = g * m,$$

for every $g \in G$ and every $m \in M$.

In Definition 6.18, it is easy to see that $\star : G \times M \to M$ really is an action of G on M (Problem 19). If no confusion threatens, we ordinarily use the same symbol $*$ (rather than $*$ and $\star$) both for the original action of G on X and for the restricted action of G on the nonempty invariant subset M of X. Thus, if $(G, X, *)$ is a TG and M is a nonempty invariant subset of X, then $(G, M, *)$ is a TG, called the TG obtained by **restricting** the action of G to the invariant subset M of X. In particular, because an orbit $M \in X/G$ is an invariant subset of X, we can restrict the action of G to M to obtain the TG $(G, M, *)$. We leave the proof of the next lemma as an exercise (Problem 21).

6.19 LEMMA

Let $(G, X, *)$ be a TG, and let $M \in X/G$. Then $(G, M, *)$ is a transitive TG. If $x \in M$, then $(G, M, *)$ is equivalent to the TG $(G, G/G_x, *)$, in which G acts by left translation on the quotient space G/G_x. If G is finite, then M is finite, and $\#M$ is the index of G_x in G.

6.20 DEFINITION

FIXED SET If $(G, X, *)$ is a TG and $g \in G$, we define

$$F_g = \{x \in X \mid g * x = x\},$$

and we refer to F_g as the **fixed set** of g. We define

$$F = \bigcap_{g \in G} F_g,$$

and we refer to F as the **fixed set** of G.

If $(G, X, *)$ is a TG, then an orbit $M \in X/G$ is said to be **trivial** if it contains only one element. If $M \in X/G$, then M is trivial with $M = \{x\}$ if and only if x belongs to the fixed set F of $(G, X, *)$ (Problem 23). Thus, F is the union of all the trivial orbits in X/G.

The following remarkable theorem, often attributed to the British algebraist William Burnside (1852–1927), says that the number of orbits of a TG is equal to the average number of elements fixed by the elements of the group.

6.21 THEOREM

BURNSIDE'S ORBIT-COUNTING THEOREM Let $(G, X, *)$ be a TG such that $\#G$ and $\#X$ are both finite. Let $c = \#(X/G)$, and let $n = \#G$. Then,

$$c = \frac{1}{n} \sum_{g \in G} \#F_g.$$

PROOF For $g \in G$ and $x \in X$, define

$$\delta(g, x) = \begin{Bmatrix} 1, & \text{if } g * x = x \\ 0, & \text{if } g * x \neq x \end{Bmatrix}. \tag{1}$$

Then, for each $g \in G$,

$$\#F_g = \sum_{x \in X} \delta(g, x), \tag{2}$$

and, for each $x \in X$,

$$\#G_x = \sum_{g \in G} \delta(g, x). \tag{3}$$

By Equations (2) and (3), we have

$$\begin{aligned} \sum_{g \in G} \#F_g &= \sum_{g \in G} \sum_{x \in X} \delta(g, x) \\ &= \sum_{x \in X} \sum_{g \in G} \delta(g, x) \\ &= \sum_{x \in X} \#G_x. \end{aligned} \tag{4}$$

Let $M_1, M_2, M_3, \ldots, M_c$ denote the distinct orbits of $(G, X, *)$. By Lemma 6.19, for each $j = 1, 2, 3, \ldots, c$,

$$x \in M_j \Rightarrow \# M_j = \#(G/G_x) = \frac{n}{\# G_x}, \tag{5}$$

and it follows that

$$x \in M_j \Rightarrow \# G_x = \frac{n}{\# M_j}. \tag{6}$$

As a consequence of (6), we have, for each $j = 1, 2, 3, \ldots, c$,

$$\sum_{x \in M_j} \# G_x = \sum_{x \in M_j} \frac{n}{\# M_j} = n. \tag{7}$$

From Equations (4) and (7) and the fact that X is the union of the disjoint sets $M_1, M_2, M_3, \ldots, M_c$, it follows that

$$\sum_{g \in G} \# F_g = \sum_{j=1}^{c} \sum_{x \in M_j} \# G_x = cn;$$

and therefore,

$$c = \frac{1}{n} \sum_{g \in G} \# F_g. \qquad \square$$

If $(G, X, *)$ is a TG, $\# G$ is finite, and $\# X$ is finite, we can summarize the results obtained in this section as follows: The number of different orbits is given by Burnside's formula (Theorem 6.21). The fixed set F is the union of all the trivial orbits. If we denote the remaining nontrivial orbits by

$$M_1, M_2, M_3, \ldots, M_k,$$

then X is a disjoint union,

$$X = F \cup M_1 \cup M_2 \cup M_3 \cup \cdots \cup M_k,$$

and each of the sets $F, M_1, M_2, M_3, \ldots, M_k$ is invariant. The action of G on X can be understood by looking at its action on each of these invariant sets.

On F, the action of G is trivial in the sense that no element of F is moved by any element of G. Consider the jth nontrivial orbit M_j for $j = 1, 2, 3, \ldots, k$. The action of G on M_j is transitive in the sense that

$$x, y \in M_j \Rightarrow \exists g \in G, g * x = y;$$

that is, any element in M_j can be moved to any other element by the action of a suitable $g \in G$. If $x \in M_j$, then the number of elements in M_j is the index in G of the isotropy subgroup

$$G_x = \{g \in G \,|\, g * x = x\}.$$

In fact, for any choice of $x \in M_j$, the TG $(G, M_j, *)$ is equivalent to the TG $(G, G/G_x, *)$ in which G is acting by left translation on the quotient space G/G_x.

PROBLEM SET 5.6

1. Let $(G, X, *)$ be a TG. For each $g \in G$, define $\sigma_g: X \to X$ by $\sigma_g(x) = g * x$ for every $x \in X$. Prove that $\sigma_g \in \mathscr{B}(X)$.
2. Let $(G, X, *)$ be a TG. For each $g \in G$, define σ_g as in Problem 1. Let $\Phi: G \to \mathscr{B}(X)$ be defined by $\Phi(g) = \sigma_g$ for each $g \in G$. Prove that the mapping $\Phi: G \to \mathscr{B}(X)$ is a group homomorphism.
3. Let G be a group, let X be a nonempty set, and let $\Phi: G \to \mathscr{B}(X)$ be a group homomorphism. (Such a homomorphism is called a **permutation representation** of G.) Define $*: G \times X \to X$ by $*((g, x)) = (\Phi(g))(x)$ for all $(g, x) \in G \times X$. Prove that $(G, X, *)$ is a TG.
4. Using the results of Problems 1, 2, and 3, write a short essay explaining the sense in which an action of the group G on a set is "essentially the same" as a permutation representation of G.
5. Define the **kernel** of a TG $(G, X, *)$ to be the set

   $$K = \{g \in G \mid g * x = x, \text{ for all } x \in X\}.$$

   (a) Show that K is a normal subgroup of G.
   (b) Show that $(G, X, *)$ is effective if and only if $K = \{1\}$.
6. Let $\Phi: G \to \mathscr{B}(X)$ be the homomorphism of Problem 2. Show that the kernel K of $(G, X, *)$ in Problem 5 is the same as the kernel of Φ.
7. (a) With the notation of Problem 5, show that

   $$K = \bigcap_{x \in X} G_x.$$

   (b) Conclude that $(G, X, *)$ is effective if and only if $\bigcap_{x \in X} G_x = \{1\}$. [See Part (b) of Problem 5.]
8. Let $(G, X, *)$ be a TG. Two elements $g, h \in G$ are said to be **disjoint** if their active sets $A_g = \{x \in X \mid g * x \neq x\}$ and $A_h = \{x \in X \mid h * x \neq x\}$ are disjoint. If g and h are disjoint and K is the kernel of $(G, X, *)$ (Problem 5), prove that $ghg^{-1}h^{-1} \in K$.
9. If G is a permutation group on X and $*$ is the corresponding natural action, prove that the TG $(G, X, *)$ is effective.
10. Let K be the kernel of the TG $(G, X, *)$ (Problem 5). Define the mapping $\star: (G/K) \times X \to X$ by $\star((gK, x)) = g * x$ for all $gK \in G/K$ and all $x \in X$.
    (a) Show that $\star$ is well-defined.
    (b) Show that $\star$ is an action of G/K on X.

(c) Show that the TG $(G/K, X, \star)$ is effective. The TG $(G/K, X, \star)$ is sometimes referred to as $(G, X, *)$ **made effective**.

11. (a) Show that every free TG is effective.
(b) Show that a TG $(G, X, *)$ is free if and only if $G_x = \{1\}$ for every $x \in X$.
(c) Give an example of an effective TG that is not free. [Hint: Look at the natural action of S_3 on $X = \{1, 2, 3\}$.]

12. Let $(G, X, *)$ be a TG, and let $(G/K, X, \star)$ be $(G, X, *)$ made effective (Problem 10). Show that $(G, X, *)$ is transitive if and only if $(G/K, X, \star)$ is transitive.

13. Let G be a group, let H be a subgroup of G, and consider the TG $(G, G/H, *)$ for which $*$ is the action of G on G/H by left translation. If K is the kernel of $(G, G/H, *)$ (Problem 5), show that $K = \bigcap_{g \in G} gHg^{-1}$.

14. With the notation of Problem 13, show that $(G, G/H, *)$ is effective if and only if H contains no subgroup $N \neq \{1\}$ such that N is a normal subgroup of G.

15. Give an example of an effective TG $(G, X, *)$ such that G is not transitive. [Hint: Try various subgroups of S_3 acting naturally on $X = \{1, 2, 3\}$.]

16. Let G be a group, let X be a nonempty set, and let $*: G \times X \to X$ satisfy Condition (i) in Definition 6.1. Suppose it also satisfies the following condition: For every $x \in X$ there exists $y \in X$ such that $1 * y = x$. Prove that $*$ is an action of G on X.

17. Let $(G, X, *)$ be a TG, suppose that $x \in X$, and let $M \subseteq X$.
(a) Show that $x \in G * x$.
(b) Show that each orbit of $(G, X, *)$ is invariant under G.
(c) If M is invariant under G, and $x \in M$, show that $G * x \subseteq M$.
(d) Show that M is invariant under G if and only if M is a union of orbits of $(G, X, *)$.
(e) Show that $(G, X, *)$ is transitive if and only if $\#(X/G) = 1$.

18. If $(G, X, *)$ is a TG, and $g \in G$ with $G = \langle g \rangle$, show that A_g, the active set of g, is invariant under G.

19. In Definition 6.18, prove that $\star: G \times M \to M$ is an action of G on M.

20. Let X be a nonempty set, let $\sigma \in \mathscr{B}(X)$, suppose A_σ is a finite nonempty set, and let $G = \langle \sigma \rangle$. Consider the TG $(G, X, *)$ for which $*$ is the natural action of G on X. By Problem 18, A_σ is invariant under G. Let $(G, A_\sigma, *)$ be the TG obtained by restricting the natural action of G to A_σ. Prove that σ is a cycle if and only if $(G, A_\sigma, *)$ is transitive.

21. Prove Lemma 6.19.

22. If G is a group, then the mapping $\star: G \times G \to G$ given by $g \star h = gh$ for all $g, h \in G$ is called the action of G on itself by **left translation**.
(a) Show that $\star: G \times G \to G$ is an action.

(b) Show that a TG $(G, X, *)$ is free and transitive if and only if it is equivalent to the TG $(G, G, \star)$.

23. If $(G, X, *)$ is a TG and F is the fixed set of $(G, X, *)$, show that $x \in F$ if and only if $\{x\} \in X/G$.

24. Let $(G, X, *)$ be a TG, and let $P = \{(g, x) \in G \times X | g * x = x\}$. For each orbit $M \in X/G$, choose one and only one representative $x_M \in M$. (Here, we use the axiom of choice!) For each $M \in X/G$, denote the isotropy group of x_M by G_M. If $M \in X/G$, we again use the axiom of choice to choose one and only one representative $\alpha_M(g) \in gG_M$ from each left coset gG_M in G/G_M. Define $\Phi: G \times (X/G) \to P$ by

$$\Phi((g, M)) = (g\alpha_M(g)^{-1}, g * x_M)$$

for all $g \in G$ and all $M \in X/G$. Prove that Φ is a bijection.

25. If G is a finite group and $(G, X, *)$ is a transitive TG, show that $\#G = \#\{(g, x) | g \in G, x \in X, g * x = x\}$. [Hint: Use Burnside's orbit-counting formula (Theorem 6.21).]

26. Use the result of Problem 24 to give an alternative proof of Burnside's orbit-counting formula (Theorem 6.21).

27. If $(G, X, *)$ is a TG, $g \in G$, and $x \in X$, show that

$$G_{g*x} = g(G_x)g^{-1}.$$

28. Let X be a finite nonempty set, and let $(G, X, \#)$ be a TG with fixed set F. Let A be a subset of X obtained by choosing one and only one element from each nontrivial orbit in X/G. Prove that

$$\#X = \#F + \sum_{a \in A} \#(G/G_a).$$

29. Let S^1 be the multiplicative group of all complex numbers of the form $e^{i\theta}$ with $\theta \in \mathbb{R}$, and let S^1 act by multiplication on the set $\mathbb{C}$ of all complex numbers to obtain the TG $(S^1, \mathbb{C}, \cdot)$.
(a) Show that the TG $(S^1, \mathbb{C}, \cdot)$ is effective.
(b) Find the fixed set F of $(S^1, \mathbb{C}, \cdot)$.
(c) Determine all of the nontrivial orbits of $(S^1, \mathbb{C}, \cdot)$.

30. Let $(G, X, *)$ be a TG, and let A be a nonempty set. Let $\mathcal{F}$ be the set of all functions $f: X \to A$. For $g \in G$ and $f \in \mathcal{F}$, define $g \star F \in \mathcal{F}$ by $(g \star f)(x) = f(g^{-1} * x)$ for all $x \in X$. Prove that $(G, \mathcal{F}, \star)$ is a TG.

31. Let $(G, X, *)$ be a TG, and denote by $\mathcal{P}(X)$ the set of all subsets of X. For $g \in G$ and $A \in \mathcal{P}(X)$, define $g \star A = \{g * a | a \in A\}$. Prove that $(G, \mathcal{P}(X), \star)$ is a TG.

32. Prove **Cayley's theorem**: If G is a group, then G is isomorphic to a group P of permutations of a set X. [Hint: Let G act on itself by left translation (Problem 22) and use Problem 2.]

*7 APPLICATIONS OF TRANSFORMATION GROUPS

We now give a few of the many applications of the theory of transformation groups developed in Section 6. One of the most interesting of these derives from the idea of letting a group G act on itself by inner automorphisms. An isomorphism $\alpha: G \to G$ is called an **automorphism** of G (Problem 19 in Problem Set 5.3). If $g \in G$, the mapping $\alpha_g: G \to G$ defined by

$$\alpha_g(x) = g^{-1}xg, \quad \text{for all } x \in G,$$

is an automorphism of G (Problem 21 in Problem Set 5.3) and is called the **inner automorphism** induced by g. The mapping

$$*: G \times G \to G$$

defined by

$$g * x = \alpha_g(x) = gxg^{-1}, \quad \text{for all } g, x \in G,$$

is an action of G on itself (Problem 1) called the **action by inner automorphisms**.

7.1 DEFINITION **CONJUGACY CLASSES** Let G be a group, and consider the TG $(G, G, *)$ for which $*$ is the action of G on itself by inner automorphisms. If $x \in G$, then the orbit $G * x = \{gxg^{-1} | g \in G\}$ is called the **conjugacy class** determined by x. Two elements $x, y \in G$ are said to be **conjugate** if and only if they belong to the same conjugacy class.

Because the orbits of a TG $(G, X, *)$ form a partition of the set X, it follows that:

The conjugacy classes of a group G decompose G into mutually exclusive and exhaustive equivalence classes.

In other words, *the relation of conjugacy is an equivalence relation on G.*

7.2 Example Let $G = \{1, a, b, c, d, f\}$ be the group in Example 2.4, page 367. Find the decomposition of G into conjugacy classes.

SOLUTION The conjugacy class determined by the neutral element 1 is given by

$$G * 1 = \{g1g^{-1} | g \in G\} = \{1\}.$$

To find $G * a$, we must compute gag^{-1} for all $g \in G$. By the multiplication table on page 367, we have

$$1a1^{-1} = a, \quad aaa^{-1} = a, \quad bab^{-1} = a$$
$$cac^{-1} = b, \quad dad^{-1} = b, \quad \text{and} \quad faf^{-1} = b.$$

Hence,

$$G * a = \{a, b\}.$$

Similar calculations (Problem 3) show that

$$G * b = \{a, b\},$$
$$G * c = \{c, d, f\},$$
$$G * d = \{c, d, f\},$$
$$G * f = \{c, d, f\}.$$

Thus, $G = \{1, a, b, c, d, f\}$ decomposes into the disjoint conjugacy classes

$$G = \{1\} \cup \{a, b\} \cup \{c, d, f\}.$$ □

If G is a group and $*$ is the action of G on itself by inner automorphisms, then, for $g, x \in G$,

$$g * x = x \Leftrightarrow gxg^{-1} = x \Leftrightarrow gx = xg.$$

Therefore, $g * x = x$ *if and only if* g *commutes with* x.

7.3 DEFINITION

CENTRALIZER Let G be a group, and consider the TG $(G, G, *)$ for which $*$ is the action of G on itself by inner automorphisms. If $x \in G$, then the stability group,

$$G_x = \{g \in G \mid g * x = x\} = \{g \in G \mid gx = xg\}$$

is called the **centralizer** of x in G.

In words, *the centralizer of* x *in* G *is the set of all elements of* G *that commute with* x.

7.4 Example For the group $G = \{1, a, b, c, d, f\}$ in Example 7.2, find the centralizer G_x of each element $x \in G$.

SOLUTION Since every element of G commutes with the neutral element 1, we have

$$G_1 = \{1, a, b, c, d, f\}.$$

From the table on page 367, we see that a commutes with 1, a, and b, but not with c, d, and f. Therefore,

$$G_a = \{1, a, b\}.$$

Similarly (Problem 4),

$$G_b = \{1, a, b\},$$
$$G_c = \{1, c\},$$
$$G_d = \{1, d\},$$
$$G_f = \{1, f\}$$

□

7.5 LEMMA If G is a finite group and $x \in G$, then the number of elements in G that are conjugate to x is equal to the index of the centralizer of x in G.

PROOF Consider the TG $(G, G, *)$ for which $*$ is the action of G on itself by inner automorphisms. Then, by Theorem 6.19 with $M = G * x$, we have

$$\#(G * x) = \#(G/G_x).$$

□

Recall that the *center* Z of a group G is the subgroup of G consisting of all elements of G that commute with every other element of G (Definition 2.5 and Theorem 2.6). By the remark prior to Definition 7.3,

$$z \in Z \Leftrightarrow g * z = z, \quad \text{for all } g \in G.$$

In other words:

The center Z of a group G is equal to the fixed set F of the action of G on itself by inner automorphisms.

A conjugacy class C in G consists of only one element, $C = \{z\}$, if and only if $z \in Z$ (Problem 11a). Such a conjugacy class is said to be **trivial**. A conjugacy class containing two or more elements is said to be **nontrivial**. If G is a finite group, then, by Lemma 7.5 and the order-index theorem:

The number of elements in a nontrivial conjugacy class of a finite group G is a proper divisor of the order of G.

(If $d \mid n$, $d \neq 1$, and $d \neq n$, then d is called a **proper divisor** of n.)

7.6 THEOREM

THE CLASS EQUATION OF A FINITE GROUP Let G be a finite noncommutative group with center Z. Let $C_1, C_2, \ldots, C_k$ be the distinct nontrivial conjugacy classes in G, and let

$$c_i = \# C_i, \quad \text{for } i = 1, 2, \ldots, k.$$

Then $\# Z, c_1, c_2, \ldots,$ and c_k are proper divisors of $\# G$, and

$$\# G = \# Z + c_1 + c_2 + \cdots + c_k.$$

PROOF G is the union of the disjoint sets

$$G = Z \cup C_1 \cup C_2 \cup \cdots \cup C_k.$$

Hence,

$$\# G = \# Z + c_1 + c_2 + \cdots + c_k.$$

Because Z is a subgroup of G, Lagrange's theorem (Theorem 2.20) implies that $\# Z \mid \# G$. Because G is noncommutative, $Z \neq G$, so $\# Z$ is a proper divisor of $\# G$. Also, each C_i is a nontrivial conjugacy class in G, so that $c_i = \# C_i$ is a proper divisor of $\# G$. □

The equation $\# G = \# Z + c_1 + c_2 + \cdots + c_k$ in Theorem 7.6 is called the **class equation** of the group G. The class equation can be used to prove the following important theorem, which is attributed to the French mathematician A. Cauchy (1789–1857).

7.7 THEOREM

CAUCHY'S THEOREM ON ELEMENTS OF PRIME ORDER If G is a finite group and p is a prime number, then G contains an element of order p if and only if p divides the order of G.

PROOF By Corollary 4.12, if G contains an element of order p, then p divides $\# G$. We prove the converse by contradiction. Suppose p is a prime and there is a finite group G such that p divides $\# G$, but G contains no element of order p. Among all such groups, choose G to be one having the *smallest* possible order.

Begin by noting that G cannot contain a *cyclic* subgroup K such that p divides $\# K$. Indeed, if p were to divide the order of such a cyclic subgroup K, then K (hence, also G) would contain an element of order p by Problem 29 in Problem Set 5.4.

Next, we show that G cannot be a commutative group. Suppose G is commutative, choose any element $a \in G$ with $a \neq 1$, and let $K = \langle a \rangle$. Since K is a cyclic subgroup of G, p cannot divide $\# K$. By the order-index theorem (Theorem 2.19),

$$\# G = \# K \cdot \#(G/K)$$

and, since p divides $\# G$ and does not divide $\# K$, it follows from Corollary 3.25 in Chapter 4 that p divides $\#(G/K)$. Since G is a commutative group, K is a normal subgroup of G, so G/K is a group. But, $\#(G/K)$ is smaller than $\# G$, and it follows from our choice of G that G/K contains an element, call it bK, of order p. If m is the order of b, then

$$(bK)^m = b^m K = 1K = K.$$

Recalling that K is the neutral element of G/K, and using Lemma 4.6, we find that p divides m. Thus, p divides the order of the cyclic subgroup $\langle b \rangle$ of G, contradicting the fact that no such cyclic subgroup can exist. Therefore, G cannot be commutative.

Now consider the class equation

$$\# G = \# Z + c_1 + c_2 + \cdots + c_k$$

of the noncommutative group G (Theorem 7.6). Suppose that p divides all of the numbers $c_1, c_2, \ldots, c_k$. Then, since p divides $\# G$, it divides $\# Z$; and because $\# Z < \# G$, it follows that Z and, hence, also G, contains an element of order p. Therefore, there exists an integer i with $1 \le i \le k$ such that p fails to divide $c_i = \# C_i$. Choose an element x in the nontrivial conjugacy class C_i. Then,

$$c_i = \# C_i = \#(G/G_x) = \# G / \# G_x;$$

so

$$\# G = c_i \cdot \# G_x.$$

Since the prime number p divides $\# G$ and does not divide c_i, it must divide $\# G_x$. Because $\# G_x < \# G$, it follows that G_x and, hence, also G, contains an element of order p. □

If G is a group, $g \in G$, and H is a subgroup of G, then,

$$\alpha_g(H) = gHg^{-1}$$

is a subgroup of G by Part (i) of Theorem 3.8. Two subgroups H and K of G are said to be **conjugate** if there exists $g \in G$ such that

$$\alpha_g(H) = gHg^{-1} = K.$$

If $\mathscr{X}$ is the set of all subgroups of G, then the mapping

$$\star : G \times \mathscr{X} \to \mathscr{X},$$

defined by

$$g \star H = \alpha_g(H) = gHg^{-1},$$

for all $g \in G$ and all $H \in \mathscr{X}$, is an action of G on $\mathscr{X}$ and $(G, \mathscr{X}, \star)$ is a TG (Problem 15). If $H \in \mathscr{X}$, then the orbit

$$G \star H = \{gHg^{-1} \mid g \in G\}$$

consists of all subgroups of G that are conjugate to H. Thus:

Conjugacy is an equivalence relation on the set of all subgroups of a group.

If $H \in \mathscr{X}$, then, for the TG $(G, \mathscr{X}, \star)$, the stability group

$$G_H = \{g \in G \mid gHg^{-1} = H\}$$

is called the **normalizer** of the subgroup H in the group G. As a consequence of Lemma 6.19:

For a finite group G, the number of subgroups that are conjugate to a subgroup H is the index of the normalizer of H in G.

Note that H is a normal subgroup of G if and only if H belongs to the fixed set of the TG $(G, \mathscr{X}, \star)$ (Problem 17). It is for this reason that normal subgroups are sometimes called *self-conjugate* subgroups.

The discussion above gives an indication of some of the applications of transformation groups to group theory itself. Another important application of transformation groups is to the study of *symmetry*. Again, we can give only an indication here of how this works.

By a **symmetry** of a geometric object, we mean, roughly, a transformation that leaves the object **invariant** in the sense that it has no discernible effect on the object. For instance, the isosceles triangle in Figure 5-6 has the same appearance after it is flipped about its vertical axis; and in this sense, the flipping leaves it invariant. Thus, we say that the transformation "flipping the triangle about its vertical axis" is a *symmetry* of the triangle, and we say that the triangle is *symmetric about its vertical axis.*

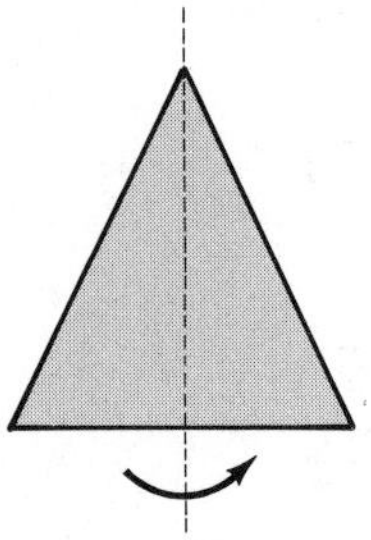

FIGURE 5-6

Suppose we have some geometric object (for instance, the triangle in Figure 5-6), and we are interested in all of its symmetries. Certainly, the identity transformation ε leaves the object invariant; so ε is a symmetry of the object. Also, if σ and τ are symmetries of the object, then transforming the object first by τ and then by σ leaves it invariant; hence the composition $\sigma \circ \tau$, which we write as $\sigma\tau$, is again a symmetry of the object. Finally, if σ is a symmetry of the object, then σ^{-1}, which simply "undoes" what σ does, also leaves the object invariant; hence, σ^{-1} is also a symmetry. Consequently:

The set of all symmetries of a geometric object forms a group.

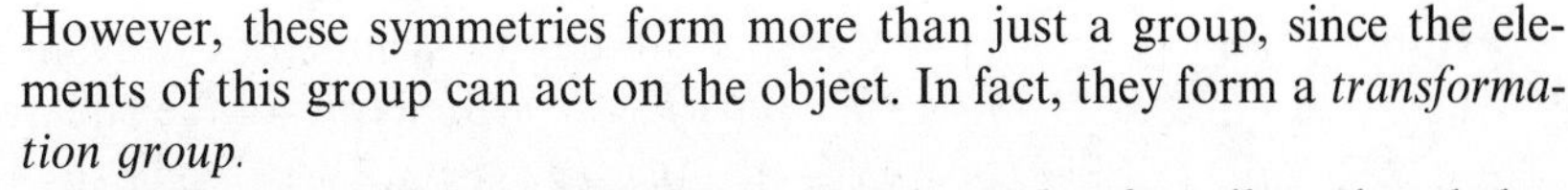

However, these symmetries form more than just a group, since the elements of this group can act on the object. In fact, they form a *transformation group.*

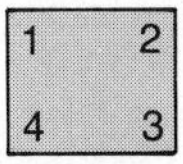

FIGURE 5-7

Imagine that the square in Figure 5-7 is made of cardboard and that it is lying flat on a table. A symmetry of the square can be thought of as a transformation effected by picking up the cardboard square, possibly rotating it, possibly turning it over, and then replacing it so that it exactly covers its original position on the table. This is the sense in which a symmetry of the square is a transformation that leaves the square invariant. Note that, under such a transformation, each vertex of the square returns to the position previously occupied by a vertex. Furthermore, the transformation is completely characterized by its effect on the four vertices.

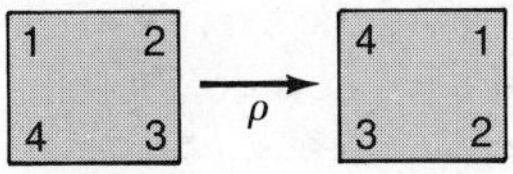

FIGURE 5-8

For instance, if we label the four vertices of the square piece of cardboard 1, 2, 3, and 4 as shown in Figure 5-8, then the effect of a 90° clockwise rotation ρ of the square is to carry vertex j into the position previously occupied by vertex $j + 1$, for $j = 1, 2,$ and 3, and to carry vertex 4 into the position previously occupied by vertex 1. In other words, on the set $X = \{1, 2, 3, 4\}$ of vertices, the effect of ρ is the same as the cyclic permutation $(1\ \ 2\ \ 3\ \ 4)$. Clearly, the 90° clockwise rotation ρ is completely characterized by the fact that it permutes the four vertices in this way; therefore, we propose to use the cyclic permutation $(1\ \ 2\ \ 3\ \ 4)$ as a label for the rotation ρ and write

$$\rho = (1\ \ 2\ \ 3\ \ 4).$$

Note that

$$\rho^2 = (1\ \ 3)(2\ \ 4)$$

corresponds to a rotation clockwise through $90° + 90° = 180°$, and

$$\rho^3 = (1\ \ 4\ \ 3\ \ 2)$$

represents a rotation clockwise through $3 \times 90° = 270°$. Of course, ρ^4, a rotation through $4 \times 90° = 360°$, carries all vertices back into their original positions, and it corresponds to the identity transformation,

$$\rho^4 = 1.$$

The four symmetries 1, ρ, ρ^2, and ρ^3 form a cyclic subgroup of the group G of all symmetries of the square, but there are additional symmetries. Consider, for instance, the effect of the transformation τ that flips the square about the axis through vertices 2 and 4 (Figure 5-9). Note that τ simply interchanges vertices 1 and 3, leaving vertices 2 and 4 fixed. Thus, representing τ as a permutation of the vertices $X = \{1, 2, 3, 4\}$, we have

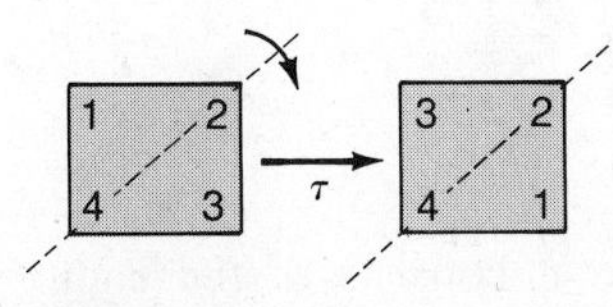

FIGURE 5-9

$$\tau = (1\ \ 3).$$

Because a product of symmetries is again a symmetry, it follows that $\tau\rho$, $\tau\rho^2$, and $\tau\rho^3$ are symmetries of the square. We have

$$\tau\rho = (1\ \ 3)(1\ \ 2\ \ 3\ \ 4) = (1\ \ 2)(3\ \ 4),$$

$$\tau\rho^2 = (1\ \ 3)(1\ \ 3)(2\ \ 4) = (2\ \ 4),$$

and

$$\tau\rho^3 = (1\ \ 3)(1\ \ 4\ \ 3\ \ 2) = (1\ \ 4)(2\ \ 3).$$

Note that $\tau\rho = (1\ 2)(3\ 4)$ corresponds to the symmetry that flips the square around the axis through the midpoint of the side containing vertices 1 and 2 and the midpoint of the side containing vertices 3 and 4 (Figure 5-10). Likewise, $\tau\rho^2 = (2\ 4)$ represents the symmetry that flips the square around the axis through vertices 1 and 3, and $\tau\rho^3 = (1\ 4)(2\ 3)$ corresponds to the symmetry that flips the square around the axis through the midpoint of the side containing vertices 1 and 4 and the side containing vertices 2 and 3.

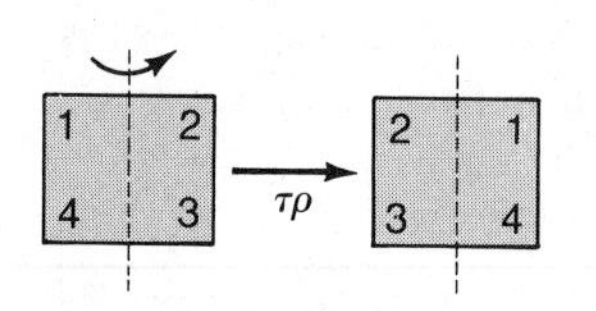

FIGURE 5-10

We have now found eight different symmetries of the square, namely,

$$1, \rho, \rho^2, \rho^3, \tau, \tau\rho, \tau\rho^2, \text{and } \tau\rho^3.$$

Are there any others? Here is where our theory of transformation groups comes in handy. The group G of all symmetries of the square contains these eight symmetries, but for all we know, there may be more symmetries in G. Now, G acts naturally on $X = \{1, 2, 3, 4\}$, producing a TG $(G, X, *)$. The elements ρ, ρ^2, and ρ^3 are sufficient to move any vertex $i \in X$ into any other vertex $j \in X$ (Problem 19a); hence, $(G, X, *)$ is a transitive TG.

Consider the stability group

$$G_1 = \{\lambda \in G \mid \lambda * 1 = 1\} = \{\lambda \in G \mid \lambda(1) = 1\}.$$

What are the symmetries in G_1? We are asking which symmetries of the square fix vertex 1. Of course, the identity does, and so does the symmetry $\tau\rho^2$ that flips the square around the axis through vertices 1 and 3. A little thought shows that these two symmetries are the *only* elements of the stability group G_1, since the only way to pick up the cardboard square and put it back down with vertex 1 in its original position is either to put it down with all vertices in their original positions or to flip it over before putting it down. Therefore, $\# G_1 = 2$. By Theorem 6.12,

$$\# X = \#(G/G_1) = \# G / \# G_1,$$

and it follows that

$$\# G = \# X \cdot \# G_1 = 4 \cdot 2 = 8.$$

Thus, G is a group of order 8, and therefore, it consists of the eight symmetries we have already found:

$$G = \{1, \rho, \rho^2, \rho^3, \tau, \tau\rho, \tau\rho^2, \tau\rho^3\}.$$

The group G is called the **octic** group. In the octic group, note that ρ is an element of order 4, τ is an element of order 2, and

$$\rho\tau = \tau\rho^3$$

(Problem 19b). This makes it relatively easy to compute the multiplication table for G (Problem 19c). For instance,

$$\rho^2(\tau\rho^2) = \rho\rho\tau\rho^2 = \rho\tau\rho^3\rho^2 = \rho\tau\rho^5 = \rho\tau\rho = \tau\rho^3\rho = \tau\rho^4 = \tau.$$

A **regular n-gon** is a plane polygon with n equal sides and n equal vertex angles. For instance, a regular 3-gon is an equilateral triangle and a regular 4-gon is a square. The group of symmetries of a regular n-gon is called the **dihedral group of degree n**, denoted D_n. For instance, D_4 is the octic group discussed above. Arguing as we did for the case $n = 4$, we can show that D_n consists of the $2n$ elements

$$1, \rho, \rho^2, \ldots, \rho^{n-1}, \tau, \tau\rho, \tau\rho^2, \ldots, \tau\rho^{n-1},$$

where ρ represents a clockwise rotation of the n-gon through $360°/n$, and τ represents a flip of the n-gon about an axis of symmetry through one of its vertices (Problem 27). In D_n, the element ρ has order n, the element τ has order 2, and the relation

$$\rho\tau = \tau\rho^{n-1}$$

holds (Problem 29). (Note that $\rho^{n-1} = \rho^{-1}$.)

The idea of a symmetry group applies to objects in three-dimensional space. For instance, the symmetries of a cube are determined by the way in which they permute the eight vertices of the cube. Again, the action is transitive. Computing the stability group of a vertex just requires observing that only a rotation about an axis through that vertex and the diametrically opposite vertex could be a symmetry; and, of these rotations, there are only four (including the identity) that will bring the cube back to fill the region of space it initially occupied. (If you have trouble visualizing these rotations, try it with an actual cube—a sugar cube or a die will do nicely!) Therefore, there are $8 \times 4 = 32$ symmetries of the cube.

PROBLEM SET 5.7

1. If G is a group, show that the mapping $*: G \times G \to G$ given by $g * x = gxg^{-1}$ for all $g \in G$ and all $x \in X$ is an action of G on itself.
2. By Definition 7.1, two elements $x, y \in G$ are conjugate if and only if there exists $g \in G$ such that $gxg^{-1} = y$. Give a direct proof, not using the theory of transformation groups, to show that the relation of being conjugate is an equivalence relation on G.
3. Complete the calculations begun in Example 7.2 by showing that $G * b = \{a, b\}$ and $G * c = G * d = G * f = \{c, d, f\}$.

4. Complete the calculations begun in Example 7.4 by showing that $G_b = \{1, a, b\}$, $G_c = \{1, c\}$, $G_d = \{1, d\}$, and $G_f = \{1, f\}$.

5. Let $\mathcal{Q} = \{1, -1, \mathbf{i}, -\mathbf{i}, \mathbf{j}, -\mathbf{j}, \mathbf{k}, -\mathbf{k}\}$ be the **quaternion group** consisting of **1**, the three quaternionic units **i**, **j**, and **k**, and the negatives of these elements (Problem 20 in Problem Set 5.2). Find the decomposition of $\mathcal{Q}$ into conjugacy classes.

6. Consider the octic group $D_4 = \{1, \rho, \rho^2, \rho^3, \tau, \tau\rho, \tau\rho^2, \tau\rho^3\}$ of symmetries of the square. Find the decomposition of D_4 into conjugacy classes.

7. For the quaternion group $\mathcal{Q}$ in Problem 5, find the centralizer of each of the eight elements.

8. For the octic group D_4, find the centralizer of each of the eight elements.

9. The quaternion group $\mathcal{Q}$ (Problem 5) and the octic group D_4 both have the same order, eight. Show that these two groups are *not* isomorphic. [Hint: Look at the orders of the elements in $\mathcal{Q}$ on the one hand and in D_4 on the other.]

10. Find the decomposition of the alternating group A_4 into conjugacy classes. [Hint: Make use of Problem 26 in Problem Set 5.5.]

11. Let G be a group, and let Z be the center of G.
    (a) If $z \in G$, show that $z \in Z$ if and only if $\{z\}$ is a conjugacy class in G.
    (b) Show that Z is the intersection of all centralizers G_x of elements $x \in G$.

12. If G is a group, prove that a subgroup N of G is normal if and only if it is a union of conjugacy classes in G.

13. If $(G, X, *)$ is a TG and if $x, y \in X$ belong to the same orbit of $(G, X, *)$, show that the stability groups G_x and G_y are conjugate subgroups of G.

14. If p is a prime and n is a positive integer, then a finite group G of order p^n is called a ***p*-group**. If G is a p-group and Z is the center of G, show that Z contains an element of order p. [Hint: Use the class equation of G to show that p divides $\#Z$, then use Cauchy's theorem.]

15. If G is a group and $\mathcal{X}$ is the set of all subgroups of G, show that the mapping $\star: G \times \mathcal{X} \to \mathcal{X}$ given by $g \star H = gHg^{-1}$ for all $g \in G$ and all $H \in \mathcal{X}$ is an action of G on $\mathcal{X}$.

16. If G is a finite group and p is a prime, show that G is a p-group (Problem 14) if and only if every element of G has order equal to some integer power of p. [Hint: For the "if" part of the proof, make an indirect argument using Cauchy's theorem.]

17. In Problem 15, show that a subgroup H of G is a normal subgroup if and only if it belongs to the fixed set of the TG $(G, \mathcal{X}, \star)$.

18. If G is a p-group (Problem 14), show that:
    (a) Every subgroup of G is a p-group.
    (b) Every homomorphic image of G is a p-group.

19. Let $D_4 = \{1, \rho, \rho^2, \rho^3, \tau, \tau\rho, \tau\rho^2, \tau\rho^3\}$ be the group of symmetries of the square (the octic group) as developed in the text.
(a) Show that the elements ρ, ρ^2, and ρ^3 are sufficient to move any vertex of the square into any other vertex.
(b) Show that $\rho\tau = \tau\rho^3$.
(c) Write out the multiplication table for the group D_4, using only the facts that ρ has order 4, τ has order 2, and $\rho\tau = \tau\rho^3$.

20. Let G be a finite group, let N be a normal subgroup of G, and suppose that N is a p-group (Problem 14). Show that G is a p-group if and only if G/N is a p-group.

21. Find the center Z of:
(a) The symmetric group S_3
(b) The octic group D_4
(c) The quaternion group $\mathcal{Q}$

22. Let G be a finite group, and let N be a normal subgroup of G. Let $(G, G, *)$ be the TG obtained by letting G act on itself by inner automorphisms, and let $(G, N, *)$ be the TG obtained by restricting the action $*$ to N.
(a) If Z is the center of G, show that the fixed set F of the TG $(G, N, *)$ is given by $F = Z \cap N$.
(b) Let $D_1, D_2, \ldots, D_q$ be the distinct nontrivial orbits of $(G, N, *)$, and let $d_i = \#D_i$ for each $i = 1, 2, \ldots, q$. Show that each d_i is a proper divisor of $\#G$ and that

$$\#N = \#F + \sum_{i=1}^{q} d_i.$$

23. Find the class equation for:
(a) The symmetric group S_3
(b) The octic group D_4
(c) The quaternion group $\mathcal{Q}$

24. If G is a p-group (Problem 14), Z is the center of G, and N is a normal subgroup of G with $N \neq \{1\}$, prove that $N \cap Z \neq \{1\}$. [Hint: Use Problem 22.]

25. If G is a group, H is a subgroup of G, and G_H is the normalizer of H in G, prove that:
(a) H is a normal subgroup of G_H.
(b) If K is a subgroup of G, and H is a normal subgroup of K, then K is a subgroup of G_H.

26. If G is a finite group and p is a prime, then a subgroup P of G is called a ***p*-subgroup** of G if P is a p-group (Problem 14). A p-subgroup P of G is called a ***p*-Sylow** subgroup of G if it is a maximal p-subgroup in the sense that there is no p-subgroup H of G such that P is a proper subgroup of H. If K is a p-subgroup of G, prove that K is contained in at least one p-Sylow subgroup P of G.

27. If D_n, the dihedral group of degree n, is the group of all symmetries of a regular n-gon, prove that D_n consists of n elements of the form ρ^j for $0 \le j \le n-1$ together with n more elements of the form $\tau\rho^j$ for $0 \le j \le n-1$, where ρ represents a clockwise rotation of the n-gon through $360°/n$, and τ represents a flip of the n-gon about an axis of symmetry through one of its vertices.

28. Let G be a finite group, let p be a prime, suppose that P is a p-Sylow subgroup of G (Problem 26), and let N be the normalizer of P in G. Prove that N/P contains no element (other than its neutral element) whose order is a power of p. [Hint: Suppose $xP \in N/P$ has order p^k with $k > 0$, let $\langle xP \rangle$ be the cyclic subgroup of N/P generated by xP, and let H be the subgroup of N obtained by taking the inverse image of $\langle xP \rangle$ under the natural homomorphism $\eta: N \to N/P$. Show that P is a normal subgroup of H, that H/P is isomorphic to $\langle xP \rangle$, and therefore, that H is a p-group (Problem 20).]

29. In Problem 27, show that $\rho\tau = \tau\rho^{n-1}$.

30. Let G be a finite group, let p be a prime, and suppose that P is a p-Sylow subgroup of G (Problem 26). Let $g \in G$, and suppose that the order of g is a power of p. If $gPg^{-1} \subseteq P$, prove that $g \in P$. [Hint: Use Problem 28.]

31. Show that D_3 is isomorphic to S_3.

32. Let P be any p-Sylow subgroup of the finite group G (Problem 26), and let $\mathcal{Y}$ be the set of all subgroups of G that are conjugate to P. Let $r = \#\mathcal{Y}$.
(a) Prove that Q is a p-Sylow subgroup of G if and only if $Q \in \mathcal{Y}$.
(b) Prove that p divides $r - 1$.
(c) Prove that r divides $\#G$. [Hint: Consider the TG $(Q, \mathcal{Y}, \star)$ in which Q acts on $\mathcal{Y}$ as in Problem 15. Write the corresponding "class equation."]

33. A **regular tetrahedron** is a solid with four congruent triangular faces, each face being an equilateral triangle. Show that the symmetry group of a regular tetrahedron has order 12.

34. It is desired to paint the four faces of a regular tetrahedron. Different faces may or may not be painted with different colors. If k different colors are available, show that this can be done in $(k^4 + 11k^2)/12$ *essentially different* ways. (Two painting schemes are essentially different if one cannot be obtained from the other simply by rotating the tetrahedron.) [Hint: Number the faces 1, 2, 3, and 4, and let $X = \{1, 2, 3, 4\}$. Begin by showing that the symmetry group of the tetrahedron is A_4 (see Problem 33). Let $C = \{c_1, c_2, \ldots, c_k\}$ denote the set of available colors. A color scheme is a function $f: X \to C$. Let S denote all color schemes. Let A_4

act on S by the rule

$$(\sigma * f)(x) = f(\sigma^{-1}(x)),$$

for $\sigma \in A_4$, $f \in S$, and $x \in X$. Show that the orbits of the resulting TG correspond to the essentially different color schemes, and use Burnside's orbit-counting formula.]

*8 RELATIONS AS GROUP MORPHISMS

A homomorphism $f: G \to H$ from a group G into a group H maps elements of G into elements of H in such a way that the group structure is preserved, and thus it effects an "algebraic connection" between G and H. In the early days of the development of group theory, people considered more general algebraic connections between groups: they studied *relations* $R \subseteq G \times H$ that preserve the group structure in the sense that

(i) $(1, 1) \in R$,
(ii) $(g_1, h_1), (g_2, h_2) \in R \Rightarrow (g_1g_2, h_1h_2) \in R$, and
(iii) $(g, h) \in R \Rightarrow (g^{-1}, h^{-1}) \in R$.

A function $f: G \to H$, regarded as a relation $f \subseteq G \times H$, satisfies (i), (ii), and (iii) if and only if it is a homomorphism (see Theorem 8.5).

After some initial experimentation with such relations, group theorists focused their attention on the special case of *functions* (that is, homomorphisms) that preserve group structure. Although the more general idea has resurfaced from time to time,† it does not seem to have found its way into the standard textbooks. Because relations that preserve group structure provide excellent illustrations of many of the ideas presented earlier, we denote this final section to a study of this topic.

Recall that a relation is a set of ordered pairs (Definition 2.3 in Chapter 2) and that a function is a special kind of relation (Definition 4.1 in Chapter 2). If R is a relation, then R^{-1}, the inverse (or converse) of R, is the relation obtained by reversing all of the ordered pairs in R. Evidently,

$$(R^{-1})^{-1} = R.$$

Because the inverse of a relation is again a relation, there is a nice symmetry in relation theory that is not generally available for functions: If f is a function, then f^{-1} is a relation, but it need not be a function.

† See, for example, Joachim Lambek, "Goursat's Theorem and the Zassenhaus Lemma," *Canadian Journal of Mathematics* **10** (1958), pp. 45–56.

Definition 5.13 in Chapter 2 extends the idea of function composition to relations. Thus, if R and S are relations,

$$R \circ S = \{(x, z) \mid \exists y, (x, y) \in S, (y, z) \in R\}.$$

Relation composition interacts nicely with the formation of inverses, and we have

$$(R \circ S)^{-1} = S^{-1} \circ R^{-1}$$

(Problem 35 in Problem Set 2.5).

By Problem 37 in Problem Set 2.5, relation composition is associative; that is, if R, S, and T are relations, then

$$(R \circ S) \circ T = R \circ (S \circ T).$$

Thus, parentheses are not really needed for compositions of three or more relations, and we can simply write

$$R \circ S \circ T$$

rather than $(R \circ S) \circ T$ or $R \circ (S \circ T)$. In Problem 1, we ask you to prove the facts in the following lemma.

8.1 LEMMA

Let R be a relation. Then:

(i) R is symmetric if and only if $R = R^{-1}$.
(ii) R is transitive if and only if $R \circ R \subseteq R$.
(iii) $R^{-1} \circ R$ is symmetric.
(iv) $R^{-1} \circ R$ is reflexive on dom(R).

If f is a function and M is a set, then Part (i) of Definition 7.1 in Chapter 2 can be rewritten in abbreviated form as

$$f(M) = \{y \mid \exists x \in M, (x, y) \in f\}.$$

This suggests the following:

8.2 DEFINITION

THE IMAGE OF A SET UNDER A RELATION If R is a relation and M is a set, we define the **image of** M **under** R, in symbols $R(M)$, by

$$R(M) = \{y \mid \exists x \in M, (x, y) \in R\}.$$

We leave the proof of the next lemma as an exercise (Problem 2).

8.3 LEMMA

Let R and S be relations and let M and N be sets. Then:

(i) $M \subseteq N \Rightarrow R(M) \subseteq R(N)$.
(ii) $\text{dom}(R) \subseteq M \Rightarrow R(M) = \text{dom}(R^{-1})$.
(iii) $(R \circ S)(M) = R(S(M))$.
(iv) $(x, y) \in R \Leftrightarrow y \in R(\{x\})$.
(v) R is a function if and only if $\# R(\{x\}) = 1$ for every $x \in \text{dom } R$.

Now we are ready to study relations that preserve group structure.

8.4 DEFINITION

RELATIONAL GROUP MORPHISM Let G and H be groups. A relation $R \subseteq G \times H$ is called a **relational group morphism** (abbreviated RGM) if it satisfies the following three conditions:

(i) $(1, 1) \in R$.
(ii) $(g_1, h_1), (g_2, h_2) \in R \Rightarrow (g_1g_2, h_1h_2) \in R$.
(iii) $(g, h) \in R \Rightarrow (g^{-1}, h^{-1}) \in R$.

The next theorem shows that the notion of an RGM really does generalize the idea of a homomorphism.

8.5 THEOREM

HOMOMORPHISMS ARE RGM'S Let G and H be groups, and let $f: G \to H$. Then, f is a homomorphism if and only if, regarded as a *relation* $f \subseteq G \times H$, f is an RGM.

PROOF First suppose that $f: G \to H$ is a homomorphism. We must show that f satisfies Conditions (i), (ii), and (iii) in Definition 8.4. By Part (i) of Lemma 3.5, $f(1) = 1$; that is, $(1, 1) \in f$, so f satisfies Condition (i). Suppose that

$$(g_1, h_1), (g_2, h_2) \in f.$$

Then,

$$f(g_1) = h_1 \quad \text{and} \quad f(g_2) = h_2.$$

Since $f: G \to H$ is a homomorphism,

$$f(g_1g_2) = f(g_1)f(g_2) = h_1h_2,$$

and it follows that

$$(g_1g_2, h_1h_2) \in f,$$

which shows that f satisfies Condition (ii). Finally, suppose that $(g, h) \in f$, so that $f(g) = h$. By Part (ii) of Lemma 3.5, $f(g^{-1}) = h^{-1}$; that is, $(g^{-1}, h^{-1}) \in f$, so f satisfies Condition (iii).

Conversely, suppose that f is an RGM, and let $g_1, g_2 \in G$. To prove that $f: G \to H$ is a homomorphism, we must show that $f(g_1 g_2) = f(g_1)f(g_2)$. Let

$$h_1 = f(g_1) \quad \text{and} \quad h_2 = f(g_2).$$

Then,

$$(g_1, h_1), (g_2, h_2) \in f,$$

and it follows from Condition (ii) in Definition 8.4 that

$$(g_1 g_2, h_1 h_2) \in f;$$

that is,

$$f(g_1 g_2) = h_1 h_2 = f(g_1)f(g_2). \qquad \square$$

In Problem 4, we ask you to prove the following:

8.6 LEMMA If G and H are groups and $R \subseteq G \times H$ is an RGM, then $R^{-1} \subseteq H \times G$ is an RGM.

Suppose that $f: G \to H$ is a homomorphism. Then, unless $f: G \to H$ is an isomorphism, we cannot meaningfully write $f^{-1}: H \to G$, much less conclude that it is a homomorphism. However, by Theorem 8.5, $f \subseteq G \times H$ is an RGM; hence, by Lemma 8.6, $f^{-1} \subseteq H \times G$ is an RGM.

In Lemma 4.21, we showed that the Cartesian product $G \times H$ of groups G and H is again a group, called the direct product of G and H, under the "coordinatewise" operations. With this idea in mind, we can rewrite Conditions (ii) and (iii) in Definition 8.4 as

(ii′) $\quad (g_1, h_1), (g_2, h_2) \in R \Rightarrow (g_1, h_1)(g_2, h_2) \in R$

and

(iii′) $\quad (g, h) \in R \Rightarrow (g, h)^{-1} \in R,$

respectively. Therefore, we have the following:

8.7 THEOREM **ALTERNATIVE CHARACTERIZATION OF AN RGM** Let G and H be groups, and let $R \subseteq G \times H$. Then R is an RGM if and only if R is a subgroup of the direct product $G \times H$.

The next theorem shows that RGM's provide a generalization of Theorem 3.6.

8.8 THEOREM

THE COMPOSITION OF RGM'S Let G, K, and H be groups, let $S \subseteq G \times K$ be an RGM, and let $R \subseteq K \times H$ be an RGM. Then $R \circ S \subseteq G \times H$ is an RGM.

PROOF Assuming that $S \subseteq G \times K$ and $R \subseteq K \times H$ are RGM's, we show that $R \circ S \subseteq G \times H$ satisfies Conditions (i), (ii), and (iii) of Definition 8.4.

Condition (i): Since $S \subseteq G \times K$ is an RGM, we have $(1, 1) \in S$. Likewise, $(1, 1) \in R$. Therefore, $(1, 1) \in R \circ S$ by the definition of relation composition.

Condition (ii): Suppose that

$$(g_1, h_1), (g_2, h_2) \in R \circ S.$$

Then, by the definition of $R \circ S$, there exist elements $k_1, k_2 \in K$ such that

$$(g_1, k_1) \in S, \quad (k_1, h_1) \in R, \quad (g_2, k_2) \in S, \quad \text{and} \quad (k_2, h_2) \in R.$$

Since S is an RGM and

$$(g_1, k_1), (g_2, k_2) \in S,$$

it follows that

$$(g_1 g_2, k_1 k_2) = (g_1, k_1)(g_2, k_2) \in S.$$

Likewise, since R is an RGM and

$$(k_1, h_1), (k_2, h_2) \in R,$$

it follows that

$$(k_1 k_2, h_1 h_2) = (k_1, h_1)(k_2, h_2) \in R.$$

By the definition of $R \circ S$ again, the facts that

$$(g_1 g_2, k_1 k_2) \in S \quad \text{and} \quad (k_1 k_2, h_1 h_2) \in R$$

imply that

$$(g_1 g_2, h_1 h_2) \in S \circ R.$$

Condition (iii): Suppose that

$$(g, h) \in R \circ S.$$

Then, by the definition of $R \circ S$, there exists $k \in K$ such that

$$(g, k) \in S \quad \text{and} \quad (k, h) \in R.$$

Because S and R are RGM's, it follows that

$$(g^{-1}, k^{-1}) \in S \quad \text{and} \quad (k^{-1}, h^{-1}) \in R.$$

Therefore, by the definition of $R \circ S$,

$$(g^{-1}, h^{-1}) \in R \circ S. \qquad \square$$

The following theorem, whose proof we leave as an exercise (Problem 9), generalizes Theorem 3.8.

8.9 THEOREM

PRESERVATION OF SUBGROUPS UNDER AN RGM Let G and H be groups, let $R \subseteq G \times H$ be an RGM, and suppose that K is a subgroup of G. Then $R(K)$ is a subgroup of H.

By symmetry, Theorem 8.9 has the following corollary:

8.10 COROLLARY

Let G and H be groups, let $R \subseteq G \times H$ be an RGM, and suppose that L is a subgroup of H. Then $R^{-1}(L)$ is a subgroup of G.

8.11 THEOREM

PRESERVATION OF NORMAL SUBGROUPS BY AN RGM Let G and H be groups, and let $R \subseteq G \times H$ be an RGM. Let K be a subgroup of G, and let N be a normal subgroup of K. Then, $R(N)$ is a normal subgroup of $R(K)$.

PROOF Let

$$h \in R(N) \quad \text{and} \quad y \in R(K).$$

We have to prove that $yhy^{-1} \in R(N)$. (See Problem 3 in Problem Set 5.3.) Since $h \in R(N)$,

$$\exists g \in N, \quad (g, h) \in R.$$

Likewise, since $y \in R(K)$,

$$\exists k \in K, \quad (k, y) \in R.$$

Because N is a normal subgroup of K, $g \in N$, and $k \in K$, we have

$$kgk^{-1} \in kNk^{-1} = N.$$

Also, $(k, y) \in R$ implies that

$$(k^{-1}, y^{-1}) \in R.$$

Now, since

$$(k, y), (g, h), (k^{-1}, y^{-1}) \in R,$$

it follows that

$$(kgk^{-1}, yhy^{-1}) = (k, y)(g, h)(k^{-1}, y^{-1}) \in R,$$

which, together with the fact that $kgk^{-1} \in N$, implies that

$$yhy^{-1} \in R(N). \qquad \square$$

If $f: G \to H$ is a homomorphism, then, by Definition 3.9,

$$\ker(f) = f^{-1}(\{1\}).$$

Therefore, we can extend the notion of a kernel to RGM's as follows:

8.12 DEFINITION

KERNEL OF AN RGM If G and H are groups and $R \subseteq G \times H$ is an RGM, we define the **kernel** of R, in symbols, $\ker(R)$, by

$$\ker(R) = R^{-1}(\{1\}).$$

Since $(R^{-1})^{-1} = R$, we have

$$\ker(R^{-1}) = R(\{1\}).$$

8.13 THEOREM

KERNELS AND DOMAINS OF RGM'S Let G and H be groups, and let $R \subseteq G \times H$ be an RGM. Then:

(i) $\ker(R)$ is a normal subgroup of $\mathrm{dom}(R)$.
(ii) $\ker(R^{-1})$ is a normal subgroup of $\mathrm{dom}(R^{-1})$.

PROOF We prove (ii) and note that (i) follows by symmetry. Since $\{1\}$ is a normal subgroup of G, it follows from Theorem 8.11 that $R(\{1\}) = \ker(R^{-1})$ is a normal subgroup of $R(G)$. But, because $\mathrm{dom}(R) \subseteq G$, Part (ii) of Lemma 8.3 implies that $R(G) = \mathrm{dom}(R^{-1})$. $\square$

8.14 LEMMA

Let G and H be groups, let $R \subseteq G \times H$ be an RGM, and let $(g, h) \in R$. Then,

$$R(\{g\}) \subseteq h \cdot \ker(R^{-1}).$$

PROOF Suppose that $h' \in R(\{g\})$. Then $(g, h') \in R$. Since $(g, h) \in R$, we have $(g^{-1}, h^{-1}) \in R$, and it follows that

$$(1, h^{-1}h') = (g^{-1}g, h^{-1}h') = (g^{-1}, h^{-1})(g, h') \in R.$$

Therefore,

$$h^{-1}h' \in R(\{1\}) = \ker(R^{-1}),$$

and so

$$h' \in h \cdot \ker(R^{-1}). \qquad \square$$

8.15 COROLLARY

Let G and H be groups, and let $R \subseteq G \times H$ be an RGM. Then R is a function if and only if $R(\{1\}) \subseteq \{1\}$.

PROOF If R is a function and $h \in R(\{1\})$, then we have both $(1, h) \in R$ and $(1, 1) \in R$; hence, $h = 1$, and so $R(\{1\}) \subseteq \{1\}$.

Conversely, suppose that $R(\{1\}) \subseteq \{1\}$; that is, $\ker(R^{-1}) \subseteq \{1\}$. Then, by Lemma 8.14,

$$(g, h) \in R \Rightarrow R(\{g\}) \subseteq h \cdot \{1\} = \{h\},$$

from which it follows that R is a function. $\square$

The following theorem gives a very powerful method for constructing homomorphisms between groups.

8.16 THEOREM

INDUCED HOMOMORPHISM THEOREM Let G, G', H, and H' be groups, let $R \subseteq G \times H$ be an RGM with $G = \text{dom}(R)$, let

$$\gamma : G \to G' \quad \text{and} \quad \eta : H \to H'$$

be homomorphisms, and suppose that $\gamma : G \to G'$ is surjective. Let $\phi \subseteq G' \times H'$ be defined by

$$\phi = \eta \circ R \circ \gamma^{-1}.$$

Then, if $R(\ker(\gamma)) \subseteq \ker(\eta)$, it follows that

$$\phi : G' \to H'$$

is a homomorphism.

PROOF Let $\phi = \eta \circ R \circ \gamma^{-1}$, and suppose that

$$R(\ker(\gamma)) \subseteq \ker(\eta).$$

Since ϕ is a composition of RGM's, it is an RGM. Because $\text{dom}(R) = G$ and $\gamma : G \to G'$ is surjective, it follows that

$$\text{dom}(\phi) = \text{dom}(\eta \circ R \circ \gamma^{-1}) = G'$$

(Problem 11). Now, since $R(\ker(\gamma)) \subseteq \ker(\eta)$, we have

$$\begin{aligned}\phi(\{1\}) &= (\eta \circ R \circ \gamma^{-1})(\{1\}) = \eta(R(\gamma^{-1}(\{1\}))) \\ &= \eta(R(\ker(\gamma))) \subseteq \eta(\ker(\eta)) = \{1\},\end{aligned}$$

and it follows from Corollary 8.15 that ϕ is a function. Therefore,

$$\phi: G' \to H'$$

is a homomorphism. □

8.17 LEMMA Let G and H be groups, and let $R \subseteq G \times H$ be an RGM. Then:

(i) $R(\ker(R)) \subseteq \ker(R^{-1})$.
(ii) $R^{-1}(\ker(R^{-1})) \subseteq \ker(R)$.

PROOF We prove (i) and note that (ii) follows by symmetry. Thus, let $h \in R(\ker(R))$. Then, there exists $k \in \ker(R)$ such that $(k, h) \in R$. Because

$$k \in \ker(R) = R^{-1}(\{1\}),$$

we have $(k, 1) \in R$, from which it follows that

$$(k^{-1}, 1^{-1}) = (k^{-1}, 1) \in R.$$

Since $(k, h), (k^{-1}, 1) \in R$, we conclude that

$$(1, h) = (kk^{-1}, h \cdot 1) = (k, h)(k^{-1}, 1) \in R;$$

so $h \in R(\{1\}) = \ker(R^{-1})$. □

8.18 THEOREM **RIGUET'S THEOREM**[†] Let G and H be groups, and let $R \subseteq G \times H$ be an RGM. Let

$$\gamma: \mathrm{dom}(R) \to \mathrm{dom}(R)/\ker(R)$$

and

$$\eta: \mathrm{dom}(R^{-1}) \to \mathrm{dom}(R^{-1})/\ker(R^{-1})$$

be the natural homomorphisms. Let

$$\phi = \eta \circ R \circ \gamma^{-1}.$$

Then,

$$\phi: \mathrm{dom}(R)/\ker(R) \to \mathrm{dom}(R^{-1})/\ker(R^{-1})$$

is an isomorphism.

[†] J. Riguet, *Bulletin de la Sociètè Mathèmatique de France, Paris* **76** (1948), pp. 114–155.

PROOF Note that $R \subseteq \text{dom}(R) \times \text{dom}(R^{-1})$. In Theorem 8.16, replace G by $\text{dom}(R)$, G' by $\text{dom}(R)/\text{ker}(R)$, H by $\text{dom}(R^{-1})$, and H' by $\text{dom}(R^{-1})/\text{ker}(R^{-1})$. Then, $\text{ker}(\gamma) = \text{ker}(R)$ and $\text{ker}(\eta) = \text{ker}(R^{-1})$; so, by Part (i) of Lemma 8.17, we have

$$R(\text{ker}(\gamma)) \subseteq \text{ker}(\eta).$$

Therefore,

$$\phi:\text{dom}(R)/\text{ker}(R) \to \text{dom}(R^{-1})/\text{ker}(R^{-1})$$

is a homomorphism. Since the natural homomorphisms are surjective, we have

$$\begin{aligned}\phi(\text{dom}(R)/\text{ker}(R)) &= \phi(\gamma(\text{dom}(R))) = (\eta \circ R \circ \gamma^{-1})(\gamma(\text{dom}(R))) \\ &= \eta(R(\gamma^{-1}(\gamma(\text{dom}(R))))) = \eta(R(\text{dom}(R))) \\ &= \eta(\text{dom}(R^{-1})) \\ &= \text{dom}(R^{-1})/\text{ker}(R^{-1}),\end{aligned}$$

and it follows that

$$\phi:\text{dom}(R)/\text{ker}(R) \to \text{dom}(R^{-1})/\text{ker}(R^{-1})$$

is surjective. Note that

$$\phi^{-1} = (\eta \circ R \circ \gamma^{-1})^{-1} = (\gamma^{-1})^{-1} \circ R^{-1} \circ \eta^{-1} = \gamma \circ R^{-1} \circ \eta^{-1}.$$

Therefore,

$$\begin{aligned}\text{ker}(\phi) &= \phi^{-1}(\{1\}) = (\gamma \circ R^{-1} \circ \eta^{-1})(\{1\}) \\ &= \gamma(R^{-1}(\eta^{-1}(\{1\}))) = \gamma(R^{-1}(\text{ker}(\eta))) \\ &= \gamma(R^{-1}(\text{ker}(R^{-1}))).\end{aligned}$$

But, by Part (ii) of Lemma 8.17,

$$R^{-1}(\text{ker}(R^{-1})) \subseteq \text{ker}(R) = \text{ker}(\gamma),$$

and it follows that

$$\text{ker}(\phi) \subseteq \gamma(\text{ker}(\gamma)) = \{1\}.$$

Of course, $\{1\} \subseteq \text{ker}(\phi)$, and so $\text{ker}(\phi) = \{1\}$. Therefore, by Theorem 3.10, ϕ is injective. Therefore,

$$\phi:\text{dom}(R)/\text{ker}(R) \to \text{dom}(R^{-1})/\text{ker}(R^{-1})$$

is both surjective and injective; hence, it is an isomorphism. □

Riguet's theorem (Theorem 8.18) is useful for establishing isomorphisms between quotient groups. Indeed, if G and H are groups, K is a normal subgroup of G, and N is a normal subgroup of H, then, to find an isomorphism of G/K onto H/N using Riguet's theorem, it is necessary only to find an RGM $R \subseteq G \times H$ with $\text{dom}(R) = G$, $\text{dom}(R^{-1}) = H$, $\ker(R) = K$, and $\ker(R^{-1}) = N$.

8.19 Example Let G be a group, let K be a normal subgroup of G, and let H be a subgroup of G. Define

$$KH = \{kh \mid k \in K, h \in H\}.$$

Prove the **Noether isomorphism theorem**: *KH is a subgroup of G, K is a normal subgroup of KH, $K \cap H$ is a normal subgroup of H, and $H/(K \cap H)$ is isomorphic to KH/K.*

SOLUTION We sketch the solution, leaving the details as an exercise (Problem 29). Define $R \subseteq G \times G$ by

$$R = \{(h, kh) \mid h \in H, k \in K\}.$$

Then R is an RGM, $\text{dom}(R) = H$, $\text{dom}(R^{-1}) = KH$, $\ker(R) = R^{-1}(\{1\}) = K \cap H$, and $\ker(R^{-1}) = R(\{1\}) = K$; so the Noether isomorphism theorem follows from Riguet's theorem (Theorem 8.18). □

8.20 DEFINITION

CONGRUENCE If G is a group, then a **congruence on** G is defined to be an RGM $E \subseteq G \times G$ that is also an equivalence relation on G.

8.21 THEOREM

A REFLEXIVE RGM IS A CONGRUENCE Let G be a group, let $E \subseteq G \times G$ be a congruence, and suppose that E is reflexive on G. Then E is a congruence on G.

PROOF We must prove that E is symmetric and transitive. Suppose $(a, b) \in E$. Then, since E is reflexive on G, $(a^{-1}, a^{-1}) \in E$, and therefore,

$$(1, a^{-1}b) = (a^{-1}a, a^{-1}b) = (a^{-1}, a^{-1})(a, b) \in E.$$

Consequently,

$$(1, b^{-1}a) = (1^{-1}, (a^{-1}b)^{-1}) \in E.$$

Again, since E is reflexive on G, $(b, b) \in E$, and it follows that

$$(b, a) = (b \cdot 1, b(b^{-1}a)) = (b, b)(1, b^{-1}a) \in E;$$

hence, E is symmetric.

To prove that E is transitive, suppose that

$(a, b), (b, c) \in E.$

Since $(b^{-1}, b^{-1}) \in E$, we have

$$(a, c) = (ab^{-1}b, bb^{-1}c) = (a, b)(b^{-1}, b^{-1})(b, c) \in E,$$

which shows that E is transitive. □

8.22 COROLLARY

Let G and H be groups, and let $R \subseteq G \times H$ be an RGM. Then $R^{-1} \circ R$ is a congruence on dom(R), and $R \circ R^{-1}$ is a congruence on dom(R^{-1}).

PROOF By Lemma 8.6, R^{-1} is an RGM; hence, by Theorem 8.8, $R^{-1} \circ R$ is an RGM. It is easy to verify that dom$(R^{-1} \circ R) =$ dom(R) (Problem 3) and that, by Part (iv) of Lemma 8.1, $R^{-1} \circ R$ is reflexive on dom(R). Hence, by Theorem 8.21, $R^{-1} \circ R$ is a congruence on dom(R). □

Suppose that G is a group and E is a congruence on G. Since E is an equivalence relation on the set G, we can form the partition G/E determined by E as in Definition 3.10 and Theorem 3.12 of Chapter 2. Also, dom$(E) = G$, and, by Part (i) of Theorem 8.13, ker(E) is a normal subgroup of G; so we can form the quotient group G/ker(E). In Problem 33, we ask you to prove the following theorem.

8.23 THEOREM

THE PARTITION DETERMINED BY A CONGRUENCE If G is a group and E is a congruence on G, then, as sets of sets,

$$G/E = G/\ker(E).$$

If N is a normal subgroup of the group G, then G/N is a partition of G by Theorem 2.15. The following theorem, whose proof we leave as an exercise (Problem 35), shows that the equivalence relation E on G determined by this partition (Definition 3.2 in Chapter 2) is a congruence on G.

8.24 THEOREM

THE CONGRUENCE DETERMINED BY A NORMAL SUBGROUP Let G be a group, let N be a normal subgroup of G, and let E be the equivalence relation on G determined by the partition G/N. Then E is a congruence on G and ker$(E) = N$.

PROBLEM SET 5.8

1. Prove Lemma 8.1.
2. Prove Lemma 8.3.
3. Let R be a relation. Prove:
   (a) $R \subseteq R \circ R^{-1} \circ R$
   (b) $\text{dom}(R) = \text{dom}(R^{-1} \circ R) = \text{dom}(R \circ R^{-1} \circ R)$
4. Prove Lemma 8.6.
5. Let G and H be groups, and define $\tau: G \times H \to H \times G$ by $\tau((g, h)) = (h, g)$ for every $(g, h) \in G \times H$.
   (a) Prove that $\tau: G \times H \to H \times G$ is an isomorphism of the direct product group $G \times H$ onto the direct product group $H \times G$.
   (b) If $R \subseteq G \times H$, show that $\tau(R) = R^{-1}$.
6. Use the results of Problem 5 together with Theorem 8.7 to give an alternative proof of Theorem 8.6.
7. Let R be a relation such that $R = R \circ R^{-1} \circ R$. Prove that $R^{-1} \circ R$ is an equivalence relation on $\text{dom}(R)$.
8. Let G and H be groups, and define $\pi: G \times H \to G$ and $\sigma: G \times H \to H$ by $\pi((g, h)) = g$ and $\sigma((g, h)) = h$ for every $(g, h) \in G \times H$.
   (a) Prove that π and σ are homomorphisms.
   (b) If $R \subseteq G \times H$ and $M \subseteq G$, prove that $R(M) = \sigma(R \cap \pi^{-1}(M))$.
9. Prove Theorem 8.9.
10. Use the results of Problem 5 together with Theorem 8.7 to give an alternative proof of Theorem 8.9.
11. Complete the proof of Theorem 8.16 by showing that $\text{dom}(\phi) = G'$.
12. Let G, K, and H be groups, let $S \subseteq G \times K$, and let $R \subseteq K \times H$. Define $\Phi: (G \times K) \times H \to G \times H$ by $\Phi(((g, k), h)) = (g, h)$ for every $((g, k), h) \in (G \times K) \times H$. Define $\Psi: G \times (K \times H) \to (G \times K) \times H$ by $\Psi((g, (k, h))) = ((g, k), h)$ for every $(g, (k, h)) \in G \times (K \times H)$.
    (a) Prove that Φ and Ψ are homomorphisms.
    (b) Prove that $R \circ S = \Phi((S \times H) \cap \Psi(G \times R))$.
    (c) Use the results in Parts (a) and (b) to give an alternative proof of Theorem 8.8.
13. Let G and H be groups, let $R \subseteq G \times H$ be an RGM, let L be a subgroup of H, and let M be a normal subgroup of L. Prove that $R^{-1}(M)$ is a normal subgroup of $R^{-1}(L)$.
14. Let $R \subseteq G \times H$ be an RGM. By Theorem 8.7, R is a subgroup of $G \times H$; hence, R is a group in its own right. Define $p: R \to \text{dom}(R)$ and $q: R \to \text{dom}(R^{-1})$ by $p((g, h)) = g$ and $q((g, h)) = h$ for every $(g, h) \in R$.
    (a) Prove that $p: R \to \text{dom}(R)$ and $q: R \to \text{dom}(R^{-1})$ are surjective homomorphisms.

(b) By Theorem 8.5, p and q are RGM's, and by Lemma 8.6, p^{-1} is an RGM. Prove that $R = q \circ p^{-1}$.

15. Prove the following stronger version of Lemma 8.14: If G and H are groups, $R \subseteq G \times H$ is an RGM, and $(g, h) \in R$, then $R(\{g\}) = h \cdot \ker(R^{-1})$.

16. Let G and H be groups, let $f: G \to H$ be a homomorphism, let K be a normal subgroup of G, and let N be a normal subgroup of H. Suppose that $\gamma: G \to G/K$ and $\eta: H \to H/N$ are the natural homomorphisms. Show that there exists a homomorphism $\phi: G/K \to H/N$ such that $\phi \circ \gamma = \eta \circ f$ if and only if $f(K) \subseteq N$.

17. Let G and H be groups, let $R \subseteq G \times H$ be an RGM, and let $C \in \operatorname{dom}(R)/\ker(R)$. If $g \in G$, prove that

$$g \in C \Leftrightarrow R(\{g\}) = R(C).$$

18. Under the hypotheses of Theorem 8.16, prove that $\phi \circ \gamma = \eta \circ R$.

19. Let G and H be groups, let $R \subseteq G \times H$ be an RGM, and let $C \in \operatorname{dom}(R)/\ker(R)$. Prove that $R(C) \in \operatorname{dom}(R^{-1})/\ker(R^{-1})$. [Hint: Combine the results in Problems 15 and 17.]

20. Let B, G', H, and H' be groups, let $\gamma: G \to G'$ be a homomorphism, and let $\eta: H \to H'$ be a surjective homomorphism. If $\phi: G' \to H'$ is a homomorphism, show that there exists an RGM $R \subseteq G \times H$ such that $\phi \circ \gamma = \eta \circ R$.

21. In Theorem 8.18, if $C \in \operatorname{dom}(R)/\ker(R)$, show that $\phi(C) = R(C)$. [See Problem 19.]

22. Suppose that G and H are groups, K is a normal subgroup of G, N is a normal subgroup of H, and $\phi: G/K \to H/N$ is an isomorphism. Prove that there exists an RGM $R \subseteq G \times H$ with $\operatorname{dom}(R) = G$, $\operatorname{dom}(R^{-1}) = H$, $\ker(R) = K$, and $\ker(R^{-1}) = N$.

23. In Parts (i) and (ii) of Lemma 8.17, show that one actually has $R(\ker(R)) = \ker(R^{-1})$ and $R^{-1}(\ker(R^{-1})) = \ker(R)$.

24. Let G and H be groups, and let $R \subseteq G \times H$ be an RGM. If K is a subgroup of G, prove that

$$R^{-1}(R(K)) = \{kg \mid k \in K \cap \operatorname{dom}(R), g \in \ker(R)\}.$$

25. Let G and H be groups, and let $f: G \to H$ be a surjective homomorphism (that is, an epimorphism). Show that Riguet's theorem (Theorem 8.18), applied to the RGM $f \subseteq G \times H$, yields the fundamental homomorphism theorem (Theorem 3.14).

26. Let G and H be groups, and let $R \subseteq G \times H$ be an RGM. If K is a subgroup of G, prove that $R^{-1}(R(K)) = K$ if and only if $\ker(R) \subseteq K \subseteq \operatorname{dom}(R)$.

27. (a) Prove **Lambek's lemma**: If G and H are groups and $R \subseteq G \times H$ is an RGM, then $R = R \circ R^{-1} \circ R$.

(b) Combine Lambek's lemma with the result of Problem 7 to give an alternative proof of Corollary 8.22.

28. Let G and H be groups, let $R \subseteq G \times H$ be an RGM, and let K be a subgroup of G with $\ker(R) \subseteq K \subseteq \mathrm{dom}(R)$. Let $L = R(K)$.
(a) Prove that $\ker(R^{-1}) \subseteq L \subseteq \mathrm{dom}(R^{-1})$.
(b) Let $S = R \cap (K \times L)$. Prove that $S \subseteq K \times L$ is an RGM, $K = \mathrm{dom}(S)$, $L = \mathrm{dom}(S^{-1})$, $\ker(S) = \ker(R)$, and $\ker(S^{-1}) = \ker(R^{-1})$.

29. Attend to the details in the solution of Example 8.19. [Hint: Since K is a normal subgroup of G, we have $Kh \subseteq hK$ for every $h \in H$; hence, if $k \in K$ and $h \in H$, then $\exists k' \in K$ such that $kh = hk'$.]

30. With the notation of Problem 28, let N be a subgroup of K with $\ker(R) \subseteq N$, and let $M = R(N)$. Prove that N is a normal subgroup of K if and only if M is a normal subgroup of L.

31. Let G and H be groups, and let $f: G \to H$ be a surjective homomorphism (that is, an epimorphism). Let N and K be subgroups of G such that $\ker(f) \subseteq N \subseteq K$.
(a) Prove that $f^{-1}(f(K)) = K$.
(b) Prove that N is a normal subgroup of K if and only if $f(N)$ is a normal subgroup of $f(K)$.
(c) If N is a normal subgroup of K, prove that K/N is isomorphic to $f(K)/f(N)$.

32. With the notation of Problems 28 and 30, suppose that N is a normal subgroup of K, so that (by Problem 30) M is a normal subgroup of L. Let $\gamma: K \to K/N$ and $\eta: L \to L/M$ be the natural homomorphisms. Let $T = \eta^{-1} \circ \eta \circ S \circ \gamma^{-1} \circ \gamma$. Prove that T is an RGM, $\mathrm{dom}(T) = K$, $\ker(T) = N$, $\mathrm{dom}(T^{-1}) = L$, and $\ker(T^{-1}) = M$. Use Riguet's theorem to conclude that K/N is isomorphic to L/M.

33. Prove Theorem 8.23.

34. Let G be a group, and let K and N be normal subgroups of G.
(a) Prove that K/N is a normal subgroup of G/N.
(b) Prove that $(G/N)/(K/N)$ is isomorphic to G/K.

35. Prove Theorem 8.24.

36. If $\mathbb{Z}$ denotes the additive group of integers, and if $R \subseteq \mathbb{Z} \times \mathbb{Z}$ is an RGM such that $\mathrm{dom}(R) = \mathrm{dom}(R^{-1}) = \mathbb{Z}$, show that R is a congruence on $\mathbb{Z}$.

37. If G is a group, and if E and F are congruences on G, show that $E \circ F$ is a congruence. [Hint: Use Theorem 8.21.]

38. Let G be a group, let K be a normal subgroup of G, and let H be a subgroup of G as in Example 8.19. Let E be the congruence on G determined by the partition G/K as in Theorem 8.24. With the notation $KH = \{kh \mid k \in K, h \in H\}$ of Example 8.19, prove that $E(H) = KH$.

39. If G is a group, and if E and F are congruences on G, show that $E \circ F = F \circ E$. [Hint: Use the result of Problem 37 to conclude that $E \circ F$ is a congruence. Then use Part (i) of Lemma 8.1 together with the fact that a congruence is symmetric.]

40. Let G be a group, let K and L be subgroups of G, and let N be a normal subgroup of K. If E is the congruence on K determined by the partition K/N as in Theorem 8.24, prove that $E(L) = (K \cap L)N$. [See Problem 38.]

41. Let G be a group, let $\mathscr{C}$ denote the set of all congruences on G, and let $E, F \in \mathscr{C}$.
(a) Prove that $E \cap F \in \mathscr{C}$.
(b) Prove that $E \circ F \in \mathscr{C}$. [See Problem 39.]
(c) In the partially ordered set $(\mathscr{C}, \subseteq)$, show that $\text{GLB}(E, F) = E \cap F$ and $\text{LUB}(E, F) = E \circ F$.

42. Let G be a group, and let K, N, K', and N' be subgroups of G such that N is a normal subgroup of K and N' is a normal subgroup of K'. Let E be the congruence on K determined by the partition K/N, and let E' be the congruence on K' determined by the partition K'/N'. [See Problem 40.] By applying Riguet's theorem to the RGM $E \circ E'$ on G, prove the **Zassenhaus lemma**: $(K \cap K')N'/(N \cap K')N'$ is isomorphic to $(K \cap K')N/(N' \cap K)N$.

43. Prove the **modular law for congruences**: If G is a group, and if E, F, and R are congruences on G with $E \subseteq R$, then $E \circ (F \cap R) = (E \circ F) \cap R$.

HISTORICAL NOTES

Affine geometry was developed by Julius Plücker (1801–1868), August Möbius (1790–1868), and Hermann Grassmann (1809–1877) in the period 1830–1845. They studied, as we have, the layout of lines and planes in three-space and, in the case of Grassmann, the higher-dimensional analogues, but their methods were rather unlike those we have used. Our methods are influenced strongly by the analogous concepts of subgroup, coset, and factor group from group theory. These ideas take a somewhat more simple form in the affine geometry setting.

Subgroups and cosets first appear in Leonhard Euler's work (1761) on the generalized congruence theorem of Pierre de Fermat (1601–1665): If $\phi(m)$ is the number of integers among $1, 2, \ldots, m$ that are relatively prime integers, then $a^{\phi(m)} \equiv 1 \pmod{m}$ or, equivalently, m divides $a^{\phi(m)} - 1$. Euler mentions neither groups nor cosets, but Euler's proof appears with only trivial cosmetic changes in group-oriented discussions of number theory today. However, couched as they are in number-theoretic language, the group theory material in Euler's work and the later work of Joseph Louis Lagrange (1736–1813) is not very apparent. The first explicit use of group theory was made by Évariste Galois (1811–1832) in his work (1831) on the solution of equations. Here, he uses a noncommutative group G and a subgroup H, and exhibits the decomposition $G = H \cup Hs_1 \cup Hs_2 \cup \cdots \cup Hs_n$ into disjoint cosets. Galois also identified the concept of normal subgroup, but did not explicitly define the factor group. Galois's group theory is concerned with groups of permutations, and the habit of working with permutation groups rather than abstract groups persists until the early 1900s. Peter Sylow's important paper (1872) on the existence of subgroups with orders that are powers of primes is still couched in permutation language, though this has nothing to do with either the subject or the technique. G. F. Frobenius (1849–1917), in 1887, proved Sylow's results in an abstract setting.

Factor groups first appear in the investigations of Pascual Jordan (1902–1980) and Otto Ludwig Hölder (1859–1937) on chains of subgroups in connection with the problem of the solvability of algebraic equations. They occur explicitly in Hölder's 1889 paper, but it took several decades for the factor group in its abstract formulation to become a generally accepted tool, and in the texts of the 1920s and early 1930s, clumsy circumlocutions were still used to avoid the concept.

It is interesting that Lagrange's theorem that the order of a subgroup divides the order of the group was distilled out of his work on the solvability of algebraic equations. Thus, Lagrange contributed a fundamental theorem of group theory over 30 years before the group concept was even defined.

Homomorphisms were defined by Jordan (1889) but were called *meriedric isomorphisms*; this was gradually replaced by our current terminology in the 1920s. The intimate relationship between homomorphisms and their kernels, the normal subgroups, was emphasized by Emmy Noether (1882–1935) in her Göttingen lectures around 1920.

Cyclic groups first gained importance in number theory, but their group-theoretical significance is due to the theorem of Kronecker (1870) that every abelian (that is, commutative) group is a direct product of certain cyclic groups.

ANNOTATED BIBLIOGRAPHY

Gallian, J. A. "The Search for Finite Simple Groups," *Mathematics Magazine* **49** (1976), 163–179.

An accessible account of the recent solution to a long-standing problem in group theory.

Infeld, Leopold. *Whom the Gods Love: The Story of Évariste Galois* (Reston, Va.: National Council of Teachers of Mathematics), 1978.

A biography of Galois, one of the principal contributors to the history of abstract algebra. An exciting story about a colorful figure who was killed in a duel at the age of 21.

Lockwood, E. H., and R. H. Macmillan. *Geometric Symmetry* (Cambridge, England: Cambridge University Press), 1978.

A beautifully printed book on groups of symmetries in the plane, with illustrations of many nice tilings.

Nový, Luboš. *Origin of Modern Algebra* (Leyden: Noordhof), 1973.

A detailed and scholarly history of abstract algebra from 1770, when Lagrange developed certain techniques to examine solvability of polynomial equations, to 1870, the close of a period of intense activity in algebra in England.

Ore, Oystein. *Niels Henrik Abel, Mathematician Extraordinary* (New York: Chelsea), 1974.

A biography of an important mathematician who solved one of the principal motivating problems in algebra—the problem of the insolvability of the quintic equation.

Rothman, Tony. "The Short Life of Évariste Galois," *Scientific American* (April 1982), 136–149.

A revisionist view of the life of Galois, making certain modifications in the romantic view espoused earlier by E. T. Bell in his popular book, Men of Mathematics. Another more scholarly account by Rothman appeared in the American Mathematical Monthly ***89*** *(1982), 84–106.*

Rotman, Joseph. *An Introduction to the Theory of Groups*, 3rd ed. (Boston: Allyn and Bacon), 1984.

An excellent graduate-level text in group theory, useful for reference.

Schattschneider, Doris. "The Plane Symmetry Groups: Their Recognition and Notation," *American Mathematical Monthly* **85** (1978), 439–450.

A very attractive article, by an authority on M. C. Escher, on groups associated with transformations in the plane that leave designs or patterns invariant.

Schattschneider, Doris. "The Pólya–Escher Connection," *Mathematics Magazine* **60** (1987), 293–298.

More on the art of Escher, this time showing the influence exerted by the twentieth century mathematician, George Pólya.

van der Waerden, B. L. *Modern Algebra*, 2 vols. (New York: Ungar), 1949, 1950.

This is the book (originally published in German) that really introduced abstract algebra into the curriculum in the United States. At that time, abstract algebra was a graduate school subject; gradually it has drifted down to the junior or even sophomore level.

Wussing, Hans. *The Genesis of the Abstract Group Concept* (Cambridge, Mass.: MIT Press), 1984.

This provides a good description of the many contributions to group theory by such mathematicians as Ruffini, Lagrange, Cauchy, Galois, Jordan, Kronecker, Klein, and Hölder, among others.

APPENDIX I

MATHEMATICAL INDUCTION

In your precalculus and calculus courses, you may have studied the method of proof known as *mathematical induction.* Since this topic is not always stressed in these courses, you might need a brief review of the method. The idea of mathematical induction is often illustrated by the so-called domino analogy. Imagine a row of dominoes standing on end as in Figure A-1. Assume that the following two conditions hold:

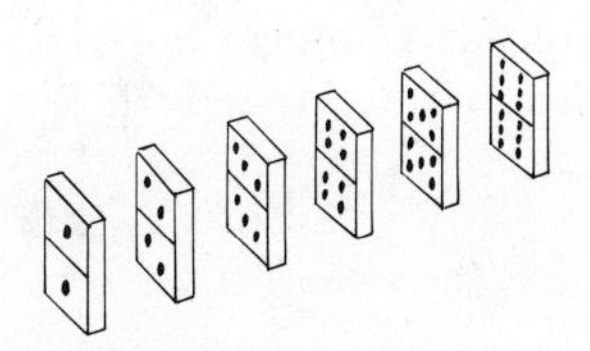

FIGURE A-1

1. The first domino is tipped over.
2. If any domino tips over, it will hit the next domino and tip it over.

Evidently, then, *all* of the dominoes tip over.

By analogy with the row of dominoes, consider a sequence of statements, or propositions,

$$P_1, P_2, P_3, P_4, P_5, P_6, \ldots,$$

each of which could be either true or false. If a proposition P_n is true, imagine that the nth domino tips over. Then the two conditions above are analogous to the following:

1. The first proposition P_1 is true.
2. If any proposition P_n is true, then the next proposition P_{n+1} is also true.

If these two conditions hold, then, by analogy with the dominoes, it appears that *all* of the propositions should be true. The **principle of mathematical induction**, which states that this argument is valid, is proved in Section 2 of Chapter 4 (see Theorem 2.4).

To prove by mathematical induction that all propositions in a sequence

$$P_1, P_2, P_3, P_4, P_5, P_6, \ldots$$

are true, begin by specifying the exact meaning of the proposition P_n for each positive integer n. Then carry out the following two steps:

Step 1. Prove that P_1 is true.

Step 2. Let n denote an arbitrary, but fixed, positive integer. Assume that P_n is true. On the basis of this assumption, prove that P_{n+1} is true.

Having done this, you are entitled to conclude, by the principle of mathematical induction, that P_n is true for every positive integer $n = 1, 2, 3, 4, 5, 6, \ldots$. This is called a proof **by induction on *n*.**

In Step 2, the assumption that P_n is true for an arbitrary, but fixed, value of n is called the **induction hypothesis**. When you make the induction hypothesis, you are not really stating that P_n is, in fact, true: you are just *supposing* it is, to see if you can prove that P_{n+1} is true on the basis of this supposition. In the domino analogy, it is as if you were checking to see if the dominoes are close enough together so that if any one of the dominoes, say the nth one, tips over, it will hit the next one and tip it over.

Example 1 If n is a positive integer, prove that the nth odd positive integer is $2n - 1$.

SOLUTION We make a proof by induction on n. The odd positive integers are 1, 3, 5, 7, 9, and so forth. Note that, by adding 2 to an odd positive integer, we obtain the next odd positive integer. For each positive integer n, let P_n be the proposition stating that the nth odd positive integer is $2n - 1$. We indicate this by writing

P_n: *The* nth *odd positive integer is* $2n - 1$.

For instance, P_3 says that

The third odd positive integer is $2(3) - 1 = 5$

(which happens to be true). The proposition P_{n+1} is obtained by replacing n by $n + 1$ in the expression for P_n; so P_{n+1} is given by

P_{n+1}: *The* $(n + 1)$st *odd positive integer is* $2(n + 1) - 1$.

To prove that P_n is true for all positive integers n, we carry out the two steps in the mathematical-induction procedure.

Step 1. P_1 is the proposition given by

P_1: *The* 1st *odd positive integer is* $2(1) - 1$;

that is, P_1 states that

The first odd positive integer is 1,

which is true.

Step 2. Let n be an arbitrary, but fixed, positive integer. We assume, for the sake of our induction argument, that P_n is true. Thus, we assume, as our induction hypothesis, that

The nth *odd positive integer is* $2n - 1$.

Then the next *odd* positive integer after $2n - 1$ is

$$(2n - 1) + 2 = 2n + 2 - 1 = 2(n + 1) - 1;$$

that is,

The $(n + 1)$st *odd positive integer is* $2(n + 1) - 1$.

This shows that P_{n+1} is true, and completes our proof by induction on n. □

Example 2 Prove that the sum of the first n odd positive integers is n^2.

SOLUTION Again, we make a proof by induction on n. For each positive integer n, let P_n be the proposition given by

P_n: *The sum of the first n odd positive integers is* n^2.

By the result of Example 1, the nth odd positive integer is $2n - 1$; so we can rewrite P_n in the equivalent form

$$P_n: 1 + 3 + 5 + \cdots + (2n - 1) = n^2.$$

Then P_{n+1} is given by

$$P_{n+1}: 1 + 3 + 5 + \cdots + (2n - 1) + [2(n + 1) - 1] = (n + 1)^2;$$

that is,

$$P_{n+1}: 1 + 3 + 5 + \cdots + (2n - 1) + (2n + 1) = (n + 1)^2.$$

Step 1. P_1 is the proposition given by

$$P_1: 1 = 1^2,$$

which is clearly true.

Step 2. Assume, as our induction hypothesis, that P_n is true for some value of n. Thus, we assume that

$$1 + 3 + 5 + \cdots + (2n - 1) = n^2 \tag{1}$$

is true. Our goal is to prove, on the basis of Equation (1), that P_{n+1} is true. A glance at the statement P_{n+1} above suggests that we add $2n + 1$ to both sides of Equation (1) to obtain

$$1 + 3 + 5 + \cdots + (2n - 1) + (2n + 1) = n^2 + (2n + 1). \tag{2}$$

Since

$$n^2 + (2n + 1) = n^2 + 2n + 1 = (n + 1)^2, \tag{3}$$

we can rewrite Equation (2) as

$$1 + 3 + 5 + \cdots + (2n - 1) + (2n + 1) = (n + 1)^2. \tag{4}$$

Thus, P_{n+1} is true, and our proof by induction is complete. □

After you become more familiar with the technique of proving theorems by induction on n, you can just keep in mind the proposition corresponding to n, rather than denoting it explicitly by P_n. The following example illustrates how this is done:

Example 3 If $f(x)$ is a real-valued function of a real variable, denote the derivative $f'(x)$ by $D_x f(x)$. Using the fact that $D_x x = 1$ and the rule for differentiating products, prove by induction on n that

$$D_x x^n = nx^{n-1}$$

holds for all positive integers n.

SOLUTION For $n = 1$, the statement to be proved becomes

$$D_x x^1 = 1 \cdot x^{1-1},$$

that is,

$$D_x x = x^0,$$

or

$$D_x x = 1,$$

which is true. Now we assume, as our induction hypothesis, that

$$D_x x^n = nx^{n-1}$$

for some arbitrary, but fixed, value of n. Our goal is to prove that

$$D_x x^{n+1} = (n + 1)x^{(n+1)-1}.$$

In other words, we want to prove that

$$D_x x^{n+1} = (n + 1)x^n.$$

But, using the facts that $D_x x = 1$ and the induction hypothesis, we have

$$\begin{aligned} D_x x^{n+1} &= D_x(x \cdot x^n) \\ &= (D_x x)x^n + x(D_x x^n) \qquad \text{(Product rule)} \\ &= 1 \cdot x^n + x(nx^{n-1}) \\ &= x^n + nx^n \\ &= (1 + n)x^n \\ &= (n + 1)x^n. \end{aligned}$$

□

PROBLEM SET FOR APPENDIX I

In Problems 1–17, use mathematical induction on n to prove that the given proposition is true for all positive integers n.

1. $1 + 2 + 3 + \cdots + n = n(n + 1)/2$; that is, the sum of the first n positive integers is $n(n + 1)/2$.
2. $1^2 + 2^2 + 3^2 + \cdots + n^2 = n(n + 1)(2n + 1)/6$; that is, the sum of the first n perfect squares is $n(n + 1)(2n + 1)/6$.

3. The sum of the first n perfect cubes is $[n(n + 1)/2]^2$.

4. $1 + 5 + 9 + \cdots + (4n - 3) = n(2n - 1)$. [Hint: In the series of numbers to be added, notice that each number is 4 more than its immediate predecessor.]

5. The sum of the squares of the first n odd positive integers is $n(2n - 1)(2n + 1)/3$.

6. $1 \cdot 2 + 2 \cdot 3 + 3 \cdot 4 + \cdots + n(n + 1) = n(n + 1)(n + 2)/3$.

7. If a and b are real numbers, then $(ab)^n = a^n b^n$.

8. $(1 \cdot 2)^{-1} + (2 \cdot 3)^{-1} + (3 \cdot 4)^{-1} + \cdots + [n(n + 1)]^{-1} = n/(n + 1)$.

9. $\cos n\pi = (-1)^n$.

10. De Moivre's theorem: If θ is a real number, then

$$(\cos \theta + i \sin \theta)^n = \cos n\theta + i \sin n\theta.$$

11. If r is a real number and $r \neq 1$, then

$$r + r^2 + r^3 + r^4 + \cdots + r^n = \frac{r(r^n - 1)}{r - 1}.$$

12. Bernoulli's inequality: If h is a real number and $h \geq -1$, then $(1 + h)^n \geq 1 + nh$.

13. $n^3 - n$ is exactly divisible by 3.

14. If $n \geq 2$, then $n^2 + 4 < (n + 1)^2$.

15. $D_x^n(x^{-1}) = (-1)^n n! x^{-(n+1)}$.

16. A polygonal region of the plane is divided into n polygonal subregions in such a way that two of the subregions overlap, if at all, only on a common edge or a common vertex. Counting an edge only once, even if it is common to two subregions, suppose there are e edges. Counting a vertex only once, even if it is common to two or more subregions, suppose there are v vertices. Then $n + v = e + 1$.

17. There are n people in a room. If every person shakes hands with every other person, show that $n(n + 1)/2$ handshakes take place.

ANSWERS AND HINTS FOR SELECTED ODD-NUMBERED PROBLEMS

CHAPTER 1 PROBLEM SET 1.1, PAGE 14

1. c, d, i, l, m, and n are propositions; a, b, g and o are not. The rest are indeterminate; e.g., for k, there could be a question of, where? When?

3. (a) If P and Q have the same truth values, and if Q and R have the same truth values, then P and R have the same truth values.
(b) Tautology 15

5. (a) True (b) False (c) False (d) True (e) True (f) False

7. (a) $P \wedge (\sim Q)$ (b) $(\sim P) \wedge Q$ (c) $Q \Rightarrow (\sim P)$ (d) $(\sim Q) \Rightarrow P$ (e) $(\sim P) \wedge Q$

9. (a) This equation does not have a solution and I am not stupid.
(b) Alfie is not studying hard or he is not doing well in his classes.

11. (a) I am not cold. (b) If I am cold, then it is snowing and I am not going home.
(c) It is snowing and I am not going home. (d) I am cold if and only if I am going home.
(e) If I am going home, then I am cold. (f) Either I am cold or I am going home, but not both.
(g) If it is snowing or I am cold, then I am going home. (h) I am going home and I am not cold.

13. (a) $P \Rightarrow R$ (b) $(P \wedge Q) \Rightarrow R$ (c) $R \Rightarrow Q$ (d) $R \Rightarrow Q$ (e) $R \Rightarrow (P \vee Q)$ (f) $\sim(P \Leftrightarrow Q)$

15.

P	Q	R	$P \vee R$	$Q \vee R$	$P \wedge R$	$Q \wedge R$	$P \wedge (Q \wedge R)$	$P \wedge (Q \vee R)$	$P \vee (Q \wedge R)$
1	1	1	1	1	1	1	1	1	1
0	1	1	1	1	0	1	0	0	1
1	0	1	1	1	1	0	0	1	1
0	0	1	1	1	0	0	0	0	0
1	1	0	1	1	0	0	0	1	1
0	1	0	0	1	0	0	0	0	0
1	0	0	1	0	0	0	0	0	0
0	0	0	0	0	0	0	0	0	0

17.

P	Q	R	$(P \wedge ((\sim Q) \vee R)) \vee Q$
1	1	1	1 1 0 1 1 1 1
0	1	1	0 0 0 1 1 1 1
1	0	1	1 1 1 1 1 1 0
0	0	1	0 0 1 1 1 0 0
1	1	0	1 0 0 0 0 1 1
0	1	0	0 0 0 0 0 1 1
1	0	0	1 1 1 1 0 1 0
0	0	0	0 0 1 1 0 0 0

19.

P	$\sim P$	$P \Leftrightarrow (\sim(\sim P))$
1	0	1 1 1
0	1	0 1 0

21.

P	Q	$(P \wedge Q) \Leftrightarrow (Q \wedge P)$
1	1	1 1 1 1 1 1 1
0	1	0 0 1 1 1 0 0
1	0	1 0 0 1 0 0 1
0	0	0 0 0 1 0 0 0

P	Q	$(P \vee Q) \Leftrightarrow (Q \vee P)$
1	1	1 1 1 1 1 1 1
0	1	0 1 1 1 1 1 0
1	0	1 1 0 1 0 1 1
0	0	0 0 0 1 0 0 0

P	Q	R	$(P \wedge Q) \wedge R \Leftrightarrow P \wedge (Q \wedge R)$
1	1	1	1 1 1 1 1 1 1 1 1 1 1
0	1	1	0 0 1 0 1 1 0 0 1 1 1
1	0	1	1 0 0 0 1 1 1 0 0 0 1
0	0	1	0 0 0 0 1 1 0 0 0 0 1
1	1	0	1 1 1 0 0 1 1 0 1 0 0
0	1	0	0 0 1 0 0 1 0 0 1 0 0
1	0	0	1 0 0 0 0 1 1 0 0 0 0
0	0	0	0 0 0 0 0 1 0 0 0 0 0

P	Q	R	$(P \vee Q) \vee R \Leftrightarrow P \vee (Q \vee R)$
1	1	1	1 1 1 1 1 1 1 1 1 1 1
0	1	1	0 1 1 1 1 1 0 1 1 1 1
1	0	1	1 1 0 1 1 1 1 1 0 1 1
0	0	1	0 0 0 1 1 1 0 1 0 1 1
1	1	0	1 1 1 1 0 1 1 1 1 1 0
0	1	0	0 1 1 1 0 1 0 1 1 1 0
1	0	0	1 1 0 1 0 1 1 1 0 0 0
0	0	0	0 0 0 0 0 1 0 0 0 0 0

23.

P	Q	R	$P \wedge (Q \vee R) \Leftrightarrow (P \wedge Q) \vee (P \wedge R)$
1	1	1	1 1 1 1 1 1 1 1 1 1 1 1 1
0	1	1	0 0 1 1 1 1 0 0 1 0 0 0 1
1	0	1	1 1 0 1 1 1 1 0 0 1 1 1 1
0	0	1	0 0 0 1 1 1 0 0 0 0 0 0 1
1	1	0	1 1 1 1 0 1 1 1 1 1 1 0 0
0	1	0	0 0 1 1 0 1 0 0 1 0 0 0 0
1	0	0	1 0 0 0 0 1 1 0 0 0 1 0 0
0	0	0	0 0 0 0 0 1 0 0 0 0 0 0 0

P	Q	R	$P \vee (Q \wedge R) \Leftrightarrow (P \vee Q) \wedge (P \vee R)$
1	1	1	1 1 1 1 1 1 1 1 1 1 1 1 1
0	1	1	0 1 1 1 1 1 0 1 1 1 0 1 1
1	0	1	1 1 0 0 1 1 1 1 0 1 1 1 1
0	0	1	0 0 0 0 1 1 0 0 0 0 0 1 1
1	1	0	1 1 1 0 0 1 1 1 1 1 1 1 0
0	1	0	0 0 1 0 0 1 0 1 1 0 0 0 0
1	0	0	1 1 0 0 0 1 1 1 0 1 1 1 0
0	0	0	0 0 0 0 0 1 0 0 0 0 0 0 0

25.

P	Q	$\sim(P \wedge Q) \Leftrightarrow (\sim P) \vee (\sim Q)$
1	1	0 1 1 1 1 0 0 0
0	1	1 0 0 1 1 1 1 0
1	0	1 1 0 0 1 0 1 1
0	0	1 0 0 0 1 1 1 1

P	Q	$\sim(P \vee Q) \Leftrightarrow (\sim P) \wedge (\sim Q)$
1	1	0 1 1 1 1 0 0 0
0	1	0 0 1 1 1 1 0 0
1	0	0 1 1 0 1 0 0 1
0	0	1 0 0 0 1 1 1 1

27.

P	Q	$(P \Rightarrow Q) \Leftrightarrow ((\sim Q) \Rightarrow (\sim P))$
1	1	1 1 1 1 0 1 0
0	1	0 1 1 1 0 1 1
1	0	1 0 0 1 1 0 0
0	0	0 1 0 1 1 1 1

29.

P	Q	$(P \Rightarrow Q) \Leftrightarrow ((\sim P) \vee Q)$
1	1	1 1 1 1 0 1 1
0	1	0 1 1 1 1 1 1
1	0	1 0 0 1 0 0 0
0	0	0 1 0 1 1 1 0

P	Q	$(P \Leftrightarrow Q) \Leftrightarrow ((P \Rightarrow Q) \wedge (Q \Rightarrow P))$
1	1	1 1 1 1 1 1 1 1 1 1 1
0	1	0 0 1 1 0 1 1 0 1 0 0
1	0	1 0 0 1 1 0 0 0 0 1 1
0	0	0 1 0 1 0 1 0 1 0 1 0

31.

P	Q	$(P \wedge Q) \Rightarrow P$
1	1	1 1 1 1 1
0	1	0 0 1 1 0
1	0	1 0 0 1 1
0	0	0 0 0 1 0

P	Q	$P \Rightarrow (Q \Rightarrow (P \wedge Q))$
1	1	1 1 1 1 1 1 1
0	1	0 1 1 0 0 0 1
1	0	1 1 0 1 1 0 0
0	0	0 1 0 1 0 0 0

P	Q	$P \Rightarrow (P \vee Q)$
1	1	1 1 1 1 1
0	1	0 1 0 1 1
1	0	1 1 1 1 0
0	0	0 1 0 0 0

33.

P	Q	$(P \wedge (P \Rightarrow Q)) \Rightarrow Q$
1	1	1 1 1 1 1 1 1
0	1	0 0 0 1 1 1 1
1	0	1 0 1 0 0 1 0
0	0	0 0 0 1 0 1 0

35. (a) If $2n + 1$ is not an odd integer, then n is not an integer.
(b) If you have not passed Math 211, then you cannot take Math 212.
(c) If $\lim_{n\to\infty} a_n \neq 0$, then the series $\sum_{k=1}^{\infty} a_k$ does not converge.
(d) If a citizen cannot vote, then he or she is not of age 18 or over.

37. 2 is prime. **39.** False; $((-4)^2 > 9$, yet $-4 \not> 3)$.

41. Note that for $n = 1$, there are exactly $2^1 = 2$ truth combinations possible. Assume that for $n = k$ there are 2^k truth combinations. By adding another propositional variable, the number of combinations grows to $2 \cdot 2^k$ since the $(k + 1)$st variable may be either true or false. Thus, since $2 \cdot 2^k = 2^{k+1}$, we conclude that for $k + 1$ propositional variables, there are 2^{k+1} truth combinations.

43. A truth table for such a connective would contain $2^2 = 4$ rows. Arguing as in Problem 41, we see that there are $2^4 = 16$ ways to place 0's and 1's in these 4 rows.

PROBLEM SET 1.2, PAGE 31

1. (a) True (b) False (c) False **3.** (a) True (b) True (c) False

5. (a) True (b) False (c) True (d) False

7. (a) $(\forall x)(T(x) \Rightarrow S(x))$, where $T(x)$ means x is a teacher and $S(x)$ means x is a sadist.
(b) There exists at least one teacher who is not a sadist.

9. (a) $(\forall x)(S(x))$, where $S(x)$ means x is a sadist. (b) There exists at least one teacher who is not a sadist.

11. (a) $(\exists x)(L(x) \wedge (\sim J(x)))$, where $L(x)$ means x is a lawyer and $J(x)$ means x is a judge.
(b) All lawyers are judges.

13. (a) $(\exists x)(\sim J(x))$, where $J(x)$ means x is a judge. (b) All lawyers are judges.

15. (a) $(\exists x)(P(x) \wedge T(x))$, where $P(x)$ means x is prime and $T(x)$ means x is divisible by 3.
(b) There does not exist a prime number that is exactly divisible by 3.

17. (a) $(\forall y)(x > y)$ means that all numbers are less than x.
(b) $(\exists x)(x > y)$ means that there is at least one number greater than y.
(c) $(\exists x)(\forall y)(x > y)$ means that there exists some number that is greater than all numbers.
(d) $(\forall y)(\exists x)(x > y)$ means that for any number, there exists a number greater than that number.

19. (a) $(\forall y)(x + y = 0)$ means that any real number will serve as an additive inverse for x.
(b) $(\exists x)(x + y = 0)$ means that there exists an additive inverse for y.
(c) $(\exists x)(\forall y)(x + y = 0)$ means that there exists some number which serves as an additive inverse for all real numbers.
(d) $(\forall y)(\exists x)(x + y = 0)$ means that for any real number, an additive inverse exists.

21. (a) True (b) True (c) False (d) True (e) True (f) True (g) True (h) True
(i) False (j) True (k) False (l) True

23. (a) $(\exists x)(\forall n)((n > x) \vee (x \geq n + 1))$ means that there exists some number x such that, for every integer n, x is less than n or x is greater than or equal to $n + 1$.
(b) $(\forall x)(\exists n)(n \leq x < n + 1)$ means that for any real number x, there exist consecutive integers n and $n + 1$ such that $n \leq x < n + 1$.

25. (a) $(\forall x)(P(x))$ (b) $(\exists x)(P(x))$

27. (a) $(\forall \varepsilon)(\exists \delta)(|x - a| < \delta \Rightarrow |f(x) - f(a)| < \varepsilon)$ (b) $(\exists \varepsilon)(\forall \delta)(|x - a| < \delta \wedge |f(x) - f(a)| \geq \varepsilon)$

29. $(\forall x)(\forall y)(x > y \Rightarrow f(x) > f(y))$

31. (1) If every person is a politician, then Joe Smith is a politician.
(2) If Joe Smith is a politician, then there is at least one politician.
(3) If every person is a politician, then there is at least one politician. And so on.

33. The statement $(\forall x)(P(x) \wedge Q(x))$ in theorem 11 on page 28 says that for all x, $P(x)$ and $Q(x)$ are simultaneously true; thus, for all x, $P(x)$ is true and for all x, $Q(x)$ is true. Conversely, $[(\forall x)(P(x)) \wedge (\forall x)(Q(x))]$ means that for any choice of x, $P(x)$ is true and for any choice of x, $Q(x)$ is true. Thus, any choice of x will suffice for both $P(x)$ and $Q(x)$ to be true.

35. By the hypothesis, $P(x)$ is true for all x or else $Q(x)$ is true for all x (or both). Suppose, for instance, that $P(x)$ is true for all x. Then, for any x, $P(x) \vee Q(x)$ is certainly true. Likewise, if $Q(x)$ is true for all x, then, for any x, $P(x) \vee Q(x)$ is again true. Hence, $(\forall x)(P(x) \vee Q(x))$ is true.

37. Similar to Problems 33 and 35. **39.** For example, if two athletes, cars, or computers are said to be "equal."

41. (a) $(\forall x)(\forall y)(\forall z)((x = y) \wedge (y = z) \Rightarrow (x = z))$
(b) If a and c are logically identical to a third object b, we have $P(a) \Leftrightarrow P(b)$ and $P(b) \Leftrightarrow P(c)$ for every propositional function $P(x)$. But, from this and tautology 15 in Section 1.1, it follows that $P(a) \Leftrightarrow P(c)$.

43. Note that $2^3 = 8$. Now if $x^3 = 8$ and $y^3 = 8$, then $x^3 - y^3 = 0$; that is, $(x - y)(x^2 + xy + y^2) = 0$. Considering $x^2 + xy + y^2 = 0$ as a quadratic equation in x (for an arbitrary, but fixed, value of y), we have a discriminant $y^2 - 4y^2 = -3y^2$. Thus, unless $y = 0$, this equation has no solution. Since $y^3 = 8$, we cannot have $y = 0$. Hence, $x^2 + xy + y^2 \neq 0$, and it follows that $x = y$.

47. (a) "For all x" means "for each and every x in the universe." If only 1 and 2 belong to the universe, then $(\forall x)(P(x))$ is equivalent to $P(1) \wedge P(2)$.
(b) "For some x..." means "there is at least one x in the universe...." If only 1 and 2 belong to the universe, then $(\exists x)(P(x))$ is equivalent to $P(1) \vee P(2)$.

49. $\sim(\forall x)(P(x)) \Leftrightarrow \sim(P(1) \wedge P(2)) \Leftrightarrow (\sim P(1)) \vee (\sim P(2)) \Leftrightarrow (\exists x)(\sim P(x))$.
Also, $\sim(\exists x)(P(x)) \Leftrightarrow \sim(P(1) \vee P(2)) \Leftrightarrow (\sim P(1)) \wedge (\sim P(2)) \Leftrightarrow \sim(\forall x)(P(x))$.

PROBLEM SET 1.3, PAGE 42

1.

P	Q	$P \Rightarrow Q$
1	1	1
0	1	1
1	0	0
0	0	1

Note that, in the second row, P is false.

3. If $P \Rightarrow Q$ is justified, then the law of contraposition states that $\sim Q \Rightarrow \sim P$ is justified. Thus, if $\sim Q$ is justified, $\sim P$ is justified by modus ponens.
5. True, by modus ponens. **7.** False, converse need not be true. **9.** False, converse need not be true.
11. Yes, by 3.5, Assertion of Hypothesis. **13.** Yes, by 3.6, Justification of Conclusion.
15. $P \Leftrightarrow \sim(\sim P)$, where P is replaced by $\cos 0 = 1$.
17. $(P \wedge (P \Rightarrow Q)) \Rightarrow Q$, where P is replaced by $\cos 0 = 1$ and Q is replaced by $\sin 0 = 0$.
19. By a truth table check, $[P \wedge (P \Rightarrow Q) \wedge (\sim Q)] \Leftrightarrow [Q \wedge (\sim Q)]$. Replace P by $x = 1$ and Q by $y = 2$.
21. Since, by tautology 18 of Section 1.1, $(P \wedge Q) \Rightarrow P$, if $(P \wedge Q)$ is justified, then, by modus ponens, P is justified as well.
23. Use substitution of equals and tautology 18 of Section 1.1.
25. Mathematically, an axiom need be neither self-evident nor recognized as being, in fact, true.
27. Suppose s is the largest positive integer. Then $s + 1 > s$, contradicting the supposition that s is the largest positive integer.
29. Keen is correct, since assuming $P \Rightarrow Q$ is false is tantamount to assuming P is true and Q is false. She must then reach a contradiction in order to show (arguing by contradiction) that $P \Rightarrow Q$ is true.
31. Assume that a second point of intersection Q exists, and show that L can no longer be perpendicular to $\overline{OP}$.
33. The proof that a number at which a function has a relative extremum is a critical number is often an indirect proof.

PROBLEM SET 1.4, PAGE 53

1. (a) $\{1, 2, 3, 4, 5, 6, 7, 8\} = A$ (b) $\{-9, -8, -7, -6, -5, -4, -3, 3, 4, 5, 6, 7, 8, 9\} = B$
(c) $\{3, 4, 5, 6, 7, 8\} = C$ (d) $\{-9, -8, -7, -6, -5, -4, -3, 1, 2, 3, 4, 5, 6, 7, 8, 9\} = D$
(e) $\{4, 8, 12, 16, 20, 24, 28, 32, 36, 40\} = E$ (f) $\{-3, 1, 2, 3\}$
3. $\{-2, 2\}$ **5.** $\varnothing$ **7.** $\{3\}$ **9.** (a) Empty (b) Nonempty (c) Empty (d) Nonempty
11. (a) $M \subseteq N$ (b) $M \subseteq N$ (c) $M \nsubseteq N$ (d) $M \subseteq N$ (e) $M \subseteq N$ (f) $M \nsubseteq N$
(g) $M \nsubseteq N$ (h) $M \nsubseteq N$
13. (a) $A = B$ (b) $A \neq B$ (c) $A = B$ (d) $A \neq B$ (e) $A = B$
15. $(A \subseteq B) \wedge (B \subseteq C) \Leftrightarrow (\forall x)(x \in A \Rightarrow x \in B) \wedge (\forall x)(x \in B \Rightarrow x \in C)$. Thus, by tautology 16 in Section 1.1, the conclusion follows.

17.

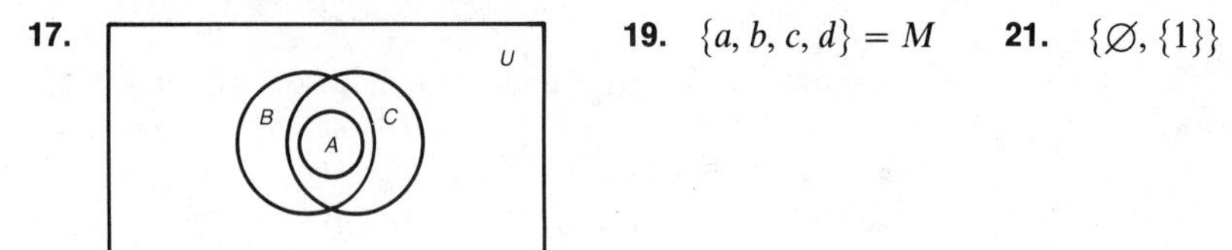

19. $\{a, b, c, d\} = M$ **21.** $\{\varnothing, \{1\}\}$

23. $\{\varnothing, \{1\}, \{2\}, \{1, 2\}\}$ **25.** $\{\varnothing, \{a\}, \{b\}, \{c\}, \{a, b\}, \{b, c\}, \{a, c\}, \{a, b, c\}\}$

27. $\varnothing, \{1\}, \{2\}, \{3\}, \{1, 2\}, \{2, 3\}, \{1, 3\}$ **29.** $\{1\}, \{2\}, \{3\}, \{1, 2\}, \{2, 3\}, \{1, 3\}$

31. False, if the set is nonempty, then $\varnothing$ is proper but trivial. **33.** Use the result of Problem 15.

PROBLEM SET 1.5, PAGE 66

1. (a) $\{1, 2, 3, 4, 5\}$ (b) $\{3, 4\}$ (c) $\{3, 4\}$ (d) $\{3, 4, 6, 7, 8, 9\}$ (e) $\{1, 2\}$ (f) $\{3, 4, 7\}$ (g) $\{1, 2\}$ (h) $\{6, 7\}$ (i) $\{1, 2, 3, 4, 6\}$ (j) $\{3, 4\}$ (k) $\{3, 4, 6\}$ (l) $\{3, 4, 5, 6\}$ (m) $\{1, 2, 3, 4, 5\}$ (n) $\{1, 2, 3, 4\}$

3. (a)

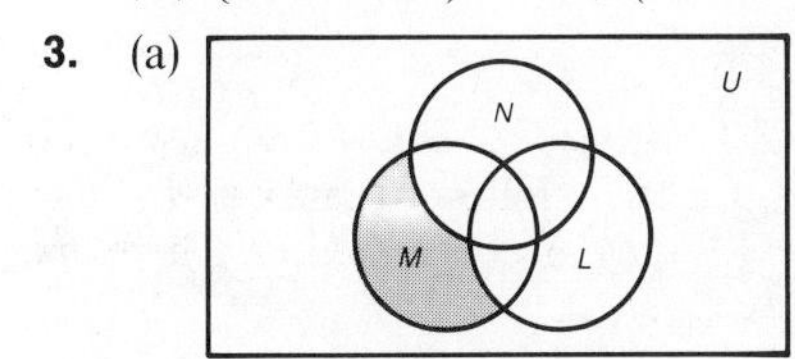

(b)

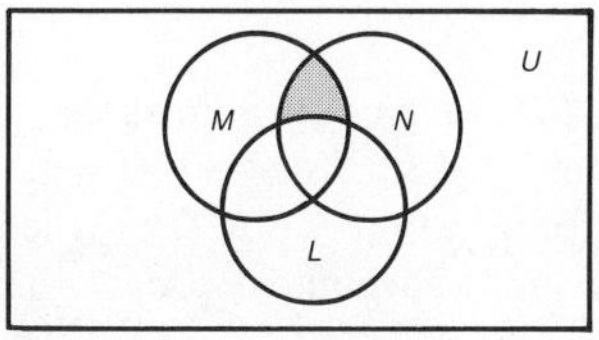

(c)

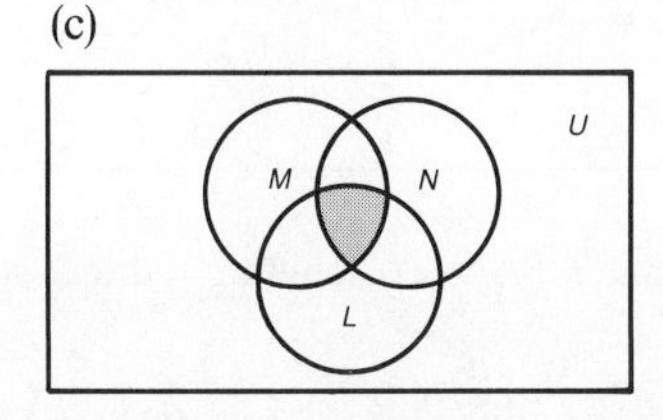

(d)

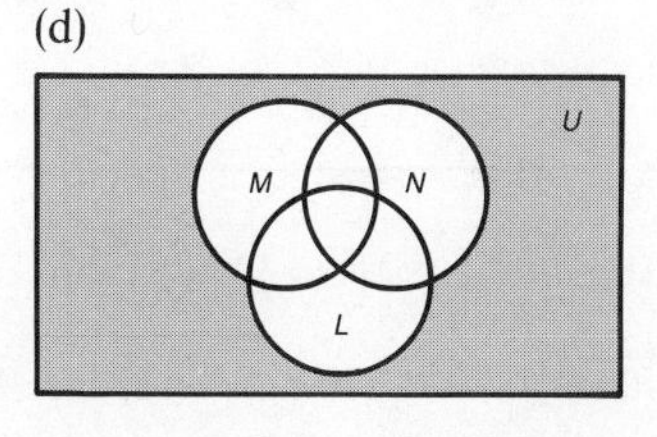

(e)

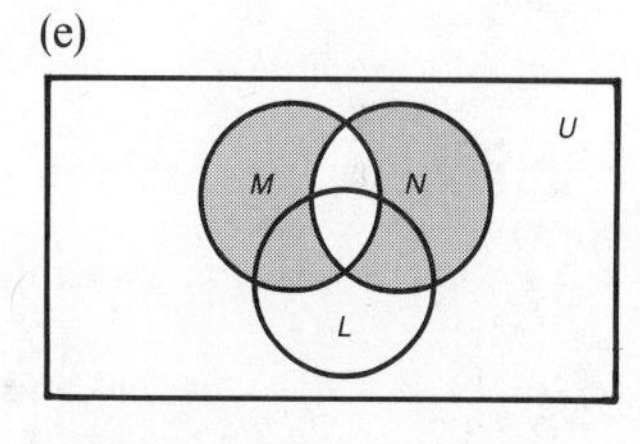

5. (a) $\{5, 6\}$ (b) $\{3\}$ (c) $\{3, 5, 6\}$ (d) $\varnothing$ (e) $\{3, 5, 6\}$ (f) $\varnothing$ (g) $\{2, 3, 4, 5, 6\}$ (h) $\{3\}$ (i) $\{5, 6\}$

7. Utilize appropriate tautology from Section 1.1. **9.** Utilize appropriate tautologies from Section 1.1.

11. (a) $(M \cap N)' = M' \cup N'$

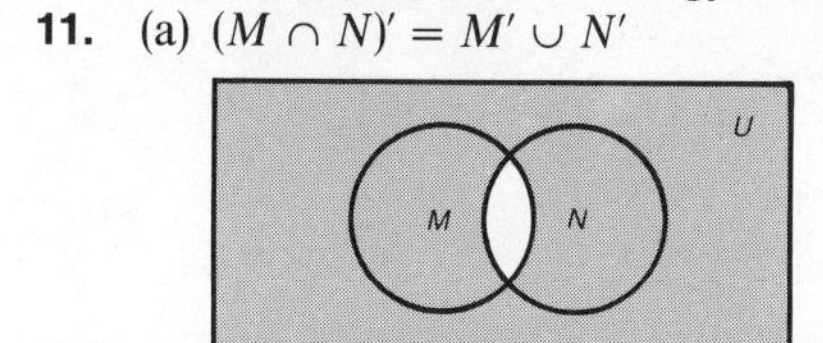

(b) $(M \cup N)' = M' \cap N'$

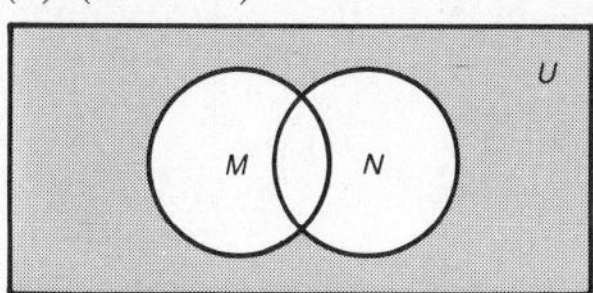

15. (a) By definition, $M \cap N = \{x | x \in M, x \in N\}$ and $M \setminus N = \{x | x \in M, x \notin N\}$. Thus, since we cannot have both $x \in N$ and $x \notin N$, we conclude that $(M \cap N) \cap (M \setminus N) = \varnothing$.

(c)

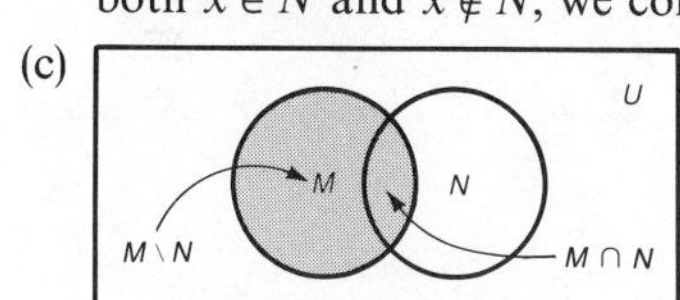

17. False

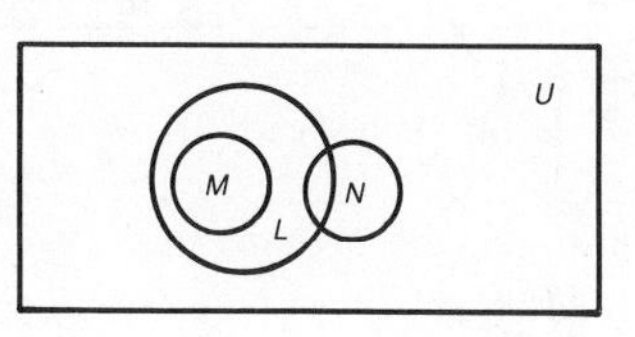

19. (a) $\{1, 4, 5\}$ (b) $\{1, 4, 5\}$ (c) $\varnothing$ (d) $\{1, 2, 3\}$

21. Give a pick-a-point proof. **23.** (a) 2 (b) 0 (c) 1 (d) 1
25. (a) 10 (b) 3 (c) 2 **27.** Note that for $N \subseteq M$, $\#(M \cap N) = \#N$.
29. Use Theorems 5.10 and 5.12. **31.** (a) $2^m - 1$ (b) $2^m - 2$
33. The 3 sets must decompose U into $2^3 = 8$ nonempty subsets as shown below.

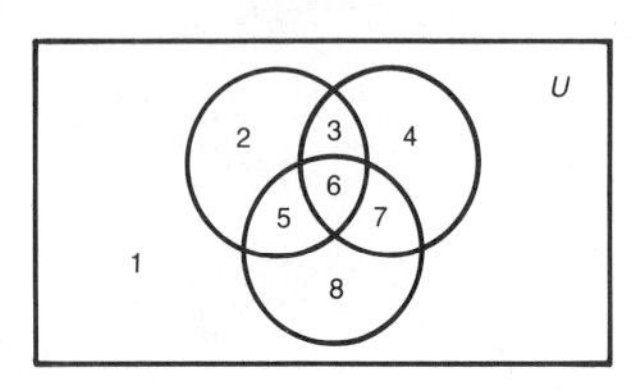

35. Since $H \subseteq H \cup M = G$ and $L \subseteq L \cup N = G$, it follows that $H \cup L \subseteq G$. To prove $G \subseteq H \cup L$, suppose $x \in G$ but $x \notin H \cup L$. Then $x \notin H$ and $x \notin L$. Since $x \in G = H \cup M$ and $x \notin H$, it follows that $x \in M$. Since $x \in G = L \cup N$ and $x \notin L$, it follows that $x \in N$. But, then, $x \in M \cap N = \varnothing$, a contradiction. Therefore, $G \subseteq H \cup L$, so $G = H \cup L$. A similar argument proves that $M \cup N = G$.

PROBLEM SET 1.6, PAGE 77

1. Use $I = \{a, b, c\}$ as the indexing set, and let $M_a = \{$Alabama, Alaska$\}$, $M_b = \{$Michigan, Ohio$\}$, and $M_c = \{$Maine, Texas$\}$.
3. Use $I = \{a, b, c, d, e\}$ as the indexing set, and let $M_a = \{1, 4, 5, 9\}$, $M_b = \{1, 3, 4, 5\}$, $M_c = \{4, 5, 7, 8\}$, $M_d = \{1, 4, 7, 9\}$, and $M_e = \{4, 5, 8, 9\}$.
5. Use $I = \{1, 2, 3, \ldots\} = \mathbb{N}$ as the indexing set, and let $M_i = \{x \cdot i \mid x \text{ is an integer, for each } i \in I\}$.
7. (a) $\{1, 3\}$ (b) $\{2, 4, 6\}$ (c) $\{4, 6, 12\}$ (d) $\{1, 3, 5, 9\}$ (e) $\{3\}$ (f) $\{1, 2, 3, 4, 5, 6, 9, 12\}$ (g) $\varnothing$
9. (a) $\{1, 2, 3, 4, 5, 7\}$ (b) $\{2\}$ **11.** (a) $\{a, b, c, d, e, f\}$ (b) $\{c, d\}$
13. (a) {red, yellow, green, blue, violet, pink, orange} (b) {green, blue}
15. (a) {right, left, up, down, front, back} (b) {right, left} **17.** (a) Problems 1 and 2 (b) None
19. (a) $[0, \infty)$ (b) $\varnothing$ **21.** (a) $[0, 1]$ (b) $\{0\}$
23. Say $p, q \in I$ with $p \neq q$. Then $\bigcap_{i \in I} M_i \subseteq M_p \cap M_q = \varnothing$. **25.** $\bigcap \mathscr{E} \subseteq \varnothing$
27. Use pick-a-point process.
29. (a) Let $x \in A$. By hypothesis, $x \in M_i$, $\forall i \in I$. By definition, $x \in \bigcap_{i \in I} M_i$.
(b) Let $x \in \bigcup_{i \in I} M_i$. By definition, $\exists i \in I$ with $x \in M_i$. By hypothesis, $x \in B$.
31. (i) $\forall x \in \varnothing$, $\exists \varepsilon > 0$ such that $(x - \varepsilon, x + \varepsilon) \subseteq \varnothing$. Thus, $\varnothing \in \mathscr{T}$.
(ii) Let $x \in \mathbb{R}$. Then $(x - 1, x + 1) \subseteq \mathbb{R}$. (iii) Choose $\varepsilon =$ minimum of $\{\varepsilon_1, \varepsilon_2\}$.
(iv) Let $x \in \bigcup \mathscr{E}$. Then $\exists G \in \mathscr{E}$ with $x \in G$. Since $\mathscr{E} \subseteq \mathscr{T}$, we have $G \in \mathscr{T}$; hence, $\exists \varepsilon > 0$ such that $(x - \varepsilon, x + \varepsilon) \subseteq G$. But $G \subseteq \bigcup \mathscr{E}$, so $(x - \varepsilon, x + \varepsilon) \subseteq \bigcup \mathscr{E}$.

CHAPTER 2 PROBLEM SET 2.1, PAGE 88

1. $x = 1, y = 1$ **3.** $(1, 2), (2, 4), (4, 3), (3, 1)$ **5.** $\{(2, 1), (2, 2), (4, 1), (4, 2), (5, 1), (5, 2)\}$
7. $\{(1, 2), (2, 2), (1, 4), (2, 4), (1, 5), (2, 5)\}$ **9.** $\varnothing$
11. $\{(2, 1), (1, 2), (2, 2), (4, 1), (1, 4), (4, 2), (2, 4), (5, 1), (1, 5), (5, 2), (2, 5)\}$
13. $\{(a, c), (a, d), (a, e), (a, f), (b, c), (b, d), (b, e), (b, f)\}$ **15.** $\{(a, c), (a, d), (b, c), (b, d), (a, e), (a, f), (b, e), (b, f)\}$

17. A prism **19.** A solid right circular cylinder **21.** 6 **23.** 78 **25.** 216

27. Let $(a, b) \in A \times B$. Then $a \in A$, and thus, $a \in X$. Similarly, $b \in Y$. Thus, $(a, b) \in X \times Y$.

29. (For first half of proof see Example 1.8.) Let $(x, y) \in (A \times C) \cup (B \times C)$. Then $(x, y) \in (A \times C)$ or $(x, y) \in (B \times C)$; that is, $y \in C \wedge (x \in A \vee x \in B)$, which is to say $(x \in A \cup B) \wedge y \in C$. Thus, $(x, y) \in (A \cup B) \times C$.

31. Use pick-a-point procedure.

33.

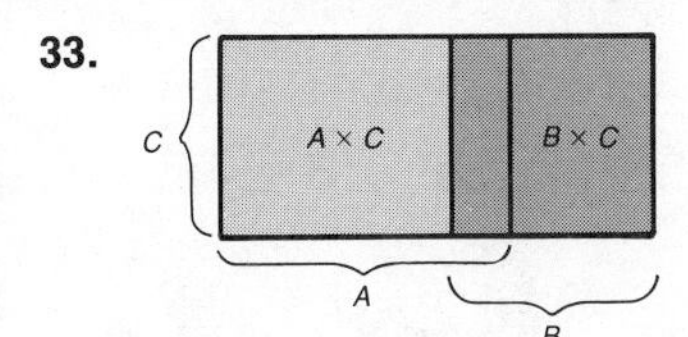

35. 2^{2m}

PROBLEM SET 2.2, PAGE 100

1. (a) Graph consists of the pairs (2, 4) and (3, 9). (b) Graph consists of the pairs (2, 3), and (3, 2). (c) Graph consists of the pairs (2, 4), (2, 3), (3, 9), and (3, 2).

3. Suppose $R \subseteq X \times X$. We must show that dom $R \subseteq X$ and codom $R \subseteq X$. Let $x \in$ dom R. Then $\exists y, (x, y) \in R$. Since $R \subseteq X \times X$, it follows that $(x, y) \in X \times X$, so $x \in X$ and $y \in X$. In particular, $x \in X$, so dom $R \subseteq X$. A similar argument shows that codom $R \subseteq X$. The proof that (dom $R \subseteq X$ and codom $R \subseteq X$) $\Rightarrow R \subseteq X \times X$ is just as easy.

5. (a) $\text{dom}(R) = \{1, 2, 3, 7\}$; $\text{codom}(R) = \{4, 5, 6, 7\}$; $R^{-1} = \{(5, 1), (5, 2), (4, 1), (6, 2), (7, 3), (6, 7)\}$
(b) $\text{dom}(R) = \{a, b, c\}$; $\text{codom}(R) = \{b, c, e, g\}$; $R^{-1} = \{(b, c), (g, b), (e, c), (b, b), (e, b), (e, a), (c, a)\}$
(c) $\text{dom}(R) = \{x \mid -3 \le x \le 3\}$; $\text{codom}(R) = \{x \mid -2 \le x \le 2\}$; $R^{-1} = \{(x, y) \in \mathbb{R} \times \mathbb{R} \mid 9x^2 + 4y^2 = 36\}$
(d) $\text{dom}(R) = \{2, 4, 5\}$; $\text{codom}(R) = \{1, 2, 4\}$; $R^{-1} = \{(4, 2), (1, 2), (4, 4), (2, 4), (2, 5)\}$

7. (a) No, since $(4, 3) \notin R$. (b) No, since $(2, 1) \in R$ and $(1, 2) \in R$ but $1 \neq 2$.

9. (a) Δ_X is reflexive since $\forall x \in \text{dom}\, \Delta_X, (x, x) \in \Delta_X$. $X \times X$ is reflexive since $\forall x \in X, (x, x) \in X \times X$.
(b) Δ_X is symmetric since $\forall (x, y) \in \Delta_X$, we have $x = y$ and thus $(y, x) \in \Delta_X$. $X \times X$ is reflexive since $\forall (x, y) \in X \times X$, we have $x \in X$, $y \in X$, and thus, $(y, x) \in X \times X$.
(c) Δ_X is transitive since, if $(x, y) \in \Delta_X$ and $(y, z) \in \Delta_X$, then $x = y$ and $y = z$. Thus, $x = z$ and $(x, z) \in \Delta_X$. $X \times X$ is transitive since, if $(x, y) \in X \times X$ and $(y, z) \in X \times X$, then $x, y, z \in X$. Thus, $(x, z) \in X \times X$.

11. Yes, since $xRy \wedge yRz \Rightarrow xRz$.

13. (a) See Problem 9. (b) $\forall x \in \mathbb{R}, x \ge x$. Also, $x \ge y$ and $y \ge z \Rightarrow x \ge z$.

15. Note that any positive integer is a multiple of itself. Also, if x is a multiple of y and y is a multiple of x, then $x = y$. Finally, if y is a multiple of x and z is a multiple of y, then z is a multiple of x.

17. (a) Symmetric (b) Reflexive, symmetric, transitive, preorder (c) Reflexive, transitive, preorder
(d) Reflexive, symmetric, transitive, preorder
(e) Reflexive, antisymmetric, transitive, preorder, partial order
(f) Reflexive, symmetric, antisymmetric, transitive, preorder, partial order
(g) Reflexive, symmetric, transitive, preorder (h) Reflexive, transitive, preorder

19. (a) Use definitions of reflexive and Δ_X. (b) Use definitions of symmetric and R^{-1}.

21. Note that for $x \in X$, $(x, x) \in (R \cup \Delta_X)$.

23. If R is reflexive on X, then (by Problem 19a) $\Delta_X \subseteq R$. Therefore, $\Delta_X \subseteq R \cup S$, so (by Problem 19a) $R \cup S$ is reflexive on X.

25. Use pick-a-point. **27.** Assume $(x, y) \in (X \times X) \setminus R$ and show $(y, x) \in (X \times X) \setminus R$.

29. Use definition of partial order. **31.** Use definition of partial order.

33. (a) 1 (b) 0 (c) 1 (d) No least element exists.

35. (a) Use definition of partial order. (b) $\sup\{M, N\} = M \cup N$; $\inf\{M, N\} = M \cap N$
37. (a) Use antisymmetric property to prove uniqueness. (b) Analogous to Part (a).

PROBLEM SET 2.3, PAGE 112

1. Partition **3.** Not a partition (7 belongs to no set in the set of sets). **5.** b, d, f, and g
7. (a) Follows from using Definition 3.5.
(b) {Los Angeles, San Francisco}, {Miami, Fort Lauderdale, Daytona Beach}, {Dallas, San Antonio}, {Chicago}
9. $\forall x \in \mathbb{R}, x^2 = x^2$. $\forall x, y \in \mathbb{R}, x^2 = y^2 \Rightarrow y^2 = x^2$. $\forall x, y, z \in \mathbb{R}, x^2 = y^2 \wedge y^2 = z^2 \Rightarrow x^2 = z^2$.
11. Show that Definition 3.5 is satisfied. **13.** $\{n\pi + (\pi/2) \mid n \text{ is an integer}\}$
15. For Part (b), $[\![x]\!]_E = \{x\} \cup \{y \mid y \text{ is a brother or sister of } x\}$. For Part (d), $[\![x]\!]_E = \{y \mid y \text{ is a triangle and } y \text{ is congruent to } x\}$. For Part (f), $[\![x]\!]_E = \{x\}$. For Part (g), $[\![x]\!]_E = \{y \mid y - x \text{ is divisible by } n\}$.
17. (a) $\{(a, a), (a, d), (d, d), (d, a), (b, b), (b, e), (e, e), (e, b), (c, c)\}$ (b) $\{\{a, d\}, \{b, e\}, \{c\}\}$
19. Let l, s, m, f, b, d, a, and c represent Los Angeles, San Francisco, Miami, Fort Lauderdale, Daytona Beach, Dallas, San Antonio and Chicago, respectively. Then $E = \{(l, l), (l, s), (s, s), (s, l), (m, m), (m, f), (m, b), (f, m), (f, f), (f, b), (b, m), (b, f), (b, b), (d, d), (d, a), (a, a), (a, d), (c, c)\}$.
25. Let $x \in X$. Then $\exists y$ such that xRy. By symmetry, yRx, and by transitivity, $xRy \wedge yRx \Rightarrow xRx$. Thus, if $X = \text{dom}(R)$, the reflexivity condition is not needed. However, if $X \neq \text{dom}(R)$, then the argument above cannot be made.
27. Let $(x, y) \in R$. By symmetry, $(y, x) \in R$, so, by antisymmetry, $x = y$. **29.** (b) $[\![x]\!]_G = [\![x]\!]_E \cap [\![x]\!]_F$
31. (a) $\{X\}$ is both exhaustive and mutually exclusive. (b) $E = X \times X$
33. Note that cRa and bRd. Thus, by the transitivity of R, $cRa \wedge aRb \wedge bRd \Rightarrow cRd$.
35. Use the fact that R is antisymmetric.
37. For $G \in \mathscr{E}$, $G \leq G$. For $G, H \in \mathscr{E}$, $G \leq H \wedge H \leq G \Rightarrow H = G$. For $G, H, K \in \mathscr{E}$, $G \leq H \wedge H \leq K \Rightarrow G \leq K$.
39. Use Theorem 3.15.
41. B_2 represents the number of equivalence relations on X if $\#X = 2$. Since the only partitions possible are $\{\{a\}, \{b\}\}$ or $\{X\}$, where $X = \{a, b\}$, we have $B_2 = 2$. Similarly, if $X = \{a, b, c\}$, the only partitions are $\{\{a\}, \{b\}, \{c\}\}$, $\{\{a, b\}, \{c\}\}$, $\{\{a\}, \{b, c\}\}$, $\{\{a, c\}, \{b\}\}$, and $\{X\}$. Thus, $B_3 = 5$. Similarly, $B_4 = 15$, $B_5 = 52$, $B_6 = 203$, $B_7 = 877$, $B_8 = 4140$.
43. The n endings to the lines may be partitioned in B_n different ways. Two line endings are in the same cell of the partition (or the same equivalence class) if and only if they rhyme.

PROBLEM SET 2.4, PAGE 126

1. b, c, and d are functions.
3. (a) $\{1, 2, 3, 5, 7\}$ (b) $\{a, b, c, d\}$ (c) c (d) d (e) b (f) c
5. (a) $\text{dom}(f) = \{1, 2, 4, 6\}$, $\text{range}(f) = \{a, b, c\}$ (b) $\text{dom}(g) = \{1, 2, 3, 5\}$, $\text{range}(g) = \{b\}$
(c) $\text{dom}(h) = \mathbb{R}$, $\text{range}(h) = \{x \mid -1 \leq x \leq 1\}$ (d) $\text{dom}(F) = \mathbb{R}$, $\text{range}(F) = \{x \mid x \geq 4\}$
(e) $\text{dom}(G) = \{x \mid x \in \mathbb{R}, x \neq (\pi/2) + n\pi \text{ for } n \text{ an integer}\}$, $\text{range}(G) = \{x \mid x \leq -1 \vee x \geq 1\}$
(f) $\text{dom}(H) = \{x \mid x \in \mathbb{R}, x \neq -2\}$, $\text{range}(H) = \{x \mid x \in \mathbb{R}, x \neq 0\}$
7. $\{(a, y), (b, x), (c, z), (d, y)\}$
9. The function under consideration is $\{(x, y) \mid x \in \mathbb{R} \wedge y \in \mathbb{R} \wedge x > -2 \wedge y = 1/\sqrt{x + 2}\}$.
11. Only Part (a) **13.** f and g are injections.
15. If f is injective, then $f(a) = f(b) \Rightarrow a = b, \forall a, b \in X$. This means that two arrows pointing to the same element of Y must start at the same element of X. That is, the two arrows are really one. The converse reverses this argument.

17. (a) f is a mapping because f satisfies the definition of function, $\text{dom}(f) = X$, and $\text{range}(f) \subseteq Y$.
(b) $\text{range}(f) = \{c, d\}$ (c) No (d) No

19. No; no arrow points to 1 or 4.

21. (a) Injective and surjective (b) Neither (c) Injective and surjective (d) Injective (e) Neither

23. Since $f(1) \neq f(2) \neq f(3) \neq f(1)$, f is injective. Since $\text{range}(f) = \{a, b, c\} = Y$, f is surjective. Thus, f is a bijection.

25. Show that $\forall a, b \in (0, 1)$, $a/(1-a) = b/(1-b) \Rightarrow a = b$ (cross multiply). Let $y \in (0, \infty)$ and show $\exists x \in (0, 1)$ such that $y = x/(1-x)$ (solve the equation).

27. (a) See Problem 21(d). (b) Let $f(x) = x^3 + x^2$. (c) See Problem 21(b) and (e).
(d) See Problem 21(a) and (c).

29. (a) $f^{-1}(x) = (2-x)/7$ (b) $g^{-1}(x) = 1/x$ (c) $h^{-1}(x) = (2x-3)/(3x-2)$
(d) $F^{-1}(x) = (x-1)^2,\ x \geq 1$ (e) $G^{-1}(x) = \ln(x + \sqrt{x^2+1})$

31. (b) Define $f: X \to Y$ by $f = \{(x, y) \mid x$ is a positive integer $\wedge\ y = 2x\}$, and show that f is bijective.

33. For each of the m objects x in X we have n choices for $f(x)$.

35. Show that an injection $f: X \to Y$ corresponds to a permutation of the n elements of Y taken m at a time.

PROBLEM SET 2.5, PAGE 140

1. (a) $\mathbb{R}$ (b) $(g \circ f)(x) = -21x - 6$ (c) $\mathbb{R}$ (d) $(f \circ g)(x) = -21x + 2$

3. (a) $\{x \mid x > -\frac{1}{3}\}$ (b) $(g \circ f)(x) = \sqrt{3x+1}$ (c) $\{x \mid x \geq -1\}$ (d) $(f \circ g)(x) = 3\sqrt{x+1}$

5. (a) $\{x \mid x \in \mathbb{R} \wedge x \neq -3\}$ (b) $(g \circ f)(x) = (3x^2 + 14x + 19)/(x^2 + 6x + 9)$ (c) $\mathbb{R}$
(d) $(f \circ g)(x) = (x^2 + 3)/(x^2 + 5)$

7. (a) $\mathbb{R}$ (b) $(g \circ f)(x) = 3\sin^2 x + 2$ (c) $\mathbb{R}$ (d) $(f \circ g)(x) = \sin(3x^2 + 2)$

9. (a) $\mathbb{R}$ (b) $(g \circ f)(x) = 3$ (c) $\mathbb{R}$ (d) $(f \circ g)(x) = 16$

11. (a) $\{(a, b), (b, c), (c, c)\}$ (b) $\{(1, 2), (2, 2), (3, 1)\}$

13. (a) $g(x) = (x+1)/(x-1)$; $f(x) = \sqrt{x}$; $\text{dom}(f \circ g) = \{x \mid x > 1 \vee x \leq -1\}$
(b) $g(x) = \sqrt{x}$; $f(x) = \cos^{-1} x$; $\text{dom}(f \circ g) = \{x \mid 0 \leq x \leq 1\}$
(c) $g(x) = x^2 - x - 6$; $f(x) = \ln x$; $\text{dom}(f \circ g) = \{x \mid x < 2 \vee x > 3\}$

15. (a) $ad + b = cb + d$ (b) Let $x = b/(1-a) = d/(1-c)$. **17.** (a) $16x - 45$ (b) $256x - 51$

19. Note that $y = (f \circ g)(x) = f(g(x))$ and the graph of $y = x$ contains the points $(g(x), g(x))$ for all x.

21. (a) $\text{dom}(g \circ f) = \{x \mid x \in \text{dom}(f) \wedge f(x) \in \text{dom}(g)\} = \{x \mid x \in X \wedge f(x) \in Y\} = X$
(b) $\text{range}(g \circ f) = \{g(y) \mid y = f(x) \wedge x \in X\} \subseteq \{g(y) \mid y \in Y\} = Z$

23. (a)
$a \quad r$
$b \to s$ (with $a \to s$)
$c \to t$
(b) $\{s, t\}$

25. (a) By hypothesis, for $f: X \to Y$, $\exists g: Y \to X$ such that $f \circ g = \Delta_Y$. Thus, $\forall y \in Y, (f \circ g)(y) = y$ or equivalently, $\forall y \in Y, f(g(y)) = y$. Thus, $\forall y \in Y, \exists x \in X$ such that $g(y) = x \wedge f(x) = y$.
(b) Let $g(a) = g(b)$. Then $f(g(a)) = f(g(b)) \Rightarrow a = b$. Thus, g is injective.

PROBLEM SET 2.6, PAGE 153

1. (a) 4 (b) 4 (c) 0 (d) -1 (e) 15 (f) $2a + b$ (g) $2a + b$ (h) a

3. (a) e (b) c (c) c (d) c (e) a (f) a

5. (a) e (b) e (c) d (d) e (e) f (f) f

7. (a) Yes (b) Yes (c) No, $\sqrt{x^2 + y^2}$ need not be an integer. (d) Yes (e) No, $(\frac{1}{2}) * (-1) \notin S$.
9. $(20 * 5) * 2 \neq 20 * (5 * 2)$ **11.** From earlier work, $(M \cup N) \cup P = M \cup (N \cup P)$.
13. $x * (y * z) = x * (y + z - yz) = x + (y + z - yz) - x(y + z - yz)$
$= x + y + z - yz - xy - xz + xyz = x + y - xy + z - (x + y - xy)z = (x * y) * z$
15. (a) $a * (b * c) = a * f = c$, $(a * b) * c = e * c = c$ (b) $f * (f * b) = f * c = b$, $(f * f) * b = e * b = b$

17. (a) See Part (b). (b)

*	1	2	3	4
1	1	1	1	1
2	1	2	2	2
3	1	2	3	3
4	1	2	3	4

(c) The table is symmetric about the diagonal from upper left to lower right.
(d) Note that $\forall x \in S$, $x * 4 = 4 * x = x$.

21.

$\circ$	e	f	g	h
e	e	f	g	h
f	f	f	h	g
g	g	g	g	g
h	h	h	h	h

Where $f(a) = b$, $f(b) = a$, $g(a) = g(b) = a$, $h(a) = h(b) = b$.

23. (a) n^n (b) $n!$ (c) 2^n
25. $\forall x, y, z \in \mathbb{R}$, $(x * y) * z = (x + y + xy) * z = x + y + xy + z + (x + y + xy)z$
$= x + y + z + yz + xy + xz + xyz = x + (y * z) + x(y * z) = x * (y * z)$.
Also, $\forall x, y \in \mathbb{R}$, $x + y + xy = y + x + yx$.
27. (a) Note that $\forall x, y, z \in S$, $(x * y) * z = x * z = x$ and $x * (y * z) = x * y = x$.
(b) No, there is no neutral element.
29. (a) $*$ is associative on S. (b) No, there is no neutral element. **31.** $x \mapsto \ln x$
33. Let $x, y, z \in S$. By $(S, *)$ associative, $(x * y) * z = x * (y * z)$, and thus, $\Phi(x * y) \star \Phi(z) = \Phi(x) \star \Phi(y * z)$ and $[\Phi(x) \star \Phi(y)] \star \Phi(z) = \Phi(x) \star [\Phi(y) \star \Phi(z)]$.
35. Since e is neutral, $e * x = x = x * e$. Thus, $\Phi(e) \star \Phi(x) = \Phi(e * x) = \Phi(x)$ and $\Phi(x) \star \Phi(e) = \Phi(x * e) = \Phi(x)$.
37. If $\Phi: \mathscr{P}(x) \to S$ is defined by $\varnothing \mapsto e$, $\{1\} \mapsto a$, $\{2\} \mapsto b$, and $\{1, 2\} \mapsto c$, show that Φ is an isomorphism.

PROBLEM SET 2.7, PAGE 163

1. (a) $\{a\}$ (b) $\{b, d\}$ (c) $\{a, b, c, d\}$ (d) $\{1, 3\}$ (e) $\{2, 5\}$ (f) $\{2, 4, 5\}$ (g) $\{4, 6\}$
3. (a) $f(X) = \{f(x) | x \in X\} = \{y | y = f(x) \wedge x \in X\} = \text{range}(f)$ (b) $f^{-1}(Y) = \{x \in \text{dom}(f) | f(x) \in Y\} = X$
(c) $f(\varnothing) = \{f(x) | x \in \varnothing\} = \varnothing$ (d) $f^{-1}(\varnothing) = \{x \in X | f(x) \in \varnothing\} = \varnothing$
5. Assume $y \in f(M)$. Then $\exists x$ with $f(x) = y$ and $x \in M$. Thus, $\exists x$ such that $x \in \text{dom}(f)$ and $x \in M$, and thus, $x \in M \cap \text{dom}(f)$ with $f(x) = y \in f(M \cap \text{dom}(f))$. For $y \in f(M \cap \text{dom}(f))$, $\exists x \in M \cap \text{dom}(f)$ with $f(x) = y$. Thus, $y \in f(M)$.
11. $y \in f(M \cup N) \Leftrightarrow y \in \{f(x) | x \in M \vee x \in N\} \Leftrightarrow y \in \{f(x) | x \in M\} \cup \{f(x) | x \in N\} \Leftrightarrow y \in f(M) \cup f(N)$
13. $f^{-1}(M \cap N) = \{x | f(x) \in M \cap N\} = \{x | f(x) \in M \wedge f(x) \in N\} = \{x | f(x) \in M\} \cap \{x | f(x) \in N\}$
$= f^{-1}(M) \cap f^{-1}(N)$
23. (a) $y = \sqrt{x}$ (b) $y = -\sqrt{x}$ (c) $y = \begin{cases} \sqrt{x}, & \text{if } 0 \leq x \leq 1 \\ -\sqrt{x}, & \text{if } x > 1 \end{cases}$

25. $f(x) = \text{Arccos } x$ **27.** Define $f: \mathbb{R} \to \{-1, 0, 1\}$ by $f(x) = \begin{cases} 1, & \text{if } x > 0 \\ 0, & \text{if } x = 0. \\ -1, & \text{if } x < 0 \end{cases}$

29. Since, by definition, $f \subseteq E$, we have $(x, y) \in f \Rightarrow xEy$. Thus, since E is an equivalence relation on X, we have $y \in X$.

31. (a) (i) Note that $\text{dom}(P) = X$ and, from elementary geometry we know that congruent triangles have equal perimeters. Thus, $xEy \Rightarrow P(x) = P(y)$.
(ii) Note that $\text{dom}(v) = X$ and, from elementary geometry we know that in congruent triangles, corresponding angles have equal measures. Thus, $xEy \Rightarrow v(x) = v(y)$.
(iii) Note that $\text{dom}(V) = X$ and, from elementary geometry corresponding angles in congruent triangles have equal measures. Thus, $xEy \Rightarrow V(x) = V(y)$.
(b) No, because $v(x) = v(y)$ and $V(x) = V(y) \not\Rightarrow xEy$ since x and y could be two similar triangles without being congruent.
(c) Yes, because $v(x) = v(y)$ and $V(x) = V(y)$ and $P(x) = P(y) \Rightarrow xEy$ since x and y must now be congruent triangles.
(d) Let $L(T)$ denote the largest side of T, etc.

33. (a) For any $x \in Z$, $x - x = 0$ is even, so xEx. If xEy, then $x - y$ is even and thus $-(x - y) = y - x$ is even, so yEx. Finally, if xEy and yEz, then $x - y = 2n$ for some $n \in \mathbb{Z}$ and $y - z = 2m$ for some $m \in \mathbb{Z}$. Thus, $x - z = 2n + 2m = 2(n + m)$, so xEz.
(b) Define $f: Z \to \{0, 1\}$ by $f(x) = \begin{cases} 1, & \text{if } x \text{ is odd} \\ 0, & \text{if } x \text{ is even.} \end{cases}$

35. There are two equivalence classes in $\mathbb{Z}/E$, the set of even numbers (which f maps to 0) and the set of odd numbers (which f maps to 1).

CHAPTER 3 PROBLEM SET 3.1, PAGE 187

1. (a) $\forall x \in \mathbb{R}$, $x + 0 - x \cdot 0 = 0 + x - 0 \cdot x = x$
(b) $a + (a^-) - a(a^-) = 0 \Rightarrow a^- = a/(a - 1) \Rightarrow a \neq 1$. Conversely, if $a \neq 1$, then $\exists s \in \mathbb{R}$ such that $s = a/(a - 1)$. Let $a^- = s$.
(c) $x + (x^-) - x(x^-) = 0 \Leftrightarrow x + (x^-)(1 - x) = 0 \Leftrightarrow x = (x^-)(x - 1) \Leftrightarrow x/(x - 1) = x^-$

3. (a) Let $l(n) = n - 1$. Then $(l \circ s)(n) = l(s(n)) = l(n + 1) = n + 1 - 1 = n$.
(b) Suppose s did have a right inverse, r. Then $(s \circ r)(1) = s(r(1)) = r(1) + 1 = 1$, and thus, $r(1) = 0$, contradicting $r \in S$.

5. $(x * y) * (x * y) = x * (y * x) * y = x * e * y = x * y$

7. $(y^-) * (x^-) * (x * y) = (y^-) * ((x^-) * x) * y = (y^-) * e * y = (y^-) * y = e$

9. (i) says $-(-x) = x$, $\forall x \in \mathbb{R}$, and (ii) says $-(x + y) = (-y) + (-x)$ for all $x, y \in \mathbb{R}$.

11. Yes, here the only element is the neutral element.

13.

*	e	a
e	e	a
a	a	e

The lower right corner is the only space not immediately filled as a result of the fact that $\forall x \in G$, $x * e = e * x = x$. Since each element appears once and only once in each row, we must have $a * a = e$.

15.

*	e	a	b
e	e	a	b
a	a	b	e
b	b	e	a

After arranging the first row and the first column in the obvious manner, the remaining spaces must be filled in such a manner that each element appears exactly once in each row and once in each column.

17. The group is AS9.

19. Note, e.g., that $a * x = a * y$ means that the same element appears in the row labeled a and in the columns labeled x and y.

21. $x * a = y * a \Rightarrow (x * a) * (a^-) = (y * a) * (a^-) \Rightarrow x * (a * (a^-)) = y * (a * (a^-)) \Rightarrow x * e = y * e \Rightarrow x = y$

23. If a solution exists, then $\exists c \in S$ such that $a * c = b$. Thus, b appears at least once in the row labeled a. Conversely, if b appears at least once in the row labeled a, then $\exists c \in S$ such that $a * c = b$. Thus, the equation $a * x = b$ has a solution in S.

29. The set S of all nonnegative integers under ordinary addition.

PROBLEM SET 3.2, PAGE 197

1. It has already been shown that $(R, +)$ is a group. Since $(0 \cdot 0) \cdot 0 = 0 = 0 \cdot (0 \cdot 0)$ and since $0 \cdot (0 + 0) = 0 \cdot 0 = 0 = 0 \cdot 0 + 0 \cdot 0$ and, similarly, $(0 + 0) \cdot 0 = 0 \cdot 0 + 0 \cdot 0$, $(R, +, \cdot)$ is a ring.

3.

+	Even	Odd
Even	Even	Odd
Odd	Odd	Even

·	Even	Odd
Even	Even	Even
Odd	Even	Odd

By earlier work, $(\mathbb{R}, +)$ is a group. Also, $(R, \cdot)$ is associative since $(\mathbb{R}, \cdot)$ is. Since $(\mathbb{R}, +, \cdot)$ obeys the left and right distributive laws, so does $(R, +, \cdot)$. Finally, it is evident that $(R, +)$ and $(R, \cdot)$ are commutative.

5. By inspection, $(Z_4, +)$ is commutative and $0 \in Z_4$ is the zero element. Note also that $-0 = 0$, $-1 = 3$, and $-2 = 2$. The associativity of $(Z_4, +)$ and $(Z_4, \cdot)$ as well as the distributive laws can be checked by direct calculation.

7. By the left distributive law, $(a + b)(c + d) = (a + b)c + (a + b)d$. By the right distributive law, $(a + b)c + (a + b)d = ac + bc + ad + bd$, and the commutativity of $+$ gives $ac + ad + bc + bd$.

9. False, since we need not have $b \cdot a = a \cdot b$.

11. $x \cdot 0 = x \cdot (0 + 0) = x \cdot 0 + x \cdot 0$; thus, $x \cdot 0 = x \cdot 0 + x \cdot 0$, so $x \cdot 0 + 0 = x \cdot 0 + x \cdot 0$, and by the left cancellation law, $0 = x \cdot 0$.

13. (a) $x(-y) + xy = x(-y + y) = x \cdot 0 = 0$; thus, by Part (i) of Theorem 2.4, $x(-y) = -(xy)$.
(b) $x + y = z \Leftrightarrow x + y + (-y) = z + (-y) \Leftrightarrow x + 0 = z - y \Leftrightarrow x = z - y$

15. By hypothesis, $0 \cdot x = x \cdot 0 = x$, $\forall x \in R$. By Theorem 2.3, $\forall x \in R$, $0 \cdot x = x \cdot 0 = 0$. Thus, $\forall x \in R$, $x = 0$, and $R = \{0\}$.

17. $\{1, -1\}$

21. By Problem 19, it suffices to show that the sum and product of Gaussian integers are again Gaussian integers.

25. $\begin{bmatrix} 1 & 0 \\ 0 & 0 \end{bmatrix} \cdot \begin{bmatrix} 0 & 0 \\ 0 & 1 \end{bmatrix} = \begin{bmatrix} 0 & 0 \\ 0 & 0 \end{bmatrix}$

PROBLEM SET 3.3, PAGE 208

1. Note that GF(2) is isomorphic to the ring $\mathbb{Z}_2$. Note that GF(2) is commutative. Thus, it suffices to show that 1 is multiplicatively invertible in GF(2). But $1 \cdot 1 = 1$.

3. Note that GF(3) is isomorphic to the commutative ring $\mathbb{Z}_3$. It suffices to show that $0 \neq x \in \text{GF}(3) \Rightarrow \exists x^{-1} \in \mathbb{Z}_3$. To see this, note that $1 \cdot 1 = 1$ and $2 \cdot 2 = 1$.

5. $a = 0 \Leftrightarrow -a = a + (-a) \Leftrightarrow -a = 0$

7.

+	0	1	2	3	4
0	0	1	2	3	4
1	1	2	3	4	0
2	2	3	4	0	1
3	3	4	0	1	2
4	4	0	1	2	3

·	0	1	2	3	4
0	0	0	0	0	0
1	0	1	2	3	4
2	0	2	4	1	3
3	0	3	1	4	2
4	0	4	3	2	1

9. (a)

+	0	1	2	3
0	0	1	2	3
1	1	2	3	0
2	2	3	0	1
3	3	0	1	2

·	0	1	2	3
0	0	0	0	0
1	0	1	2	3
2	0	2	0	2
3	0	3	2	1

(b) Note that $2 \cdot 2 = 0$, but $2 \neq 0$.

15. (a) $\frac{2}{3} = 2 \cdot 3^{-1} = 2 \cdot 2 = 4$ (b) $-(\frac{2}{3}) = -4 = 1$ **17.** (a) $\frac{1}{4} = 1 \cdot 4^{-1} = 4^{-1} = 4$ (b) $-(\frac{1}{4}) = 1$

19. Idea of the proof: $ab = c \Leftrightarrow abb^{-1} = cb^{-1} \Leftrightarrow a = c/b$

21. Idea of the proof: $a/b = c/d \Leftrightarrow (a/b)(bd) = (b/d)(cd) \Leftrightarrow ad = bc$

23. Idea of the proof: $(a/b) - (c/d) = (ad/bd) - (cb/bd) = (ad/bd) - (bc/bd) = (ad - bc)/bd$

25. Idea of the proof: $(a/b)/(c/d) = (a/b \cdot d/c)/(c/d \cdot d/c) = (ad/bc)/1 = ad/bc = a/b \cdot d/c$

27. Do by fixing n and performing induction on m for $m \geq 0$. Then take care of the case $m < 0$ by using $a^m = (a^{-1})^{-m}$.

29. Do by induction on n for $n \geq 0$. Then take care of $n < 0$.

31. Use Problem 27, let $m = -n$, and use the fact that $a^{-n} = (a^{-1})^n$.

33. $0^2 + 0 + 1 \neq 0$ and $1^2 + 1 + 1 = 1 \neq 0$.

35. (a) $\mathbb{Q}$ and $\mathbb{Q}(\sqrt{2})$ are subfields of $\mathbb{R}$ (b) $\mathbb{C}$ is an extension field of $\mathbb{R}$.

PROBLEM SET 3.4, PAGE 220

1. $(x/y) \in P \Rightarrow (x/y)y^2 \in P$ (since $y \neq 0 \Rightarrow y^2 \in P$)

3. (i) $x \in P \Leftrightarrow x - 0 \in P \Leftrightarrow 0 < x$ (ii) $-x \in P \Leftrightarrow 0 - x \in P \Leftrightarrow x < 0$
(iii) $x < y \Leftrightarrow y - x \in P \Rightarrow y - x \neq 0 \Rightarrow x \neq y$

5. Apply the law of trichotomy to $y - x$. **9.** Follows from the law of trichotomy.

11. In GF(2), $1 = -1$. Thus, if GF(2) were an ordered field, 1 would be both positive and negative.

13. Consider separately the cases $y < z$ and $y = z$.

17. To prove $x + z < y + z \Rightarrow x < y$, use Part (i) of Theorem 4.11 to add $-z$ to both sides.

19. Consider separately the cases $0 = z$ and $0 < z$. **21.** Case analysis **33.** $(x + y - |x - y|)/(1 + 1)$

37. $xy \leq 0$

CHAPTER 4 PROBLEM SET 4.1, PAGE 260

1. $[0, 1)$ **3.** Note that $x = y \Leftrightarrow -x = -y$.

5. By Theorem 4.12(i) in Chapter 3, $1 < b \Rightarrow b < b^2$. **7.** Use the fact that $\mathbb{R}$ is complete and Lemma 1.4.

9. Since b is the *greatest* lower bound for S and $b < b + \varepsilon$, it follows that $b + \varepsilon$ cannot be a lower bound for S; hence, $\exists s \in S$ such that $s < b + \varepsilon$. Since b is a lower bound for S, we also have $b \leq s$.

11. Let δ be the smaller of the two numbers 1 and $\varepsilon/5$, or their common value if they are equal.
13. By Definition 1.11, $\sqrt{4} = 2$. **15.** Yes. Both $\sqrt{a}$ and $-\sqrt{a}$ satisfy the equation.
17. Use Problem 13 in Problem Set 3.3. **19.** $(y^2 - x^2) - (x - y)^2 = 2x(y - x) \geq 0$
21. Apply Problem 19 separately to the two cases $x \leq y$ and (with x and y interchanged in Problem 19) $y < x$.
23. In Problem 21, replace x by $\sqrt{x}$ and y by $\sqrt{y}$. **25.** Given $\varepsilon > 0$, let $\delta = \varepsilon^2$ and use Problem 23.

PROBLEM SET 4.2, PAGE 272

1. Since $1 \in \mathbb{R}$ and $1 > 0$, we have $1 \in P$. Also, $\forall n \in \mathbb{R}$, $n \in P \Rightarrow n > 0 \Rightarrow n + 1 > 0 \Rightarrow n + 1 \in P$.
3. Since $1 \in \mathbb{R}$ and $1 \geq 1$, we have $1 \in J$. Also, $\forall n \in \mathbb{R}$, $n \in J \Rightarrow n \geq 1 \Rightarrow n + 1 \geq 2 \Rightarrow n + 1 \geq 1 \Rightarrow n + 1 \in J$.
5. By Lemma 2.3, if I is an inductive subset of $\mathbb{R}$, then $\mathbb{N} \subseteq I$. Thus, if $I \subseteq \mathbb{N}$, we have $I = \mathbb{N}$.
7. False. Let $I = \mathbb{N}$ and let $J = \mathbb{N} \cup \{1/2\}$.
9. By Theorem 2.7, $m < n$ or $m > n + 1$. Since $m > n$, we cannot have $m < n$ and so $m > n + 1$.
11. Fix $m \in \mathbb{N}$. Then $m \cdot 1 = m \in \mathbb{N}$. Assume the induction hypothesis $mn \in \mathbb{N}$. Then, $m(n + 1) = mn + m \in \mathbb{N}$ by Theorem 2.9.
13. By Theorem 2.11, $\exists m \in \mathbb{N}$, $m\varepsilon > 1$. Thus, $\varepsilon > 1/m$. Now, if $n > m$, then $1/m > 1/n$, so $\varepsilon > 1/m$.
15. Use Theorem 2.7.
17. Since $n \leq x$ and $x < m + 1$, we have $n < m + 1$. Also, since $m \leq x$ and $x < n + 1$, we have $m < n + 1$. Thus, $m + 1 < n + 2$, so $n < m + 1 < n + 2$ and, from Problem 15, $m + 1 = n + 1$, from which it follows that $m = n$.
19. (i) $m + 1 = n + 1 \Rightarrow m = n$. (ii) $1 = x + 1 \Rightarrow x = 0$, contradicting $0 \notin \mathbb{N}$. (iii) Use Problem 5.
21. $X/n\varepsilon$ always has a positive leading coefficient, namely $1/n\varepsilon$. **23.** Use Theorem 2.14 and Lemma 1.4.
25. (a) If n is even, then $n = 2k$ for some $k \in \mathbb{N}$. If $n + 1$ were also even, we would have $n + 1 = 2j$ for some $j \in \mathbb{N}$. Then, with $m = j - k$, we would have $1 = 2m$, $m = 1/2$. Thus, $m > 0$, so $j > k$, and it would follow from Theorem 2.8 that $1/2 = m \in \mathbb{N}$, contradicting Example 2.5.
(b) Suppose there were a number n such that both n and $n + 1$ are odd. Then, by Theorem 2.14, there would be a smallest such n. Since 2 is even, $n \neq 1$, hence, $n - 1 \in \mathbb{N}$ by Lemma 2.6. By our choice of n, we cannot have both $n - 1$ and n odd; hence, $n - 1$ is even, so $\exists k \in \mathbb{N}$ with $n - 1 = 2k$. Then $n + 1 = 2k + 2 = 2(k + 1)$, contradicting the fact that $n + 1$ is odd.
(c) Obvious, since odd means not even.
(d) Let m and n be odd. Without loss of generality, we can suppose $m, n > 1$. By Part (b), both $m - 1$ and $n - 1$ are even, so $\exists j, k \in \mathbb{N}$ such that $m - 1 = 2j$ and $n - 1 = 2k$, it follows that $mn = (2j + 1)(2k + 1) = 2q + 1$, where $q = 2jk + j + k \in \mathbb{N}$. Since $2q$ is even, it follows from Part (a) that mn is odd.
(e) Argue by contradiction, using Part (d).
27. (a) Use Theorem 2.14. (b) Follow the hint given.
(c) Since $n^2 = 2m^2$ and $n = 2k$, we have $(2k)^2 = 2m^2$, $4k^2 = 2m^2$, and so $m^2 = 2k^2$. Now apply the reasoning in Part (b) to m.
(d) Substituting $m = 2r$ into $m^2 = 2k^2$ and canceling the factor 2 from both sides, we obtain $k^2 = 2r^2$. That $k < n$ follows from $n = 2k$.

PROBLEM SET 4.3, PAGE 286

1. Follows from the facts that $m > 0$ for every $m \in \mathbb{N}$ and $m \leq 0$ for every $m \in \{0\} \cup \{-n \mid n \in \mathbb{N}\}$.
3. First prove $k \in \mathbb{Z} \Rightarrow -k \in \mathbb{Z}$ by a case analysis. Then replace k by $-k$ to obtain $-k \in \mathbb{Z} \Rightarrow -(-k) \in \mathbb{Z}$, from which the converse implication $-k \in \mathbb{Z} \Rightarrow k \in \mathbb{Z}$ follows.

5. Suppose $mn = 1$ with $m, n \in \mathbb{Z}$. We consider the case in which $m > 0$. (Evidently, $m \neq 0$, and the case in which $m < 0$ is handled similarly.) Then $m \in \mathbb{N}$, so $m \geq 1$ by Example 2.5. If $m = 1$, then $n = 1$ and we are done. But if $m > 1$, then $0 < 1/m < 1$, so $0 < n < 1$, contradicting Theorem 3.4.

7. $x^* \cdot 10^n = [\![x \cdot 10^n + 0.5]\!]$, so $x^* \cdot 10^n \leq x \cdot 10^n + 0.5 < 10^n x^* + 1$. Subtracting $x^* \cdot 10^n + 0.5$ from all members of the last inequality, we find that $-0.5 \leq (x - x^*) \cdot 10^n < 0.5$, so that $-(10^{-n})/2 \leq x - x^* < 10^{-n}/2$. Consequently, $|x - x^*| \leq 10^{-n}/2$.

9. (i) $m = 1 \cdot m$ (ii) $0 = 0 \cdot m$ (iii) $n = n \cdot 1$ (iv) $n = (-n)(-1)$

11. (v) $m = nq \Rightarrow m = (-n)(-q)$

(vi) Suppose $n = mq$ and $m = nr$ with $q, r \in \mathbb{Z}$. If $n = 0$, then $m = 0$ and so $n = m$. Thus, we can assume $n \neq 0$. Now, $n = mq = (nr)q = n(rq)$, and, since $n \neq 0$, it follows that $rq = 1$. Now use Problem 5 to conclude that $r = q = 1$ or $r = q = -1$, and hence that $n = \pm m$.

(vii) $m = nq \Rightarrow mk = n(qk)$

(viii) $m = nq$ and $k = nr \Rightarrow m + k = n(q + r)$

(ix) $m = nq$ and $k = mr \Rightarrow k = n(qr)$

13. (a) Suppose $m = nq$ and let $j \in \mathbb{Z}m$. Then $\exists k \in \mathbb{Z}$ with $j = km$. Thus, $j = k(nq) = (kq)n \in \mathbb{Z}n$. Conversely, suppose $\mathbb{Z}m \subseteq \mathbb{Z}n$. Then $m = 1 \cdot m \in \mathbb{Z}m$ implies that $m \in \mathbb{Z}n$, so $\exists q \in \mathbb{Z}$ with $m = qn$. Thus, $m = nq$, so $n|m$.

(b) $h = jn$ and $k = mn \Rightarrow h + k = (j + m)n$

(c) $h = jn \Rightarrow hk = (jn)k = (jk)n$

15. (i) Let $h, k \in I + J$, so $\exists a, b \in I$ and $\exists c, d \in J$ with $h = a + c$ and $k = b + d$. Since I and J are ideals, we have $a + b \in I$ and $c + d \in J$, from which it follows that $h + k = (a + c) + (b + d) = (a + b) + (c + d) \in I + J$.

(ii) Let $h \in I + J$, so $\exists a \in I$ and $\exists b \in J$ with $h = a + b$. Let $k \in \mathbb{Z}$. Since I and J are ideals, $ak \in I$ and $bk \in J$, so it follows that $hk = (a + b)k = ak + bk \in I + J$.

17. $m = 10$, $h = 4$, $k = 5$

19. Prove that if the prime p does not divide the integer h, then p and h are relatively prime; then use Lemma 3.22.

21. First suppose that $\text{GCD}(m, n) = 1$. Then, by Part (ii) of Theorem 3.18, $\exists a, b \in \mathbb{Z}$ such that $am + bn = 1$. Conversely, suppose $\exists a, b \in \mathbb{Z}$ such that $am + bn = 1$. Let $g = \text{GCD}(m, n)$. By Part (ii) of Theorem 3.18, $\exists j, k \in \mathbb{Z}$ such that $m = gj$ and $n = gk$. Therefore, $a(gj) + b(gk) = 1$, from which it follows that $1 = (aj + bk)g$; that is, $g|1$. By Part (i) of Theorem 3.11, $1|g$; hence, by Part (vi) of the same theorem, $g = \pm 1$. By Definition 3.17, $g \geq 0$; hence, $g = 1$.

23. By Part (i) of Theorem 3.10, $\exists j, k \in \mathbb{Z}$ with $m = jg$ and $n = kg$. Since $m, n \neq 0$, it follows that $g \neq 0$. By Part (ii) of the same theorem, $\exists a, b \in \mathbb{Z}$ with $g = am + bn$; hence, $g = ajg + bkg = (aj + bk)g$. Since $g \neq 0$, we have $1 = aj + bk$; hence, j and k are relatively prime by Problem 21. But $j = m/g$ and $k = n/g$.

25. Suppose $k = ma = nb$ for $a, b \in \mathbb{Z}$. If $k = 0$, then $c|k$ is automatic, so we can assume that $k \neq 0$. Consequently, we must have $m, n \neq 0$. Thus, c is the smallest natural number such that $m|c$ and $n|c$. By Theorem 3.8, $\exists q, r \in \mathbb{Z}$ such that $k = cq + r$ and $0 \leq r < c$. Thus, $r = k - cq$. Since $m|k$ and $m|c$, it follows that $m|r$. Likewise, $n|r$. Since $r < c$ and c is the smallest natural number divisible by both m and n, it follows that $r = 0$; hence, $k = cq$.

27. Let $g = \text{GCD}(m, n)$. Because $g|m$ and $g|n$, $\exists j, k \in \mathbb{N}$ such that $m = gj$ and $n = gk$. By Problem 23, j and k are relatively prime. Let $c = mk$. Evidently, c is an integer multiple of m. Also, $c = (gj)k = j(gk) = jn$, so c is also an integer multiple of n. We claim that $c = \text{LCM}(m, n)$. To prove this, we must prove that if $r \in \mathbb{Z}$ and r is an integer multiple of both m and n, then $c \leq r$. Thus, suppose $r = ma$ and $r = nb$ with $a, b \in \mathbb{N}$. Now $gja = ma = r = nb = gkb$, so $gja = gkb$, and it follows that $ja = kb$. Therefore, $j|kb$. Since j and k are relatively prime, Lemma 3.22 implies that $j|b$. Thus, $\exists s \in \mathbb{N}$ with $b = sj$, and therefore $r = nb = nsj = sjn = sc$. From $r = sc$ with $s \in \mathbb{N}$, it follows that $c \leq r$, establishing the fact that $c = \text{LCM}(m, n)$. Finally, $mn = gjgk = ggjk = gmk = gc = \text{GCD}(m, n) \cdot \text{LCM}(m, n)$.

29. If $a, b \in \mathbb{N}$, then $ab = \text{GCD}(a, b) \cdot \text{LCM}(a, b) = 1 \cdot \text{LCM}(a, b) = \text{LCM}(a, b)$ by Problem 27, and the result follows from Problem 25. To settle the general case, just replace a and b by $|a|$ and $|b|$.

31. Let C be the set of all integers that divide both m and n, and let $g = \text{GCD}(m, n)$. Then $g \in C$ by Part (i) of Theorem 3.18. Suppose $k \in C$. We must prove that $k \le c$. By Definition 3.17, $g \ge 0$, so we need only consider the case in which $k > 0$. Since k divides both m and n, Part (iii) of Theorem 3.18 implies that $k|g$; hence, $k \le g$.

33. No; $\mathbb{Z}$ is certainly a nonempty subset of itself, but it has no smallest element.

PROBLEM SET 4.4, PAGE 294

1. $\sqrt{2} + (-\sqrt{2}) = 0 \in \mathbb{Q}$, $\sqrt{2} - \sqrt{2} = 0 \in \mathbb{Q}$, $\sqrt{2} \cdot \sqrt{2} = 2 \in \mathbb{Q}$, and $\sqrt{2}/\sqrt{2} = 1 \in \mathbb{Q}$

3. $\frac{98}{219}$ **5.** $-1, -\frac{1}{3}, \frac{2}{3}$

7. By Theorem 4.7, if $x = h/k$ is a rational number in reduced form, and if x is a solution of the equation, then $k|1$, and it follows that $k = 1$.

9. Let $k = 2$ in Example 4.9.

11. Let x denote the cube root of 2. Then x is a solution of the equation $x^3 - 2 = 0$; hence, if x were rational, it would be an integer by Corollary 4.8. But 2 is not the cube of any integer.

13. Let $\sqrt{m/n} = a/b$, where the rational number a/b is in reduced form, so that $\text{GCD}(a, b) = 1$. Let $g = \text{GCD}(a^2, b^2)$. If $g > 1$, there exists a prime number p such that $p|g$. Then, $p|a^2$, so $p|a$ by Corollary 3.25. Likewise, $p|b$, contradicting $\text{GCD}(a, b) = 1$. Hence, $g = 1$, so a^2/b^2 is in reduced form. Because $m/n = a^2/b^2$, Lemma 4.5 implies that $m = a^2$ and $n = b^2$.

15. $x = 10^n x/10^n$ and $10^n x$ is an integer.

17. $[\![10^{n-1}x]\!] \le 10^{n-1}x < [\![10^{n-1}x]\!] + 1$ implies that $0 \le y < 1$. To obtain the decimal expansion of $10^{n-1}x$ from the decimal expansion of x, you move the decimal point $n - 1$ places to the right, which is where the repetition begins. Subtracting the integer part $10^{n-1}x$ from $10^{n-1}x$ leaves only the repeating portion.

19. Note that $10^k y - y = m$. Thus, $(10^k - 1)y = m$, so $y = m/(10^k - 1)$.

21. Let y be determined from x as in Problem 17. By Problem 19, y is rational. Therefore, $y + [\![10^{n-1}x]\!]$ is rational, and it follows that $x = 10^{1-n}(y + [\![10^{n-1}x]\!])$ is rational.

23. $\frac{392{,}413}{124{,}875}$

PROBLEM SET 4.5, PAGE 307

1. (a) $(-i)^2 + 1 = i^2 + 1 = 0$

(b) If $i = -i$, then $2i = 0$, so $i = 0$, contradicting $i^2 + 1 = 0$.

(c) Suppose $j \in F$ and $j^2 + 1 = 0$. Then $j^2 = i^2 = -1$, so $j^2 - i^2 = 0$. Therefore, $(j - i)(j + i) = 0$; hence, $j - i = 0$ or $j + i = 0$. Thus, $j = i$ or $j = -i$.

3. By direct calculation, check each of the postulates for a commutative ring with unit.

5. Follow the hint given. **7.** (a) $\phi^{-1}(x) = (x, 0)$ (b) $y \ne 0 \Rightarrow \phi^{-1}((x, y)) = (x, y)$

9. By Definition 5.6, $\mathbb{R} \subseteq \mathbb{C}$; by Theorem 5.9, $\mathbb{C}$ is a field; and, by Definitions 5.4 and 5.8, the operations of addition and multiplication in $\mathbb{R}$ are restrictions of the corresponding operations in $\mathbb{C}$.

11. (i) $x + (a, b) = \phi[\phi^{-1}(x) + \phi^{-1}((a, b))] = \phi[(x, 0) + (a, b)] = \phi((x + a, 0 + b)) = \phi((x + a, b)) = (x + a, b)$ because $b \ne 0$.

(ii) $y \cdot (a, b) = \phi[\phi^{-1}(y) \cdot \phi^{-1}((a, b))] = \phi[(y, 0) \cdot (a, b)] = \phi[(ya - 0 \cdot b, yb + 0 \cdot a)] = \phi((ya, yb)) = (ya, yb)$ because $yb \ne 0$.

13. (i) Let $z = x + yi$, $w = u + vi$. Then $z + w = (x + u) + (y + v)i$ and $(z + w)^* = (x + u) - (y + v)i = (x + u) + (-y - v)i = x + (-y)i + u + (-v)i = x - yi + u - vi = z^* + w^*$.

(ii) Make a similar calculation.

15. (a) $\frac{17}{21} + \frac{1}{4}i$ (b) $\frac{11}{21} + \frac{9}{4}i$ (c) $\frac{113}{84} - \frac{41}{84}i$ (d) $-\frac{679}{600} + \frac{497}{600}i$
(e) $-\frac{1164}{2023} - \frac{852}{2023}i$ (f) $-\frac{97}{84} + \frac{71}{84}i$ (g) $\frac{17}{12}$

17. Let $z = x + yi$ with $x, y \in \mathbb{R}$.
(a) $z \in \mathbb{R} \Rightarrow z = x + 0 \cdot i \Rightarrow z = z^*$. On the other hand, if $z = z^*$, then $x + yi = x - yi$, so $yi = -yi$; hence, $2yi = 0$. Since $i \neq 0$, $y = 0$, so $z = x$, and therefore $z \in \mathbb{R}$.
(b) $|z| = \sqrt{zz^*} = \sqrt{(x + yi)(x - yi)} = \sqrt{x^2 + y^2} \geq 0$. If $|x| = 0$, then $x^2 + y^2 = 0$ and, since x and y are real numbers, it follows that $x = y = 0$.
(c) Since $|z| = \sqrt{zz^*}$, it follows that $|z|^2 = zz^*$.

19. Look at the vectors from the origin to the points z and w. The vector from the origin to $z + w$ (respectively, $z - w$) is the sum (respectively, difference) of these vectors. The vector from the origin to $-z$ is the negative of the vector from the origin to z, and $|z|$ is the length of this vector.

21. $|z/w| = \sqrt{(z/w)(z/w)^*} = \sqrt{(z/w)(z^*/w^*)} = \sqrt{zz^*/ww^*} = \sqrt{zz^*}/\sqrt{ww^*} = |z|/|w|$

23. (a) We have $a^2 + b^2 \geq a^2$. Taking square roots, we find that $\sqrt{a^2 + b^2} \geq |a|$. Hence, since $|a| \geq -a$, $\sqrt{a^2 + b^2} \geq -a$. Therefore, $r \geq -a$, so $r + a \geq 0$.
(b) The argument is the same as in Part (a), except we use the fact that $|a| \geq a$.
(c) $(x + yi)^2 = x^2 - y^2 + 2xyi = (r + a)/2 - (r - a)/2 + 2xyi = a + 2\sqrt{(r + a)(r - a)/4}\,\text{sgn}(b)i = a + \sqrt{r^2 - a^2}\,\text{sgn}(b)i = a + \sqrt{b^2}\,\text{sgn}(b)i = a + |b|\,\text{sgn}(b)i = a + bi$. A similar calculation shows that $(x - yi)^2 = a + bi$.

25. Let $p(x) = a_n x^n + a_{n-1}x^{n-1} + \cdots + a_1 x + a_0$. Then, $a_n z^n + a_{n-1}z^{n-1} + \cdots + a_1 z + a_0 = 0$. Take the conjugate on both sides of the last equation, noting that $(z^k)^* = (z^*)^k$ and that each of the real coefficients is its own conjugate.

27. Proof by contradiction: Suppose there were a polynomial of degree 1 or more that could not be so factored. Let $p(z)$ be such a polynomial whose degree is minimal. (Here, we use the fact that the natural numbers are well-ordered.) By the fundamental theorem of algebra, $\exists z_1 \in \mathbb{Z}$ such that $p(z_1) = 0$. Using long division, we divide $p(z)$ by $z - z_1$ to obtain a quotient polynomial $q(z)$ and a remainder r, which, since the divisor $z - z_1$ has degree 1, is a complex number. Thus, $p(z) = (z - z_1)q(z) + r$. Putting $z = z_1$, we find that $r = 0$, so $p(z) = (z - z_1)q(z)$. If $q(z)$ had degree 0, it would be a constant complex number q, and $p(z) = qz - qz_1$ would be factored, contrary to our supposition. Therefore, $q(z)$ has degree 1 or more. However, the degree of $q(z)$ is less than the degree of $p(z)$, so $q(z)$ must factor into a product of first-degree polynomials. Thus, $p(z)$ also factors into such a product, contrary to our supposition.

29. Factor the polynomial $p(x)$ as a product of a constant k and linear factors of the form $x - z_i$, where $z_1, z_2, \ldots, z_n$ are the complex solutions of $p(x) = 0$. Group the nonreal z_i's in pairs of complex conjugates, and multiply the corresponding linear factors. Note that $(x - z_i)(x - z_i^*) = x^2 + bx + c$, where $b = -(z_i + z_i^*)$ and $c = z_i z_i^*$ are real numbers.

31. (a) If $|z| = 1$ and $|w| = 1$, then $|zw| = |z|\,|w| = 1$.
(b) If $|z| = 1$, then $|1/z| = 1/|z| = 1$. Also, $|1| = 1$. Therefore, $1 \in S^1$, and S^1 is closed under multiplication and the formation of multiplicative inverses. It is now an easy matter to verify the postulates for a commutative group.
(c) Visualize S^1 as a circle of radius 1 with center at the origin in the Gaussian plane.

CHAPTER 5 PROBLEM SETS

We omit answers and hints for selected odd-numbered problems in Chapter 5, because the material in the chapter is optional.

INDEX

SYMBOL	MEANING	PAGE
$f(M)$	The image of the set M under the function f	156
$g\|_X$	The restriction of a function g to the set X	169
x^-	The inverse of x in a monoid $(S, *)$	180
$(R, +, \cdot)$	A ring R with addition operation $+$ and multiplication operation $\cdot$	189
$\mathbb{Z}_4$	The ring of integers modulo 4	191
$a \equiv_n b$	$b - a$ is an integer multiple of n	199
$\equiv_n$	Congruence modulo n	199
$(\mathbb{Z}_n, +, \cdot)$	Ring of integers modulo n	199
$(F, +, \cdot)$	A field F with addition operation $+$ and multiplication operation $\cdot$	200
$\mathbb{R}$	The field of real numbers	201
$\mathbb{Q}$	The field of rational numbers	201
$\mathbb{C}$	The field of complex numbers	201
$\mathbb{Q}(\sqrt{2})$	The field obtained by adjoining $\sqrt{2}$ to $\mathbb{Q}$	201
$\mathbb{R}(X)$	The field of rational functions in X with coefficients in $\mathbb{R}$	201
$\mathrm{GF}(p)$	The Galois field of order p	201
$F \setminus \{0\}$	The group of invertible elements in a field F	202
$\mathrm{sgn}: F \to F$	The signum function for the ordered field F	218
$\mathrm{sgn}(x)$	The signum of x	218
$+: V \times V \to V$	Vector addition	222
F^n	Coordinate n-space over the field F	223
F^X	The set of all functions $f: X \to F$ from a fixed set X into a fixed field F	223
$F^{[X]}$	The subset of F^X consisting of all finitely nonzero functions $f: X \to F$	223
$C(D)$	The set of all continuous functions $f: D \to \mathbb{R}$ with domain $D \subseteq \mathbb{R}$	223
$\mathbf{e}_1, \mathbf{e}_2, \mathbf{e}_3, \ldots, \mathbf{e}_n$	Standard basis vectors for F^n	227
$\dim_F(V)$	Dimension of V over F	231
$\mathrm{lin}(M)$	Set of all linear combinations of finite sets of vectors in M	236
$P \oplus Q$	Direct sum of subspaces P and Q of a vector space V	237
$T(P)$	The image of a subspace P of a vector space V under a linear transformation T	241
$T^{-1}(Q)$	The inverse image of a subspace Q of a vector space W under a linear transformation T	241
$\ker(T)$	Kernel of a linear transformation T	242
$\mathrm{range}(T)$	Range of a linear transformation T	243
INT	The greatest integer function (BASIC)	276
CINT	The closest integer function (BASIC)	276
$x \mapsto [\![x]\!]$	The greatest integer function	276
$\mathrm{GCD}(m, n)$	The greatest common divisor of the integers m and n	282
$\mathrm{LCM}(m, n)$	The least common multiple of the integers m and n	287
$(\mathbb{Q}, +, \cdot)$	Field of rational numbers	289